How to Listen to the Audio Files

The audio files that accompany this book are available streaming online at
www.petersonbirdsounds.com

How to Look Up an Unfamiliar Bird Sound

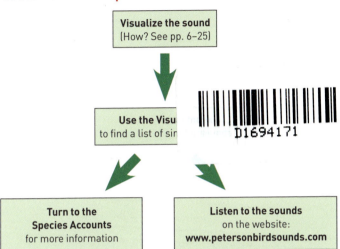

Visualize the sound
(How? See pp. 6–25)

Use the Visu
to find a list of sir

D1694171

**Turn to the
Species Accounts**
for more information

Listen to the sounds
on the website:
www.petersonbirdsounds.com

How to Use the Visual Index

The index organizes sounds by pattern, providing a quick reference
to soundalikes and a way to look up unfamiliar sounds.

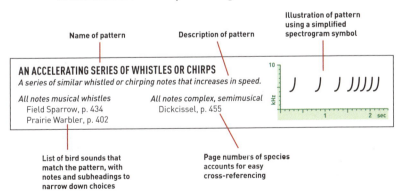

Name of pattern

Description of pattern

Illustration of pattern
using a simplified
spectrogram symbol

AN ACCELERATING SERIES OF WHISTLES OR CHIRPS
A series of similar whistled or chirping notes that increases in speed.

All notes musical whistles
Field Sparrow, p. 434
Prairie Warbler, p. 402

All notes complex, semimusical
Dickcissel, p. 455

List of bird sounds that
match the pattern, with
notes and subheadings to
narrow down choices

Page numbers of species
accounts for easy
cross-referencing

PETERSON FIELD GUIDE TO

BIRD SOUNDS

of Eastern North America

THE PETERSON FIELD GUIDE SERIES

PETERSON FIELD GUIDE TO

BIRD SOUNDS
of Eastern North America

Nathan Pieplow

HOUGHTON MIFFLIN HARCOURT
BOSTON   NEW YORK   2017

Sponsored by the National Wildlife Federation
and the Roger Tory Peterson Institute

Address requests for permission to make copies of Houghton Mifflin Harcourt
material to trade.permissions@hmhco.com or Permissions, Houghton Mifflin Harcourt
Publishing Company, 3 Park Avenue, 19th Floor, New York, NY 10016.

www.hmhco.com

PETERSON FIELD GUIDES and PETERSON FIELD GUIDE SERIES are
registered trademarks of Houghton Mifflin Harcourt Publishing Company.

Library of Congress Cataloging-in-Publication Data is available.
ISBN 978-0-547-90558-7

Book design by Eugenie S. Delaney

Printed in China

SCP 10 9 8 7 6 5 4 3 2 1

*This book is dedicated to
my grandparents, for sparking
my interest in birds
and
to Molly, for her love, support,
and tremendous patience*

Continuing the work of Roger Tory Peterson through Art, Education, and Conservation

In 1984, the Roger Tory Peterson Institute of Natural History (RTPI) was founded in Peterson's hometown of Jamestown, New York, as an educational institution charged by Peterson with preserving his lifetime body of work and making it available to the world for educational purposes.

RTPI is the only official institutional steward of Roger Tory Peterson's body of work and his enduring legacy. It is our mission to foster understanding, appreciation, and protection of the natural world. By providing people with opportunities to engage in nature-focused art, education, and conservation projects, we promote the study of natural history and its connections to human health and economic prosperity.

Art—Using Art to Inspire Appreciation of Nature

The RTPI Archives contains the largest collection of Peterson's art in the world—iconic images that continue to inspire an awareness of and appreciation for nature.

Education—Explaining the Importance of Studying Natural History

We need to study, firsthand, the workings of the natural world and its importance to human life. Local surroundings can provide an engaging context for the study of natural history and its relationship to other disciplines such as math, science, and language. Environmental literacy is everybody's responsibility—not just experts and special interests.

Conservation—Sustaining and Restoring the Natural World

RTPI works to inspire people to choose action over inaction, and engages in meaningful conservation research and actions that transcend political and other boundaries. Our goal is to increase awareness and understanding of the natural connections between species, habitats, and people—connections that are critical to effective conservation.

For more information, and to support RTPI, please visit rtpi.org.

CONTENTS

INTRODUCTION

A New Approach to Bird Sounds

Bird sounds fill the wilderness, echo between city skyscrapers, and penetrate the windows of speeding cars. Few other natural phenomena are as ubiquitous, and few are as rich in beauty, variety, and meaning.

They have been called "nature's music," but bird sounds are perhaps better described as nature's language. Like human speech, they carry complex messages from singer to listener. This book is a dictionary for the language of the birds.

Identifying the species of a bird by sound, also called "birding by ear" or "ear-birding," has long been important to naturalists. In fact, some say that experienced birders detect and identify ten times as many birds with their ears as with their eyes. But listening to a bird can reveal far more than simply the identity of the singer. It can also reveal what the bird is doing and why, what it is communicating and to whom.

Listening to birds is one of the best ways to connect to the natural world. Whether out in the wilderness or in between the skyscrapers, birds can help remind us that we humans are not the only organisms on Earth capable of complex social interaction, communication, drama, and even artistry.

This book, the most comprehensive reference to North American bird sounds ever produced, takes a new approach to the language of the birds based on visualizing and indexing sounds.

Learning to visualize sounds takes only a short time, but revolutionizes listening and hearing. Picturing sounds makes it possible to understand at a glance what a bird will sound like, even before hearing the accompanying recording. It makes it possible to search this book visually for a bird song heard in the field.

Thus, this book makes it possible, for the first time, to look up an unfamiliar sound, just as one can look up an unfamiliar word in the dictionary.

Whenever possible, this book also seeks to illuminate the meanings of sounds—that is, their behavioral context, including whether they are made by males, females, or both; whether by adults, juveniles, or birds of all ages; whether in association with alarm, courtship, territorial defense, or the need to keep a flock together. Our knowledge of the sounds in this book is incomplete, and readers might add to that knowledge with their own careful observations of bird sounds and bird behaviors.

We have much else to learn about birds as well. Perhaps the most pressing question is how the birds are faring, and how they will continue to fare, on a planet increasingly repurposed and redesigned for human ends. This book cannot answer that question, but it does aim to bring bird sounds out of the background noise, making them more comprehensible, more meaningful, more enjoyable, and thereby more beautiful, and more difficult to ignore.

Organization of the book

The introductory section of this book explains how to listen to and visualize bird sounds, and discusses some aspects of their production, function, and development. The species accounts, arranged family-by-family in taxonomic order, depict the sounds of each species spectrographically. The visual index lists similar sounds in groups by pattern, with reference to the page number of the species account in which each sound is discussed.

Scope of the book

This book covers 520 species of birds regularly found in the eastern United States and southeastern Canada (hereafter, "the East"), south of about the 50th parallel and west to roughly the 100th meridian. Certain species that occur regularly only at the fringes of this area are omitted. Most pelagic species are omitted.

The book primarily treats sounds known to be given in the East. Thus, the breeding-season sounds of some species that nest in the Arctic, such as certain sandpipers, have been omitted, unless they are known to be given in migration or winter.

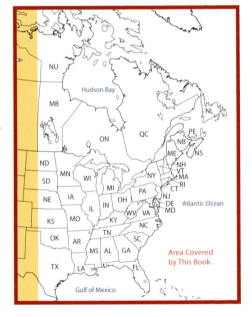

Extinct species are included when recordings of their voices exist. Extinct species for which no recordings exist are mentioned in the family introductions.

A number of exotic species are included in the guide, when wild or feral individuals can regularly be encountered in the East. Whenever possible, recordings of these species come from within the geographic scope of this book, because in some cases wild or feral birds may sound different than their wild ancestors.

Notes on the audio files

This book is accompanied by more than 5,400 audio files of bird sounds, which are available streaming on the Web at www.petersonbirdsounds.com.

The recordings were made by over 250 different recordists. Many of the recordings have been edited, sometimes extensively, in order to amplify the target sound, reduce interference from background sounds, and/or shorten intervals between vocalizations. Some recordings are of natural vocalizations, while others are of birds responding to playback. A few recordings are of captive birds.

The website lists information about each audio file, including the name of the recordist, the date and location of the recording, and any other bird species that can be heard in the background, sometimes with additional notes on the behavioral context of the sounds and the circumstances of the recording. The website also hosts links to additional resources as well as video tutorials on how to read spectrograms.

Notes on the spectrograms

All the spectrograms in this book were created from the audio files on the accompanying website using the Raven Pro software developed by the Cornell Laboratory of Ornithology. The spectrograms were generated using Hann windows, typically with a window size of 512 samples and an overlap of 90 percent. Window size was sometimes adjusted for clarity. All spectrograms were then modified using graphics editing software, to remove visible traces of echoes and background sounds and to improve contrast.

In most cases the spectrogram shown is of the first vocalization on the sound file, but some spectrograms were generated from later sections of a recording. The website explains which part of the sound file each spectrogram corresponds to.

How Birds Produce Sound

Sounds produced by a bird's vocal tract are called **vocal sounds** or **phonations**, while those produced in other ways are called **nonvocal sounds, mechanical sounds,** or **sonations.** Most of the sounds in this book are vocal sounds.

Vocal sounds

The vocal organ of birds is called a **syrinx** (plural **syringes**). The birds with the most complex vocalizations—hummingbirds, parrots, and passerines—tend to be those with the most complex musculature around the syrinx. Black and Turkey Vultures lack most or all syringeal muscles, so their vocal repertoires are limited to hisses and grunts.

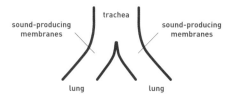

FIG. 1. *The avian syrinx*

In many species, the syrinx can produce two independently controlled voices at once, one from each lung. The result is called a **polyphonic** sound. Many birds have this capability but do not use it; they may produce all their sounds with one side of the syrinx, or they may use both sides, but not at the same time—for example, to produce different notes in the same song.

Nonvocal sounds

1. Beating the bill against a hard surface: Perhaps the most familiar type of mechanical bird sound is the drumming that woodpeckers make by striking hard surfaces with their bills. The initial sounds made by a displaying male Ruddy Duck may also qualify, as the bird apparently makes those sounds by slapping its bill rapidly downward against its own chest feathers.

2. Snapping the bill shut: In most birds, the sound of the bill snapping shut is not very loud, but some species, particularly owls, flycatchers, and gnatcatchers, use it in close-range aggressive displays. A few birds, notably Greater Roadrunner and Wood Stork, make clattering sounds by beating their mandibles together many times in quick succession.

3. Clapping the wings together: Among North American birds, this type of sound production typically takes place in flight. Some species, such as Long-eared and Short-eared Owls, clap their wings together below their bodies, while others, such as pigeons, clap their wings together above their backs.

4. Moving feathers through air: The wings of most birds can be heard in flight, at least at close range, but some species are specialized for sound production with their wings or tail. Sounds produced by feathers include the drumming of Ruffed Grouse, the winnowing of Wilson's Snipe, the high-pitched whistles of Mourning Doves flushing from a perch, and many hummingbird sounds.

5. Inflating body cavities with air: All birds have air sacs inside their body, but only a few species have specially modified air sacs that are used in sound production. Displaying Greater and Gunnison Sage-Grouse make popping sounds by expelling air from their air sacs explosively. The bizarre song of the American Bittern apparently involves the inflation of the esophagus, although the mechanism of production of this sound, as with many bird sounds, is not completely understood.

In some cases, an inflated air sac or body cavity may serve as a resonating chamber for a vocal sound. Birds such as Trumpeter Swans and Whooping Cranes have tracheas that are elongated or even looped inside the body, apparently to amplify or modify the voice (see p. 119).

Visualizing Sound

Visualizing bird sounds makes it easier to identify them, because the aspects of bird sound that are important for visualization are the same ones that are important for identification: pitch pattern, speed, repetition, pauses, and tone quality. Creating a mental image of the sound makes it possible to look up the sound in the visual index of this book (p. 495), where similar sounds are grouped by their visual pattern.

How to see it

The best way of depicting bird sounds visually is the spectrogram (often called the Sonagram), a computer-generated graph of sound frequencies across time. Below, compare how the same familiar tune appears in two different visual depictions: standard musical notation on the left, and a spectrogram on the right:

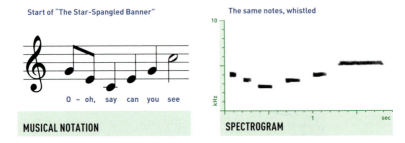

Musical notation places notes on a staff, while spectrograms measure the frequency of sounds in kilohertz (kHz), but the basic principle is the same: they both read from left to right, with high notes near the top of the chart and low notes near the bottom. On a spectrogram, the more horizontal space a note takes up, the longer it lasts in time.

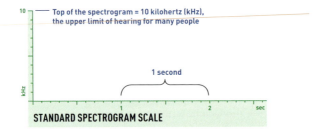

Most spectrograms in this book conform to the scale above, with the top of the spectrogram at 10 kHz (near the upper limit of hearing in most adults) and numbers across the bottom marking intervals of one second.

Real spectrograms vs. spectrogram symbols

This book uses two types of visualizations: real spectrograms in the species accounts, where accuracy and detail are important; and spectrogram symbols in the visual index, where basic patterns and similarities are more important than detail.

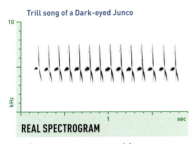

REAL SPECTROGRAM

- **Computer-generated from an audio recording**
- **Shows fine details, even some not audible to the human ear**
- **Used in the species accounts**

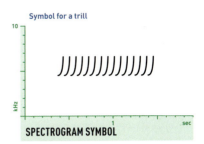

SPECTROGRAM SYMBOL

- **Artistic approximation of a real spectrogram**
- **Emphasizes basic patterns rather than details**
- **Used in the visual index**

The five basic pitch patterns

Unlike music, bird sound identification does not require attention to the precise pitch of notes; more important is *how the pitch changes*. All bird sounds can be described with just five basic pitch patterns (or combinations thereof), which can be visualized this way:

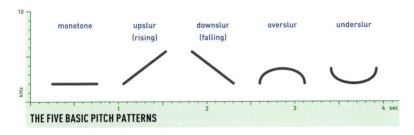

THE FIVE BASIC PITCH PATTERNS

- **Monotone sounds** do not change in pitch, and appear horizontal on the spectrogram.
- **Upslurred sounds** rise in pitch, and appear tilted upward.
- **Downslurred sounds** fall in pitch, and appear tilted downward.
- **Overslurred sounds** rise and then fall in pitch, appearing and sounding highest in the middle.
- **Underslurred sounds** fall and then rise, appearing and sounding lowest in the middle.

The four basic patterns of repetition and speed

The key to hearing and visualizing the repetition and speed of bird sounds lies in two questions:

1. Does the bird ever sing the same note twice?
2. Are the notes slow enough to count, or too fast to count?

Together, these two questions make it possible to identify four basic patterns of bird sound: **phrases, series, warbles,** and **trills.** *Phrases* and *series* are slower sounds, with individual notes slow enough to count; phrases contain unique notes that are not repeated, while series consist of one note repeated over and over. *Warbles* and *trills* are faster versions of phrases and series, with notes too fast to count (faster than about eight notes per second). At this speed, unique notes run together into a single warbled sound, and similar notes merge into a trill.

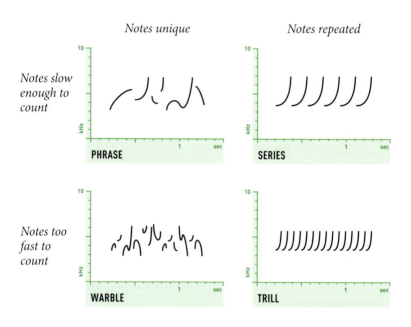

Even tremendously long and complex bird songs can be described as a combination of phrases, series, warbles, and trills. These four patterns form the most important building blocks of the visual index in this book.

The examples on the following page illustrate these four basic patterns with real spectrograms, generated from recordings that you can listen to on the accompanying website **www.petersonbirdsounds.com.**

Examples of phrases (unique notes, slow enough to count)

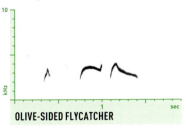

OLIVE-SIDED FLYCATCHER

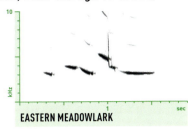

EASTERN MEADOWLARK

Examples of series (repeated notes, slow enough to count)

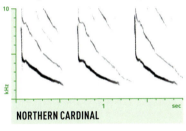

NORTHERN CARDINAL

AMERICAN GOLDFINCH

Examples of warbles (unique notes, too fast to count)

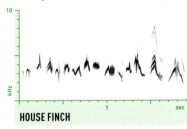

HOUSE FINCH

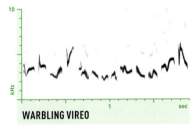

WARBLING VIREO

Examples of trills (repeated notes, too fast to count)

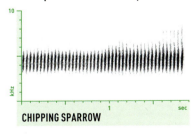

CHIPPING SPARROW

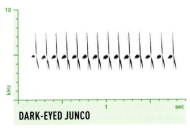

DARK-EYED JUNCO

Changes in speed and pitch of whole songs

Some bird sounds consist of series that change in speed. If the elements in a series are more closely spaced on the spectrogram as you move from left to right, then they are growing more closely spaced in time, which means that the series accelerates. If the elements grow farther apart, the series decelerates.

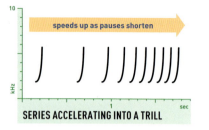

SERIES ACCELERATING INTO A TRILL

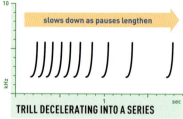

TRILL DECELERATING INTO A SERIES

Phrases, series, warbles, and trills can also change in pitch. For example, a warble might sound upslurred if it shows an overall trend toward higher notes. Similarly, a series might fall in pitch if each note starts slightly lower than the last, regardless of the pitch pattern of the individual notes.

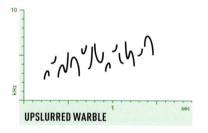

UPSLURRED WARBLE

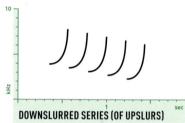

DOWNSLURRED SERIES (OF UPSLURS)

The song of the Canyon Wren is a downslurred, decelerating series. Notice that the individual notes in this particular song are not downslurred, however; most of them are underslurred. As they progress from left to right, they slow down. They also become fainter on the spectrogram, meaning they are not as loud.

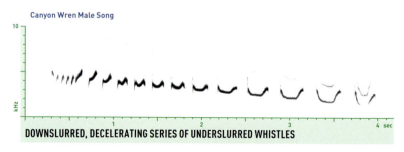

DOWNSLURRED, DECELERATING SERIES OF UNDERSLURRED WHISTLES

These songs both accelerate. The Field Sparrow's song begins as a series and speeds up into a trill; in this example the series is downslurred. The Horned Lark's song begins as a phrase and becomes a warble, rising in pitch at the end.

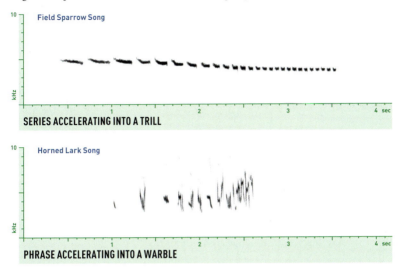

SERIES ACCELERATING INTO A TRILL

PHRASE ACCELERATING INTO A WARBLE

Complex series

Sometimes the repeated elements in a series themselves consist of multiple notes. A **couplet series** sounds like a 2-syllable word repeated, such as "peter peter peter"; a **triplet series** sounds like a 3-syllable word repeated, such as "teakettle teakettle teakettle." Series of 1-syllabled notes can be called **simple series**.

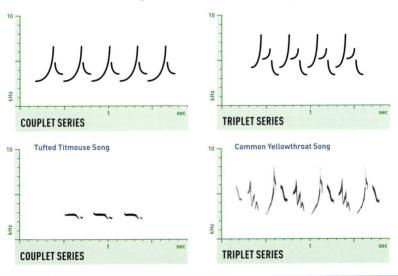

COUPLET SERIES

TRIPLET SERIES

COUPLET SERIES

TRIPLET SERIES

Pauses

A very important question to ask about bird sounds is whether and when the bird stops to "take a breath." Some birds can sing for 30 seconds or more without any noticeable pause. The song of the American Goldfinch, for example, often consists of several short series strung seamlessly together:

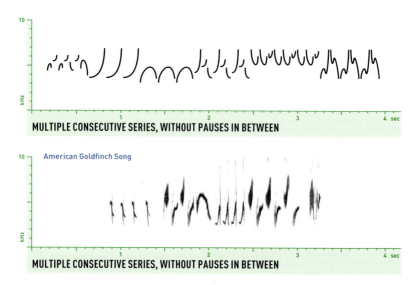

MULTIPLE CONSECUTIVE SERIES, WITHOUT PAUSES IN BETWEEN

American Goldfinch Song

MULTIPLE CONSECUTIVE SERIES, WITHOUT PAUSES IN BETWEEN

Compare the song pattern of this Northern Mockingbird, in which the bird takes short "breaths" between almost all of the series in its song:

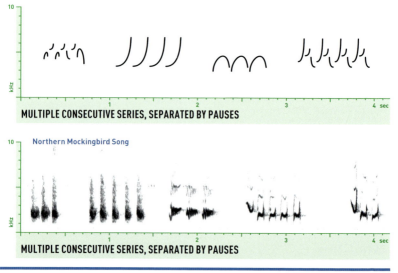

MULTIPLE CONSECUTIVE SERIES, SEPARATED BY PAUSES

Northern Mockingbird Song

MULTIPLE CONSECUTIVE SERIES, SEPARATED BY PAUSES

Some birds, especially many vireos and flycatchers, sing very short unique phrases separated by relatively long pauses:

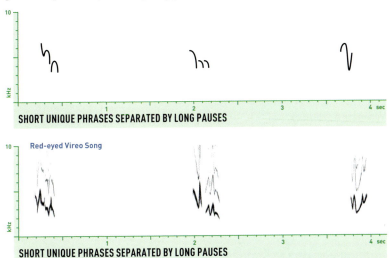

SHORT UNIQUE PHRASES SEPARATED BY LONG PAUSES

Red-eyed Vireo Song

SHORT UNIQUE PHRASES SEPARATED BY LONG PAUSES

Other singers, including the American Robin, Rose-breasted and Black-headed Grosbeaks, and many tanagers, group short unique phrases into clusters that are separated by longer pauses:

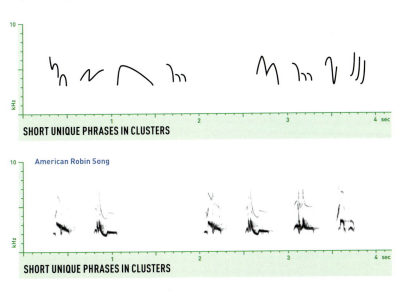

SHORT UNIQUE PHRASES IN CLUSTERS

American Robin Song

SHORT UNIQUE PHRASES IN CLUSTERS

The seven basic tone qualities

The distinctive voice of a bird sound, also called its tone quality, can be an important clue to its identification. Many species have unique tone qualities, and although the differences are easy to perceive, many people have found them difficult to descibe. However, all bird voices are composed of variations or combinations of just seven basic tone qualities, illustrated below and explained in more detail on the following pages.

Whistled sounds   p. 16

Whistles are the most basic sounds, appearing on the spectrogram as simple nonvertical lines. Examples include typical human whistling and the sounds of flutes and piccolos. Birds with whistled songs include Black-capped Chickadee, Yellow Warbler, and Northern Cardinal.

Hooting and cooing sounds   p. 16

The hoots of large owls and the coos of doves are just extremely low-pitched whistles, less than 1 kHz in frequency. Because they are so low, they are illustrated in special spectrograms with a vertical scale of just 0 to 2 kHz. These low-pitched spectrograms are marked with an owl symbol.

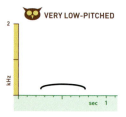

VERY LOW-PITCHED

Ticking sounds   p. 17

Instantaneous bursts of noise sound like ticks, snaps, or knocks, and appear on the spectrogram as vertical lines. The ticking of a clock, the drumming of a woodpecker's bill against a tree, the bill snap of an angry flycatcher, and the ticking song of Yellow Rail all fall into this category.

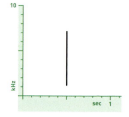

Burry and buzzy sounds   p. 19

When a sound rises and falls very rapidly in pitch, it forms a squiggly line on the spectrogram and sounds trilled, like a referee whistle. If the squiggles are tall and fast enough, they sound less musical, more like an electric buzzer. These sounds are common among birds such as warblers and sparrows.

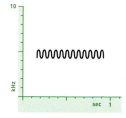

Noisy sounds   p. 20

Noisy sounds contain noise—that is, random sound at multiple frequencies, which looks like television static on the spectrogram and sounds like static to the ear. Very noisy bird sounds have a rough or harsh quality, like the alarm chatters of wrens and the hissing of angry swans and geese.

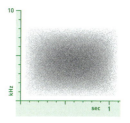

Nasal sounds   p. 20

Nasal sounds are composed of whistles stacked vertically that the human ear interprets as a single sound. Examples include police sirens, the whine of mosquito wings, and the sounds of oboes and violins. Birds with nasal voices include Red-breasted Nuthatch, Black-billed Magpie, and Pinyon Jay.

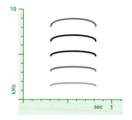

Polyphonic sounds   p. 23

Birds can make two whistled or nasal sounds simultaneously, creating a distinctive sound with a spectrographic pattern of lines that cross and/or stack without exactly matching. Such sounds may be dissonant and "whiny," like the calls of goldfinches, or metallic, like the songs of many thrushes.

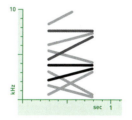

These seven tone qualities interact and combine in various ways. A few of the more common combinations are listed here.

- **Nasal and noisy sounds** include shrieks and screeches.
- **Nasal, noisy, and burry sounds** include the caws of crows and the quacks of ducks.
- **Polyphonic and noisy sounds** include the agitated whines of vireos and the harsh metallic cries of scrub-jays. Many of these sounds are also burry.

Whistles and hoots

Tone quality changes with pitch

The pitch of a sound has a significant impact on its tone quality. This is true of all sounds, but it is particularly true of whistles.

Hoots and coos are low-pitched whistles

The hooting of owls and the cooing of doves are just whistles that are lower than about 1 kHz. Similar low whistles can be made by blowing across the top of a large bottle. Extremely low-pitched sounds like these are illustrated using a different vertical scale than the rest of the spectrograms in this book, and marked with the symbol of an owl.

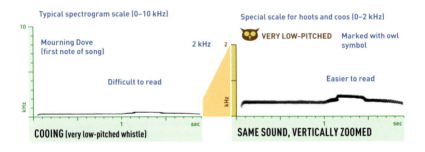

Pitch and whistle quality

As whistles rise to a pitch above hooting and cooing sounds, they first take on a mellow quality. As they continue to rise, above 3 kHz or so, they gradually become thinner and more penetrating. Whistles above 6 kHz tend to sound sibilant, almost like the hiss of air escaping a tire.

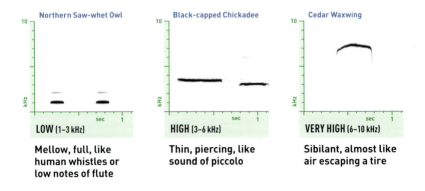

Ticking sounds and sharp whistles

As explained on page 14, nonvertical lines on the spectrogram represent whistles, and vertical lines represent ticks. As whistles become **sharper** (more vertical on the spectrogram), they gradually lose their musical quality and become more like ticks.

THE CONTINUUM OF INCREASING SHARPNESS FROM WHISTLES TO TICKS

The song of the Black-chinned Sparrow provides an excellent example of this phenomenon. The song is an accelerating series of whistles in which each successive note is slightly sharper and therefore less musical. The first half of the song is musical because the whistles are closer to the horizontal, but as the song continues, the whistles approach the vertical and become decidedly ticklike. The last few notes just sound like a toneless buzz.

ACCELERATING SERIES OF WHISTLES (gradually changing from musical to unmusical)

The Whit calls of *Empidonax* flycatchers and the Chip calls of warblers are examples of very sharp, unmusical whistles. Whits are sharp upslurs; Chips are downslurs.

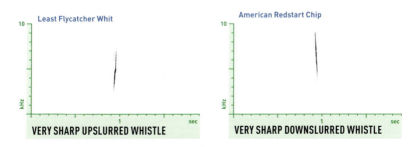

VERY SHARP UPSLURRED WHISTLE

VERY SHARP DOWNSLURRED WHISTLE

The quality of a trill depends on the individual note

The principles explained on the previous page apply to trills as well as individual notes. A trill of musical notes sounds musical, while a trill of unmusical notes sounds unmusical.

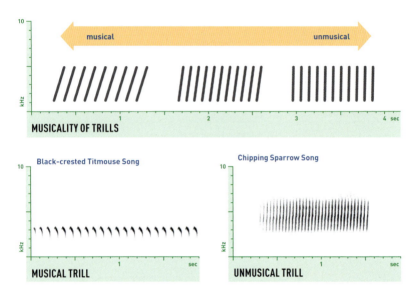

Musical trumps unmusical

Our perception of a sound's musicality tends to be determined by its most musical elements. For example, when a whistle on a spectrogram has both steep and flat sections, we tend to hear the flat, musical sections more strongly. This applies to trills as well.

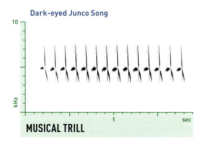

This junco's song consists of two alternated notes—one musical, one unmusical—repeated so fast that we hear only a single trill. Because musical sounds tend to dominate, the trill sounds rather musical to the ear.

Burry and buzzy sounds

These sounds rise and fall very rapidly in pitch, creating up-and-down squiggles, or **beats,** on the spectrogram. **Burrs** are more musical, and tend to take up less vertical space on the spectrogram; **buzzes** are less musical, and tend to take up more vertical space. In **coarse** burrs and buzzes, the squiggles are farther apart; in **fine** burrs and buzzes, they are closer together.

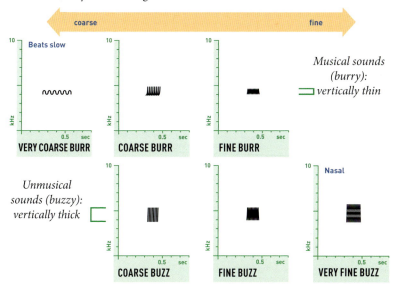

Tremolos are burrs so coarse that the individual beats are countable, or nearly countable. **Peents** are buzzes so fine that they take on a nasal quality.

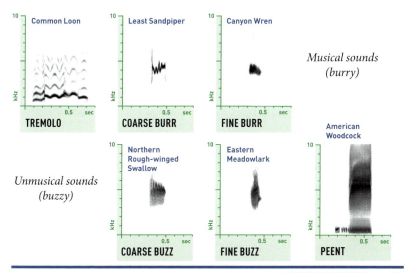

Noisy sounds

Darker is louder

The darker a note appears on the spectrogram, the louder it is.

In noisy sounds, the pitch and pitch pattern are determined by the placement and shape of the darkest (and thus the loudest) parts of the noise.

If the darkest part of the sound is closer to the top of the spectrogram, the noise will sound higher in pitch. If the darkest part rises toward the end of the spectrogram, the noise will sound upslurred; if it falls, the noise will sound downslurred.

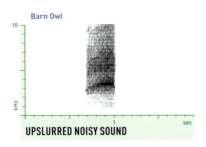

UPSLURRED NOISY SOUND

DOWNSLURRED NOISY SOUND

Nasal sounds

Many bird sounds are actually combinations of multiple simultaneous whistles on different pitches that the human brain typically perceives as a single sound. The individual whistles that make up a complex sound of this kind are called **partials.** The first (lowest) partial is often called the **fundamental.**

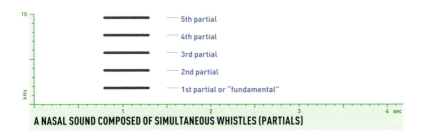

A NASAL SOUND COMPOSED OF SIMULTANEOUS WHISTLES (PARTIALS)

All nasal sounds are stacks of partials, but not all stacks of partials sound nasal. The tone quality of a complex sound depends on which partials are loudest—that is, darkest on the spectrogram.

The higher the darkest line in a stack, the more nasal the sound

If the fundamental or the second partial is the loudest (darkest), then the call will not sound nasal at all, but whistled. Once the darkest band climbs as high as the third partial, the call will start to sound nasal, and the higher it climbs, the more nasal the tone quality, as the following graph illustrates.

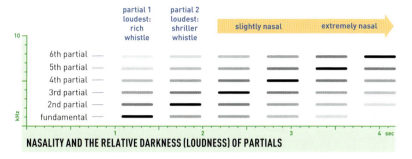

NASALITY AND THE RELATIVE DARKNESS (LOUDNESS) OF PARTIALS

The six bird sounds below are arranged in order of increasing nasality; the higher the loudest (darkest) partial, the more nasal the sound. Note that the counting of partials starts at the fundamental even if it is too faint to appear on the spectrogram. This is because in nasal sounds, the numbering of the partials is determined by the mathematical relationship between their frequencies: the frequency of the second partial is twice that of the fundamental, the frequency of the third partial is three times that of the fundamental, and so on.

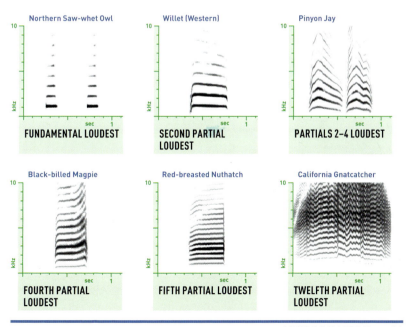

Northern Saw-whet Owl — **FUNDAMENTAL LOUDEST**

Willet (Western) — **SECOND PARTIAL LOUDEST**

Pinyon Jay — **PARTIALS 2–4 LOUDEST**

Black-billed Magpie — **FOURTH PARTIAL LOUDEST**

Red-breasted Nuthatch — **FIFTH PARTIAL LOUDEST**

California Gnatcatcher — **TWELFTH PARTIAL LOUDEST**

The farther apart the stacked lines, the higher the pitch

The key to gauging the pitch of a nasal sound is **the spacing between the partials.** The farther apart the partials, the higher the pitch of the sound. If the partials stretch to the top and bottom of the spectrogram, counting them is helpful in estimating pitch. High sounds with three to five visible partials have a distinctive seminasal tone quality.

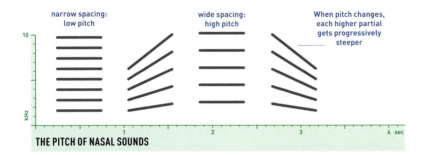

THE PITCH OF NASAL SOUNDS

The voices of large birds often break

The voices of large birds, such as geese, hawks, gulls, shorebirds, and woodpeckers, may jump suddenly to a higher or lower pitch. In nasal sounds, voice breaks appear on the spectrogram as vertical "fault lines" along which the stripes of the partials don't match up.

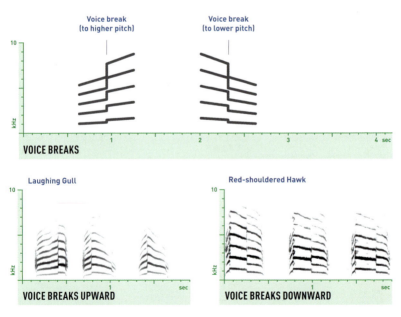

VOICE BREAKS

VOICE BREAKS UPWARD

VOICE BREAKS DOWNWARD

Polyphonic sounds

Many birds can produce two separate sounds simultaneously, one from each lung. When birds use this ability, the two original sounds blend into one polyphonic sound. Polyphonic sounds generally look similar to nasal sounds on the spectrogram, but they usually show one or more of these telltale signs, illustrated below:

1. Simultaneous rising and falling partials;
2. Stacks of partials with dissimilar shapes;
3. Partials that cross each other;
4. Partials that are irregularly spaced.

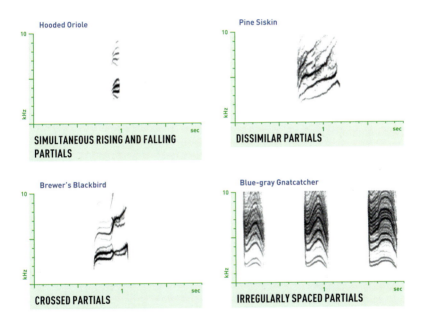

Polyphonic sounds are diverse, but with practice, they can be consistently distinguished from all other types of sounds. Most often, they sound either distinctively metallic or distinctively whiny.

If the polyphonic notes are very brief or contain monotone segments, they tend to sound metallic, like the Hooded Oriole call above, certain calls of Redwinged Blackbird, and the jangling melodies of thrushes like the Veery.

If the polyphonic notes do not contain any monotone segments, they tend to have a whiny quality, like the Pine Siskin and Blue-gray Gnatcatcher calls above, as well as the common calls of House Finch.

The Naming of Bird Sounds

Traditionally, bird sounds have been named either for their biological function ("alarm call," "flight call," "begging call," "song") or for their sound ("chewink," "chip," "buzz," "rattle").

Naming sounds for their function is often difficult, because little is known about the functions of many bird sounds. Even more important, one sound frequently serves several different functions, and different sounds can serve the same function. But assigning names based on sound can also be difficult, especially for sounds that are complex, highly variable, or plastic (see p. 28).

No method of naming sounds is perfect. This book seeks to strike a balance by using the name "Song" for primary breeding-season vocalizations that likely serve in mate attraction and territorial defense, and using sound-based names in most other cases. Mnemonic phrases are listed for some songs, when they are deemed useful.

In this book, the names of vocalizations are capitalized.

Transcribing bird sounds into human speech

One way to describe bird sounds is by transcribing them into the English alphabet. Bird voices and human voices are vastly different, and no standard method of transcription exists, but people do tend to follow certain patterns when imitating bird sounds in speech.

Vertical = consonant; horizontal = vowel

On the spectrogram, many bird sounds start and end with brief steep sections. The shape of the steep part tends to determine the consonant sound that many people hear, as follows:

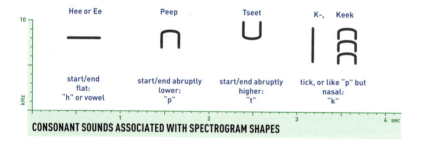

CONSONANT SOUNDS ASSOCIATED WITH SPECTROGRAM SHAPES

Listeners do not always agree on consonants, especially at the end of a sound. There tends to be more agreement on vowels.

Certain consonants correspond to tone qualities

In addition to the consonants that correspond to spectrogram shapes, some consonants are often used to indicate the quality of a sound, as follows:

consonant	often denotes a sound that is
s	very high-pitched
ch, sh	noisy
r	burry (but see below)
z	buzzy or polyphonic
l	broken

Different vowels represent different pitches

Linguists describe vowels as "high" or "low" based on the height of the tongue in the mouth during pronunciation. Low-pitched bird sounds are often transcribed using low vowels, and high-pitched bird sounds are often transcribed using high vowels.

The letters r, w, and y are vowel-like consonants (or "semivowels") that also correspond to pitches: **r** and **w** are low, while **y** is high.

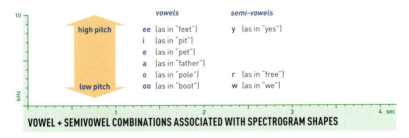

VOWEL + SEMIVOWEL COMBINATIONS ASSOCIATED WITH SPECTROGRAM SHAPES

Vowel combinations represent changes in pitch

Monotone bird sounds are typically written with a single vowel sound. Bird sounds that change in pitch are typically written with a combination of vowels and semivowels that mirrors the change in pitch. For example, upslurred sounds are usually transcribed with a low vowel or semivowel, followed immediately by a higher one.

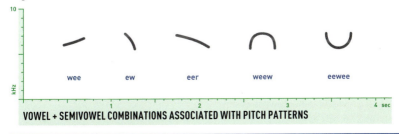

VOWEL + SEMIVOWEL COMBINATIONS ASSOCIATED WITH PITCH PATTERNS

The Biology of Bird Sounds

Innate vs. learned sounds

Some sounds are unlearned, or innate, meaning that birds can produce them without ever having heard them before. The ability to produce innate sounds is inborn and genetically controlled.

Some bird sounds are learned, meaning that in order to produce them properly, a young bird must hear them from an adult tutor of the same species. Tutors do not actively "teach"; instead, young birds learn by listening from a distance. The tutors are often the neighboring males on a young bird's first summer territory.

As far as we know, the only North American birds that can learn sounds are the hummingbirds, the parrots, and the passerines (excluding the flycatchers). Even in these groups, many sounds are innate.

Songs vs. calls

In 1961, W. H. Thorpe, the pioneering British ornithologist who originated the spectrographic analysis of bird sounds, laid out a set of criteria to distinguish between the songs and the calls of birds. According to his scheme, songs tend to be complex, learned, and given principally by males in prolonged bouts in the breeding season to establish a territory and attract a mate. Calls tend to be simple, innate, and used by both sexes in more general contexts, such as to raise an alarm, maintain contact between flock members, or beg for food.

The traditional distinction between songs and calls is blurred in many species. For example, the "Che-bek" of Least Flycatcher is innate, rather simple, and regularly given by females—but because it is given incessantly by territorial males and resembles the more complex songs of related flycatcher species, it is traditionally considered a song.

Ultimately, drawing the line between songs and calls is a matter of personal judgment. In drawing that line, this book seeks to strike a balance between traditional descriptions of bird sounds and the most recent research into their structure and function.

Mislearned songs

In captivity, some young birds will learn the songs of other species if they do not hear their own species' song. In a very few cases, birds in the wild will do the same. This happens most frequently among closely related species like Eastern and Western Meadowlarks, but even then it is rare.

Abnormalities may affect a whole song or just part of a song. One Song Sparrow from British Columbia sang a song that began with the notes of a Northern Waterthrush but ended with those of a typical Song Sparrow. In the wild, birds that sing abnormally are usually considered less likely to attract mates and pass their genes on to the next generation, meaning that most serious song abnormalities are selected out of the gene pool.

The learning process: subsong, plastic song, and crystallization

Young birds learning to sing usually start by producing **subsong,** a soft, disorganized jumble of notes that sometimes bears little resemblance to adult song. Subsong is typically heard between late summer and early spring, often from birds hidden in dense vegetation. Many subsongs include imitations of other species.

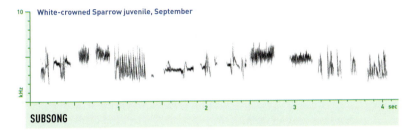

White-crowned Sparrow juvenile, September

SUBSONG

Eventually, young birds begin singing versions of their own species' songs that are recognizable, but variable and poorly stereotyped. This **plastic song** is typical of many young birds in their first spring. Adults in some species also give plastic song from fall through spring, as vocal centers in the brain undergo seasonal atrophy to reduce energy needs.

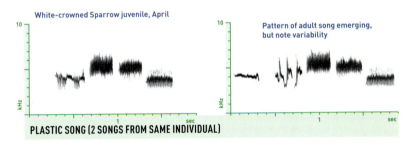

White-crowned Sparrow juvenile, April

Pattern of adult song emerging, but note variability

PLASTIC SONG (2 SONGS FROM SAME INDIVIDUAL)

The final stage of song development is marked by a process called **crystallization.** At this stage, changes in the bird's brain "lock in" stereotyped versions of the song or songs that it has learned. In many species, after crystallization, no further learning ever occurs.

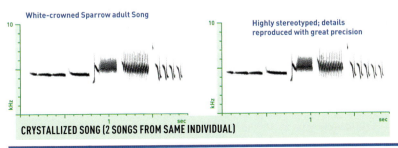

White-crowned Sparrow adult Song

Highly stereotyped; details reproduced with great precision

CRYSTALLIZED SONG (2 SONGS FROM SAME INDIVIDUAL)

Song differences within individuals

Plastic vs. stereotyped sounds

Plastic sounds differ slightly each time they are produced, even by the same bird. Stereotyped sounds are always the same when made by the same bird. Plastic songs are typical of young birds, but some sounds of certain species remain plastic into adulthood, like several calls of the House Finch.

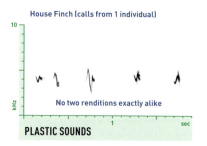

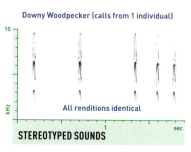

In adult birds, plasticity may vary with situation and season. Highly stereotyped versions of song tend to be associated with courtship. Plastic versions may be heard in winter, when levels of breeding hormones are lower.

Repertoires

In some species, individual birds learn only a single version of their species' song. In other species, individual birds learn multiple versions. These versions are called **songtypes**, and the set of all the songtypes that an individual bird can sing is called its **song repertoire.**

Birds with multiple songtypes may cycle through their repertoire, singing each song just once, or they may repeat one songtype multiple times before switching to another. If consecutive songtypes differ, a bird is said to be singing with **immediate variety.** If consecutive songtypes tend to be the same, it is said to be singing with **eventual variety.** Many intermediate patterns occur, and some birds change pattern depending on the situation.

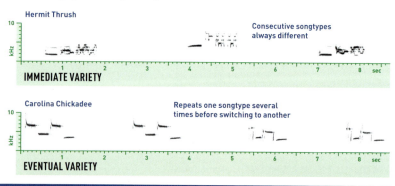

Song differences between individuals

Innovation, dialects, and individual recognition

Not all young birds copy songs from a tutor. Some invent their own songtypes according to the typical pattern of their species. In some species, like Song Sparrows, this is a rare behavior shown by just a few individuals. In other species it is a common strategy. Young American Robins may innovate up to 80 percent of their song phrases, learning the remaining 20 percent from tutors. Sedge Wrens may invent all their songtypes.

In species that learn from tutors, some young birds typically end up with slightly different versions than their tutors sang. The difference may be as minor as a slight modification to one note, or as major as a mixing of two previously distinct songtypes. As these birds get older they may serve as tutors for other young birds, establishing the modified song as a local variant. When birds breeding in a particular area sound similar to each other but different from members of the same species elsewhere, they are said to have a regional **dialect.**

Even within a dialect group, songs of individual birds may vary slightly. This variation can serve a biological purpose, allowing individual members of some species to recognize one another by song. Females may also use subtle distinctions between male songs to determine which males might make the fittest mates.

Variable vs. uniform sounds

This book uses the word **variable** to describe sounds that show high levels of individual variation, and **uniform** to describe sounds with low levels of individual variation. Variable sounds differ from one bird to the next; uniform sounds are similar in all members of a species.

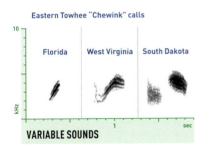

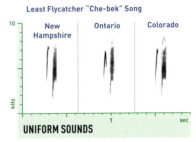

Imitation

Some bird species incorporate the sounds of other bird species, or even man-made sounds, into their own songs. This practice has often been called "mimicry," but many biologists prefer the terms "imitation" or "appropriation."

Why some birds appropriate others' songs is not well understood. However, the purpose is likely not to impersonate the imitated species. Most birds that imitate always signal their true identity by arranging the notes of other species in diagnostic patterns, mixing them with distinctive call notes, or both. Blue Jays that imitate hawks may be an exception. It has been hypothesized that jays may impersonate hawks as a way of drawing attention to the presence of a silent hawk or in order to frighten other birds, but more study is needed.

Species that regularly imitate include some vireos, corvids, gnatcatchers, mimids, European Starling, mynas, Phainopepla, Yellow-breasted Chat, western forms of Fox Sparrow, some orioles, and several species of cardueline finch, particularly Pine Grosbeak and Lesser Goldfinch. Many other birds imitate during the song-learning process.

Duets and choruses

Some bird species regularly perform duets, in which two birds sing together. Duetting birds are often mated pairs. In synchronized duets, the notes of the two birds are carefully timed, and the performance stereotyped. In unsynchronized duets, the timing and the performance are variable.

Duets occur in several North American families, including cranes, grebes, loons, chachalacas, and icterids. When more than two birds participate, the result is called a chorus. In this book, duets and choruses are indicated on the spectrograms by contrasting colors.

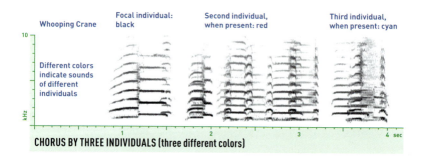

CHORUS BY THREE INDIVIDUALS (three different colors)

Sounds and species boundaries

Because females in many species choose mates by listening to the precise characteristics of each male's vocalizations, vocal differences can be a key factor keeping members of different species from pairing with one another—a reproductive isolating mechanism. In other words, differences in vocalizations can be not merely a symptom of being different species, but actually a mechanism that drives populations apart and keeps them separate. For this reason, taxonomists consider differences in song to be one important indicator of species boundaries.

Vocal differences can reproductively isolate populations that seem otherwise nearly identical. Such populations are often called **cryptic species.** For example, Willow and Alder Flycatchers were thought to be one species until the 1960s, when fieldwork showed that they often nest in the same areas without interbreeding, owing in large part to differences in their songs and calls.

Not all vocal differences are taxonomically important. Demonstrating that different-sounding populations are actually cryptic species requires careful fieldwork and genetic analysis to provide convincing evidence of reproductive isolation.

Sounds of hybrid birds

When birds of two different species pair and produce offspring, the resulting hybrid birds often give songs and calls that show characteristics of both parent species. This is generally true of birds with innate sounds, such as flycatchers, but can also occur in birds with learned sounds, such as sparrows.

In a few species pairs, such as Blue-winged and Golden-winged Warblers, hybrid offspring typically sound like one parent species or the other, not a mixture of both.

Insects and mammals that sound like birds

Some animals can sound quite similar to birds, especially frogs, toads, chipmunks, squirrels, and singing insects such as crickets, katydids, and cicadas. Such species are beyond the scope of this guide, but a few of the nonbird sounds most likely to cause confusion are mentioned in parentheses in the visual index alongside the bird sounds they resemble, so as to warn listeners of the possibility of misidentification.

Listeners should also be aware that recordings of birds in distress and/or birds of prey are sometimes used to repel pigeons, starlings, and gulls from roosting on buildings. Such recordings can sometimes be mistaken for the sounds of wild birds.

Recording Bird Sounds and Making Spectrograms

Creating one's own recordings and spectrograms is an excellent way to learn about bird sounds, and it has never been easier.

Many people can record audio with devices they already own, such as smartphones and digital cameras. Many different pocket voice recorders are now available, some for less than $100, that work quite well to record sounds that are reasonably close and loud.

Creating high-quality audio recordings of birds requires a high-quality microphone and a dedicated audio recorder. Many recordists use a shotgun microphone or a parabolic reflector dish to amplify the sound of a target bird relative to the background noise. The price of excellent audio recording equipment is comparable to the price of excellent photography equipment, but far fewer people own recording equipment. Good recordings have enormous potential to advance our knowledge of bird behavior, identifcation, and taxonomy.

From the 1960s through the 1990s, the only way to make spectrograms was using a Sona-Graph machine, manufactured by the Kay Elemetrics Company. The resulting "Sonagrams" were printed on paper, often in high-contrast black-and-white, which was not always ideal for showing the fine details of the sound.

Modern computer software makes it much easier to generate detailed and comprehensible pictures of bird sounds. Free software such as Raven Lite, available for download from the Cornell Laboratory of Ornithology website (**http:// birds.cornell.edu**), allows anyone to create spectrograms like those used in this book and customize them in myriad ways.

Playback and birding ethics

Improvements in technology have made it increasingly easy to play bird sound recordings back to birds in the field, usually in an attempt to draw them closer so that people can get a better view. This practice has long been controversial.

Playback changes the behavior of wild birds. It simulates a territorial challenge, or in some cases a threat from a predator, creating a situation to which birds must respond aggressively. What birders may call "cooperative" individual birds may in fact be extremely agitated.

Some have argued that playback may negatively impact the breeding success of the responding birds. The evidence on this point is scanty and mixed. However, playback that is extremely loud, long, and/or disruptive to the birds could be considered harassment, which is unethical in all cases, and illegal in the case of threatened and endangered species.

In deciding whether to use playback, consider these questions:

- **Can I achieve my purpose without playback?** If you can, perhaps you should.
- **Would the use of playback violate any regulations at this site?** Many state and national parks, national wildlife refuges, and other sites prohibit playback. Know and follow the relevant rules.

- **How often will these same individual birds likely be the target of playback?** If you think the answer is more than two to three times per season, consider moving to another area to target different individuals, or avoiding playback altogether.
- **Would playback disturb other people?** If you are birding with others, or near others, seek consensus on the use of playback before beginning.
- **How loud is too loud, and how long is too long?** The answers may vary, but consider erring on the side of quietude and brevity.

When in doubt about the appropriateness of playback or any other behavior, check the American Birding Association Code of Ethics, **www.aba.org/about/ethics.html**, for general guidelines.

Additional resources

Online bird sound libraries
In addition to the sounds that accompany this book online at **www.peterson birdsounds.com**, these websites host streaming audio of vast numbers of bird sounds:

Xeno-Canto: **www.xeno-canto.org**
Macaulay Library: **www.macaulaylibrary.org**
Borror Laboratory of Bioacoustics: **drc.ohiolink.edu/handle/2374 .OX/30658**
Florida Museum of Natural History: **www.flmnh.ufl.edu/bird-sounds**

Nocturnal flight call resources
Flight Calls of Migratory Birds (CD-ROM), by Bill Evans and Michael O'Brien, available online at **www.oldbird.org**

Books
Two books in particular provide good comprehensive introductions to the biology of bird sounds and the reading of spectrograms:

The Singing Life of Birds by Donald Kroodsma, 2007, Houghton Mifflin, Boston, MA
The Sound Approach to Birding by Mark Constantine and the Sound Approach, 2006, The Sound Approach, Poole, Dorset, UK

Blog
Earbirding.com: Nathan Pieplow (author of this book) and Andrew Spencer blog about recording, identifying, and interpreting bird sounds.

BIRDS BY SONG: SPECIES ACCOUNTS

WATERFOWL (Family Anatidae)

This family includes the geese, the swans, and the ducks, all of which are adapted to floating on the surface of the water. Many species are brightly and distinctively colored, and some have been domesticated.

WHISTLING-DUCKS (Family Anatidae, Subfamily Dendrocygninae)

In size and shape, these species are intermediate between ducks and geese, with long legs and long necks. Formerly called "tree ducks," they are partly arboreal. Their vocalizations, consisting mostly of breathy whistles, are apparently innate.

FULVOUS WHISTLING-DUCK

Dendrocygna bicolor

A tropical species that expanded into our area in the nineteenth century. U.S. population has fluctuated ever since; closely tied to flooded rice fields.

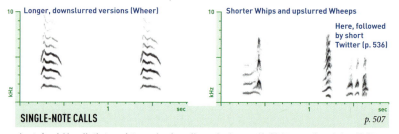

Flocks often give Twitter of wheezy Pips, punctuated by Ki-wheers

Typical version: clear, nasal

Wheezy version

Multinote version (uncommon; p. 554)

KI-WHEER

p. 551

All year; most common and distinctive call, likely by both sexes, often in flight. Fairly variable and plastic. Clear and wheezy versions may correspond to differences in sex, age, or function; more study needed. Multinote version recalls Black-bellied Whistling-Duck, but lower and more nasal.

Longer, downslurred versions (Wheer)

Shorter Whips and upslurred Wheeps

Here, followed by short Twitter (p. 536)

SINGLE-NOTE CALLS

p. 507

A set of variable calls that may intergrade; given all year, in alarm and in flight, sometimes in long Twitters. Most versions are 1-noted, like single syllables of Ki-wheer; same note often given twice in quick succession. Quality nasal to slightly polyphonic to fairly noisy. Usually given with Ki-wheers.

BLACK-BELLIED WHISTLING-DUCK

Dendrocygna autumnalis

Highly social and vocal, often gathering in loud flocks in shallow wetlands or adjacent fields. Range is expanding northward.

Flocks often give constant Twitter punctuated by Kip-whee-witters

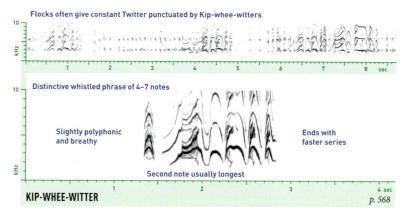

Distinctive whistled phrase of 4–7 notes

Slightly polyphonic and breathy

Ends with faster series

Second note usually longest

KIP-WHEE-WITTER
p. 568

All year; most common and distinctive call, likely by both sexes, often in flight. Fairly variable and plastic; individuals seem to have distinctive versions. Function little known.

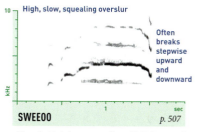

High, slow, squealing overslur

Often breaks stepwise upward and downward

SWEEOO p. 507

Given by birds in groups, often with Twitters and Kip-whee-witters; function unknown.

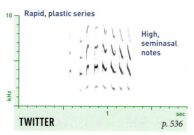

Rapid, plastic series

High, seminasal notes

TWITTER p. 536

Given by birds in groups, perhaps in contact. Highly plastic.

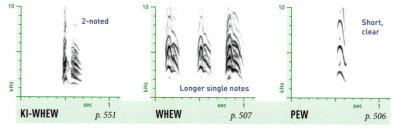

2-noted

KI-WHEW p. 551

Longer single notes

WHEW p. 507

Short, clear

PEW p. 506

A set of variable calls that may intergrade; given all year, in alarm and in flight, sometimes in long series. Whew and Pew are most common; Ki-whew recalls Ki-wheer of Fulvous Whistling-Duck, but shorter. Always at least slightly polyphonic, but some short versions are nearly whistled.

GEESE (Family Anatidae, Subfamily Anserinae)

Geese are familiar to all as the larger, longer-necked cousins of ducks. The voices and displays of the North American species are surprisingly little studied, but on average, their vocal communication appears to be simpler than that of ducks.

All goose sounds are apparently innate; most are plastic variations on a basic Honk, Yap, or Grunt. High levels of variation and plasticity make it difficult to determine the degree to which different sounds serve different functions, but at least some species have several discrete types of calls and displays.

Domestic and Feral Geese and Ducks

Many farms and urban parks host flocks of geese and ducks of domestic origin, which come in a variety of shapes, sizes, and colors. Many are entirely or partially white, with orange bills and legs; others may be mostly or partly black. Domestic geese are mostly derived from Eurasian species: the Greylag Goose (*Anser anser*), the Swan Goose (*Anser cygnoides*), or hybrids between them. Domestic ducks are mostly derived from the Mallard (p. 48) or the Muscovy Duck (p. 44).

These feral birds can be highly vocal; their voices generally resemble those of their wild ancestors, but generations of selective breeding and hybridization can alter their sounds.

EGYPTIAN GOOSE

Alopochen aegyptiacus

Native to Africa; now feral in parts of Florida and Texas, and escapes occur elsewhere. Nests in large trees or on ground, often in urban parks.

Display, often by duetting pairs: Grunt Series, sometimes accelerating into rapid Chatter (p. 547)

Female grunts

Male snarls

GRUNT SERIES   *p. 548*   **SNARL SERIES**   *p. 547*

All year. Only slightly variable, but calls vary subtly with behavioral context. Male gives hissing Snarl Series, while female gives louder Grunt Series, sometimes nasal or quacking. Series often continue for many minutes. Pairs and groups display loudly, often bowing heads or spreading wings.

Snow Goose

Chen caerulescens

Frequents marshes and farmland. Most are white with black wingtips. Dark morph, called "Blue Goose" (shown), is most common mid-continent.

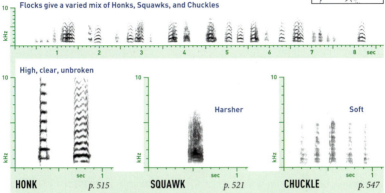

Flocks give a varied mix of Honks, Squawks, and Chuckles

High, clear, unbroken

Harsher

Soft

HONK   *p. 515*   **SQUAWK**   *p. 521*   **CHUCKLE**   *p. 547*

All sounds variable and plastic, heard year-round. Sound of large flocks giving high clear Honks in alarm is distinctive, higher than Canada Goose and lower than Greater White-fronted. Soft Chuckles, often given along with other sounds by flying flocks, nearly unique to this species and Ross's Goose.

Ross's Goose

Chen rossii

Smaller than Snow Goose, with shorter neck and stubby, triangular bill with blue warty base, lacking Snow's black "grinning patch."

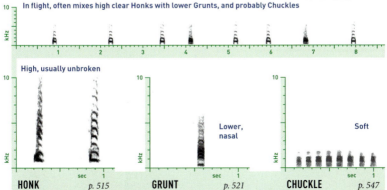

In flight, often mixes high clear Honks with lower Grunts, and probably Chuckles

High, usually unbroken

Lower, nasal

Soft

HONK   *p. 515*   **GRUNT**   *p. 521*   **CHUCKLE**   *p. 547*

Voice little studied, but very similar to Snow Goose; averages lower-pitched with shorter notes, but probably not reliably separable. All sounds variable and plastic, likely given all year. Sole available recording of Chuckles is from nest site, but flying flocks probably also give a version, as in Snow Goose.

Canada Goose

Branta canadensis

Abundant and familiar; now resident in many areas where it formerly occurred only in winter or migration. Subspecies differ greatly in size.

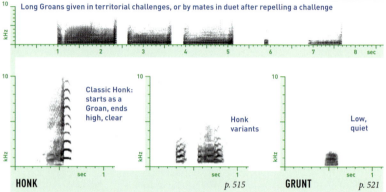

Long Groans given in territorial challenges, or by mates in duet after repelling a challenge

Classic Honk: starts as a Groan, ends high, clear

Honk variants

Low, quiet

HONK

p. 515

GRUNT

p. 521

All sounds highly variable and plastic; larger-bodied individuals have lower-pitched voices. Dominant birds in group are far more vocal than submissive birds. Subtle differences in Honks associated with different behaviors and displays; Grunts given in close contact. When threatened, adults give Hiss (p. 522).

Cackling Goose

Branta hutchinsii

Like a small Canada Goose, and formerly considered part of that species. Subspecies vary in size, with larger birds nearly identical to small Canadas.

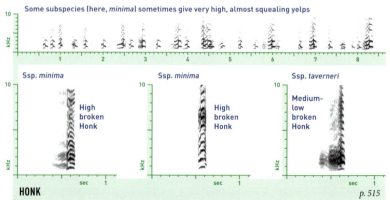

Some subspecies (here, *minima*) sometimes give very high, almost squealing yelps

Ssp. *minima*

High broken Honk

Ssp. *minima*

High broken Honk

Ssp. *taverneri*

Medium-low broken Honk

HONK

p. 515

All sounds highly variable and plastic, varying between subspecies, individuals, and situations. Larger subspecies (like Alaskan *taverneri*) can sound like Canada Goose; smaller subspecies (like Pacific Northwest *minima*) make some sounds that are higher than any Canada. Full repertoires poorly known.

Brant

Branta bernicla

Winters in flocks in salt water. Eastern birds have pale bellies and western birds dark bellies; a gray-bellied form breeds in the central Canadian Arctic.

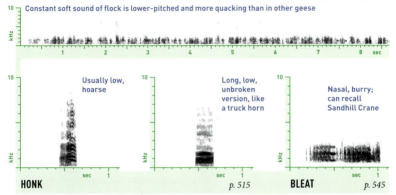

Constant soft sound of flock is lower-pitched and more quacking than in other geese

Usually low, hoarse

HONK

Long, low, unbroken version, like a truck horn

p. 515

Nasal, burry; can recall Sandhill Crane

BLEAT　　p. 545

All sounds variable and plastic. Generally lower-pitched and hoarser than sounds of other geese; bleating sounds are distinctive. If vocal differences exist between paler Atlantic population (top) and western "Black Brant" population (left, center, and right), they are subtle.

Greater White-fronted Goose

Anser albifrons

Medium-sized, with conspicuous orange legs. Orange-billed birds from Greenland occasionally wander to the East Coast. Vocally distinctive.

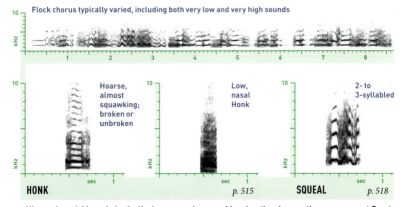

Flock chorus typically varied, including both very low and very high sounds

Hoarse, almost squawking; broken or unbroken

HONK

Low, nasal Honk

p. 515

2- to 3-syllabled

SQUEAL　　p. 518

All sounds variable and plastic. Honks average lower and harsher than in any other goose except Brant. Most distinctive sound is a 2- to 3-syllabled Squeal, high and hoarse, that can give away even a single individual of this species in a large flock of other geese. Rarely gives Chuckle like Snow Goose.

SWANS (Family Anatidae, Subfamily Anserinae)

Currently classified in the same subfamily as geese, the swans are a distinctive group. The largest of the waterfowl, they dip their extremely long necks under the water to forage on submerged plants. All three of our species have primarily white plumage as adults and gray plumage as juveniles.

Sounds are apparently innate and tend to be relatively simple and gooselike; pair and group displays may be accompanied by loud, unsynchronized series of calls in duet or chorus. In flight, the wings of all three species make distinctive hums or rattles that can carry for considerable distances, possibly serving a communicative function.

MUTE SWAN

Cygnus olor

The familiar swan of European parks and gardens, introduced in our area. Found on protected coastal waterways and freshwater ponds. Not mute at all.

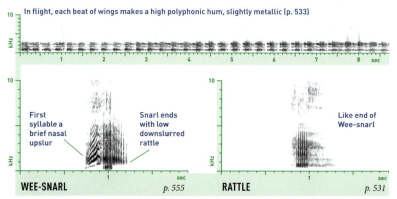

In flight, each beat of wings makes a high polyphonic hum, slightly metallic (p. 533)

WEE-SNARL *p. 555*

First syllable a brief nasal upslur

Snarl ends with low downslurred rattle

RATTLE *p. 531*

Like end of Wee-snarl

Wee-snarl and Rattle given all year, especially in breeding season, by adults, ranging from soft to fairly loud. Snarl sometimes much longer than shown. Variety of calls described in different situations, perhaps mostly variations on Rattle and Wee-snarl; vocal behavior needs more study.

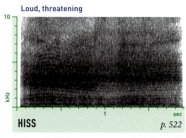

Loud, threatening

HISS *p. 522*

All year, in high aggression, especially by adults in summer defending nests or young.

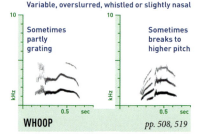

Variable, overslurred, whistled or slightly nasal

Sometimes partly grating

Sometimes breaks to higher pitch

WHOOP *pp. 508, 519*

At least Oct.–Jan., by juveniles, often in flight. Like Tundra Swan Whoop, but usually higher. Variable, plastic; likely develops into Wee-snarl.

TRUMPETER SWAN

Cygnus buccinator

Our largest swan. Hunted to near-extinction in the nineteenth century and extirpated from the East. Wild and re-introduced populations now increasing.

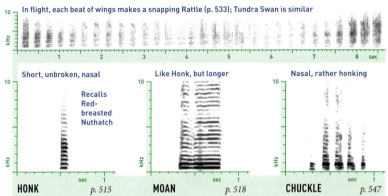

In flight, each beat of wings makes a snapping Rattle (p. 533); Tundra Swan is similar

Short, unbroken, nasal

Recalls Red-breasted Nuthatch

HONK   *p. 515*

Like Honk, but longer

MOAN   *p. 518*

Nasal, rather honking

CHUCKLE   *p. 547*

All sounds variable and plastic, mostly variations on a low, nasal Honk, with quality recalling a paper party horn. Lower and more nasal than Tundra Swan, with almost no overlap. Slow Chuckles are typical at the end of Long Calls given during displays by pairs or families when reuniting or after aggressive encounters.

TUNDRA SWAN

Cygnus columbianus

More common than Trumpeter Swan in the East. Winters in large flocks on coastal estuaries. At a distance, best distinguished from Trumpeter by voice.

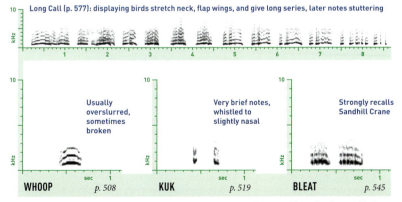

Long Call (p. 577): displaying birds stretch neck, flap wings, and give long series, later notes stuttering

Usually overslurred, sometimes broken

WHOOP   *p. 508*

Very brief notes, whistled to slightly nasal

KUK   *p. 519*

Strongly recalls Sandhill Crane

BLEAT   *p. 545*

All sounds variable and plastic, mostly variations on a high clear Whoop, with quality that recalls Snow Goose. More varied than Trumpeter; rare low Moans, perhaps given only on breeding grounds, most similar to Trumpeter. Long Call (top) given by pairs and families when reuniting or after aggressive encounters.

DUCKS (Family Anatidae, Subfamily Anatinae)

The "true" ducks in the subfamily Anatinae are a diverse group. As far as is known, all duck sounds are innate. Many sounds are sex-specific.

In many species, courtship occurs primarily on the wintering grounds, starting as early as October. Especially in the dabbling ducks, pairs are thought to migrate northward together in spring, and the pair bond typically lasts only until the eggs hatch, with new pairs forming the following winter.

Courtship generally takes place in groups on the water, with several males simultaneously courting one or more females. Many species have multiple courtship displays, some of which include vocalizations. In this context, some behaviors that may seem casual—such as preening, flapping the wings, and/or turning away from the female—can actually represent highly stereotyped courtship behaviors.

Courtship in many ducks is associated with a high degree of aggression, especially by males toward females. Many female ducks direct soft Chuckles or Growls (sometimes termed "inciting calls") at males they prefer, and louder Screeches (sometimes termed "repulsion calls") at males they are trying to fight off. Females may take to the air with males following in pursuit flights, one or both sexes often vocalizing loudly.

Females of all dabbling duck species give Long Calls (or "Decrescendo Calls"), a series of quacks that usually accelerates. Given in long-distance contact, especially to a bird's mate, Long Calls often incite other females to respond in kind. The Long Call of Mallard is by far the most frequently heard; in many species it is rare. In Cinnamon Teal and both wigeons, no definitively identified recordings are available.

MUSCOVY DUCK

Cairina moschata

Feral birds vary greatly in plumage; often seen in city parks, especially in Florida. Wild birds are rare, local, and shy, found along wooded waterways.

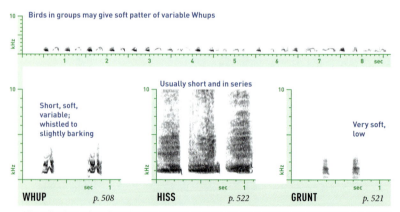

All vocalizations quite soft and infrequent. Whups little known; apparently given mostly by females. Hisses given by males in excitement. Both sexes reported to vocalize during tail-wagging, head-bobbing display. Grunts given in alarm, usually singly.

Wood Duck

Aix sponsa

Found on secluded, wooded rivers and ponds, usually in pairs. Nests in tree cavities or nest boxes. Generally quiet and fairly inconspicuous.

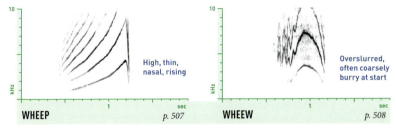

High, thin, nasal, rising

WHEEP *p. 507*

Overslurred, often coarsely burry at start

WHEEW *p. 508*

All year, by males only. Audible only at close range. Wheep is given frequently in a variety of situations. Wheew is associated mostly with alarm and the urge to fly. In courtship, males give a variety of plastic notes, all with similar distinctive thin, nasal quality, easily mistaken for polyphonic.

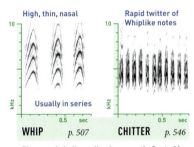

High, thin, nasal

Rapid twitter of Whiplike notes

Usually in series

WHIP *p. 507*

CHITTER *p. 546*

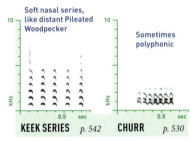

Soft nasal series, like distant Pileated Woodpecker

Sometimes polyphonic

KEEK SERIES *p. 542*

CHURR *p. 530*

These and similar calls given mostly Sept.–May, by males in courtship, along with a very soft, low, coughing Whup (not shown).

These and similar calls given all year by females, in courtship and in family contact. Somewhat variable and plastic; never loud.

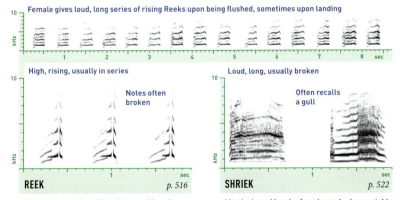

Female gives loud, long series of rising Reeks upon being flushed, sometimes upon landing

High, rising, usually in series

Notes often broken

REEK *p. 516*

Loud, long, usually broken

Often recalls a gull

SHRIEK *p. 522*

All year, by female only, in alarm and in pair contact. Plastic and variable but distinctive; by far the species' most common call.

Mostly Aug.–May, by females only, in courtship and in long-distance pair contact. Highly plastic. Usually given singly at long intervals.

GADWALL

Anas strepera

Breeds in dense wetland vegetation or on islands; widespread in migration and winter. Note white patch on trailing edge of wing.

Short version of Long Call

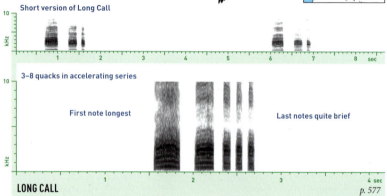

3–8 quacks in accelerating series

First note longest

Last notes quite brief

LONG CALL *p. 577*

Apparently all year, in long-distance contact by females only, but rarely given. Somewhat variable and plastic. Quality typically quite nasal, slightly harsh, without much change in pitch between notes. When 5 or more notes are present, note rapid pace at end.

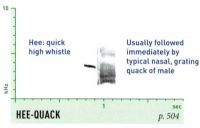

Hee: quick high whistle

Usually followed immediately by typical nasal, grating quack of male

HEE-QUACK *p. 504*

Most common sound of courting males, given while rearing up out of water with bill pressed to chest. Somewhat plastic.

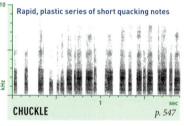

Rapid, plastic series of short quacking notes

CHUCKLE *p. 547*

Given in courtship, reportedly by both sexes. Much like Mallard Chuckle.

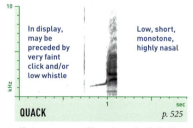

In display, may be preceded by very faint click and/or low whistle

Low, short, monotone, highly nasal

QUACK *p. 525*

Most common sound by males; quite distinctive.

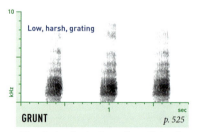

Low, harsh, grating

GRUNT *p. 525*

Given by females in alarm. Generally harsher than Mallard female Quack.

AMERICAN WIGEON

Anas americana

Found in shallow ponds, bays, estuaries, and city parks, sometimes in large flocks. Sometimes called "Baldpate" because of the male's white forehead.

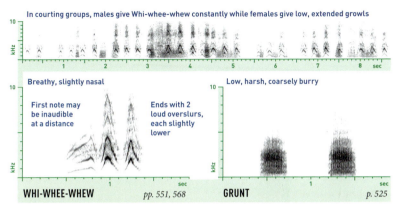

In courting groups, males give Whi-whee-whew constantly while females give low, extended growls

Breathy, slightly nasal

First note may be inaudible at a distance

Ends with 2 loud overslurs, each slightly lower

WHI-WHEE-WHEW   *pp. 551, 568*

Low, harsh, coarsely burry

GRUNT   *p. 525*

Whi-whee-whew given all year by males, in alarm, contact, flight, and courtship. Slightly plastic, but highly distinctive. Most versions are 3-noted; some have 2 or 4 notes. Grunts given only by females, in alarm or flight. Females rarely gives a 3- to 5-note Long Call like other ducks; quality apparently like typical Grunts.

EURASIAN WIGEON

Anas penelope

The Old World counterpart of American Wigeon, and closely related to it; the two occasionally hybridize. Rare in winter, usually with American Wigeons.

In courting groups, males give Wheew constantly while females give low, extended growls

Breathy, slightly nasal

First note audible only at close range

1 loud overslur, longer and higher than American's

WHEEW   *pp. 508, 551*

Low, very coarse; may average harsher than American

GRUNT   *p. 525*

Wheew given all year by males, in alarm, contact, flight, and courtship. Slightly individual and plastic, but highly distinctive. Grunts given only by females, in alarm or flight. Female rarely gives a 3- to 5-note Long Call like other ducks; quality apparently like typical Grunts.

Mallard

Anas platyrhynchos

The world's most common and wide-spread duck. As in many ducks, females tends to be more vocal than males. Late summer males resemble females.

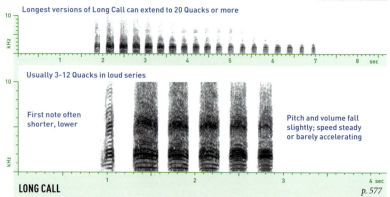

Longest versions of Long Call can extend to 20 Quacks or more

Usually 3-12 Quacks in loud series

First note often shorter, lower

Pitch and volume fall slightly; speed steady or barely accelerating

LONG CALL *p. 577*

All year, especially Sept.–Apr., by females only. Somewhat variable and plastic, especially in number of notes. Given in long-distance contact with distant mate or with other groups of Mallards. Tone quality variable, but usually strongly nasal and only slightly harsh.

Brief, nearly monotone whistle

Often sounds inhaled; followed by faint grunt

HEE *p. 504*

Quality like typical male Quack

SHORT QUACK *p. 525*

Much faster, more stuttering than male

CHUCKLE *p. 547*

Sept.–Mar., by males displaying in small groups. Slightly variable and plastic.

Mostly Sept.–Mar., in courtship; similar soft calls when feeding. Short Quacks given by males in plastic series. Chuckle given by females courted or pursued by males, including on the wing.

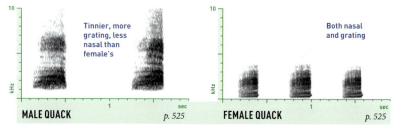

Tinnier, more grating, less nasal than female's

MALE QUACK *p. 525*

Both nasal and grating

FEMALE QUACK *p. 525*

All year; most common call, given loudly in alarm and contact, often upon takeoff or in flight. Usually in slow series. Male and female versions differ, often quite noticeably. Females Quack loudly and persistently for long periods in early spring when searching for nest sites.

American Black Duck

Anas rubripes

Closely related to Mallard; interbreeds with it frequently. Both sexes resemble very dark female Mallard, but male has yellower bill than female.

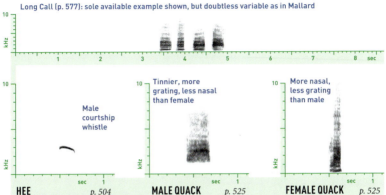

Long Call (p. 577): sole available example shown, but doubtless variable as in Mallard

Male courtship whistle

HEE   *p. 504*

Tinnier, more grating, less nasal than female

MALE QUACK   *p. 525*

More nasal, less grating than male

FEMALE QUACK   *p. 525*

All vocalizations and displays apparently similar to those of Mallard, but few recordings available. As in Mallard, female Quacks rather variable and plastic, ranging from nasal to fairly harsh; male Quacks less nasal and more tinny, lacking bass register.

Mottled Duck

Anas fulvigula

Closely related to Mallard; interbreeds with it frequently. Both sexes resemble slightly dark female Mallard with buffy face; black spot borders base of bill.

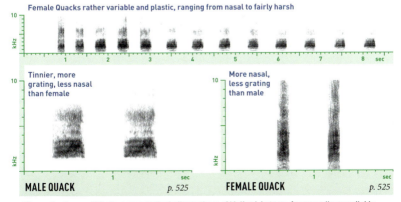

Female Quacks rather variable and plastic, ranging from nasal to fairly harsh

Tinnier, more grating, less nasal than female

MALE QUACK   *p. 525*

More nasal, less grating than male

FEMALE QUACK   *p. 525*

All vocalizations and displays apparently similar to those of Mallard, but very few recordings available.

BLUE-WINGED TEAL

Anas discors

A small duck of shallow ponds and wetlands. Blue shoulder patches, visible on
both sexes in flight, are shared with
Cinnamon Teal and Northern Shoveler.

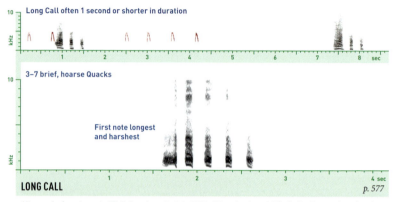

Long Call often 1 second or shorter in duration

3–7 brief, hoarse Quacks

First note longest
and harshest

LONG CALL *p. 577*

All year, by females only, likely in pair contact. Individual females have fairly distinctive versions, but these
are plastic, especially in number of notes. Shortness of notes can be useful in identification; Long Calls of
Gadwall and Northern Shoveler can be quite similar.

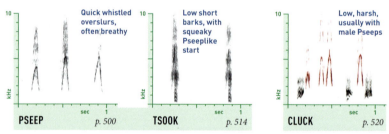

Quick whistled
overslurs,
often breathy

Low short
barks, with
squeaky
Pseeplike
start

Low, harsh,
usually with
male Pseeps

PSEEP *p. 500*   **TSOOK** *p. 514*   **CLUCK** *p. 520*

Pseep given all year, by males, in courtship and other contexts,
sometimes in series. Tsook apparently by males, possibly during
territorial or courtship conflicts; may grade into Pseep.

By courting females. Soft
and rather plastic; never
particularly fast.

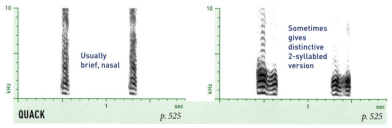

Usually
brief, nasal

Sometimes
gives
distinctive
2-syllabled
version

QUACK *p. 525*   *p. 525*

All year, by females only, in contact and alarm. Somewhat variable and quite plastic; often short, but may
become longer and rougher in agitation. Long series of Quacks sometimes given by females in spring, especially at dusk.

Cinnamon Teal

Anas cyanoptera

Found on shallow ponds with emergent vegetation. Closely related to Blue-winged. Females similar, but Cinnamon usually has larger bill and plainer face.

Female gives loud Quacks in series upon flushing; male gives series of Rattles

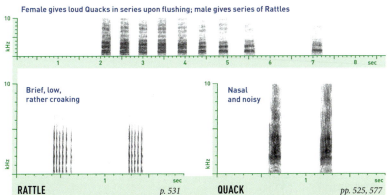

Brief, low, rather croaking

RATTLE *p. 531*

Nasal and noisy

QUACK *pp. 525, 577*

Rattle is typical male sound, given in courtship, in flight, and in alarm; completely unlike any other dabbling duck sound. Quacks, by females in alarm, are variable and plastic, but average medium-harsh. Females reported to give infrequent Long Call like Northern Shoveler; no recordings available.

Northern Shoveler

Anas clypeata

Found on shallow ponds. Named for its very large bill, which is immediately distinctive in both sexes. Closely related to Blue-winged and Cinnamon Teal.

Long Call (p. 577): 3-8 notes, first one usually longest, the rest in rapid series

Quick, swallowed squeaky bark

TSOOK *p. 514*

Generally long, nasal, slightly harsh

Sometimes distinctively 2-syllabled

QUACK *p. 525*

Tsook is typical male sound, given in courtship, in flight, and in alarm; some versions more nasal and barking. In flight, male wings make a low beating sound. Female Quacks rather variable and plastic, ranging from nasal to harsh; sometimes given in long series in spring, especially at dusk. Long Call infrequent.

Northern Pintail

Anas acuta

A slender, elegant duck, named for the male's elongated central tail feathers. Common on shallow water. Note all-dark bill and unmarked face of female.

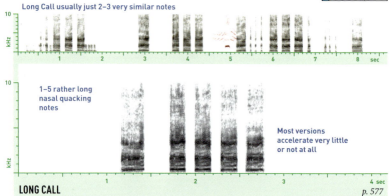

Long Call usually just 2–3 very similar notes

1–5 rather long nasal quacking notes

Most versions accelerate very little or not at all

LONG CALL *p. 577*

All year, by females only, likely in pair contact. Slightly variable and somewhat plastic; in some versions the individual notes are squeakier, vaguely 2-noted. Most versions are distinctively slow. Little distinction in this species between brief Long Calls and short series of Quacks.

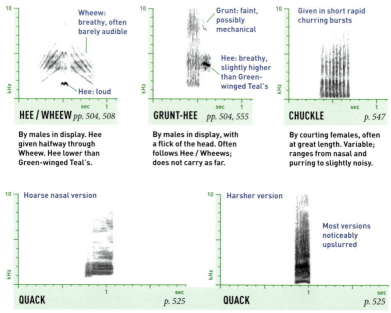

Wheew: breathy, often barely audible

Hee: loud

HEE / WHEEW *pp. 504, 508*

By males in display. Hee given halfway through Wheew. Hee lower than Green-winged Teal's.

Grunt: faint, possibly mechanical

Hee: breathy, slightly higher than Green-winged Teal's

GRUNT-HEE *pp. 504, 555*

By males in display, with a flick of the head. Often follows Hee / Wheews; does not carry as far.

Given in short rapid churring bursts

CHUCKLE *p. 547*

By courting females, often at great length. Variable; ranges from nasal and purring to slightly noisy.

Hoarse nasal version

QUACK *p. 525*

Harsher version

Most versions noticeably upslurred

QUACK *p. 525*

All year, by females only, in contact and alarm. Somewhat variable and quite plastic. Generally less frequent than Chuckle in this species, but long series of Quacks sometimes given by females in spring. Most versions as rough as or rougher than Mallard Quack.

Green-winged Teal

Anas crecca

Our smallest dabbling duck. Eurasian form, rare on coasts, has horizontal white stripe on wing, no vertical white bar on side; sounds similar.

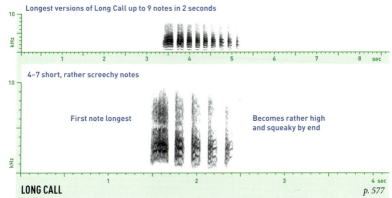

Longest versions of Long Call up to 9 notes in 2 seconds

4–7 short, rather screechy notes

First note longest

Becomes rather high and squeaky by end

LONG CALL   *p. 577*

All year, by females only, likely in pair contact. Slightly variable and somewhat plastic. On average the highest Long Call of any duck in our area; squeaky tone of last few notes distinctive when present.

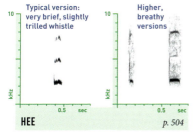

Typical version: very brief, slightly trilled whistle

Higher, breathy versions

HEE   *p. 504*

Mostly Sept.–Mar.; plastic. Given by courting males; different versions apparently accompany different courtship displays.

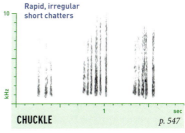

Rapid, irregular short chatters

CHUCKLE   *p. 547*

Mostly Sept.–Mar.; plastic. Given by courting females, usually with male Hees. Male gives a similar call during bill-up display.

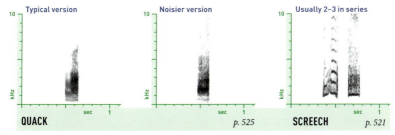

Typical version

Noisier version

Usually 2–3 in series

QUACK   *p. 525*

SCREECH   *p. 521*

All year, by females only, in alarm and contact. Somewhat plastic; may become rougher in agitation. Sometimes vaguely 2-syllabled. Long series of Quacks sometimes given by females in spring. Screeches given in aggression against males; may grade into Long Call.

REDHEAD

Aythya americana

Breeds mostly in prairie marshes. As in many other ducks, females often lay eggs in nests of other Redheads, or even other duck species.

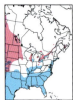

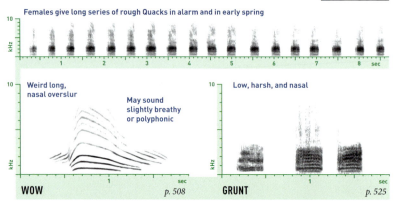

Females give long series of rough Quacks in alarm and in early spring

Weird long, nasal overslur

May sound slightly breathy or polyphonic

WOW   *p. 508*

Low, harsh, and nasal

GRUNT   *p. 525*

Wow is male display sound, given frequently Feb.–May, with a backward toss of the head. Distinctive, uniform, and stereotyped; longer than any similar sound by other Aythya ducks. Male may give some other sounds in display; more study needed. Female Quacks plastic, but average very rough.

CANVASBACK

Aythya valisineria

Larger than Redhead, with an elegant, sloping bill profile. Male has a striking white back. Population has fluctuated in recent years.

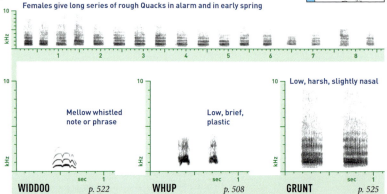

Females give long series of rough Quacks in alarm and in early spring

Mellow whistled note or phrase

WIDDOO   *p. 522*

Low, brief, plastic

WHUP   *p. 508*

Low, harsh, slightly nasal

GRUNT   *p. 525*

Not very vocal. Widdoo is one of male's display sounds, given mostly Feb.–May. Highly plastic, possibly grading into other, poorly known types of display sounds. Highly plastic Whups also given during courtship; unknown whether by males or females. Female Quacks plastic, but average very rough.

LESSER SCAUP

Aythya affinis

Very similar to Greater Scaup, but note slight peak on crown just behind eye. Generally more common than Greater, especially inland.

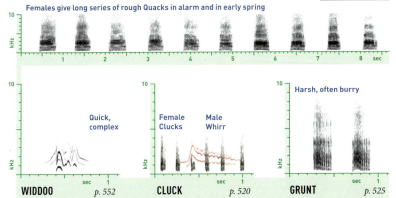

Females give long series of rough Quacks in alarm and in early spring

Quick, complex	Female Clucks	Male Whirr
		Harsh, often bury
WIDDOO *p. 552*	**CLUCK** *p. 520*	**GRUNT** *p. 525*

Widdoo given mostly Feb.–May, by males displaying with backward toss of head. Whirr (p. 507) is likely one of several other male courtship sounds; more study needed. Courting females give Cluck, sometimes run into slow Chuckles. All sounds soft except for Grunt, which is sometimes more of a Quack, given by females.

GREATER SCAUP

Aythya marila

Resembles Lesser Scaup, but slightly larger, with more rounded head shape. In flight, wing stripe more extensively white. Most winter on salt water.

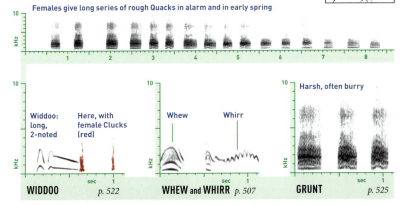

Females give long series of rough Quacks in alarm and in early spring

Widdoo: long, 2-noted	Here, with female Clucks (red)	Whew / Whirr
		Harsh, often bury
WIDDOO *p. 522*	**WHEW and WHIRR** *p. 507*	**GRUNT** *p. 525*

Widdoo given mostly Feb.–May, by males displaying with backward toss of head. Whew and Whirr are likely other male courtship sounds, but more study needed. Females give Clucks (p. 520) in courting groups. All sounds fairly soft except for Grunt, given by females.

RING-NECKED DUCK

Aythya collaris

Often found on wooded ponds. Note distinctive bill markings, peaked head shape, and, in males, bright white vertical stripe on side of breast.

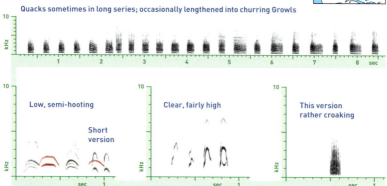

Quacks sometimes in long series; occasionally lengthened into churring Growls

WHEW — Low, semi-hooting / Short version — *p. 507*

PSEEP — Clear, fairly high — *p. 500*

GRUNT — This version rather croaking — *p. 525*

Several different sounds apparently by males in courtship, but more study needed. Whew like those of scaup, but sometimes longer. Pseeps recall Blue-winged Teal, but lower. Grunts, by females, vary from croaking to fairly clear, nasal, and quacking.

COMMON EIDER

Somateria mollissima

Our largest duck, found almost exclusively on salt water. Often forms large flocks just offshore. Like scoters, dives to feed on mollusks and crustaceans.

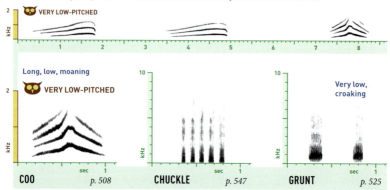

Male Coos may be up- or overslurred; different versions may serve different functions

VERY LOW-PITCHED

COO — Long, low, moaning — VERY LOW-PITCHED — *p. 508*

CHUCKLE — *p. 547*

GRUNT — Very low, croaking — *p. 525*

Coo given starting in fall by courting males, usually accompanied by a dramatic forward or backward toss of the head. Chuckle given by courting females in response. Grunts given by females in alarm and likely in some courtship displays.

Long-tailed Duck

Clangula hyemalis

Breeds in the Arctic; winters primarily along seacoasts. Rare inland away from the Great Lakes. Molt schedule complex, resulting in a variety of plumages.

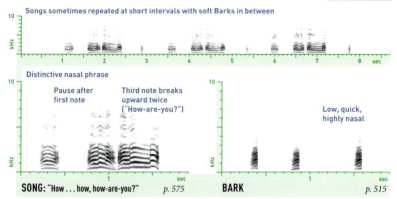

Songs sometimes repeated at short intervals with soft Barks in between

Distinctive nasal phrase

Pause after first note

Third note breaks upward twice ("How-are-you?")

Low, quick, highly nasal

SONG: "How . . . how, how-are-you?"   *p. 575*

BARK   *p. 515*

Song given all year, especially Jan.–June, reportedly only by males. Rather uniform; variation between males is mostly in final syllable. Sometimes given in flight. Barks generally soft, given in alarm or contact. Very soft Crackle (p. 523), possibly nonvocal, sometimes given with Barks in alarm.

Harlequin Duck

Histrionicus histrionicus

Strikingly patterned and social. Breeds along clear, fast-flowing streams; winters on rocky seacoasts, in or near the crashing surf. Rare inland.

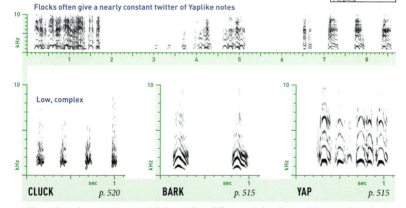

Flocks often give a nearly constant twitter of Yaplike notes

Low, complex

CLUCK   *p. 520*

BARK   *p. 515*

YAP   *p. 515*

All sounds are frequent, given mostly in interactions. Differences in function poorly known; voices of sexes may differ. Yaps are most frequent sounds; highly plastic, grading into low Barks and Clucks. Also gives soft, low Moans (not shown; p. 518).

Surf Scoter

Melanitta perspicillata

Our most strikingly patterned scoter. Nests on lakes in boreal forest; winters in flocks along seacoasts. Rare inland. Rarely vocalizes.

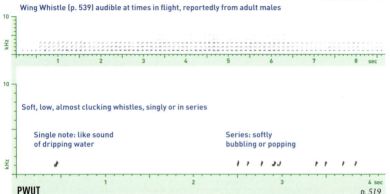

Wing Whistle (p. 539) audible at times in flight, reportedly from adult males

Soft, low, almost clucking whistles, singly or in series

Single note: like sound of dripping water

Series: softly bubbling or popping

PWUT

p. 519

Pwut given at least Mar.–June, by displaying males, but does not carry far. Displaying male stretches neck out, sometimes bringing bill to chest. Single notes and series likely accompany different displays. Females reportedly give crowlike Croaks in display and in nest defense; no recordings available.

White-winged Scoter

Melanitta fusca

Our largest scoter. White wing patch distinctive. Nests on lakes in boreal forest; winters in flocks along seacoasts. Rare inland. Rarely vocalizes.

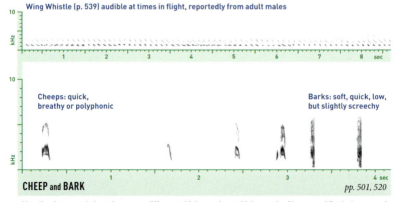

Wing Whistle (p. 539) audible at times in flight, reportedly from adult males

Cheeps: quick, breathy or polyphonic

Barks: soft, quick, low, but slightly screechy

CHEEP and BARK

pp. 501, 520

Vocalizations rarely heard; sources differ on which sex gives which sounds. Cheeps and Barks known only from group interactions on breeding grounds; both sounds audible only at close range. Cheep may accompany chin-lift display. Females may croak or gurgle in nest defense; no recordings available.

BLACK SCOTER

Melanitta americana

Our smallest scoter. Nests on lakes in boreal forest; winters in flocks along seacoasts. Rare inland. Males vocalize frequently.

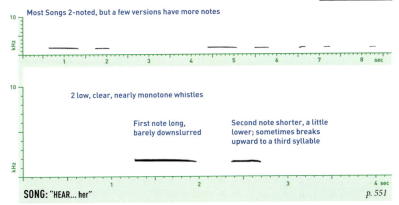

Most Songs 2-noted, but a few versions have more notes

2 low, clear, nearly monotone whistles

First note long, barely downslurred

Second note shorter, a little lower; sometimes breaks upward to a third syllable

SONG: "HEAR... her"

p. 551

Song given at least Nov.–June, by males; carries for long distances. Plaintive quality distinctive. Sometimes given in flight by flushed birds. Females reportedly give nasal or growling calls; may give harsh Quacks like scaup. Wing Whistle (p. 539) audible at times in flight, reportedly from adult males.

BUFFLEHEAD

Bucephala albeola

Our smallest diving duck. Nests in tree cavities, usually old flicker nests, in mixed woods. Winters on fresh and salt water. Rarely vocalizes.

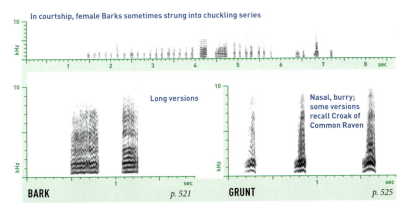

In courtship, female Barks sometimes strung into chuckling series

Long versions

Nasal, burry; some versions recall Croak of Common Raven

BARK p. 521

GRUNT p. 525

Male is nearly silent in display, except for soft slapping sounds made by flicking wings. Courting female gives plastic Barks, often in series, while swimming behind a displaying male. In alarm, female gives Quacks, some versions harsh and grating. Wings do not whistle in flight.

COMMON GOLDENEYE

Bucephala clangula

Breeds in tree cavities in boreal forests; winters on both fresh and salt water. Forehead slopes back from bill to peak above eye; bill of female mostly dark.

Wing Whistle (p. 539): in flight, male's wings make loud, clear whistled series

2-note nasal buzz, all on one pitch

Second note louder, longer

Soft, low ticking trills

PI-PEENT *p. 553*

RATTLE *p. 531*

Mostly Dec.–May, by males in courting groups on the water. Pi-peent is loudest and most distinctive sound; accompanies dramatic backward head throw. Rattles given with head stretched forward, nearly touching water.

Rapid, often long

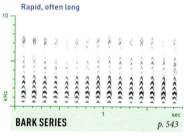

BARK SERIES *p. 543*

Mostly Apr.–June, by female, possibly in courtship, sometimes in flight. Recalls Pileated Woodpecker Keek Series, but lower, more nasal.

High, soft, monotone

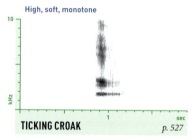

TICKING CROAK *p. 527*

All year, by males, in alarm and in flight; may also play a role in courtship displays. Audible mostly at close range.

Harsh, nasal, burry

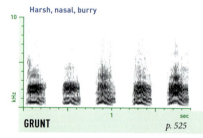

GRUNT *p. 525*

Low, cootlike

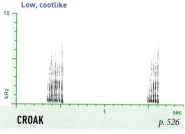

CROAK *p. 526*

By females. Grunts all year, in alarm and in flight; much like calls of *Aythya* ducks. Croaks not well known; possibly given mostly in nest defense. Both calls apparently plastic; may grade into one another and into Bark Series.

Barrow's Goldeneye

Bucephala islandica

Breeds in tree cavities in boreal and montane forests. Winters along coast in small numbers. Peak of head is forward of eye; bill of female mostly yellow.

Wing Whistle (p. 539): slightly lower than Common Goldeneye's

Much softer than Common's Pi-peent, less nasal

Second note as short as first, or shorter

PEENT-IP   *p. 553*

Soft ticks in couplets

TICK-IT SERIES   *p. 556*

Mostly Dec.–May, by males in courting groups on the water. Peent-ip accompanies dramatic backward head throw; sometimes preceded by soft short Hiss. Tick-it Series apparently given with head stretched forward, nearly touching water, or during head throw. Both sounds difficult to hear from a distance, especially Tick-it Series.

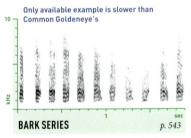

Only available example is slower than Common Goldeneye's

BARK SERIES   *p. 543*

Mostly Apr.–June, by females, possibly in courtship, sometimes in flight.

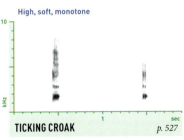

High, soft, monotone

TICKING CROAK   *p. 527*

All year, by males, in alarm and in flight; may also play a role in courtship displays. Audible mostly at close range.

Harsh, nasal, burry

GRUNT   *p. 525*

Low, cootlike

Here, followed by short nasal barks

CROAK   *p. 526*

By females. Grunts all year, in alarm and in flight; much like calls of *Aythya* ducks. Croaks not well known; possibly given mostly in nest defense. Both calls apparently plastic; may grade into one another and into Bark Series.

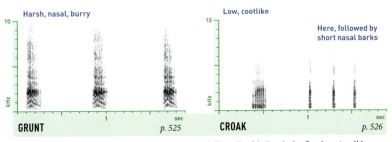

Common Merganser

Mergus merganser

Nests on lakes and along clear rivers; winters on both fresh and salt water. Like other mergansers, dives to catch fish with serrated bill.

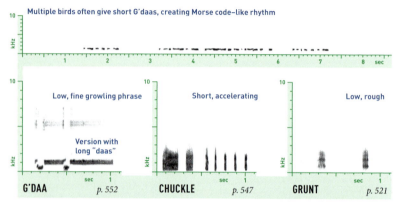

Multiple birds often give short G'daas, creating Morse code–like rhythm

Low, fine growling phrase

Short, accelerating

Low, rough

Version with long "daas"

G'DAA *p. 552*   **CHUCKLE** *p. 547*   **GRUNT** *p. 521*

G'daa given mostly Dec.–May by males in courting groups on the water, while stretching neck and/or tossing head. Several different variants may correspond to different displays. Chuckle given infrequently by females during courtship. Grunts given all year, at least by females, in alarm, often in flight.

Red-breasted Merganser

Mergus serrator

Breeds on lakes and rivers; winters primarily on salt water, but some winter inland. Slightly smaller than Common; both sexes have ragged crest.

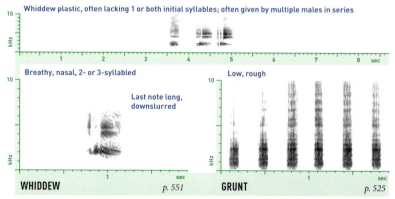

Whiddew plastic, often lacking 1 or both initial syllables; often given by multiple males in series

Breathy, nasal, 2- or 3-syllabled

Last note long, downslurred

Low, rough

WHIDDEW *p. 551*   **GRUNT** *p. 525*

Whiddew given mostly Dec.–May by males in courting groups on the water, with brief vertical stretch of head and bill. Distinctive; from a distance, may sound catlike. Females reported to growl in courtship; no recordings available. Grunts given all year, at least by females, in alarm, often in flight.

HOODED MERGANSER

Lophodytes cucullatus

Nests in tree cavities around wooded ponds. Male's white head patch is a large semicircle when crest is raised, a thin stripe when it is lowered.

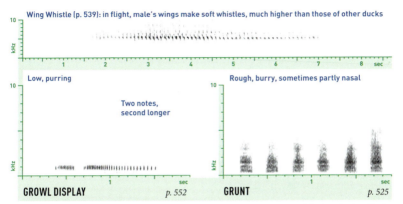

Wing Whistle (p. 539): in flight, male's wings make soft whistles, much higher than those of other ducks

Low, purring

Two notes, second longer

GROWL DISPLAY   *p. 552*

Rough, burry, sometimes partly nasal

GRUNT   *p. 525*

Growl Display given mostly Dec.–May by males in courting groups on the water, while throwing head backward to touch the back. Does not carry far. Females reported to make low croaks in courtship; no recordings available. Grunts given all year, at least by females, in alarm, often in flight.

RUDDY DUCK

Oxyura jamaicensis

Breeds in marshes and on ponds with a mix of emergent vegetation and open water. Aggressive in behavior; highly aquatic, almost unable to walk on land.

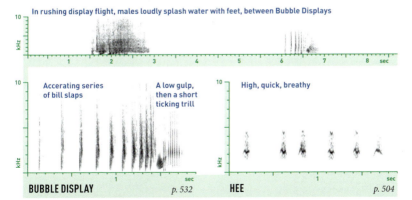

In rushing display flight, males loudly splash water with feet, between Bubble Displays

Accerating series of bill slaps

A low gulp, then a short ticking trill

BUBBLE DISPLAY   *p. 532*

High, quick, breathy

HEE   *p. 504*

Unlike other ducks, displays only on breeding grounds. Bubble Display mostly May–July, by courting males with tail vertical; bill slapped rapidly down against chest, frothing the water. Whistles given by courting females. Females also reportedly give nasal calls in alarm; no recordings available.

CHICKENLIKE BIRDS (Order Galliformes)

This diverse order contains several families of terrestrial birds ranging from medium-sized to very large. Often known as "game birds," they tend to fly only short distances, usually when flushed by a predator.

Most birds in this group have well-developed vocal repertoires. All vocalizations appear to be innate. Many species, especially peacocks, turkeys, and grouse, have striking courtship displays that involve fanned tails, wing flaps, or elaborate dances, often with distinctive sounds. Some species, including chachalacas and some quail, perform synchronized duets or choruses.

Domestic and feral game birds

Several species in this group have been domesticated, and most are hunted; many have been introduced to North America for hunting purposes, including Ring-necked Pheasant and Gray Partridge. In several species, some populations continue to be maintained or supplemented by released birds. Even some native species, such as Northern Bobwhite, are sometimes released far from their native range in order to be hunted.

The Domestic Chicken, Indian Peafowl, and Helmeted Guineafowl are among the most common bird species kept in captivity, and all three sometimes roam free in fields, yards, and even towns. Although they have not established self-sustaining feral populations in North America, their vocalizations are a significant component of the soundscape in many areas.

DOMESTIC CHICKEN

Gallus gallus

Ubiquitous and familiar. Domestic birds derived from the Red Junglefowl of southeast Asia, but show huge variation in plumage; vocalizations apparently still strongly resemble those of wild forms.

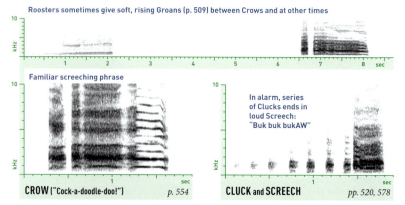

Roosters sometimes give soft, rising Groans (p. 509) between Crows and at other times

Familiar screeching phrase

In alarm, series of Clucks ends in loud Screech: "Buk buk bukAW"

CROW ("Cock-a-doodle-doo!") *p. 554*

CLUCK and SCREECH *pp. 520, 578*

Crows given mostly by roosters, rarely by hens, often starting well before dawn. Crows usually 4-syllabled, sometimes 2- or 3-syllabled, ranging from very screechy to nearly clear. Clucks given in alarm, sometimes in long series, often culminating in loud Screech.

Indian Peafowl

Pavo cristatus

Common in zoos, often wandering into adjacent neighborhoods. Some apparently breed in the wild in parts of urban southern California and possibly south Florida.

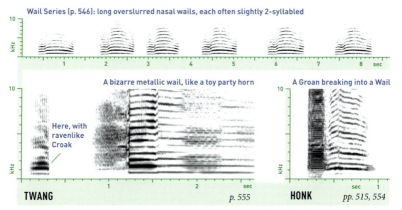

Wail Series (p. 546): long overslurred nasal wails, each often slightly 2-syllabled

A bizarre metallic wail, like a toy party horn

Here, with ravenlike Croak

TWANG   *p. 555*

A Groan breaking into a Wail

HONK   *pp. 515, 554*

Wail Series (top) is most familiar call; loud, frequent, and distinctive. Male with tail spread in display often silent, but may give faster, lower version of Wail Series, and less frequently Twang, sometimes with a soft Croak. Honks and similar calls may be given in alarm or long-distance contact.

Helmeted Guineafowl

Numida meleagris

Native to Africa; widely kept in captivity, sometimes roaming free. A few pairs may breed in the wild in Florida. Domestic birds may have white or pied plumage.

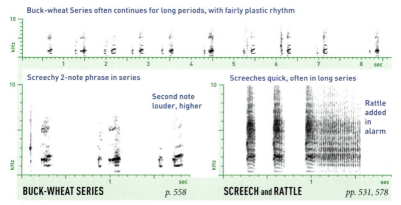

Buck-wheat Series often continues for long periods, with fairly plastic rhythm

Screechy 2-note phrase in series

Second note louder, higher

BUCK-WHEAT SERIES   *p. 558*

Screeches quick, often in long series

Rattle added in alarm

SCREECH and RATTLE   *pp. 531, 578*

All calls loud and generally unmusical. Buck-wheat Series given all year, reportedly only by female; variable and plastic, tone varying from screechy to nearly whistled. Screeches and Rattles reportedly given by both sexes; male often gives long series of quick Screeches. Rattles given by agitated birds.

PLAIN CHACHALACA

Ortalis vetula

A pheasant-sized, mostly arboreal bird of thorn scrub and woodlands. Dominates the soundscape with its raucous, trumpeting choruses.

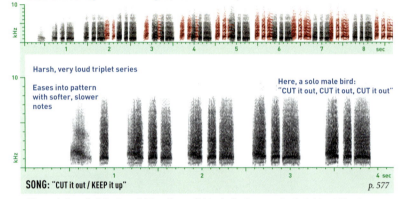

Duet: female (red) joins male with higher, clearer "KEEP it up!" series

Harsh, very loud triplet series

Eases into pattern with softer, slower notes

Here, a solo male bird: "CUT it out, CUT it out, CUT it out"

SONG: "CUT it out / KEEP it up" *p. 577*

All year, but mostly Feb.–June. Fairly uniform; slighty plastic. Song usually started by adult male on conspicuous perch, then joined by female lower in same tree. Other individuals often join in, creating a truly ear-splitting chorus. Most frequent at dawn and dusk; sometimes given at night.

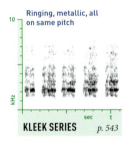

Ringing, metallic, all on same pitch

KLEEK SERIES *p. 543*

All year, in high alarm. Loud, plastic series of screechy 1- or 2-syllabled chirps.

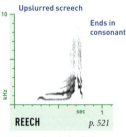

Upslurred screech

Ends in consonant

REECH *p. 521*

All year. Loud, plastic, in high alarm; associated with Kleek Series.

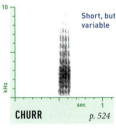

Short, but variable

CHURR *p. 524*

All year; most common call, often run into steady chatter. Often introduces Song.

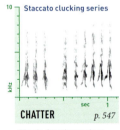

Staccato clucking series

CHATTER *p. 547*

Plastic; function unclear. Has an unmusical, machine-gun quality.

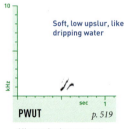

Soft, low upslur, like dripping water

PWUT *p. 519*

All year, in close contact; inaudible at a distance. Plastic.

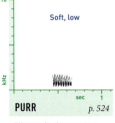

Soft, low

PURR *p. 524*

All year, in close contact, often with Pwut. Slightly plastic.

WILD TURKEY

Meleagris gallopavo

Found in wooded areas and adjacent fields. Range shrank in nineteenth century owing to hunting; now larger than ever thanks to reintroductions.

Some versions of Yelp Series are clear, almost whistled

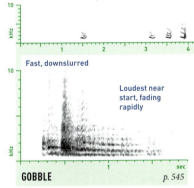

Fast, downslurred

Loudest near start, fading rapidly

GOBBLE   p. 545

Unmistakable. By males, often from a tree, to attract a mate; often echoed by other males. Male in strutting display nearly silent.

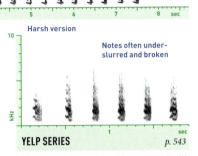

Harsh version

Notes often underslurred and broken

YELP SERIES   p. 543

All year, by both sexes. Variable and plastic; different versions may serve functions from alarm to reassembly of scattered flocks.

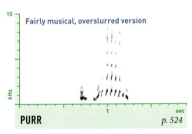

Fairly musical, overslurred version

PURR   p. 524

All year. Variable, ranging from musical trills to harsh rattles; often nasal. Used to maintain spacing in flocks, in threats, and in alarm.

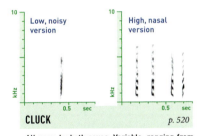

Low, noisy version

High, nasal version

CLUCK   p. 520

All year, by both sexes. Variable, ranging from soft, noisy versions in contact to high, fast, nasal Keklike versions in high alarm.

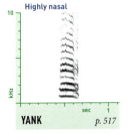

Highly nasal

YANK   p. 517

All year. Highly plastic; given during flock interactions, usually with Purrs.

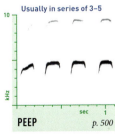

Usually in series of 3–5

PEEP   p. 500

As early as May, by young birds separated from hen; grades into Yelp by fall.

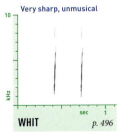

Very sharp, unmusical

WHIT   p. 496

As early as May, by young birds in alarm, often with Peeps.

SCALED QUAIL

Callipepla squamata

Found in dry, grassy habitats with some cover, sometimes including xeriscaped residential areas. Quite social, forming coveys outside the breeding season.

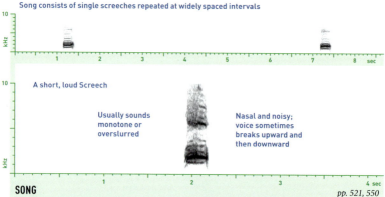

Song consists of single screeches repeated at widely spaced intervals

A short, loud Screech

Usually sounds monotone or overslurred

Nasal and noisy; voice sometimes breaks upward and then downward

SONG *pp. 521, 550*

Mostly Apr.–Aug., by unmated males, usually from a conspicuous perch at mid-height, such as a fencepost. Slightly variable and plastic. Kuk-curr calls sometimes interspersed. Can resemble notes from Yelp Series of Northern Bobwhite, but harsher and given at longer intervals.

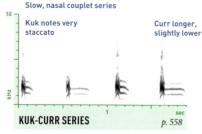

Slow, nasal couplet series

Kuk notes very staccato

Curr longer, slightly lower

KUK-CURR SERIES *p. 558*

All year, by both sexes, when separated from other quail. Series often starts very softly and grows louder; can continue for long periods.

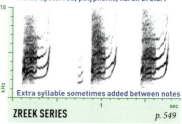

Notes upslurred, polyphonic, harsh at start

Extra syllable sometimes added between notes

ZREEK SERIES *p. 549*

All year, but mostly during breeding season, by males in aggressive display, with head toss; rare from females.

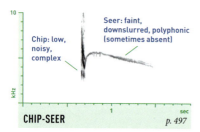

Chip: low, noisy, complex

Seer: faint, downslurred, polyphonic (sometimes absent)

CHIP-SEER *p. 497*

All year, by both sexes, in alarm. Sometimes given upon flushing. Whistled portion often audible only at close range.

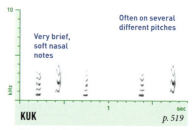

Very brief, soft nasal notes

Often on several different pitches

KUK *p. 519*

All year, by both sexes, in close contact. Quite plastic; never loud. Low, rough versions sometimes given upon flushing.

NORTHERN BOBWHITE

Colinus virginianus

A social quail of forest edges, shelter-belts, open woods, and savannas. In many areas, large numbers are raised and released to be hunted.

Duet (p. 568): "Hoyp-woo . . . Bob . . . WHITE?," possibly by mated pairs

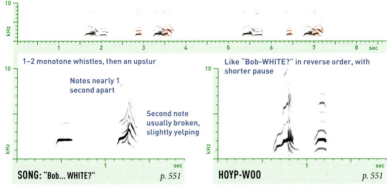

1–2 monotone whistles, then an upslur

Notes nearly 1 second apart

Second note usually broken, slightly yelping

SONG: "Bob... WHITE?" p. 551

Highly distinctive. Mostly Apr.–Aug., by males, those either without mates or separated from mates. Females sing rarely and softly.

Like "Bob-WHITE?" in reverse order, with shorter pause

HOYP-WOO p. 551

All year, in series of 4–5, by flockmates reassembling or by females responding to male Song. Quite variable, somewhat plastic.

Broken, overslurred, rather musical

End lower, sometimes sounding like another syllable

YELP SERIES p. 543

All year, in similar situations to Hoyp-woo; also at dawn by waking coveys. Variable, somewhat plastic.

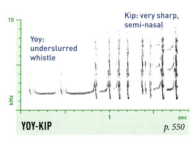

Kip: very sharp, semi-nasal

Yoy: underslurred whistle

YOY-KIP p. 550

All year, in alarm. Yoy notes soft, sometimes nasal; Kips multiply as agitation rises, sometimes running into a sharp, chipping chatter.

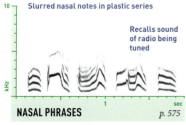

Slurred nasal notes in plastic series

Recalls sound of radio being tuned

NASAL PHRASES p. 575

All year. Soft and plastic. Given mostly by males in aggressive encounters; also possibly in alarm.

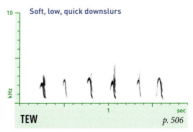

Soft, low, quick downslurs

TEW p. 506

All year, in close contact, often with short wailing or squealing calls. Plastic; not very loud.

MONTEZUMA QUAIL

Cyrtonyx montezumae

An elusive quail, most common in open, grassy oak woodlands, but ranging locally from shrubby deserts all the way up to pine forests.

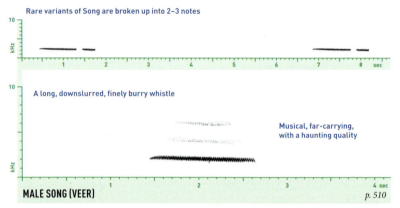

Rare variants of Song are broken up into 2–3 notes

A long, downslurred, finely burry whistle

Musical, far-carrying, with a haunting quality

MALE SONG (VEER)
p. 510

Mostly Feb.–Sept., by males. May be heard at any time of day, but most frequent at dawn and dusk. May serve a territorial function; also sometimes given as birds attempt to reunite after covey is flushed.

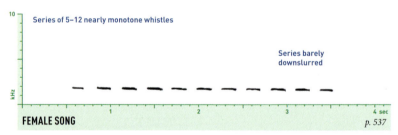

Series of 5–12 nearly monotone whistles

Series barely downslurred

FEMALE SONG
p. 537

Medium-loud version given mostly Feb.–Sept., primarily at dawn and dusk, probably by unmated females. Causes nearby males to sing and approach. Quieter version given year-round by scattered birds, including immatures, seeking to reunite.

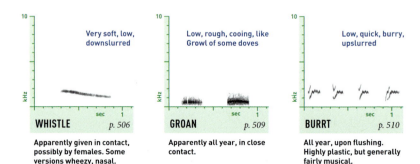

Very soft, low, downslurred

WHISTLE *p. 506*

Apparently given in contact, possibly by females. Some versions wheezy, nasal.

Low, rough, cooing, like Growl of some doves

GROAN *p. 509*

Apparently all year, in close contact.

Low, quick, burry, upslurred

BURRT *p. 510*

All year, upon flushing. Highly plastic, but generally fairly musical.

Ring-necked Pheasant

Phasianus colchicus

Native to Asia. Prefers agricultural areas with some brushy or wooded cover nearby. In many areas, large numbers are raised and released to be hunted.

Flushing birds give loud, rapid series of Kuttuks, along with loud wingbeats

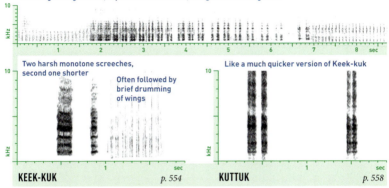

Two harsh monotone screeches, second one shorter

Often followed by brief drumming of wings

KEEK-KUK
p. 554

Like a much quicker version of Keek-kuk

KUTTUK
p. 558

Keek-kuk given all year, especially Mar.–June, by males only, from ground. Loud and far-carrying; some individual variation. Kuttuk given all year, by males only, in alarm. Rather plastic. Flushing females give only soft calls, or are silent. Chicks give high seerlike whistles (not indexed).

Gray Partridge

Perdix perdix

Native to Europe. Found in grasslands and agricultural fields, especially of wheat and other grasslike crops. In flight, note short, rusty-colored tail.

Territorial males give Kee-raw at regular intervals

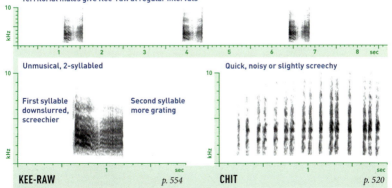

Unmusical, 2-syllabled

First syllable downslurred, screechier

Second syllable more grating

KEE-RAW
p. 554

Quick, noisy or slightly screechy

CHIT
p. 520

Kee-raw given all year, especially Feb.–June, by males. Somewhat variable. Often given in flight after flushing; these versions often shorter. Chit given all year, in alarm; given singly from ground, but often in series in high alarm, especially right before or after flushing.

RUFFED GROUSE

Bonasa umbellus

A bird of dense forests, especially young mixed forests and aspen groves; drumming of males is a characteristic sound of spring throughout range.

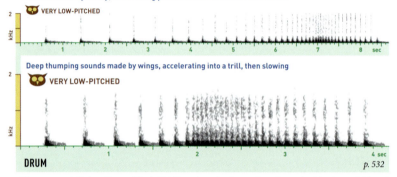

Drum starts very slowly; accelerating portion can last more than 10 seconds

🦉 VERY LOW-PITCHED

Deep thumping sounds made by wings, accelerating into a trill, then slowing

🦉 VERY LOW-PITCHED

DRUM *p. 532*

All year, especially Apr.–May and Sept.–Oct. By males only, atop preferred log or boulder, in courtship and territorial defense. Unmistakable, far-carrying. Loudest at low frequencies; often felt in the chest as much as heard. Poorly reproduced by most speakers. Bird also flushes with thumping wingbeats.

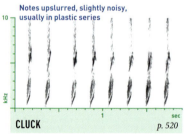

Plastic, sometimes slightly burry

WAIL *p. 518*

Mostly May–Sept., by females in high alarm, especially when chicks are threatened. Squealing versions sometimes given in distraction display.

Notes upslurred, slightly noisy, usually in plastic series

CLUCK *p. 520*

Likely all year, in alarm. Can strongly recall alarm notes of Red Squirrel. Often given in conjunction with Purrs.

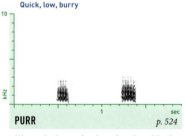

Quick, low, burry

PURR *p. 524*

All year, in alarm, often in conjunction with other calls. Fairly plastic.

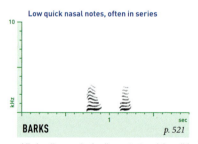

Low quick nasal notes, often in series

BARKS *p. 521*

Likely all year, in family contact and in mild alarm. Plastic, grading into Wails. Rather soft.

SPRUCE GROUSE

Falcipennis canadensis

A tame but quiet and elusive resident of pine barrens and spruce bogs. Female shorter-tailed and shorter-crested than Ruffed Grouse, with finer barring below.

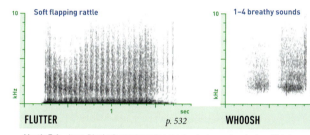

FLUTTER   Soft flapping rattle   *p. 532*

WHOOSH   1–4 breathy sounds   *p. 523*

Mostly Feb.–June. Displaying male makes mechanical sounds only, inaudible beyond about 50 meters. Male flies up into tree with audible wingbeats, then flies down, giving Flutter during brief hover. Whoosh made by opening and closing spread tail. Also gives soft single Thump with wings against body.

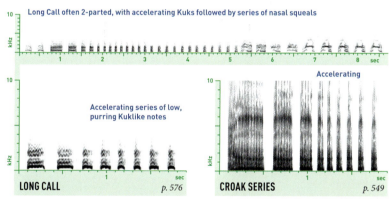

Long Call often 2-parted, with accelerating Kuks followed by series of nasal squeals

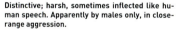

LONG CALL   Accelerating series of low, purring Kuklike notes   *p. 576*

CROAK SERIES   Accelerating   *p. 549*

Mostly Mar.–May, by females only, in songlike fashion from trees at dawn and dusk, as well as in territorial encounters.

Distinctive; harsh, sometimes inflected like human speech. Apparently by males only, in close-range aggression.

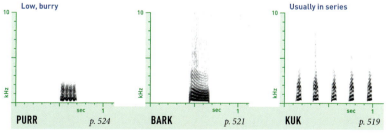

PURR   Low, burry   *p. 524*

BARK   *p. 521*

KUK   Usually in series   *p. 519*

A variety of quick low nasal notes given in alarm and brood contact, possibly only by females. Purr given singly or followed by Kuks. Barks given singly; quite plastic, possibly grading into other calls.

SHARP-TAILED GROUSE

Tympanuchus phasianellus

Local in grasslands and shrub steppes, sometimes in bogs. Range has shrunk and numbers have declined, but still our most widespread prairie grouse.

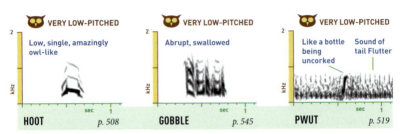

👀 VERY LOW-PITCHED	👀 VERY LOW-PITCHED	👀 VERY LOW-PITCHED
Low, single, amazingly owl-like	Abrupt, swallowed	Like a bottle being uncorked / Sound of tail Flutter
HOOT *p. 508*	**GOBBLE** *p. 545*	**PWUT** *p. 519*

Hoot given by lekking males, with neck sacs fully inflated and tail raised. Gobbles given in aggression toward other lekking males. Pwuts given with Tsooks in high-intensity dancing display for female, with wings spread; male shakes tail and stamps feet in rapid unison, creating a rattling Flutter (p.532).

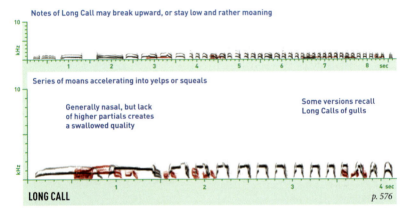

Notes of Long Call may break upward, or stay low and rather moaning

Series of moans accelerating into yelps or squeals

Generally nasal, but lack of higher partials creates a swallowed quality

Some versions recall Long Calls of gulls

LONG CALL *p. 576*

Given during lekking displays by males, mostly during aggressive face-offs or fights with other males. Highly plastic, but usually not divided into distinct segments; and short snippets or single notes are frequently heard.

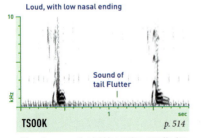

Loud, with low nasal ending

Sound of tail Flutter

TSOOK *p. 514*

Given by aggressive lekking males, usually during or after dancing, possibly to interfere with sound of other males' Pwuts.

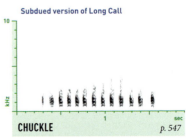

Subdued version of Long Call

CHUCKLE *p. 547*

All year, upon flushing. Nearly identical to Greater Prairie-Chicken Chuckle.

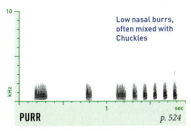

Low nasal burrs, often mixed with Chuckles

10

kHz

1   sec

PURR   *p. 524*

One of a variety of similar notes given by alarmed hens with broods; prairie-chickens likely give similar calls.

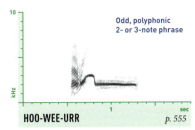

Odd, polyphonic 2- or 3-note phrase

10

kHz

1   sec

HOO-WEE-URR   *p. 555*

Poorly known. Apparently Aug.–Sept., by nearly independent young separated from parents. Prairie-chickens likely give similar calls.

LEKS AND LEKKING

All five North American species of "prairie grouse" (Sharp-tailed Grouse, Greater and Lesser Prairie-Chickens, and Greater and Gunnison Sage-Grouse) exhibit a unique mating behavior called lekking. Males gather each spring on traditional display sites, called leks. Beginning before first light, the male grouse perform elaborate display dances, inflating their throat sacs, spreading their tails, and making a variety of unique sounds. These displays are given for a few hours around dawn, and sometimes in the evening. They sometimes occur in fall, as well as in spring.

By displaying and fighting, males establish a strict dominance hierarchy, and each claims a small part of the lek as his territory, with the highest-ranking males occupying the largest territories near the center. After the male hierarchy is established, females arrive to mate primarily with the high-ranking males.

All North American prairie grouse have declined dramatically in population over the past century, primarily because of human-caused changes in habitat. Lek mating may make these species particularly vulnerable to disturbances, and the tendency for a few high-ranking males to do most of the mating can result in low genetic diversity. The Heath Hen, an East Coast subspecies of Greater Prairie-Chicken, has been extinct since 1932. The Gulf coast subspecies, "Attwater's" Prairie-Chicken, is critically endangered.

Typical display postures of prairie-chickens

GREATER PRAIRIE-CHICKEN

Tympanuchus cupido

Uncommon and local in mid- to tall-grass prairie and oak savanna; populations have disappeared from 12 states and provinces. Neck sac yellowish.

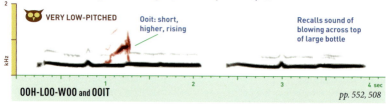

Ooh-loo-woo: low, 3-syllabled moaning coo, all on nearly the same pitch

VERY LOW-PITCHED

Ooit: short, higher, rising

Recalls sound of blowing across top of large bottle

OOH-LOO-WOO and OOIT

pp. 552, 508

Ooh-loo-woo is main display sound of lekking males, given with neck sacs fully inflated, tail raised, and pinnae feathers erected into a pointed crest behind the head, often while stamping feet. Ooit given infrequently, when females present, almost always during another male's Ooh-loo-woo.

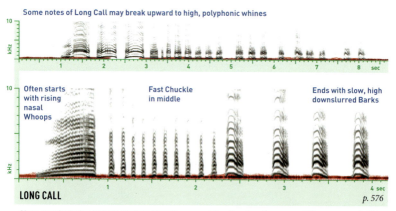

Some notes of Long Call may break upward to high, polyphonic whines

Often starts with rising nasal Whoops

Fast Chuckle in middle

Ends with slow, high downslurred Barks

LONG CALL

p. 576

Given mostly during lekking displays by males, especially when females appear. Often accompanied by a flutter-jump display, in which the bird leaps into the air briefly, flapping its wings. Quite plastic; often 3-parted, as shown above, but each part of the Long Call call also given separately at times.

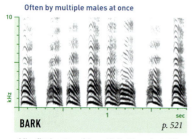

Often by multiple males at once

BARK

p. 521

Like final notes of Long Call. Given by lekking males, possibly in antiphonal duets with other males; rising nasal cries often mixed in.

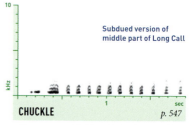

Subdued version of middle part of Long Call

CHUCKLE

p. 547

All year, upon flushing. Much like Chuckle of Lesser Prairie-Chicken, but pitch slightly lower.

LESSER PRAIRIE-CHICKEN

Tympanuchus pallidicinctus

Very uncommon and local in sandsage prairie and oak shrublands; range has contracted by over 75 percent since the 1960s. Note pinkish neck sac.

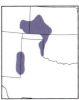

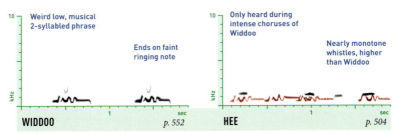

WIDDOO   p. 552

Weird low, musical 2-syllabled phrase

Ends on faint ringing note

HEE   p. 504

Only heard during intense choruses of Widdoo

Nearly monotone whistles, higher than Widdoo

Widdoo is main display sound of lekking males, given with neck sacs fully inflated, tail raised, and pinnae feathers erected into a pointed crest behind the head, often with a flick of the tail and a bob of the head. Sometimes males face each other and give Widdoos at high rate, with Hees interspersed.

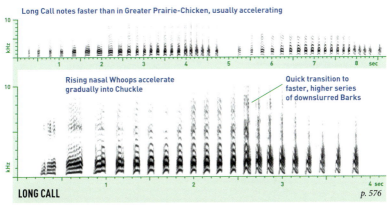

Long Call notes faster than in Greater Prairie-Chicken, usually accelerating

Rising nasal Whoops accelerate gradually into Chuckle

Quick transition to faster, higher series of downslurred Barks

LONG CALL   p. 576

Given mostly during lekking displays by males, especially when females appear. Often accompanied by a flutter-jump display, in which the bird leaps into the air briefly, flapping its wings. Quite plastic; often 2-parted, as shown above, but each part of the Long Call also given separately at times.

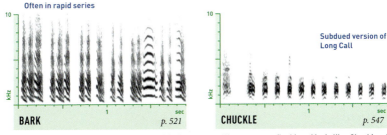

Often in rapid series

BARK   p. 521

Like final notes of Long Call. Given rapidly by lekking males, possibly in antiphonal duets with other males; longer notes often mixed in.

Subdued version of Long Call

CHUCKLE   p. 547

All year, upon flushing. Much like Chuckle of Greater Prairie-Chicken, but slightly higher.

GREBES (Family Podicipedidae)

Highly aquatic, grebes have lobed feet attached far back on the body, limiting their mobility on land. Vocalizations are likely innate. Vocal repertoires are often complex; some species engage in synchronized duets. Courtship displays of some species include elaborately choreographed water dances.

CLARK'S GREBE

Aechmophorus clarkii

Formerly considered a color morph of Western Grebe, but the two breed side-by-side with little hybridization. Rare in Midwest; accidental in East.

Songs sometimes mixed with Dziks, especially when calling to chicks

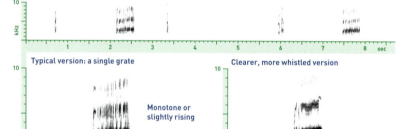

Typical version: a single grate

Monotone or slightly rising

Clearer, more whistled version

SONG: "Kreeek"

p. 525

All year, to attract potential mates and to maintain contact with distant family members; sometimes repeated in slow series. Variable and plastic; female versions average slightly higher. In areas where both species breed, Clark's and Western generally do not respond to one another's Songs.

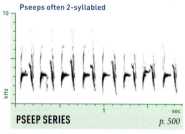

Pseeps often 2-syllabled

PSEEP SERIES

p. 500

Mostly July–Nov., by begging juveniles. Varies individually and with age. Loud and often incessant, carrying far across the water.

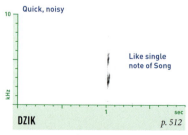

Quick, noisy

Like single note of Song

DZIK

p. 512

All year, in alarm and family contact. Highly plastic. Similar notes, singly or in series, accompany certain courtship displays.

WESTERN GREBE

Aechmophorus occidentalis

Shares many displays with Clark's, including the famous rushing courtship display: 2–3 birds rise out of water, run silently across it together, then dive.

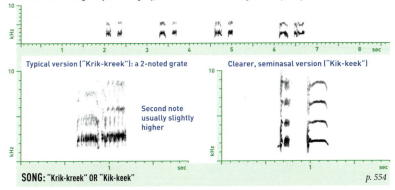

Consecutive Songs may differ slightly, but individuals often recognizable by Song

Typical version ("Krik-kreek"): a 2-noted grate

Second note usually slightly higher

Clearer, seminasal version ("Kik-keek")

SONG: "Krik-kreek" OR "Kik-keek"

p. 554

All year, to attract potential mates and to maintain contact with distant family members; sometimes repeated in slow series. Variable and plastic; beware occasional 1-note versions similar to Song of Clark's Grebe. Female versions average slightly higher. Occasionally heard at night.

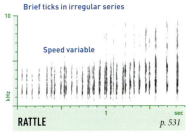

Brief ticks in irregular series

Speed variable

RATTLE *p. 531*

Mostly May–June, in Ratchet-pointing display: birds face off with necks low, occasionally shaking bill in water, often before Rushing.

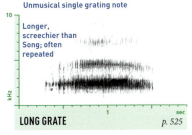

Unmusical single grating note

Longer, screechier than Song; often repeated

LONG GRATE *p. 525*

May–June, in Barge-trill display: two competing males rise out of water as if to Rush, but move slowly, giving ritual head turns.

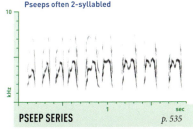

Pseeps often 2-syllabled

PSEEP SERIES *p. 535*

July–Nov., by begging juveniles. Varies individually and with age. Loud and often incessant, carrying far across the water.

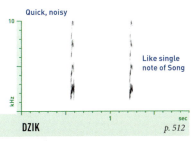

Quick, noisy

Like single note of Song

DZIK *p. 512*

All year, in alarm and family contact. Highly plastic. Similar notes, singly or in series, accompany certain courtship displays.

EARED GREBE

Podiceps nigricollis

Nests on shallow ponds with emergent vegetation, often in dense colonies. In migration and winter, gathers in vast numbers on saline lakes in the West.

 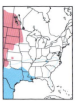

Song often given in series; here, answered by Twitter of another bird

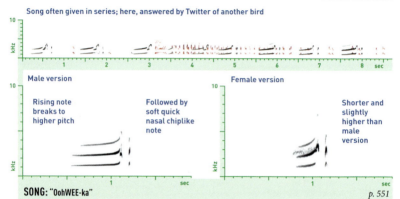

Male version

Rising note breaks to higher pitch

Followed by soft quick nasal chiplike note

Female version

Shorter and slightly higher than male version

SONG: "OohWEE-ka" *p. 551*

Most common vocalization; given mostly Mar.–July, by solo birds seeking mates, or by pairs in courtship or when feeding young. Male and female versions similar but often separable by ear. At start of breeding season, Song may be followed by elaborate dancing courtship displays by pair.

Often a short burr followed by a series of shrill Pips

Some versions faster, more trilling

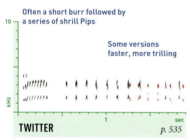

TWITTER *p. 535*

Mostly Apr.–June, by both sexes, in courtship. Aggressive versions faster, trilled. Rarely downslurred and decelerating, recalling Sora.

Long breathy, burry whistle

Monotone or slightly rising

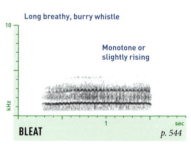

BLEAT *p. 544*

Mostly May–June, by both sexes, in close courtship. Does not carry far. Often followed by copulation, accompanied by clearer Wail.

Slightly nasal; usually sounds upslurred

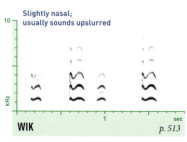

WIK *p. 513*

Likely mostly Apr.–June. Not well understood, but likely given in agitation or alarm. Begging juveniles give Pseep Series (p. 535).

Soft, abrupt

Sometimes in 2-note pattern: Pik-up

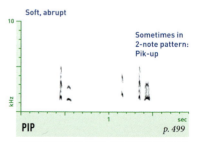

PIP *p. 499*

Likely all year, in alarm and contact. Does not carry far. Like single note of Twitter; grades into both Wik and Twitter.

Red-necked Grebe

Podiceps grisegena

Large, strikingly patterned in breeding season, with long yellowish bill. Breeds on ponds with emergent vegetation; winters mostly along coasts.

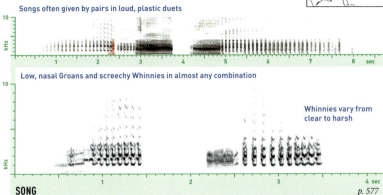

Songs often given by pairs in loud, plastic duets

Low, nasal Groans and screechy Whinnies in almost any combination

Whinnies vary from clear to harsh

SONG   *p. 577*

All year, but especially Jan.–Aug., by lone birds of either sex to attract a mate, and by duetting pairs in courtship or after successful defense of territories. Little difference between male and female versions. Extremely plastic, but distinctive. Long, burry versions of Groan given during copulation.

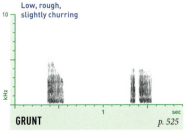

Chuckle: low, harsh, stuttering

CHUCKLE   *p. 547*

Whinny: high, screechy

WHINNY   *p. 546*

A set of similar vocalizations that intergrade. All year, but especially Jan.–Aug., by both sexes. Some versions resemble Whinnies from Song; other versions are noisier, more chattering. Some versions may be given during elaborate courtship displays prior to pairing.

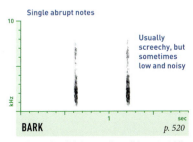

Low, rough, slightly churring

GRUNT   *p. 525*

Single abrupt notes

Usually screechy, but sometimes low and noisy

BARK   *p. 520*

Function not entirely clear; may be given in high alarm. Plastic. Begging juveniles likely give Pseep Series (no recordings).

All year, in mild alarm and possibly contact. Like single note of Whinny from Song; sometimes given in slow, irregular series.

Horned Grebe

Podiceps auritus

Breeds on shallow ponds with emergent vegetation, usually not in colonies. Winters on both salt and fresh water. Head flatter, less peaked than Eared's.

In greeting, pairs duet with plastic, trilled Kwirrs and wailing sounds (p. 577)

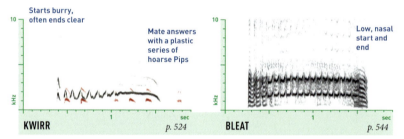

Slightly screechy wail, with burry end

Downslurred or overslurred

Harsher version

SONG   *p. 525*

All year, but especially Jan.–Aug., by lone birds seeking a mate, and by courting pairs, starting in late winter and extending through spring migration. Little difference between male and female versions. Variable and plastic.

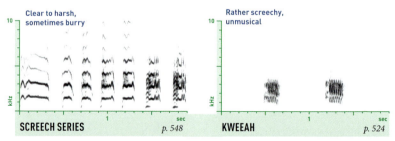

Starts burry, often ends clear

Mate answers with a plastic series of hoarse Pips

Low, nasal start and end

KWIRR   *p. 524*   **BLEAT**   *p. 544*

Highly plastic vocalizations given in close courtship and high excitement. Kwirrs often precede copulation; Bleat given during copulation. Both may grade into Song. Often given by pairs in unsynchronized duets. Hoarse Pips given by male near nest, grading into Screech Series.

Clear to harsh, sometimes burry

Rather screechy, unmusical

SCREECH SERIES   *p. 548*   **KWEEAH**   *p. 524*

Mostly May–Aug., in high alarm. Extremely plastic. Usually in series. Some versions clear, not screechy. Juveniles give Pseep Series (p. 535).

Likely mostly May–July, in alarm near nest or young. Fairly uniform. May grade into Screeches.

PIED-BILLED GREBE

Podilymbus podiceps

Prefers marshy ponds. Outlandish voice is often heard but frequently given from dense cover, and casual observers often do not connect it with this species.

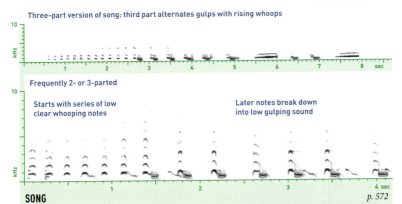

Three-part version of song: third part alternates gulps with rising whoops

Frequently 2- or 3-parted

Starts with series of low clear whooping notes

Later notes break down into low gulping sound

SONG *p. 572*

All year, especially Apr.–June. Highly variable; usually contain 2–3 contiguous series, each slower than the last and with more complex notes. Any series may last up to 15–20 notes; first and second series often given by themselves. Both sexes sing, but apparently only males give 3-part songs.

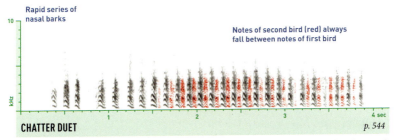

Rapid series of nasal barks

Notes of second bird (red) always fall between notes of first bird

CHATTER DUET *p. 544*

All year. Chatters sometimes given by solo birds, but mostly in duets between members of a pair on territory, resulting in a distinctive rapid laughing couplet series. One bird (female?) often sounds significantly higher than the other. Often introduces Song, or given simultaneously with mate's Song.

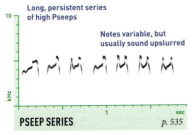

Long, persistent series of high Pseeps

Notes variable, but usually sound upslurred

PSEEP SERIES *p. 535*

Mostly May–Aug., by begging juveniles. Often long and loud. Sometimes given while young ride on parent's back during brooding.

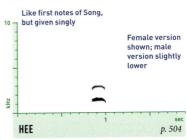

Like first notes of Song, but given singly

Female version shown; male version slightly lower

HEE *p. 504*

At least Nov.–May, by birds exploring occupied territories, or in mild alarm. Recalls Keek of Song.

LEAST GREBE

Tachybaptus dominicus

Our smallest grebe, widespread in the tropics on densely vegetated ponds. Black chin of breeding birds becomes white in winter.

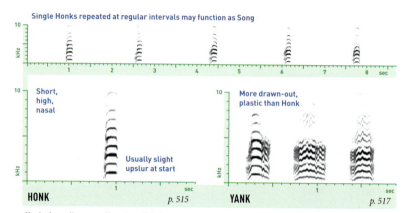

Single Honks repeated at regular intervals may function as Song

HONK p. 515

Short, high, nasal

Usually slight upslur at start

YANK p. 517

More drawn-out, plastic than Honk

Honk given all year; uniform but slightly plastic. Used by mated pairs when separated, or in alarm; possibly also to advertise for mates. Often accompanied by erect posture, bird sitting high in water, with neck stretched vertically and head feathers fluffed. Longer, plastic Yanks apparently given in alarm.

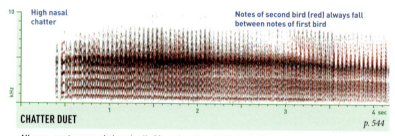

High nasal chatter

Notes of second bird (red) always fall between notes of first bird

CHATTER DUET p. 544

All year; most commonly heard call. Often given by pairs in duet, alternating notes so rapidly that they sound like a Churr from a single bird; apparently to reinforce pair bond and/or maintain territory. Plastic; speed can change irregularly.

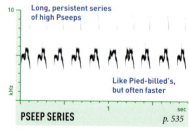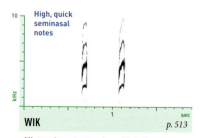

Long, persistent series of high Pseeps

Like Pied-billed's, but often faster

PSEEP SERIES p. 535

High, quick seminasal notes

WIK p. 513

From dependent fledglings. Mostly May–Sept., but the species can breed at any time of year in south Texas if weather is mild.

All year, in contact or alarm. Variable; some versions much higher and more piercing.

PIGEONS AND DOVES (Family Columbidae)

Often common even in cities, pigeons and doves are some of our most widespread and familiar birds. Many have adapted well to human changes to the landscape; some species from the dry Southwest, such as White-winged and Inca Doves, have expanded their ranges north and east in the past century. Species from Europe and Asia, such as the Rock Pigeon and the Eurasian Collared-Dove, have spread across the continent. Not every species has fared well. The Passenger Pigeon (*Ectopistes migratorius*), which likely numbered in the billions when Europeans began colonizing North America, became extinct in 1914.

The sounds of doves are innate, varying little between or within individuals. Most are soft and low-pitched, with a characteristic cooing or hooting quality that makes them easy to mistake for owls. Low growling or purring sounds are typical of courtship. Voice breaks are common.

Many dove species also make sounds with their wings, including wing whistles on takeoff that may warn other birds of danger (see p. 539) and wing claps or wing rattles that may serve territorial or courtship functions (see p. 532).

ROCK PIGEON

Columba livia

The familiar pigeon of cities and parks, native to Europe but now widespread in North America. Extremely variable in plumage color and pattern.

Song lower than those of other doves; not very loud

VERY LOW-PITCHED

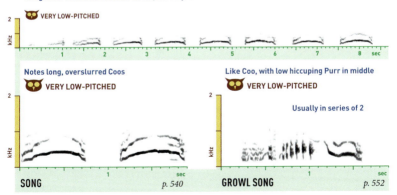

Notes long, overslurred Coos
VERY LOW-PITCHED

Like Coo, with low hiccuping Purr in middle
VERY LOW-PITCHED

Usually in series of 2

SONG   *p. 540*

GROWL SONG   *p. 552*

Song given all year, as the species nests year-round, by both sexes near nest or potential nest site. Single Coos given singly in alarm. Growl Song given all year, by males bowing with puffed-out throats and fanned tails, to court females and to repel other males. Does not carry far. Wings often slap loudly on takeoff (p. 533).

WHITE-CROWNED PIGEON

Patagioenas leucocephala

A large fruit-eating pigeon of the Caribbean, common but shy in dense forests in extreme south Florida. Populations at risk in much of its range.

Song: overslurred Coo, then a quadruplet series ("WHOOO! Ca-REER but POOR, ca-REER but POOR")

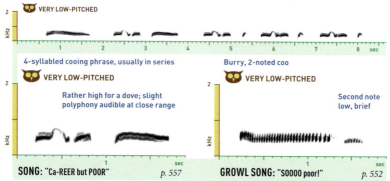

🦉 VERY LOW-PITCHED

4-syllabled cooing phrase, usually in series
🦉 VERY LOW-PITCHED

Rather high for a dove; slight polyphony audible at close range

Burry, 2-noted coo
🦉 VERY LOW-PITCHED

Second note low, brief

SONG: "Ca-REER but POOR"   p. 557

GROWL SONG: "SOOOO poor!"   p. 552

Song given all year by both sexes, but especially by male, May–Aug. No other dove in range gives quadruplet series. Growl Song given mostly May–Aug., from both sexes on or near the nest. Distinctive; often interspersed with Songs.

RED-BILLED PIGEON

Patagioenas flavirostris

A large, dark, shy, fruit-eating bird of the Central American lowlands. In our area, uncommon in summer and rare in winter in woodlands with tall trees.

Song: overslurred Coo, then a quadruplet series ("WHOOO!... UP cuppa COO... UP cuppa COO...")

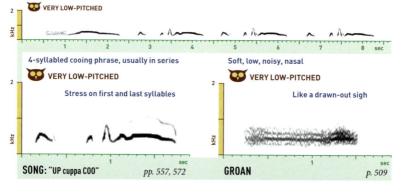

🦉 VERY LOW-PITCHED

4-syllabled cooing phrase, usually in series
🦉 VERY LOW-PITCHED

Stress on first and last syllables

Soft, low, noisy, nasal
🦉 VERY LOW-PITCHED

Like a drawn-out sigh

SONG: "UP cuppa COO"   pp. 557, 572

GROAN   p. 509

Song given all year, by both sexes, but especially by male, May–Aug. No other dove in range gives quadruplet series. Groan given all year. Most common call, but audible only at close range.

EURASIAN COLLARED-DOVE

Streptopelia decaocto

Native to Eurasia, but since the 1980s has spread rapidly across North American cities, towns, and farmsteads. In many areas, nests nearly year-round.

Songs given singly or strung together in long triplet series

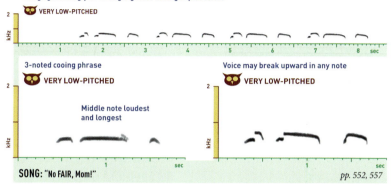

VERY LOW-PITCHED

3-noted cooing phrase

VERY LOW-PITCHED

Middle note loudest and longest

Voice may break upward in any note

VERY LOW-PITCHED

SONG: "No FAIR, Mom!"

pp. 552, 557

Nearly all year. Highly variable, but triplet pattern of Coos is distinctive. All notes typically on the same pitch, the triplets either given singly or strung together in long series. Rare 2-noted variant omits the final note.

Hoarse, groaning version of typical Song

VERY LOW-PITCHED

Very soft

GROWL SONG
p. 552

Nearly all year; from courting birds near potential nest sites, often well hidden. Never in series. Does not carry far.

Nasal, finely burry overslur

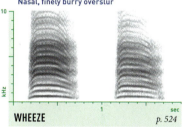

WHEEZE
p. 524

All year. Distinctive; usually given on the wing, including in wing-clapping display flight. Often in series of 2–3.

AFRICAN COLLARED-DOVE

Streptopelia roseogrisea

Domesticated form, "Ring-necked Dove" or "Ringed Turtle-Dove" (*S. risoria*), escapes widely. Resembles Eurasian Collared-Dove; hybrids sound intermediate.

VERY LOW-PITCHED

Repeated 2-noted burry phrase

SONG: "WHIP brooOOm"
pp. 552, 557

Falling, slowing nasal series

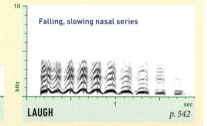

LAUGH
p. 542

INCA DOVE

Columbina inca

Originally restricted to deserts, but has become a familiar bird of residential areas throughout the Southwest, often nesting in yards and visiting feeders.

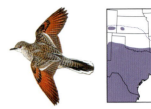

Song repeated at short intervals for long periods

🦉 **VERY LOW-PITCHED**

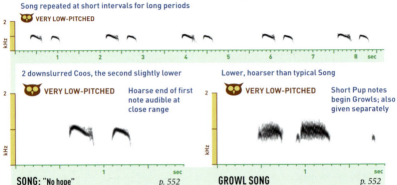

2 downslurred Coos, the second slightly lower

🦉 **VERY LOW-PITCHED** Hoarse end of first note audible at close range

Lower, hoarser than typical Song

🦉 **VERY LOW-PITCHED** Short Pup notes begin Growls; also given separately

SONG: "No hope" p. 552

GROWL SONG p. 552

Song given all year, but mostly Feb.–Aug., by males. Repeated incessantly, even in the heat of the day; highly uniform. Growl Song mostly Feb.–Oct., in close courtship or aggressive interactions. Soft; rhythm varies with excitement, but above pattern is typical. Wing Flutter (p. 532) given on takeoff and in aggression.

COMMON GROUND-DOVE

Columbina passerina

Fairly common in dry scrub and second growth. Also found in residential areas, but has declined as suburbs expand, unlike some other dove species.

Song repeated in slow steady series for long periods

🦉 **VERY LOW-PITCHED**

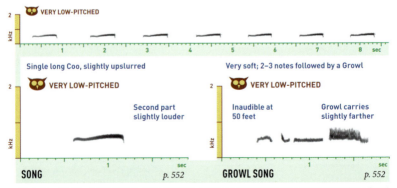

Single long Coo, slightly upslurred

🦉 **VERY LOW-PITCHED** Second part slightly louder

Very soft; 2–3 notes followed by a Growl

🦉 **VERY LOW-PITCHED** Inaudible at 50 feet Growl carries slightly farther

SONG p. 552

GROWL SONG p. 552

Song given all year, especially Jan.–Sept., by males. Sings even in the heat of the day; highly uniform. Growl Song given mostly Jan.–Sept., in close courtship and possibly some territorial encounters; rarely noticed because of low volume. Wing Flutter (p. 532) given on takeoff.

White-tipped Dove

Leptotila verreauxi

A large, pot-bellied dove of dense underbrush. Easy to hear, but shy and often difficult to see except in fast flight across woodland clearings.

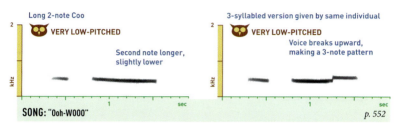

Long 2-note Coo

VERY LOW-PITCHED

Second note longer, slightly lower

2 kHz

1 · sec

SONG: "Ooh-WOOO"

3-syllabled version given by same individual

VERY LOW-PITCHED

Voice breaks upward, making a 3-note pattern

2 kHz

1 · sec

p. 552

All year, especially Mar.–May, by both sexes, often in duet. Uniform. Soft and difficult to localize, but far-carrying. Typical version shown at left. 3-syllabled version less common, but can be repeated in long series; function unknown, but may be given at potential nest site.

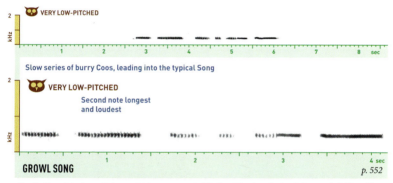

VERY LOW-PITCHED

2 kHz

1 2 3 4 5 6 7 8 sec

Slow series of burry Coos, leading into the typical Song

VERY LOW-PITCHED

Second note longest and loudest

2 kHz

1 2 3 4 sec

GROWL SONG

p. 552

Mostly Mar.–July, by male in courtship, accompanied by head-bowing display. Infrequent. Quite uniform and sterotyped. Begins with 2 monotone, cooing trills like a burry version of the typical Song; then a few shorter, softer trills; then the typical 2-note Song.

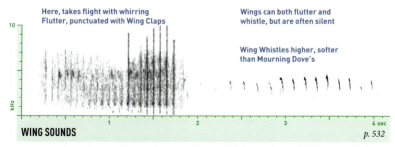

Here, takes flight with whirring Flutter, punctuated with Wing Claps

10 kHz

Wings can both flutter and whistle, but are often silent

Wing Whistles higher, softer than Mourning Dove's

kHz

1 2 3 4 sec

WING SOUNDS

p. 532

As in other doves, wing sounds vary. Often takes flight with brief, noisy slapping Flutter, rather like that of Inca Dove. Wings can also make soft rapid series of high-pitched Whistles, each sharply downslurred, like Mourning Dove Wing Whistle but higher, recalling Common Goldeneye.

Mourning Dove

Zenaida macroura

Our most familiar native dove, common in a huge variety of habitats across the continent. One of the first birds that many people learn to identify by sound.

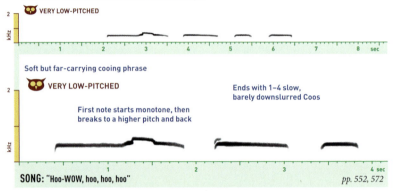

Number of Hoo notes at the end varies, but usually 3

🦉 VERY LOW-PITCHED

Soft but far-carrying cooing phrase

🦉 VERY LOW-PITCHED

First note starts monotone, then breaks to a higher pitch and back

Ends with 1–4 slow, barely downslurred Coos

SONG: "Hoo-WOW, hoo, hoo, hoo"   pp. 552, 572

Mostly Apr.–Aug. A well-known and evocative song, the sad sound of which gives the bird its common name. Sometimes mistaken for an owl, but generally given only during the day. Highly uniform.

Like first note of Song, but often hoarser

🦉 VERY LOW-PITCHED

Voice breaks to a higher pitch and back

GROWL SONG   p. 508

Mostly Mar.–Aug. Given by male at nest or prospective nest site, often well concealed.

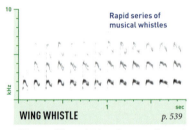

Rapid series of musical whistles

WING WHISTLE   p. 539

All year. Often mistaken for a vocal sound. Mostly heard at takeoff, but takeoff can be silent, and whistle can occur in sustained flight.

PIGEON AND DOVE WING SOUNDS

All species of North American doves make sounds with their wings, and many of these sounds serve communicative functions.

Whistles: Most familiar is the loud musical whistle of the Mourning Dove's wings. Mourning Doves may be able to control this sound, and it can provoke alarm responses in other birds, leading some to hypothesize that the Wing Whistle is an intentional alarm signal. The wings of Rock Pigeon, White-winged Dove, and Eurasian Collared-Dove sometimes whistle in a similar fashion.

Flutters: Several species of doves can produce a loose, slapping Flutter on takeoff, instead of or in addition to Wing Whistles. Inca Doves also give Flutters while perched in aggressive displays toward other members of their species.

Claps: Most dove and pigeon species in our area give Wing Claps in courtship flight displays. In all such displays, a male bird sallies out from a high perch, claps its wings together above its back 5–10 times, then goes into a stiff-winged downward glide, usually to a different perch.

White-winged Dove

Zenaida asiatica

Vocally and visually conspicuous in a variety of habitats. Native southwestern range has expanded north and east. Introduced and established in Florida.

Fewer repeating phrases than in other long dove songs

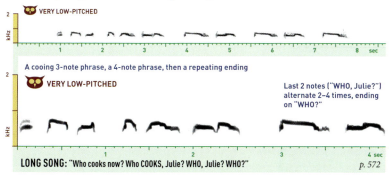

2 kHz — **VERY LOW-PITCHED**

A cooing 3-note phrase, a 4-note phrase, then a repeating ending

2 kHz — **VERY LOW-PITCHED**

Last 2 notes ("WHO, Julie?") alternate 2–4 times, ending on "WHO?"

LONG SONG: "Who cooks now? Who COOKS, Julie? WHO, Julie? WHO?"   *p. 572*

Nearly all year, but mostly Jan.–Aug. Highly distinctive; uniform and stereotyped. Given by males upon first landing on a song perch, after courtship flights, and when rival males approach. If Long Song is not answered, male often switches to Short Song or alternates the two song forms.

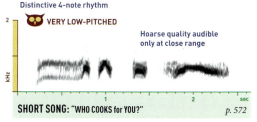

Distinctive 4-note rhythm

2 kHz — **VERY LOW-PITCHED**

Hoarse quality audible only at close range

SHORT SONG: "WHO COOKS for YOU?"   *p. 572*

Nearly all year, but mostly Jan.–Aug. Never in series. Higher than Barred Owl's "Who cooks for you?" Burrier 2- and 3-note versions, possibly from female, given in some courtship duets.

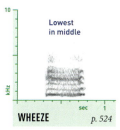

Lowest in middle

WHEEZE   *p. 524*

All year, in encounters with conspecifics. Quality like air escaping balloon.

CUCKOOS AND RELATIVES (Family Cuculidae) *following pages*

The cuckoo family is a diverse, cosmopolitan group. Sounds are apparently innate. The familiar 2-note "COO-coo" song popularized by the cuckoo clock is not heard in North America; it is the song of the European Cuckoo. The North American cuckoos in the genus *Coccyzus*, slender and secretive birds of dense woods, tend to give mellow coos and mechanical-sounding knocks and clucks. They usually build their own nests, but sometimes lay eggs in the nests of other cuckoos or other bird species. The roadrunner, a terrestrial desert species, also gives coos, among other sounds. The anis, large-billed black birds of tropical open country, have generally higher-pitched, squeakier voices.

YELLOW-BILLED CUCKOO

Coccyzus americanus

A striking but reclusive bird of dense deciduous woods; more often heard than seen. Eats mostly large insects, especially caterpillars.

Slow series of mellow, slightly nasal downslurred Coos or Barks

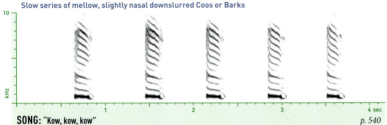

SONG: "Kow, kow, kow"   *p. 540*

Mostly Apr.–June, by unmated males seeking mates. Given with bill closed, throat inflating with each note. Quality varies from clear, rather dovelike cooing to low, swallowed barking; some versions polyphonic. Length of series plastic, but often 9–12 notes; last few often lower, softer, more 2-syllabled.

Initial rattle sometimes omitted

Classic version: rapid knocking rattle, then slower, decelerating, couplet or triplet series of Clucks

Loudest in middle

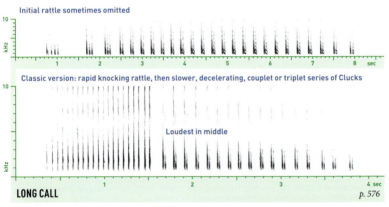

LONG CALL   *p. 576*

Mostly Apr.–July, possibly only by males. Most common vocalization, but usually given at very long intervals of 10 or more minutes. Often prompts other cuckoos to respond with Long Call. May function to reunite pairs and to maintain territorial boundaries. Very distinctive, but rhythm is variable and plastic.

Rattle: like first part of Long Call

Chortle: a Cluck joined to a low Growl

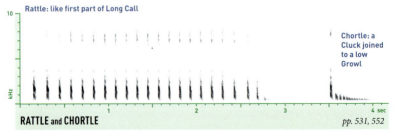

RATTLE and CHORTLE   *pp. 531, 552*

Rattle given by both sexes in pair contact, mostly in breeding season. Can be given rather frequently near nest. Rattle may end in Chortle; Chortle also given separately, sometimes as short as 2 notes. Night migrants give all sounds, but mostly Chortle, often in series.

BLACK-BILLED CUCKOO

Coccyzus erythropthalmus

Generally uncommon near the edges of deciduous and mixed forests. Similar to Yellow-billed Cuckoo, but wing and tail patterns differ; eye-ring red, not yellow.

Low, cooing toots in quick series of 3–4

Notes clear, usually monotone; sometimes downslurred

Series typically repeated about every 2 seconds

SONG: "Po-po-po"

p. 539

Mostly May–July, reportedly by both sexes. Can be given at any time of day or night on the breeding grounds, sometimes in nocturnal song flights; also occasionally by night migrants. Sometimes gives up to 9 notes in a series, especially if excited. Recalls Least Bittern Song, but higher and clearer.

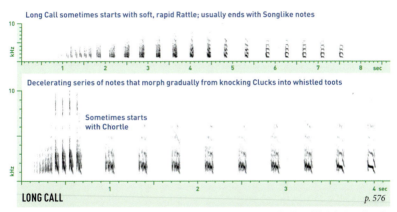

Long Call sometimes starts with soft, rapid Rattle; usually ends with Songlike notes

Decelerating series of notes that morph gradually from knocking Clucks into whistled toots

Sometimes starts with Chortle

LONG CALL

p. 576

Mostly May–July. Function little known, but likely similar to Long Call of Yellow-billed Cuckoo. Usually given at very long intervals of 10 or more minutes. May introduce bouts of Song. Pattern variable, especially at start. Gradual transition to clear, musical notes is highly distinctive.

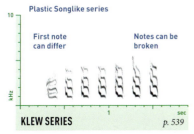

Plastic Songlike series

First note can differ

Notes can be broken

KLEW SERIES

p. 539

Much like Song, but series longer, more plastic. Perhaps a vocalization intermediate between Long Call and Song.

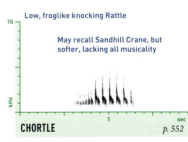

Low, froglike knocking Rattle

May recall Sandhill Crane, but softer, lacking all musicality

CHORTLE

p. 552

Distinctive. Given in night flights on breeding grounds and in migration; rarely during the day. Some versions shorter, more monotone.

MANGROVE CUCKOO

Coccyzus minor

An uncommon resident of mangroves
and of tropical hardwood hammocks.
Secretive, but when encountered, can
be surprisingly approachable.

Series of odd, low, quacking Barks

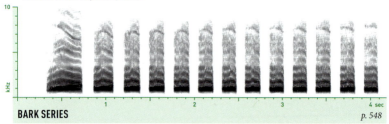

BARK SERIES   *p. 548*

Mostly Apr.–Aug.; most common vocalization. Variable; speed ranges from 3 to 6 notes/second. First note
sometimes longer; last 2–3 notes often strikingly different, slower, lower, and more growling. Function
unstudied, but may serve to establish territory or attract mate.

Long Call: Often starts with accelerating Rattle, ends with slowing, fading Growls

Accelerating Rattle, transitioning into Bark Series

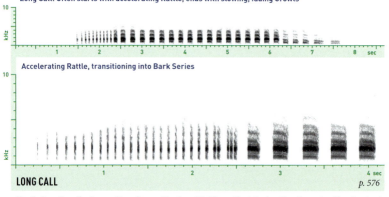

LONG CALL   *p. 576*

Mostly Apr.–Aug. Can be considered a combination of Rattle and Bark Series. Function not well understood.
Individuals give both Bark Series and Long Call.

Accelerating series of noisy Keks

Start often lower, softer

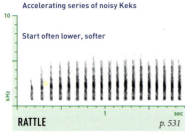

RATTLE   *p. 531*

Mostly Apr.–Aug., between Bark Series or when
foraging. Highly variable; often ends in a swallowed growl. Much like Yellow-billed's Rattle.

Slightly decelerating

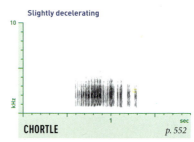

CHORTLE   *p. 552*

Little known; only available example given in response to playback. Noisier, more chattering
than Black-billed Cuckoo Chortle.

GREATER ROADRUNNER

Geococcyx californianus

Unmistakable. Usually solitary; prowls arid places in search of snakes, lizards, and other prey. Can run up to 20 miles per hour; flies infrequently but well.

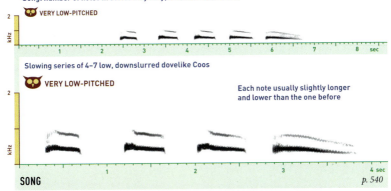

Song: number of notes in series may vary, but most often 5 or 6

🦉 VERY LOW-PITCHED

Slowing series of 4–7 low, downslurred dovelike Coos

🦉 VERY LOW-PITCHED

Each note usually slightly longer and lower than the one before

SONG *p. 540*

Mostly Feb.–June, reportedly only by males. Uniform. Seems soft but carries far; often given from an exposed perch with bill pointed downward, head bobbing forward with each note. Final notes often distinctly 2-syllabled, with a downward voice break. Very soft 1- and 2-note versions sometimes given.

Polyphonic, finely burry, usually rising

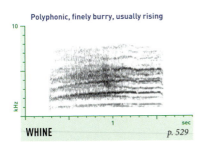

WHINE *p. 529*

By both sexes, in alarm near nest and when pairs forage together; also in crouching display with complex head motions. Highly plastic.

Low, purring, audible only at close range

🦉 VERY LOW-PITCHED

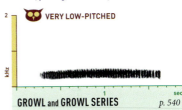

GROWL and GROWL SERIES *p. 540*

Variable. Long single Growls given by courting males; short Growls or Grunts given in series by both sexes, in courtship and at nest.

Loud, rapid, snapping

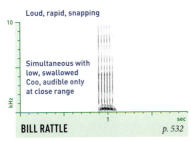

Simultaneous with low, swallowed Coo, audible only at close range

BILL RATTLE *p. 532*

All year, by both sexes, in alarm and contact. Plastic; varies from 1 snap to many. Series of pops made with wings also reported.

Low, clear, nasal; slows slightly at end

First and last few notes lower

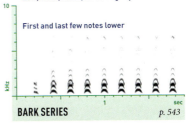

BARK SERIES *p. 543*

Mostly Feb.–June, reportedly only by female, often in response to male Song or Growl.

Groove-billed Ani

Crotophaga sulcirostris

Local in pastures and open thorn scrub. Like Smooth-billed Ani, breeds in communal groups of up to 5 pairs, all the females laying eggs in the same nest.

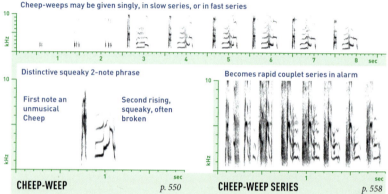

Cheep-weeps may be given singly, in slow series, or in fast series

Distinctive squeaky 2-note phrase

First note an unmusical Cheep

Second rising, squeaky, often broken

Becomes rapid couplet series in alarm

CHEEP-WEEP *p. 550*

CHEEP-WEEP SERIES *p. 558*

All year; most common call. Quite plastic and somewhat variable, but distinctive rangewide. Different variants, repeated at different speeds, apparently correspond with different behavioral contexts. Stress may fall on first or second note.

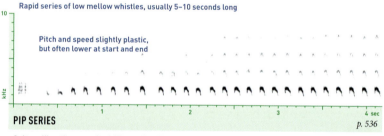

Rapid series of low mellow whistles, usually 5–10 seconds long

Pitch and speed slightly plastic, but often lower at start and end

PIP SERIES *p. 536*

Quite unlike other sounds of this species. Given by solo birds from prominent perches, sometimes repeated for long periods, reportedly in attempt to join or rejoin a group. Version shown recalls Northern Flicker or Ferruginous Pygmy-Owl. Some versions slower, with longer notes, more like Smooth-billed Weep Series.

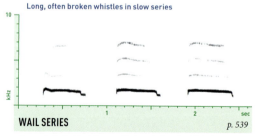

Long, often broken whistles in slow series

WAIL SERIES *p. 539*

Somewhat plastic. Infrequently heard; function little known, but may communicate anxiety or distress. Compare Gray Hawk Wail Series.

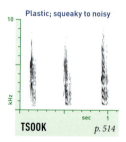

Plastic; squeaky to noisy

TSOOK *p. 514*

All year, in close contact and mild alarm; may grade into Cheep-weep. Quite soft.

SMOOTH-BILLED ANI

Crotophaga ani

Widespread in the tropics, but rare and local in our area in pastures and open second growth; breeding population in Florida nearly extirpated.

Weelips may be given singly, in slow series, or in fast series

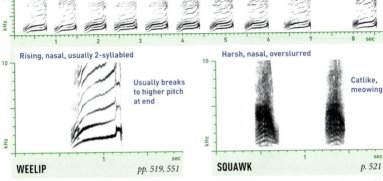

Rising, nasal, usually 2-syllabled

Usually breaks to higher pitch at end

WEELIP *pp. 519, 551*

All year; most common call. Plastic and variable, but distinctive. Different variants apparently correspond with different behaviors.

Harsh, nasal, overslurred

Catlike, meowing

SQUAWK *p. 521*

All year, in response to aerial predators and in other high alarm situations. Recalls harsh Mew of Gray Catbird.

Long rapid series of low mellow whistles

Individual notes may sound upslurred

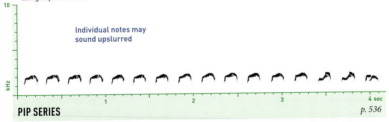

PIP SERIES *p. 536*

Like Weep Series of Groove-billed Ani, and apparently given in similar situations, by birds separated from flockmates. Variable; notes range from Weeplike in slower versions (shown) to Piplike in faster versions.

Long, often broken whistles in slow series

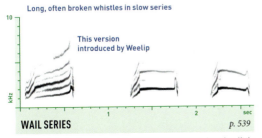

This version introduced by Weelip

WAIL SERIES *p. 539*

Somewhat plastic; grades into Weelip. Infrequent; function little known.

Often churring

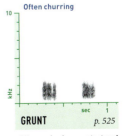

GRUNT *p. 525*

All year, in close contact and mild alarm. Soft, variable, plastic.

NIGHTJARS (Order Caprimulgiformes, Family Caprimulgidae)

Active mostly at dawn and dusk, nightjars have large eyes and wide mouths lined with bristles to capture flying insects in low light. Cryptic plumage lets them avoid detection during the day, when they roost on the ground or lengthwise along tree branches.

Nightjars give simple, often repetitive vocalizations which appear to be innate (but see notes under Common Pauraque Song, p. 553). Many species vocalize for long periods from perches, usually at dawn or dusk, less often in the middle of the night, and very rarely during daylight hours. Some species are named for their evocative, repetitive calls; several species give wing-clapping displays.

The nighthawks (genus *Chordeiles*) are more aerial than other nightjars and more likely to be active by day.

COMMON NIGHTHAWK

Chordeiles minor

Widespread; seen mostly at dusk and dawn, but also sometimes at midday. Often vocal on the wing. Nests on open sandy ground or flat urban rooftops.

Peents given regularly in flight; Fooms generally infrequent

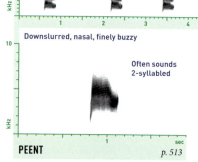

Downslurred, nasal, finely buzzy

Often sounds 2-syllabled

PEENT *p. 513*

Mostly Apr.–Aug. A familiar sound of summer evenings across North America; usually given in flight with a rapid wing flutter.

Low, breathy, slightly nasal whooshing

Almost always with simultaneous Peent

FOOM *p. 522*

May–Aug. Produced in flight at bottom of shallow dive in vicinity of nest; may be made by wings. Sounds soft, but carries far.

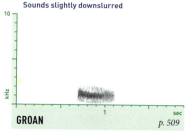

Sounds slightly downslurred

GROAN *p. 509*

May–Aug. Given from perch by male during courtship, usually in presence of female, and often in conjunction with Peent.

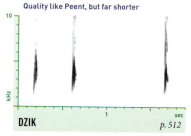

Quality like Peent, but far shorter

DZIK *p. 512*

May–Aug., in flight chases; infrequent. Buzzier, louder than Lesser Nighthawk Kuk; may not be separable from Antillean Nighthawk Dzik.

LESSER NIGHTHAWK

Chordeiles acutipennis

A desert bird, usually seen only at dusk and dawn. Often silent in flight, but can be quite vocal on predawn perches and during flight chases.

Trill may continue for long periods, but regularly punctuated by Chortle or by brief pauses

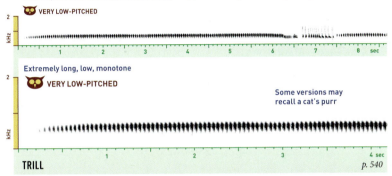

VERY LOW-PITCHED

Extremely long, low, monotone

VERY LOW-PITCHED

Some versions may recall a cat's purr

TRILL   *p. 540*

Mar.–Aug., from ground or bush, usually before dawn or after dusk; rarely in flight. Uniform and stereotyped. Like screech-owl Trills but much longer, sometimes continuing for many minutes; soft and easily overlooked or mistaken for a distant mechanical sound or the similar trilling of some toad species.

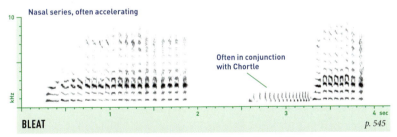

Nasal series, often accelerating

Often in conjunction with Chortle

BLEAT   *p. 545*

Mostly Mar.–Aug. Frequently given in flight chases; possibly also in courtship and alarm. Plastic; intergrades with Kuk, and sometimes appended to Trill. Distinctive in our area.

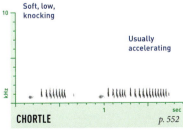

Soft, low, knocking

Usually accelerating

CHORTLE   *p. 552*

Mostly Mar.–Aug. Usually heard as part of Trill, but also given separately on the wing. Function unknown.

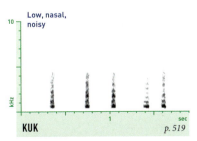

Low, nasal, noisy

KUK   *p. 519*

All year. Like single notes of Bleat, but usually lower-pitched, briefer, and noisier. Never grating or buzzy.

ANTILLEAN NIGHTHAWK

Chordeilis gundlachii

Virtually identical to Common Night-hawk, but voice distinctive. Found in Florida mostly in summer, foraging over savannas, clearings, and beaches.

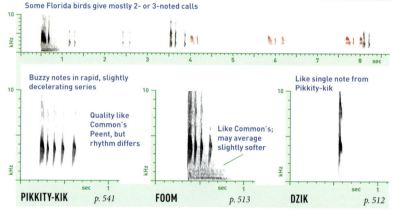

Some Florida birds give mostly 2- or 3-noted calls

Buzzy notes in rapid, slightly decelerating series

Quality like Common's Peent, but rhythm differs

Like Common's; may average slightly softer

Like single note from Pikkity-kik

PIKKITY-KIK   *p. 541*   **FOOM**   *p. 513*   **DZIK**   *p. 512*

Pikkity-kik is most common sound, given frequently on the wing by foraging birds. Usually 4–5 notes; some 2- or 3-note versions may come from hybrids with Common, but this is not confirmed. Foom given at bottom of stiff-winged dive, usually with simultaneous Pikkity-kik. Dzik given infrequently.

COMMON POORWILL

Phalaenoptilus nuttallii

Our smallest nightjar. Fairly common in scrubby canyons, rocky areas, and open coniferous forests, but nocturnal and rarely seen.

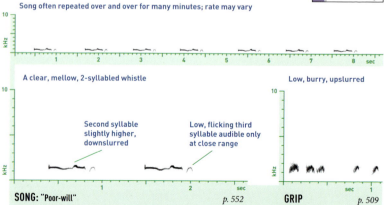

Song often repeated over and over for many minutes; rate may vary

A clear, mellow, 2-syllabled whistle

Second syllable slightly higher, downslurred

Low, flicking third syllable audible only at close range

Low, burry, upslurred

SONG: "Poor-will"   *p. 552*   **GRIP**   *p. 509*

All year, but mostly Mar.–Aug., around dawn and dusk and on moonlit nights, but also on rare occasions during the day. Highly uniform and stereotyped. Both sexes reported to sing. Song is easily imitated by human whistling. Grip given all year; most common call, but infrequent. Somewhat plastic.

EASTERN WHIP-POOR-WILL

Antrostomus vociferus

Locally common in deciduous woods and second growth, but nocturnal and rarely seen. Named for its distinctive, incessant nighttime call.

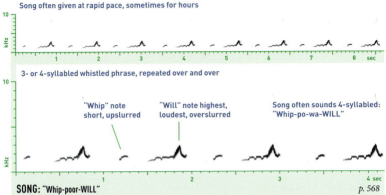

Song often given at rapid pace, sometimes for hours

3- or 4-syllabled whistled phrase, repeated over and over

"Whip" note short, upslurred

"Will" note highest, loudest, overslurred

Song often sounds 4-syllabled: "Whip-po-wa-WILL"

SONG: "Whip-poor-WILL"   *p. 568*

All year, but especially Apr.–June, around dawn and dusk and on moonlit nights, but also on rare occasions during the day. Not known whether females sing. Quite uniform and stereotyped. Final "will" note usually sounds very slightly burry. Last bouts before sunrise often frantic in pace, without pauses.

Long, low, repetitive growling series

Growls alternated with soft Clucks; Clucks may be doubled but sound single

Often associated with slow series of Wing Claps given in flight (p. 532)

GROWL-CLUCK SERIES   *p. 557*

Apparently given mostly during breeding season by excited birds, possibly during interactions, but function is little known. Associated with both Song and with wing-clapping displays; Wing Claps, at least, may be territorial. Individual Growl may sometimes be given singly, without Clucks. Audible only at close range.

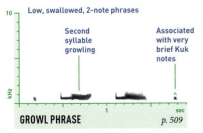

Low, swallowed, 2-note phrases

Second syllable growling

Associated with very brief Kuk notes

GROWL PHRASE   *p. 509*

Apparently May–June. Little known; sole available recording is introduced by Song, Growl-Cluck Series, and Wing Claps.

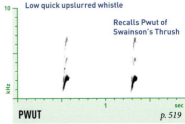

Low quick upslurred whistle

Recalls Pwut of Swainson's Thrush

PWUT   *p. 519*

All year; most common call. Given in pair contact and in alarm, including distraction display near nest, along with loud hissing.

CHUCK-WILL'S-WIDOW

Antrostomus carolinensis

Fairly common in woodlands and second growth. Even larger than Common Nighthawk. Plumage is often reddish, but some are grayer.

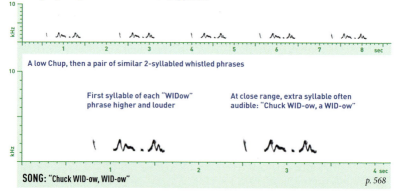

Song may be repeated over and over for long periods

A low Chup, then a pair of similar 2-syllabled whistled phrases

First syllable of each "WIDow" phrase higher and louder

At close range, extra syllable often audible: "Chuck WID-ow, a WID-ow"

SONG: "Chuck WID-ow, WID-ow"
p. 568

Mostly Apr.–Aug., around dawn and dusk and on moonlit nights, but also on rare occasions during the day. Not known whether females sing. Quite uniform and stereotyped. Initial "chuck" may be inaudible at a distance. Quality like Eastern Whip-poor-will Song, but slightly lower.

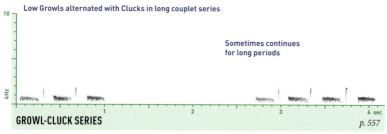

Low Growls alternated with Clucks in long couplet series

Sometimes continues for long periods

GROWL-CLUCK SERIES
p. 557

Given mostly during breeding season by excited birds, apparently during interactions, but function is little known. Associated with both Song and wing-clapping displays (p. 532). Sometimes given in flight, generally without Clucks. Audible only at close range.

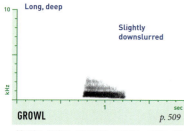

Long, deep

Slightly downslurred

GROWL
p. 509

Mostly during breeding season, apparently mostly in territorial disputes. Given singly or in conjunction with Grip.

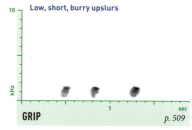

Low, short, burry upslurs

GRIP
p. 509

All year; most common call. Given upon being flushed and in various interactions; often introduces bouts of Song.

COMMON PAURAQUE

Nyctidromus albicollis

Widespread in the tropics. In Texas, found in thorn scrub and mesquite-oak savanna. Active mostly just before dawn and after dusk.

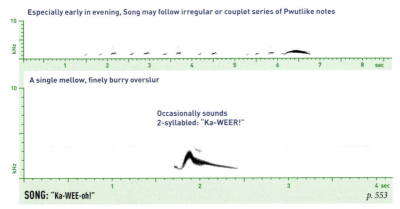

Especially early in evening, Song may follow irregular or couplet series of Pwutlike notes

A single mellow, finely burry overslur

Occasionally sounds 2-syllabled: "Ka-WEER!"

SONG: "Ka-WEE-oh!"
p. 553

Mostly Mar.–July, around dawn and dusk and on moonlit nights, rarely during the day. Unlike other nightjars, males reportedly have multiple songtypes, repeating one many times before switching, and sometimes matching songtypes of rival males; this requires study, as it suggests Song may be learned.

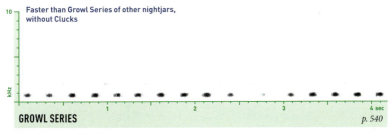

Faster than Growl Series of other nightjars, without Clucks

GROWL SERIES
p. 540

Possibly all year. Function little known; appears to be associated with Song and possibly with short-range interactions. Audible mostly at close range.

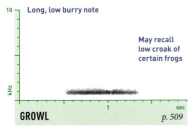

Long, low burry note

May recall low croak of certain frogs

GROWL
p. 509

All year, possibly in interactions. Plastic; various similar notes may be versions of this call, or serve different functions.

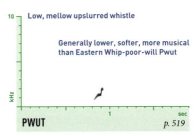

Low, mellow upslurred whistle

Generally lower, softer, more musical than Eastern Whip-poor-will Pwut

PWUT
p. 519

All year; most common call. Plastic. Given upon taking flight and in various other contexts; often introduces bouts of Song.

SWIFTS (Order Apodiformes, Family Apodidae)

HUMMINGBIRDS (Order Apodiformes, Family Trochilidae)

Though these two families differ greatly in shape and habits, anatomical details show them to be related lineages that developed from a common ancestor, and they share the order Apodiformes.

Swifts are highly aerial birds that pursue flying insects on the wing. They rarely perch on anything other than a vertical surface. In some species, some individuals may stay aloft for months at a time, even sleeping on the wing. Vocalizations tend to be rather simple and highly plastic; they are apparently innate.

Hummingbirds are adapted to hovering in front of flowers in order to drink nectar. Our smallest birds, they live at a frenetic pace, their wings a blur. Combative and pugnacious, they sometimes drive away birds many times their size. Vocal repertoires vary from fairly simple to very well-developed; some species sing exceedingly complex songs. At least some sounds in some species are learned. Most of the North American species perform elaborate flight displays, often accompanied by mechanical sounds produced by the tail and/or wings.

CHIMNEY SWIFT

Chaetura pelagica

Often seen over towns. Note dark color, cigar-shaped body, and stiff wingbeats. Once nested in hollow trees, but now primarily in chimneys.

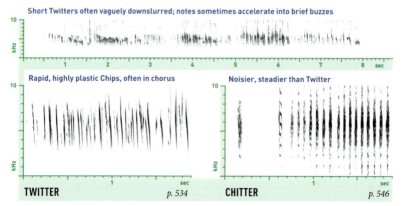

Short Twitters often vaguely downslurred; notes sometimes accelerate into brief buzzes

Rapid, highly plastic Chips, often in chorus

Noisier, steadier than Twitter

TWITTER   *p. 534*

CHITTER   *p. 546*

Twitter given all year, by adults, generally in flight. Mostly associated with social behavior, but function unstudied. Extremely plastic; variations in speed and tone may correspond to different behaviors. Chitter given in chimneys by young birds; by 5 weeks of age, Chitter starts becoming Twitter.

Buff-bellied Hummingbird

Amazilia yucatanensis

Fairly large for a hummingbird. Breeds in south Texas, where a few stay all year; after breeding, some wander northeast along the Gulf Coast.

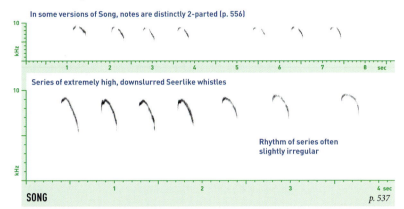

In some versions of Song, notes are distinctly 2-parted (p. 556)

Series of extremely high, downslurred Seerlike whistles

Rhythm of series often slightly irregular

SONG *p. 537*

Mostly given during breeding season in early morning. Birds reportedly sing in horizontal flight, sometimes flying repeatedly along the same path, in what may serve as this species' flight display. Also possibly by perched birds.

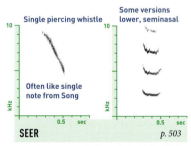

Single piercing whistle

Some versions lower, seminasal

Often like single note from Song

SEER *p. 503*

Possibly all year. Various single high whistles regularly heard, from perched birds or prior to chases; function unclear.

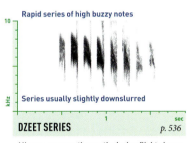

Rapid series of high buzzy notes

Series usually slightly downslurred

DZEET SERIES *p. 536*

All year, apparently mostly during flight chases. Variable in pitch and highly plastic.

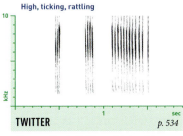

High, ticking, rattling

TWITTER *p. 534*

All year, often in flight, possibly in mild aggression. Length highly plastic. Grades into Smacks.

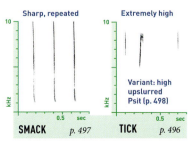

Sharp, repeated

Extremely high

Variant: high upslurred Psit (p. 498)

SMACK *p. 497*　　**TICK** *p. 496*

All year, when foraging and in contact. Many variations; some may serve different functions. More study needed.

RUBY-THROATED HUMMINGBIRD

Archilochus colubris

Breeds in deciduous and mixed forests and residential areas. Males resemble Black-chinned Hummingbird in poor light; females nearly identical.

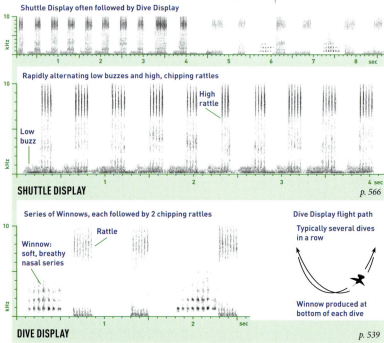

Shuttle Display often followed by Dive Display

Rapidly alternating low buzzes and high, chipping rattles

High rattle

Low buzz

SHUTTLE DISPLAY *p. 566*

Series of Winnows, each followed by 2 chipping rattles

Rattle

Winnow: soft, breathy nasal series

Dive Display flight path

Typically several dives in a row

Winnow produced at bottom of each dive

DIVE DISPLAY *p. 539*

Two types of flight display given during the breeding season, apparently only by males, including immature males. Shuttle Display given mostly in courtship: male buzzes back and forth along a short horizontal path, very close to the female; rate about 2 shuttles/second. Dive Display given in aggression or courtship, often directed at non-hummingbirds; rate about 1 dive every 1.5–2 seconds. Both displays faster than in Black-chinned Hummingbird, with distinctly different sounds. All display sounds likely nonvocal.

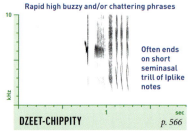

Rapid high buzzy and/or chattering phrases

Often ends on short seminasal trill of Iplike notes

DZEET-CHIPPITY *p. 566*

All year, by both sexes, in aggressive interactions. Highly plastic; the note types shown here combine into many patterns.

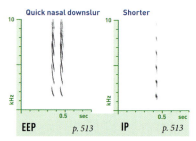

Quick nasal downslur   **Shorter**

EEP *p. 513*   **IP** *p. 513*

All year; most common calls. In interactions, often gives loud repeating series of 2–3 Eeps; solo birds give short, single Ips.

BLACK-CHINNED HUMMINGBIRD

Archilochus alexandri

Very closely related to Ruby-throated; sometimes hybridizes with it. Breeds in riparian woods, pinyon-juniper forests, and residential areas.

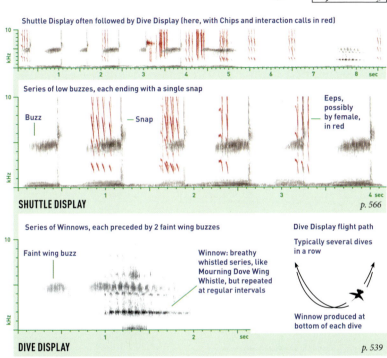

Shuttle Display often followed by Dive Display (here, with Chips and interaction calls in red)

Series of low buzzes, each ending with a single snap

Buzz

— Snap

Eeps, possibly by female, in red

SHUTTLE DISPLAY *p. 566*

Series of Winnows, each preceded by 2 faint wing buzzes

Faint wing buzz

Winnow: breathy whistled series, like Mourning Dove Wing Whistle, but repeated at regular intervals

Dive Display flight path

Typically several dives in a row

Winnow produced at bottom of each dive

DIVE DISPLAY *p. 539*

Two types of flight display given during the breeding season, apparently only by males, including immature males. Shuttle Display given mostly in courtship: male buzzes back and forth along a short horizontal path, very close to the female; rate about 3 shuttles in 2 seconds. Dive Display given in aggression or courtship, often directed at non-hummingbirds; rate about 1 dive every 3–4 seconds. Both displays slower than in Ruby-throated Hummingbird, with distinctly different sounds. All display sounds likely nonvocal.

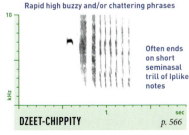

Rapid high buzzy and/or chattering phrases

Often ends on short seminasal trill of Iplike notes

DZEET-CHIPPITY *p. 566*

All year, by both sexes, in aggressive interactions. Highly plastic; the note types shown here combine into many patterns.

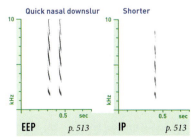

Quick nasal downslur   Shorter

EEP *p. 513*   **IP** *p. 513*

All year; most common calls. In interactions, often gives loud repeating series of 2–3 Eeps; solo birds give short, single Ips.

RAILS (Family Rallidae)

Rails are related to Limpkin and to cranes; the three families share the order Gruiformes. The rails are medium-sized to small birds that live in dense marsh vegetation. They are much more often heard than seen. In many species, vocal repertoires are complex, and duets are common. Sounds are apparently innate.

LIMPKIN (Family Aramidae)

The sole member of its family, this odd bird of the Florida swamps is intermediate in size and shape between rails and cranes, and eats almost exclusively apple snails. Its haunting, far-carrying sounds are apparently innate.

CRANES (Family Gruidae)

The cranes have played symbolic roles in many cultures worldwide owing to their graceful beauty, their elaborate mating dances, and their loud trumpeting vocalizations. Vocal duetting is well developed in our two species. As far as is known, sounds are innate.

YELLOW RAIL

Coturnicops noveboracensis

A tiny rail, highly local in summer in sedge marshes and in winter in salt marshes. Very difficult to see. Prefers to run from danger rather than fly.

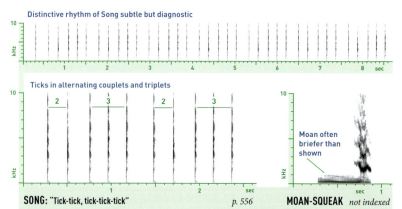

Distinctive rhythm of Song subtle but diagnostic

Ticks in alternating couplets and triplets

SONG: "Tick-tick, tick-tick-tick" *p. 556*

Moan often briefer than shown

MOAN-SQUEAK *not indexed*

Song given May–Aug., by males, mostly after dark; rarely by day early in season. Northern Cricket Frog has similar song, but with different rhythm. Moan-Squeak given in high alarm; Moan and Squeak sometimes given separately. Also a descending Cackle rarely in migration; no recordings.

BLACK RAIL

Laterallus jamaicensis

Very local in salt marshes and certain inland freshwater marshes. Tiny and rarely seen. Note that juveniles of other rail species are small and black.

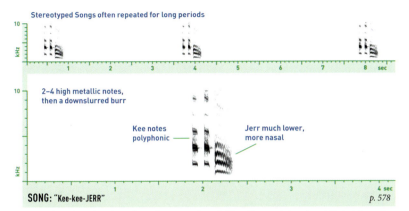

Stereotyped Songs often repeated for long periods

2–4 high metallic notes, then a downslurred burr

Kee notes polyphonic

Jerr much lower, more nasal

SONG: "Kee-kee-JERR"
p. 578

Mostly Mar.–June, but given all year in some regions. Fairly uniform and stereotyped. Up to 4 Kee notes, but usually 2; up to 3 Jerr notes, but almost always just 1. Given mostly at night, especially in East, but occasionally during day.

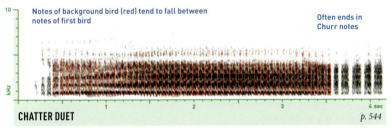

Notes of background bird (red) tend to fall between notes of first bird

Often ends in Churr notes

CHATTER DUET
p. 544

Mostly during breeding season; function little known. Often given by pairs in duet, alternating notes so rapidly that they sound like a single bird. Sometimes mixed with Churrs; one bird may give Churrs in response to partner's Chatter. Strikingly similar to call given by duetting Least Grebes.

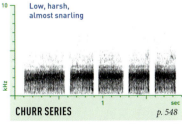

Low, harsh, almost snarling

CHURR SERIES
p. 548

All year, but especially in breeding season, in agitation or alarm. Notes fairly constant in pitch, but series plastic in speed.

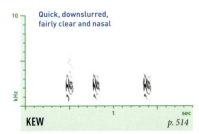

Quick, downslurred, fairly clear and nasal

KEW
p. 514

All year, by both sexes. Possibly a contact call. Higher versions in faster series may be associated with alarm.

Clapper Rail

Rallus longirostris

A large rail of salt marshes. Very closely related to King Rail. Atlantic Coast birds much duller than King; Gulf Coast birds intermediate.

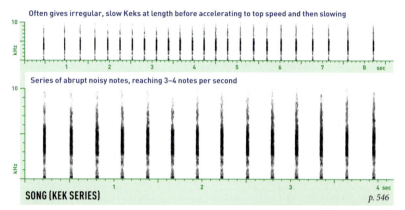

Often gives irregular, slow Keks at length before accelerating to top speed and then slowing

Series of abrupt noisy notes, reaching 3–4 notes per second

SONG (KEK SERIES)   *p. 546*

Mostly Mar.–June, apparently by unmated males advertising for a mate. Usually delivered in irregular bouts lasting 4–10 seconds, accelerating and then decelerating, but may be given continuously. Speed of series increases with level of agitation; excited birds can temporarily exceed 5 notes per second.

Rapid series of harsh grunts, 4–6 notes per second, often accelerating at end

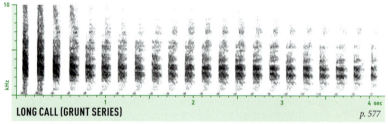

LONG CALL (GRUNT SERIES)   *p. 577*

All year, often by mates in unsynchronized duets, in courtship and to maintain pair bond; also used in territorial disputes. Pair duets tend to be longer and less accelerating than solo versions. Averages faster than King Rail Grunt Series, but much overlap; only fastest versions safely identifiable.

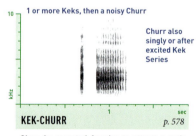

1 or more Keks, then a noisy Churr

Churr also singly or after excited Kek Series

KEK-CHURR   *p. 578*

Given by unmated females to attract a male. Churr without Kek given by both sexes, perhaps as alarm or contact call near nest.

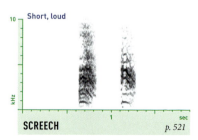

Short, loud

SCREECH   *p. 521*

Given singly but repeatedly by agitated birds. Quite plastic; often nasal, but varies from rather grunting to rather squeaky.

King Rail

Rallus elegans

Breeds in freshwater marshes. Hybridizes regularly with Clapper Rail along narrow zone of overlap in brackish coastal wetlands.

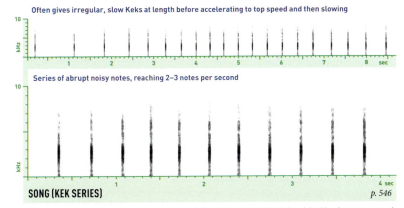

Often gives irregular, slow Keks at length before accelerating to top speed and then slowing

Series of abrupt noisy notes, reaching 2–3 notes per second

SONG (KEK SERIES) *p. 546*

Given in same situations and patterns as Clapper Rail Kek Series; best distinguished by slower top speed, especially during steady, extended calling bouts. Fast Kings overlap with slow Clappers, especially when Kings are excited, but many birds are identifiable if sufficient time is spent listening.

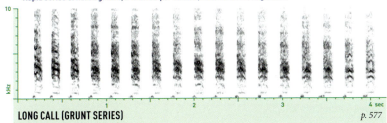

Rapid series of harsh grunts, 3–5 notes per second, often accelerating at end

LONG CALL (GRUNT SERIES) *p. 577*

Given in same situations as Grunt Series of other rails, including in response to loud noises. This and Clapper Rail's Grunt Series higher, screechier, and longer than Virginia Rail's, up to 10 seconds or more (Virginia Rail Grunt Series usually 5 seconds or less).

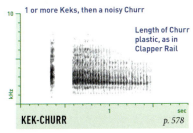

1 or more Keks, then a noisy Churr

Length of Churr plastic, as in Clapper Rail

KEK-CHURR *p. 578*

Probably not safely separable from Clapper Kek-churr; likely given in similar circumstances, but more study needed.

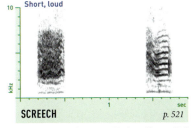

Short, loud

SCREECH *p. 521*

Probably not safely separable from Clapper Rail Screech. Highly plastic. Both species also give a rare, very low Hoot.

VIRGINIA RAIL

Rallus limicola

Common but secretive. Breeds in dense cattails or bulrushes, rarely in salt marshes. Only two-thirds the size of Clapper and King Rails.

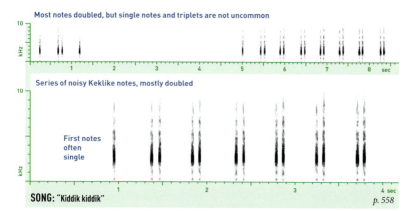

Most notes doubled, but single notes and triplets are not uncommon

Series of noisy Keklike notes, mostly doubled

First notes often single

SONG: "Kiddik kiddik"
p. 558

Mostly Apr.–June, possibly only from males; apparently functions primarily in attracting a mate, though also given after eggs hatch. Consistent doubling of notes diagnostic, but occasional birds give 10 or more single notes before doubling. Higher than calls of Clapper and King Rails, lacking the bass register.

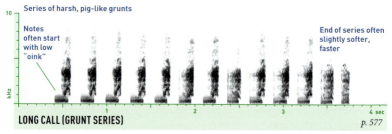

Series of harsh, pig-like grunts

Notes often start with low "oink"

End of series often slightly softer, faster

LONG CALL (GRUNT SERIES)
p. 577

All year, often by mates in unsynchronized duets, apparently to maintain pair bond; also in territorial disputes. Pair duets tend to be longer and less accelerating than solo versions. Averages shorter than Grunt Series of Clapper and King Rails, with lower, more growling, more disyllabic notes.

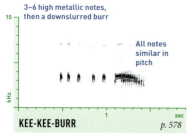

3–6 high metallic notes, then a downslurred burr

All notes similar in pitch

KEE-KEE-BURR
p. 578

In early spring, perhaps by females. Introductory notes and final burry note may be given separately.

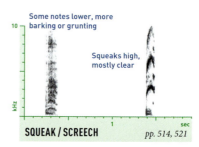

Some notes lower, more barking or grunting

Squeaks high, mostly clear

SQUEAK / SCREECH
pp. 514, 521

Highly plastic; Squeaks grade into longer Screeches and sometimes lower Barks. Given singly in alarm by both sexes.

SORA

Porzana carolina

Breeds mostly in freshwater marshes; winters in brackish and salt marshes also. Rather bold for a rail, but still more often heard than seen.

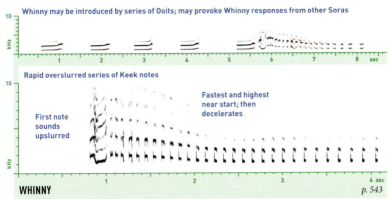

Whinny may be introduced by series of Ooits; may provoke Whinny responses from other Soras

Rapid overslurred series of Keek notes

First note sounds upslurred

Fastest and highest near start; then decelerates

WHINNY *p. 543*

Mostly in spring migration and breeding season, apparently in territorial defense and pair contact; often associated with chases. May be provoked by sounds of other Soras, Virginia Rails, or any loud noise. Females often respond to mate's Whinny, often starting halfway through his call (see top).

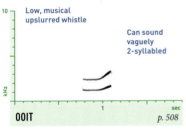

Low, musical upslurred whistle

Can sound vaguely 2-syllabled

OOIT *p. 508*

Mostly May–June, often in long bouts, perhaps to attract mate. Similar calls reportedly given singly by night migrants.

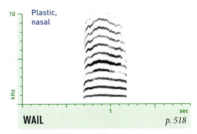

Plastic, nasal

WAIL *p. 518*

Mostly in fall, especially by juveniles; possibly also by night migrants. Length and pitch pattern highly plastic; grades into Ooit.

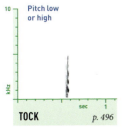

Pitch low or high

TOCK *p. 496*

Mostly in contact between family members near nest. Like coot calls, but softer.

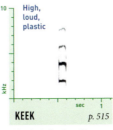

High, loud, plastic

KEEK *p. 515*

Mostly during breeding, perhaps in alarm; sometimes in long regular bouts.

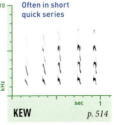

Often in short quick series

KEW *p. 514*

Variable, plastic; grades into Keek. May be related to female version of Whinny.

PURPLE SWAMPHEN

Porphyrio porphyrio

An Old World species, established and expanding in Florida. Nearly twice the size of Purple Gallinule. A few have more blue on head than shown.

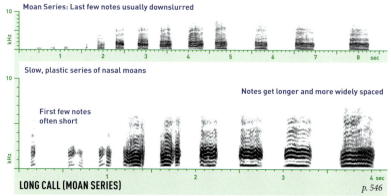

Moan Series: Last few notes usually downslurred

Slow, plastic series of nasal moans

Notes get longer and more widely spaced

First few notes often short

LONG CALL (MOAN SERIES) *p. 546*

Highly variable and plastic, but characteristically lower and slower than Long Calls of Purple and Common Gallinules, with nasal, sometimes groaning or bleating quality. Little studied, but apparently given in agitation, sometimes by pairs or groups. Individual notes can be given singly.

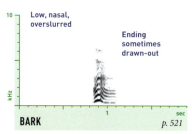

Low, nasal, overslurred

Ending sometimes drawn-out

BARK *p. 521*

Highly variable and plastic; function little known. Grades into other calls.

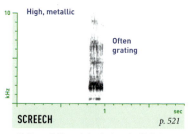

High, metallic

Often grating

SCREECH *p. 521*

Highly variable and plastic; function little known. Grades into other calls.

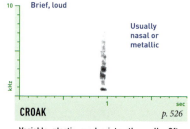

Brief, loud

Usually nasal or metallic

CROAK *p. 526*

Variable, plastic; grades into other calls. Often like Croak of American Coot.

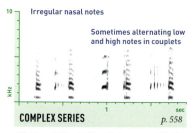

Irregular nasal notes

Sometimes alternating low and high notes in couplets

COMPLEX SERIES *p. 558*

Highly variable, plastic; note types vary. Rhythm often unsteady.

Purple Gallinule

Porphyrio martinicus

Spectacularly colored, but often inconspicuous. Found primarily around the edges of vegetated freshwater ponds and in mature rice fields.

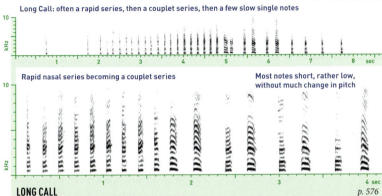

Long Call: often a rapid series, then a couplet series, then a few slow single notes

Rapid nasal series becoming a couplet series

Most notes short, rather low, without much change in pitch

LONG CALL *p. 576*

All year, but mostly Mar.–Aug., when disturbed and possibly in territorial defense and mate contact; often given in flight. Variable and plastic. Sometimes precedes or grades into Keek-a Series, creating a confusing variety of combination sounds. Some versions are brief, quick Kek Series (p. 546).

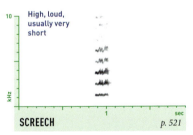

High, loud, usually very short

SCREECH *p. 521*

Highly variable and plastic; grades into other calls. Usually shorter, clearer than Wail of Common Gallinule.

High, nasal

Sometimes broken

WAIL *p. 518*

Highly plastic. Usually clear, but occasionally slightly grating; can strongly resemble Limpkin Wail or single-note calls of gulls.

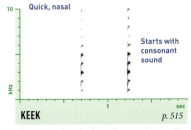

Quick, nasal

Starts with consonant sound

KEEK *p. 515*

All year, perhaps in alarm. Rare versions grating or croaking. Averages lower than Common Gallinule Keek, more like Black-necked Stilt.

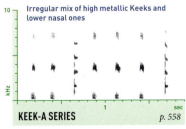

Irregular mix of high metallic Keeks and lower nasal ones

KEEK-A SERIES *p. 558*

Variable, plastic. Irregular low-high pattern nearly diagnostic when present. May be a variant of the Long Call.

Common Gallinule

Gallinula galeata

Common in southern marshes; local farther north. Not particularly shy, but often inconspicuous. Formerly lumped with the Common Moorhen of Eurasia.

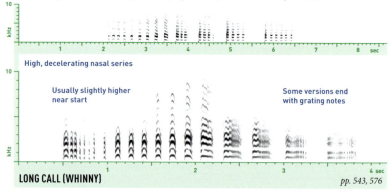

Some Long Calls break into couplet series, triplet series, or even more complex series

High, decelerating nasal series

Usually slightly higher near start

Some versions end with grating notes

LONG CALL (WHINNY)

pp. 543, 576

All year, but mostly Apr.–June. Believed to be used by males to advertise territory; perhaps also in mate contact. Often given in response to mates or neighbors. Versions that end in grates or complex series fairly distinctive, but some overlap with pattern of Purple Gallinule Long Call.

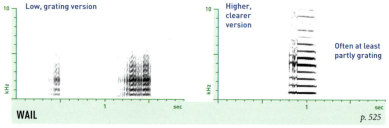

Low, grating version

Higher, clearer version

Often at least partly grating

WAIL

p. 525

Exceedingly variable and plastic category of sounds; often loud. Varies in pitch and tone, but strained quality rather distinctive when present, as though bird is trying and failing to keep voice clear of grating sounds. Function and behavioral context poorly understood.

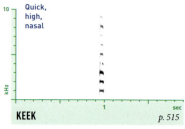

Quick, high, nasal

KEEK

p. 515

All year, possibly in alarm. Rare versions grating or clucking Tocks (p. 596), or low Barks (p. 515). Averages higher than Purple Gallinule Keek.

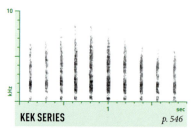

KEK SERIES

p. 546

Not well known. Kek Series variable, plastic; speed can be irregular. Note type can change during series, but does not alternate.

AMERICAN COOT

Fulica americana

Abundant on vegetated ponds as well as larger lakes. In migration and winter, occasionally found on salt water. Swims and dives more than other rails.

Long Call: here, nasal couplet version followed by grating version

Grating version: slow series of grating Croaks and/or Wails

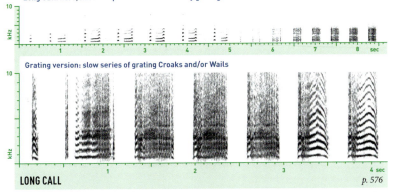

LONG CALL *p. 576*

Highly plastic and variable. Often introduced by single-note calls. Two main types, possibly corresponding to different sexes or purposes. More common version consists mostly of long Grates. Another version includes nasal couplet series, with second note of each couplet often long, rising.

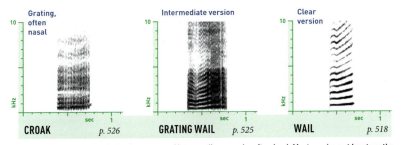

Grating, often nasal

Intermediate version

Clear version

CROAK *p. 526* **GRATING WAIL** *p. 525* **WAIL** *p. 518*

Exceedingly variable and plastic category of intergrading sounds; often loud. Most versions at least partly croaking or grating, but some are clear and nasal. Function and behavioral context need more study. Begging juveniles give whistled Screech (p. 521) or Pseep Series (p. 535).

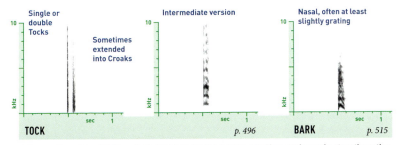

Single or double Tocks

Sometimes extended into Croaks

Intermediate version

Nasal, often at least slightly grating

TOCK *p. 496* **BARK** *p. 515*

Gives a wide variety of highly variable and plastic single-note sounds. Most versions at least partly grating. Function and behavioral context need more study.

Sandhill Crane

Antigone canadensis

Breeds in bogs and grassy wetlands. Locally common; most populations stable or increasing. In migration, often stages for weeks in enormous flocks.

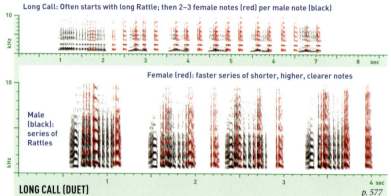

Long Call: Often starts with long Rattle; then 2–3 female notes (red) per male note (black)

Female (red): faster series of shorter, higher, clearer notes

Male (black): series of Rattles

LONG CALL (DUET)   *p. 577*

All year, by mated pairs, sometimes joined by their older offspring. Variable, but often highly synchronized, given with display postures. Used in aggressive encounters with other pairs or family groups; also apparently to reinforce pair bond. Birds separated from mates may give their half of the duet alone.

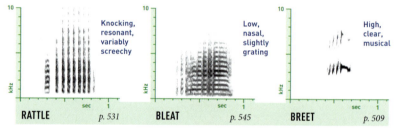

Knocking, resonant, variably screechy

Low, nasal, slightly grating

High, clear, musical

RATTLE   *p. 531*   **BLEAT**   *p. 545*   **BREET**   *p. 509*

Rattle given all year, in mild alarm, in aggressive encounters, and in contact, especially on the wing. Variable and plastic, but distinctive, frequent, and loud. Calls of pairs and family groups often overlap. Bleats infrequent; function unclear. Juveniles give Breet, grading into Rattle around the age of 9 months.

CONFUSING SOUNDS OF GALLINULES

Coots, gallinules, and swamphens make a tremendous variety of sounds, which in most species have not been well described or studied, in part because of the difficulty of determining the source of a call in dense marsh vegetation. Although each species gives certain distinctive sounds, there is a great deal of overlap. In areas where more than one species occurs, caution is warranted when attempting to identify birds by voice alone. Be aware that other marsh birds can sometimes make rail-like sounds, especially Least Bittern (p. 188).

WHOOPING CRANE

Grus americana

Exceedingly rare. A few hundred live in the wild thanks to conservation, captive breeding, and training juveniles to migrate by following ultralight aircraft.

Here, Long Call is initiated by female (black), then joined by two males (red and cyan)

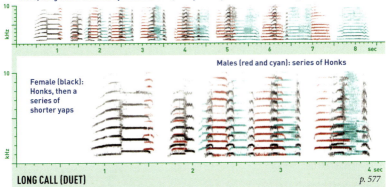

Males (red and cyan): series of Honks

Female (black): Honks, then a series of shorter yaps

LONG CALL (DUET) *p. 577*

All year, by mated pairs, sometimes joined by their older offspring. Variable, but often highly synchronized, given with display postures. Used in aggressive encounters with other pairs or family groups; also apparently to reinforce pair bond. Birds separated from mates may give their half of the duet alone.

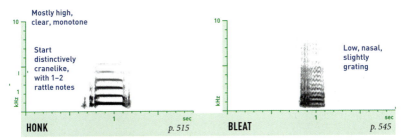

Mostly high, clear, monotone

Start distinctively cranelike, with 1–2 rattle notes

Low, nasal, slightly grating

HONK *p. 515*

BLEAT *p. 545*

Honk given all year, in mild alarm, in aggressive encounters, and in contact, especially on the wing. Variable and plastic, but highly distinctive. Calls of pairs and family groups often overlap. Bleat softer; function unclear. Juveniles give Breet (p. 509) like Sandhill Crane; no recordings available.

ELONGATED TRACHEAS

The distinctively resonant vocalizations of some birds may be connected to the unique anatomy of their tracheas, which are extremely long, looping around inside the body and sometimes penetrating the sternum (breastbone) before entering the lungs. Some have argued that the elongation of the trachea serves to lower the pitch of the calls, while others believe its primary purpose is to make the calls louder by using the sternum as a resonator. More study is needed. The North American birds known to possess this anatomy include some of our loudest species: Limpkin, Sandhill and Whooping Cranes, Trumpeter and Tundra Swans, and Plain Chachalaca.

LIMPKIN

Aramus guarauna

Local in marshes and swamps, where it forages for apple snails, leaving tellale piles of shells. Named for its bent-legged, head-pumping "limping" gait.

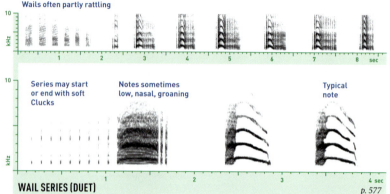

Wails often partly rattling

Series may start or end with soft Clucks

Notes sometimes low, nasal, groaning

Typical note

WAIL SERIES (DUET) *p. 577*

All year, especially by unpaired males, often for long periods, even at night. Variable and plastic, but loud and distinctive. Grades into Kreew. Females duet with mate, giving Kewlike calls at slower pace, leading to characteristic irregular rhythm. In flight, wings sometimes Rattle with each wingbeat (p. 533).

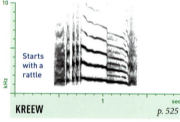

Overslurred rattling wail, usually in series

Starts with a rattle

KREEW *p. 525*

Like Wails, but longer, in slower series, with cranelike rattling quality. Apparently by males in many contexts, often in chorus.

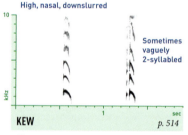

High, nasal, downslurred

Sometimes vaguely 2-syllabled

KEW *p. 514*

All year, apparently by females, in a variety of circumstances. Rather variable and plastic.

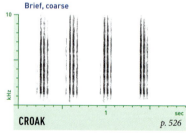

Brief, coarse

CROAK *p. 526*

All year, by both sexes, in alarm and possibly other contexts. Some versions as short as a single Tock (p. 496), like American Coot.

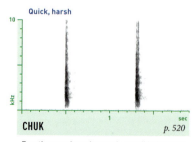

Quick, harsh

CHUK *p. 520*

Function poorly understood; possibly best considered a harsh version of Kew, with which it may intergrade.

SHOREBIRDS

This diverse group includes several related families of birds. The majority of species are adapted to feeding along shorelines, where they search for invertebrate prey in various ways, from probing in mud to wading in water. Some species are regularly found far from water. Despite great variation in size and shape, many shorebirds share a distinctive basic body plan, often with long legs, a long neck, and a long bill. Many species have extensive vocal repertoires.

STILTS AND AVOCETS (Family Recurvirostridae)

This family includes two species in our area, both with long legs and striking plumage. Their vocalizations are loud and frequent, but much simpler than those of most other shorebirds. All sounds are presumed to be innate.

OYSTERCATCHERS (Family Haematopodidae)

The oystercatchers are large, strikingly patterned birds with distinctive long, bright red bills that are deep but quite thin side-to-side, which they use in foraging for bivalves, worms, and other marine invertebrates. Our species inhabits beaches and barrier islands, especially those made almost entirely of mollusk shells. All sounds are presumed to be innate.

Black-necked Stilt

Himantopus mexicanus

Unmistakable, with bold plumage and ridiculously long legs. Breeds on shallow briny ponds; winters in flooded fields and salt marshes.

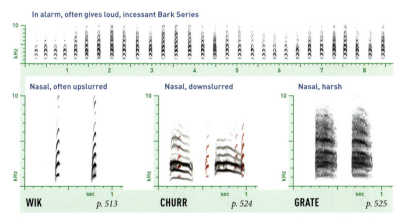

In alarm, often gives loud, incessant Bark Series

Nasal, often upslurred   Nasal, downslurred   Nasal, harsh

WIK   *p. 513*   CHURR   *p. 524*   GRATE   *p. 525*

Highly vocal at all times of year. Vocalizations plastic, grading into one another fairly smoothly according to level of agitation. Wik is most common call, given in alarm and in flight; plastic. Often noticeably upslurred, but some versions are Barks (p. 515). Grates given in high alarm; Churrs intermediate.

American Avocet

Recurvirostra americana

Beautiful and elegant. Rusty plumage turns white in winter. Often feeds by sweeping upturned bill back and forth just under the surface of the water.

In circle display, 3–4 avocets face each other with lowered heads and give rapid Klee Series (p. 543)

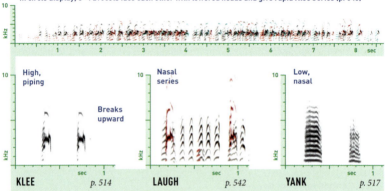

KLEE — High, piping / Breaks upward — *p. 514*

LAUGH — Nasal series — *p. 542*

YANK — Low, nasal — *p. 517*

Klee is most common call, given all year in alarm and in flight, and in loud rapid chorus during circle display, apparently between 2 mated pairs or a mated pair and a single bird. Some versions are 2-syllabled Ka-lees. Laugh little known. Yank given during wing-raising distraction display.

American Oystercatcher

Haematopus palliatus

Often vocal and conspicuous on coastal mudflats and shellfish beds, but overall population is estimated to be low. True to its name, forages mostly on bivalves.

Peeps sometimes run into loud Twitter (p. 536), often by 2 or more birds at once

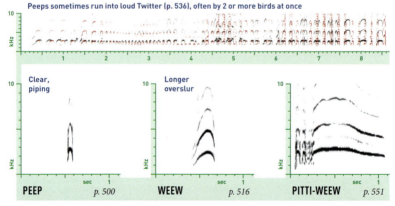

PEEP — Clear, piping — *p. 500*

WEEW — Longer overslur — *p. 516*

PITTI-WEEW — *p. 551*

All calls highly plastic, grading into one another. Peeps are most common, in contact and alarm. In courtship and territorial display, Peeps run into long Twitter with sudden changes in pitch and speed. Weew and combinations like Pitti-weew may function in pair bonding.

PLOVERS (Family Charadriidae)

The plovers are small to medium-sized shorebirds with short, stout bills and fairly long legs. Rather than probing in mud, they forage by alternately running and pausing, using their large eyes to spot prey. They are well known for their distraction displays, often feigning a broken wing or other injury in order to lead potential predators, including humans, away from nest or young. Sometimes these displays are accompanied by distinctive vocalizations.

All plover sounds are likely innate; most are whistles, with occasional trills or burrs. In several species, a flight display, with associated songlike sounds, forms an important part of courtship.

SNOWY PLOVER

Charadrius nivosus

Uncommon to rare; highly local. Nests on open sandy beaches and the edges of alkaline lakes. Populations fluctuate, but have declined in many areas.

Ter-weet and Peer-purr often given together, in plastic series (p. 574)

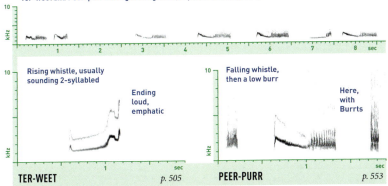

Rising whistle, usually sounding 2-syllabled

Ending loud, emphatic

TER-WEET   *p. 505*

Falling whistle, then a low burr

Here, with Burrts

PEER-PURR   *p. 553*

Mostly Jan.–Sept. Ter-weet given by males excavating nest scrapes in courtship, or in territorial confrontations, and by both sexes in response to potential predators. Peer-purr is given by males in courtship, possibly also in other contexts and by females. Both calls uniform but plastic.

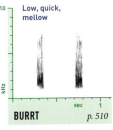

Low, quick, mellow

BURRT   *p. 510*

All year, in alarm. Plastic; in high alarm, can become longer, coarser.

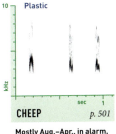

Plastic

CHEEP   *p. 501*

Mostly Aug.–Apr., in alarm, often in flight, including by flocks.

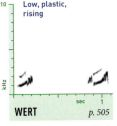

Low, plastic, rising

WERT   *p. 505*

Possibly all year. May simply be a short, plastic version of Ter-weet.

WILSON'S PLOVER

Charadrius wilsonia

A medium-sized plover of beaches, dunes, and salt flats; rarely found inland or on mud. Note single breast band and heavy black bill.

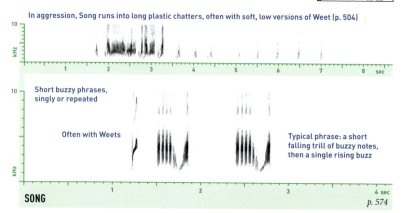

In aggression, Song runs into long plastic chatters, often with soft, low versions of Weet (p. 504)

Short buzzy phrases, singly or repeated

Often with Weets

Typical phrase: a short falling trill of buzzy notes, then a single rising buzz

SONG
p. 574

Mostly in spring, by both sexes in territorial displays and aggressive chases. Sometimes given by male chasing another male in flight, but the species may lack a stereotyped flight display. Male in close courtship reported to give a faint Moan like that of Mountain Plover; no recordings available.

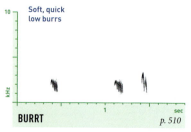

Soft, quick low burrs

BURRT
p. 510

Not well known; given by family groups with chicks, possibly in contact.

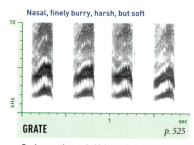

Nasal, finely burry, harsh, but soft

GRATE
p. 525

During nesting and chick rearing, by both sexes in broken-wing distraction display.

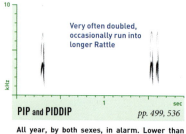

Very often doubled, occasionally run into longer Rattle

PIP and PIDDIP
pp. 499, 536

All year, by both sexes, in alarm. Lower than Weet.

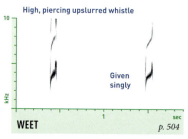

High, piercing upslurred whistle

Given singly

WEET
p. 504

All year, in alarm, mostly by males. Often mixed with Pips and Piddips.

Semipalmated Plover

Charadrius semipalmatus

Nests on patches of gravel or sand near water. In migration and winter, found on mudflats and beaches. Small, with one breast band, dark back.

In flight display, long series of Weeiks are occasionally punctuated by a Laugh and 3–4 Growls

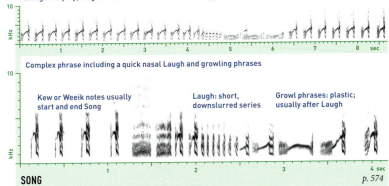

Complex phrase including a quick nasal Laugh and growling phrases

Kew or Weeik notes usually start and end Song

Laugh: short, downslurred series

Growl phrases: plastic; usually after Laugh

SONG   p. 574

All year. Highly stereotyped version given in flight display, mostly on breeding grounds but occasionally on spring migration. Softer, more plastic versions given frequently in migration and winter in aggressive interactions with other members of the same species, often with single rising Weeik notes.

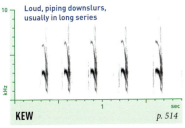

Loud, piping downslurs, usually in long series

KEW   p. 514

All year, in anxiety, possibly in aggression. Plastic; grades into Weeik notes of Song and likely also into Peeplike notes.

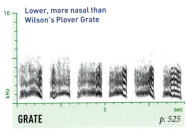

Lower, more nasal than Wilson's Plover Grate

GRATE   p. 525

Mostly June–Aug., in high-intensity distraction display near nest or young, while feigning injury.

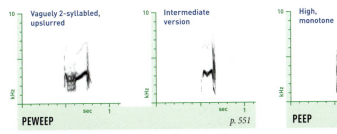

Vaguely 2-syllabled, upslurred

Intermediate version

High, monotone

PEWEEP   p. 551   **PEEP**   p. 500

All year. Most common calls, often given in flight. Also given in alarm, in low-intensity distraction display, and in conjunction with Kew calls. Highly plastic; 2-syllabled Peweep and 1-syllabled Peep are typical, but gives a range of intermediates.

PIPING PLOVER

Charadrius melodus

Uncommon and highly local; nests on open sand or gravel near water. Shares orange legs and bill with Semipalmated Plover, but much paler.

Song may continue for 10 seconds or more

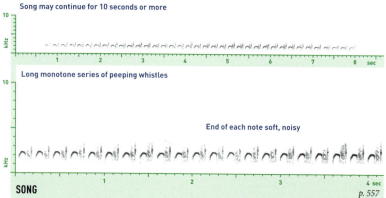

Long monotone series of peeping whistles

End of each note soft, noisy

SONG
p. 557

Apr.–July, by males, usually in slow-flapping display flight over territory, or from ground after display flight ends. Uniform and stereotyped. Often introduced by a long series of Weets. Female often responds to male Song with Peeps.

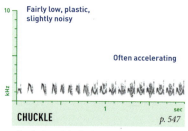

Fairly low, plastic, slightly noisy

Often accelerating

CHUCKLE
p. 547

Mostly Apr.–July, by both sexes, in aggression, often accompanied by horizontal threat posture. Peeping version given from nest scrape.

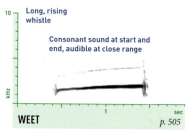

Long, rising whistle

Consonant sound at start and end, audible at close range

WEET
p. 505

Possibly all year, by both sexes, in alarm and aggression. Also associated with male Song, and sometimes given during flight display.

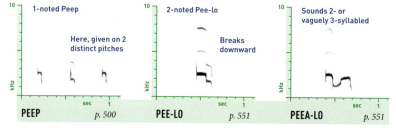

1-noted Peep

Here, given on 2 distinct pitches

PEEP
p. 500

2-noted Pee-lo

Breaks downward

PEE-LO
p. 551

Sounds 2- or vaguely 3-syllabled

PEEA-LO
p. 551

A set of intergrading calls, given all year by both sexes in contexts ranging from contact to alarm; different versions have slightly different functions. 1-note Peeps run into rapid series by males excavating nest scrapes in courtship; some drawn-out versions given in broken-wing distraction display.

KILLDEER

Charadrius vociferus

Large, with two breast bands. Found in open fields and areas with short grass, not necessarily near water. Highly vocal, often calling nearly nonstop.

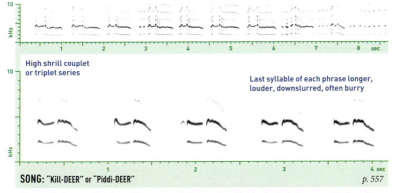

Excited versions of Song often burrier and more plastic, grading into Trill or other sounds

High shrill couplet or triplet series

Last syllable of each phrase longer, louder, downslurred, often burry

SONG: "Kill-DEER" or "Piddi-DEER"
p. 557

The species' namesake, given all year, especially in spring. Highly plastic, but geographically uniform; typically delivered on the wing, sometimes nonstop for many minutes. Some versions are quadruplet series. Associated with courtship, but also given in alarm and excitement.

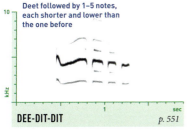

Deet followed by 1–5 notes, each shorter and lower than the one before

DEE-DIT-DIT
p. 551

Excited version of Deet, ranging from 2-noted Dee-dit to longer versions that grade into the Twitter.

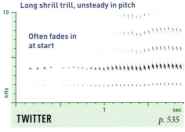

Long shrill trill, unsteady in pitch

Often fades in at start

TWITTER
p. 535

Plastic; up to 10 seconds long. Given in close courtship and in high alarm, including in broken-wing distraction displays.

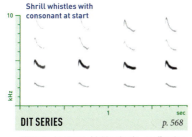

Shrill whistles with consonant at start

DIT SERIES
p. 568

All year, in alarm; highly plastic, grading into other sounds, especially Deet. Note length and shape highly variable.

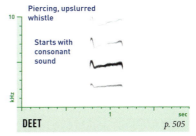

Piercing, upslurred whistle

Starts with consonant sound

DEET
p. 505

All year, in many situations. Highly plastic, especially in length; typical version shown. Grades into all other sounds.

MOUNTAIN PLOVER

Charadrius montanus

Breeds in well-grazed patches of arid shortgrass prairie, agricultural fields, and prairie dog towns. Population has apparently declined.

Song can last 5 seconds or more

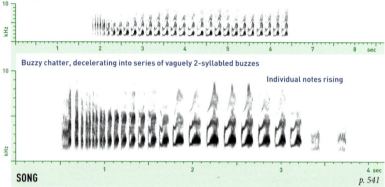

Buzzy chatter, decelerating into series of vaguely 2-syllabled buzzes

Individual notes rising

SONG
p. 541

Mostly Mar.–June, by both sexes, in territorial flight display, from the ground, or during aggressive interactions. Occasionally given in fall migration. Somewhat variable and slightly plastic; often truncated. Compare Dunlin Song.

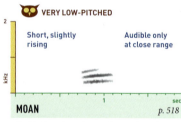

🦉 VERY LOW-PITCHED

Short, slightly rising

Audible only at close range

MOAN
p. 518

Mostly Apr.–June, by courting male and sometimes female, at potential nest scrape, bowing head with fanned tail.

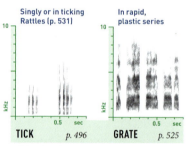

Singly or in ticking Rattles (p. 531)

In rapid, plastic series

TICK *p. 496*   **GRATE** *p. 525*

Ticks given in threat display to potential nest predators; Grates and soft, clear Wails given in injury-feigning distraction display.

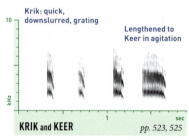

Krik: quick, downslurred, grating

Lengthened to Keer in agitation

KRIK and KEER *pp. 523, 525*

Apparently all year, by both sexes, in alarm and in interactions, often on the wing. Like Keer of Forster's Tern, but usually shorter.

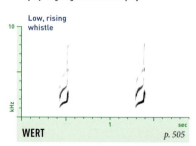

Low, rising whistle

WERT *p. 505*

Apparently all year, by both sexes, in mild alarm, often with Keer. Somewhat plastic.

Black-bellied Plover

Pluvialis squatarola

Breeds in Arctic; winters mostly along coasts. Generally prefers mudflats to fields. In flight, shows large black patch at base of underwing.

Song (p. 568): Usually starts with trill, then 1–2 slurred, broken whistles, like Long-billed Curlew

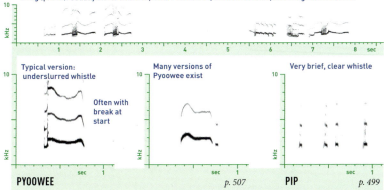

Typical version: underslurred whistle

Often with break at start

PYOOWEE

Many versions of Pyoowee exist

p. 507

Very brief, clear whistle

PIP   *p. 499*

Only sounds heard in migration and winter are shown. Typical Pyoowee is highly distinctive, but variable and plastic; many kinds of whistles may be given. Pips soft, given during foraging, perhaps in contact. Songs given during interactions; nonbreeding versions are plastic and often truncated.

American Golden-Plover

Pluvialis dominica

Breeds in Arctic; winters in southern South America. In migration, frequents fields, away from water. Lacks Black-bellied's white tail and wing stripe.

Away from breeding grounds, gives plastic, songlike phrases of short notes (p. 568)

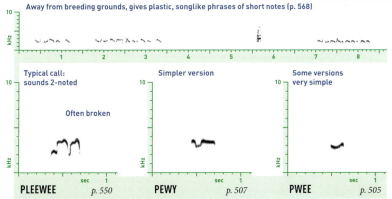

Typical call: sounds 2-noted

Often broken

PLEEWEE   *p. 550*

Simpler version

PEWY   *p. 507*

Some versions very simple

PWEE   *p. 505*

Only sounds heard in migration and winter are shown. Pleewee and Pewy are typical of the variable, plastic calls given in flight and alarm; most versions are brief and complex, without much change in pitch. Plastic songlike phrases often quite soft; quality may recall song of Willet.

SANDPIPERS (Family Scolopacidae)

The sandpipers are a highly diverse family of shorebirds. Many species breed in the high Arctic, where they perform elaborate courtship displays, drawing on large, complex vocal repertoires. In many species, these breeding sounds are not heard south of the Arctic, and are therefore omitted from this book.

Snipe and woodcock are terrestrial sandpipers with short necks and legs and extremely long bills. They give aerial flight displays accompanied by mechanical sounds made by the tail (snipe) or the wings (woodcock).

The phalaropes swim more than other shorebirds, often spinning in tight circles to raise food to the surface. Females are brightly colored, while males incubate the eggs and raise the young. Vocalizations tend to be fairly simple.

UPLAND SANDPIPER

Bartramia longicauda

A bird of tallgrass prairie. Related to curlews; note similar vocalizations and plumage. Flies on stiff wings, held high for a moment after landing.

Song sometimes interspersed with low, soft Groans

A rising, accelerating trill, then a broken overslurred whistle

Often preceded by Chuckle

Notes of trill start buzzy, become whistled

SONG   *p. 568*

All year, by both sexes, but mostly Apr.–Aug., by males, often in gliding flight display. Sometimes given at night. Variable and somewhat plastic; frequently truncated, especially in late summer. Much like some versions of Long-billed Curlew Song, but averages more complex.

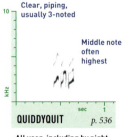

Clear, piping, usually 3-noted

Middle note often highest

QUIDDYQUIT   *p. 536*

All year, including by night migrants. Plastic, especially in length.

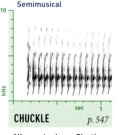

Semimusical

CHUCKLE   *p. 547*

All year, in alarm. Plastic; grades into Quiddyquit. Rarely shortened to 1 Cluck.

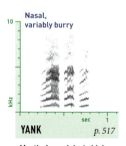

Nasal, variably burry

YANK   *p. 517*

Mostly June–July, in high alarm, by parents with chicks.

WHIMBREL

Numenius phaeopus

Breeds on Arctic tundra. In migration and winter, found on tidal flats and beaches; a few remain on wintering range year-round.

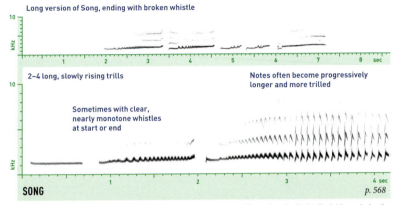

Long version of Song, ending with broken whistle

2–4 long, slowly rising trills

Notes often become progressively longer and more trilled

Sometimes with clear, nearly monotone whistles at start or end

SONG   *p. 568*

All year, but especially in aerial display on breeding grounds. Often given in flight. Variable and plastic; versions given in migration and winter are often truncated, sometimes with shorter and clearer notes.

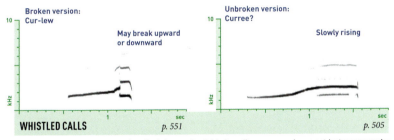

Broken version:
Cur-lew

May break upward or downward

Unbroken version:
Curree?

Slowly rising

WHISTLED CALLS   *p. 551*   *p. 505*

Probably all year, but apparently quite infrequent outside breeding season, when most instances may be better described as truncated songs. More study needed. Variable and plastic, but apparently averages longer than similar calls of Long-billed Curlew.

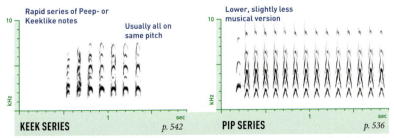

Rapid series of Peep- or Keeklike notes

Usually all on same pitch

Lower, slightly less musical version

KEEK SERIES   *p. 542*   **PIP SERIES**   *p. 536*

All year, in alarm. Most common call outside breeding season, given whenever birds are disturbed, often in flight. Somewhat variable in tone quality and plastic in length.

LONG-BILLED CURLEW

Numenius americanus

Large, with an amazing bill. Breeds on dry shortgrass prairie; in migration and winter, found on mudflats and in wet meadows. Populations have declined.

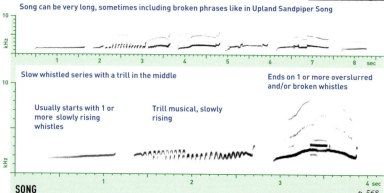

Song can be very long, sometimes including broken phrases like in Upland Sandpiper Song

Slow whistled series with a trill in the middle

Ends on 1 or more overslurred and/or broken whistles

Usually starts with 1 or more slowly rising whistles

Trill musical, slowly rising

SONG *p. 568*

All year, by both sexes, especially during the early breeding season. Highly variable and plastic. Often given from ground, especially after alighting; also given in circling flight and during territorial chases on breeding grounds.

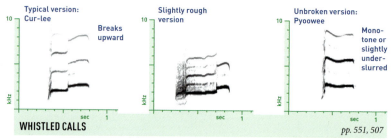

Typical version: Cur-lee

Breaks upward

Slightly rough version

Unbroken version: Pyoowee

Mono-tone or slightly under-slurred

WHISTLED CALLS *pp. 551, 507*

All year, by both sexes; most common call, given in contact or alarm. Highly variable and plastic. Typical version is a clear, upward-breaking Cur-lee, but many variations exist. Given singly or in rapid series, from ground or in flight.

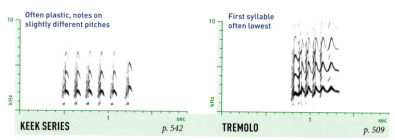

Often plastic, notes on slightly different pitches

First syllable often lowest

KEEK SERIES *p. 542*

TREMOLO *p. 509*

All year, especially during breeding season, by both sexes, in agitation, usually in flight. Sometimes used to mob predators, including humans. Highly plastic, grading into whistled calls. Some versions are Tremolos, some are Trills, and some are repeated screechy series of 2–3 notes.

Marbled Godwit

Limosa fedoa

Nests in well-grazed grasslands, often far from water; in migration and winter, frequents mudflats. Resembles Long-billed Curlew, but bill shape differs.

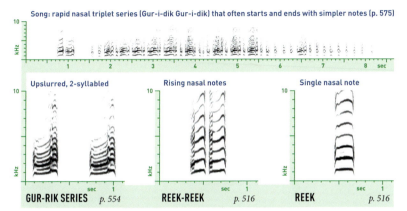

Song: rapid nasal triplet series (Gur-i-dik Gur-i-dik) that often starts and ends with simpler notes (p. 575)

Upslurred, 2-syllabled

Rising nasal notes

Single nasal note

GUR-RIK SERIES *p. 554*

REEK-REEK *p. 516*

REEK *p. 516*

All vocalizations plastic, grading into one another, and given all year. Most distinctive sound is Song, a rapid series of Gur-i-dik phrases, given during interactions. Reek and Reek-reek given in alarm, often in flight. Gur-rik Series given repeatedly by males in circling courtship flights; also in other contexts.

Hudsonian Godwit

Limosa haemastica

Smaller than Marbled Godwit. Migrates mostly along East Coast in fall, through Great Plains in spring; winters in southern South America.

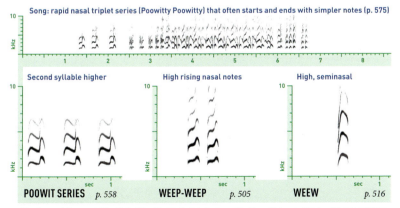

Song: rapid nasal triplet series (Poowitty Poowitty) that often starts and ends with simpler notes (p. 575)

Second syllable higher

High rising nasal notes

High, seminasal

POOWIT SERIES *p. 558*

WEEP-WEEP *p. 505*

WEEW *p. 516*

Vocal repertoire much like that of Marbled Godwit, but all vocalizations higher and faster. Weew and Weep-weep given in alarm and in flight; often shortened to quick soft Wip. Poowit Series and Song possibly not often heard away from breeding grounds. Also gives soft Growls (not indexed).

Ruddy Turnstone

Arenaria interpres

Plumage and short, upturned bill are unique. Found mostly on rocky or sandy shores, occasionally mudflats. Uncommon to rare inland in migration.

Flocks in flight may give various, intergrading versions of Keer and Cheep

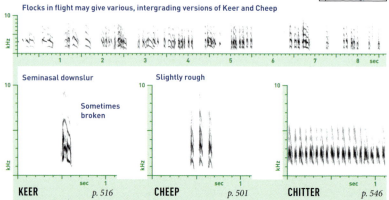

Seminasal downslur

Sometimes broken

Slightly rough

KEER *p. 516*

CHEEP *p. 501*

CHITTER *p. 546*

Vocal all year; most sounds are plastic variants of those shown. Keer usually given singly, in alarm. Grades into Cheeps, given in short quick series in flight and in flocks. Chitter given in aggression, often with tail raised. Tone often varies within a Chitter; some versions are rattling.

Red Knot

Calidris canutus

Larger than other *Calidris* sandpipers. Winters on sandy beaches; uncommon inland in migration. North American population has declined rapidly.

Pip sometimes run into Twitter by agitated flocks

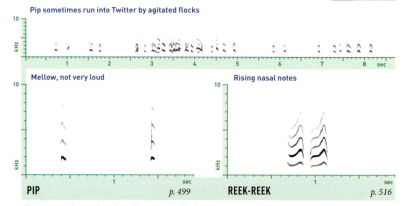

Mellow, not very loud

Rising nasal notes

PIP *p. 499*

REEK-REEK *p. 516*

Not very vocal away from breeding grounds. Pip is most common call in migration and winter; may sometimes run into a rapid mellow Twitter. Reek-reek given all year in alarm; sometimes single or in longer series. Complex flight song of slow whistles not known away from breeding grounds.

STILT SANDPIPER

Calidris himantopus

In migration and winter, found mostly on freshwater ponds, where it forages by wading up to its belly. Note slightly downcurved bill, white back and rump.

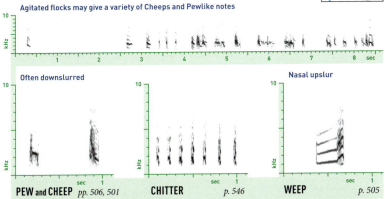

Agitated flocks may give a variety of Cheeps and Pewlike notes

Often downslurred

Nasal upslur

PEW and CHEEP *pp. 506, 501*   **CHITTER** *p. 546*   **WEEP** *p. 505*

Cheep is most common call in migration and winter, given in alarm or upon taking flight. Some versions are clear single Pews recalling Lesser Yellowlegs. Screechy Chitter may be given in interactions; never very fast. Weep given in alarm, singly or in short series. All calls plastic.

SANDERLING

Calidris alba

Winters on sandy beaches, habitually chasing the outgoing waves and fleeing the incoming ones. Uncommon inland in migration.

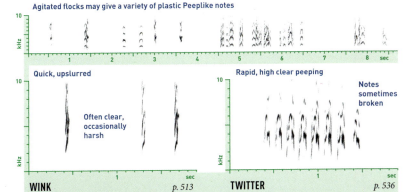

Agitated flocks may give a variety of plastic Peeplike notes

Quick, upslurred

Often clear, occasionally harsh

Rapid, high clear peeping

Notes sometimes broken

WINK *p. 513*   **TWITTER** *p. 536*

Wink is most common call in migration and winter, given in mild alarm or upon taking flight. Twitter apparently given during interactions. Both calls highly plastic, grading into various Peeplike notes.

Dunlin

Calidris alpina

Winters on mudflats, mostly near coast, often in flocks of thousands. Forages with sewing-machine motion in shallow water. Bill long, slightly downcurved.

Song: series of semimusical burrs, higher at start, notes often becoming vaguely 2-syllabled (p. 541)

Distinctively screechy, often slightly downslurred

GRATE p. 525

TWITTER p. 536

Grate given in flight or alarm; classic version is quite distinctive, long and scratchy. Twitter may be given in interactions. All calls are exceedingly plastic; excited flocks make many sounds, from single Peeps to short trilled or noisy notes. Song sometimes given in migration.

Purple Sandpiper

Calidris maritima

Winters farther north than other sandpipers, on rocky shores and jetties exposed to the surf. Very rare inland, except locally on the Great Lakes.

Cheep highly plastic, especially on takeoff; occasionally given in chittering series in flight

Slightly upslurred

Harsher versions

Clear

CHEEP

PEEP p. 500

p. 501

Not very vocal away from the breeding grounds. Most sounds in migration and winter are variations on the Cheep, given when flushed. Cheep highly plastic; becomes harsh in high alarm. Gives various sounds during interactions, including occasional clear Peeps and whistled notes.

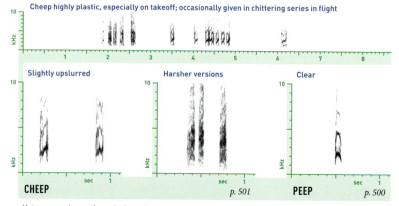

PECTORAL SANDPIPER

Calidris melanotos

Fairly common migrant on mudflats. Resembles Least Sandpiper, but larger. Breeding male develops an inflatable throat sac used in hooting displays.

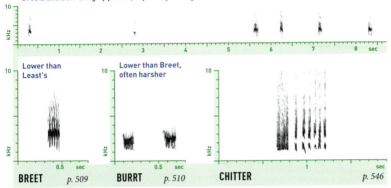

Breets and Burrts highly plastic, especially in length

BREET — Lower than Least's — *p. 509*

BURRT — Lower than Breet, often harsher — *p. 510*

CHITTER — *p. 546*

Breet and Burrt given all year, in alarm and in flight, Breet possibly by females and Burrt by males. At least on breeding grounds, males give low, harsh Snarls (p. 526) without burriness. Chitter given infrequently all year, during aggressive interactions, sometimes with high soft squeaking notes.

BUFF-BREASTED SANDPIPER

Calidris subruficollis

Our only sandpiper in which males display on leks (p. 75). Once abundant, now uncommon. Migrants use mown fields and dry, sandy edges of ponds.

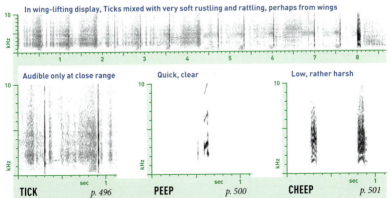

In wing-lifting display, Ticks mixed with very soft rustling and rattling, perhaps from wings

TICK — Audible only at close range — *p. 496*

PEEP — Quick, clear — *p. 500*

CHEEP — Low, rather harsh — *p. 501*

Very quiet for a sandpiper. Soft Ticks given only by male in display, lifting one wing at a time; mostly on breeding grounds but occasionally in spring migration. Peep known from a single recording, from male in display. Cheep given all year in alarm or in flight; plastic, but low for a sandpiper.

Baird's Sandpiper

Calidris bairdii

Abundant migrant on the Great Plains; rare on both coasts. Larger and longer-winged than Semipalmated. Prefers slightly drier ground than other peeps.

Flock gives Breets at slighly varying pitches when flushed or alarmed; rather quiet while feeding

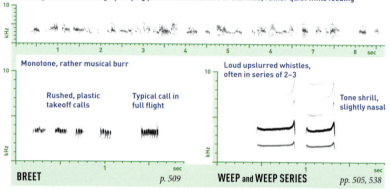

Monotone, rather musical burr

Rushed, plastic takeoff calls

Typical call in full flight

BREET   *p. 509*

Loud upslurred whistles, often in series of 2–3

Tone shrill, slightly nasal

WEEP and WEEP SERIES   *pp. 505, 538*

Breet given all year, in alarm and in flight. Lower and more monotone than Least Sandpiper's; slightly higher and more musical than Pectoral's. Weep and Weep Series given in alarm, mostly on breeding grounds, but sometimes on migration and in winter; distinctive. Also gives peeping Twitter (p. 536).

White-rumped Sandpiper

Calidris fuscicollis

Size and shape like Baird's. Migrates mostly through Great Plains in spring, down East Coast in fall. Vocalizations distinctive, standing out in mixed flocks.

Highly vocal in flocks, keeping up nearly constant high Twitter

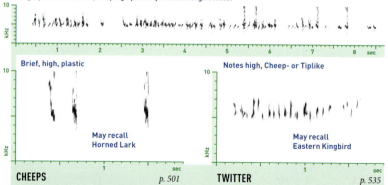

Brief, high, plastic

May recall Horned Lark

CHEEPS   *p. 501*

Notes high, Cheep- or Tiplike

May recall Eastern Kingbird

TWITTER   *p. 535*

Cheeps plastic, but consistently higher than any other sandpiper call; sometimes likened to squeaks of mice or bats. Each note very short. Twitter highly plastic; the end of a continuum of increasingly rapid Cheeps as excitement level rises.

LEAST SANDPIPER

Calidris minutilla

The smallest sandpiper in the world. Note yellow legs. Migrants found on mudflats, often feeding on the periphery of mixed shorebird flocks.

Highly plastic peeping Twitter (p. 536) often given in flight, when feeding, or in mild alarm

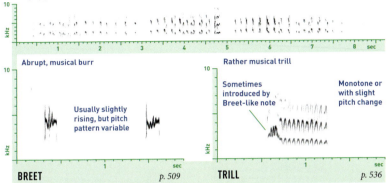

Abrupt, musical burr

Usually slightly rising, but pitch pattern variable

BREET   *p. 509*

Rather musical trill

Sometimes introduced by Breet-like note

Monotone or with slight pitch change

TRILL   *p. 536*

Breet given all year, by both sexes, usually in full flight. Plastic but fairly distinctive; higher than Breet of Baird's Sandpiper. Trill given in high alarm near nest or young; rare in migration and winter. Plastic and variable; averages lower than Western Sandpiper Trill, but much overlap.

PEEPING TWITTERS OF SANDPIPERS

Least, Semipalmated, and Western Sandpipers give highly plastic peeping notes when in groups, and sometimes while feeding or while flying; large flocks can generate a high-pitched muddle of constant peeping. Similar peeping Twitters are given by both dowitchers, Sanderling, Dunlin, and Baird's Sandpiper. White-rumped Sandpiper gives a Twitter that is distinctively high in pitch.

TAKEOFF CALLS OF SANDPIPERS

Many sandpipers give a slightly modified version of the typical call upon taking flight in alarm. These "takeoff calls" tend to be shorter, harsher, somewhat lower, and repeated in brief rapid series. They often last only a few seconds before the bird reverts to more typical calls.

The takeoff calls of Long-billed Dowitcher can recall the Pew Series of Short-billed Dowitcher. The takeoff calls of Least and, especially, Western Sandpipers can resemble the Cheep of Semipalmated. When listening to sandpipers flushing, pay particular attention to the calls made after the initial burst, once the birds are completely aloft.

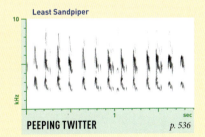

Least Sandpiper

PEEPING TWITTER   *p. 536*

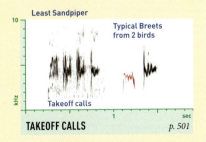

Least Sandpiper

Typical Breets from 2 birds

Takeoff calls

TAKEOFF CALLS   *p. 501*

SEMIPALMATED SANDPIPER

Calidris pusilla

One of the most common small sandpipers in the East during migration. Can
be difficult to separate from the closely
related Western Sandpiper; note voice.

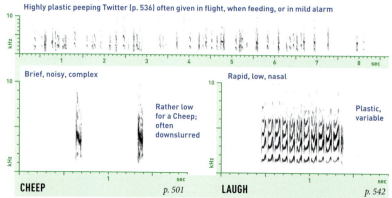

Highly plastic peeping Twitter (p. 536) often given in flight, when feeding, or in mild alarm

Brief, noisy, complex

Rather low
for a Cheep;
often
downslurred

CHEEP   *p. 501*

Rapid, low, nasal

Plastic,
variable

LAUGH   *p. 542*

Cheep given all year, by both sexes, in flight or in alarm. Western Sandpiper calls average higher, but can
match Semipalmated. Laugh given in alarm near nest, in interactions on migration. Lower, more nasal, and
far more often heard than Trills of Least and Western.

WESTERN SANDPIPER

Calidris mauri

Bill averages longer, thinner, and more
downcurved than Semipalmated's, but
with much overlap. Much more likely
than Semipalmated to occur in winter.

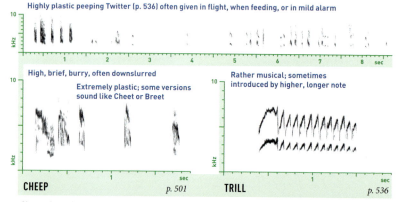

Highly plastic peeping Twitter (p. 536) often given in flight, when feeding, or in mild alarm

High, brief, burry, often downslurred

Extremely plastic; some versions
sound like Cheet or Breet

Rather musical; sometimes
introduced by higher, longer note

CHEEP   *p. 501*

TRILL   *p. 536*

Cheep given all year, mostly in full flight. Classic version notably higher than Semipalmated Sandpiper
Cheep, but low, noisy versions can match Semipalmated. Trill given in high alarm near nest or young; rare
in migration and winter. Plastic and variable; some versions match Trill of Least.

LONG-BILLED DOWITCHER

Limnodromus scolopaceus

Feeds in flocks, wading up to its belly and probing mud with a rapid sewing-machine motion of its bill. Prefers fresh water, often with emergent vegetation.

Song: Rattles, then 3 or more complex burry phrases in series (p. 574)

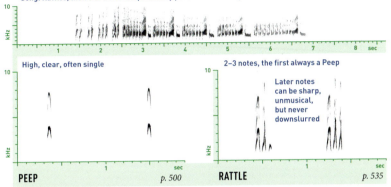

High, clear, often single

2–3 notes, the first always a Peep

Later notes can be sharp, unmusical, but never downslurred

PEEP   *p. 500*

RATTLE   *p. 535*

Peep given all year, even when calm, as well as in flight. Varies little in pitch and quality. Excited birds run Peeps together into long Twitters. Peeping Rattle given by alarmed birds, on the ground or upon flushing. Faster than Short-billed calls; consistently different in quality. Song sometimes given in migration.

SHORT-BILLED DOWITCHER

Limnodromus griseus

Very similar to Long-billed Dowitcher, and sometimes flocks with it. Best identified by voice. Generally prefers coastlines and mudflats.

Song: Pews, then rattling phrases, each ending with 2–4 upslurred burrs (p. 574)

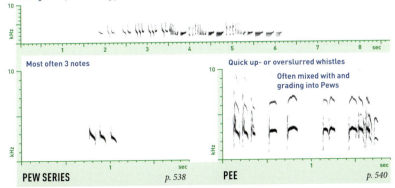

Most often 3 notes

Quick up- or overslurred whistles

Often mixed with and grading into Pews

PEW SERIES   *p. 538*

PEE   *p. 540*

Pews somewhat plastic; like calls of yellowlegs, but slightly faster and higher, more likely to sound like Tew. Can become slightly noisy in alarm. Pee calls given by excited birds on the ground. Highly plastic, but generally longer and more musical than Long-billed Peeps. Song sometimes given in migration.

Wilson's Snipe

Gallinago delicata

Common but rather secretive in dense wetlands. Occasionally sits up on low perches such as fenceposts, often when giving Bark Series or Kik-a Series.

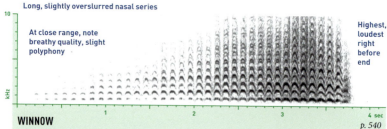

Long, slightly overslurred nasal series

At close range, note breathy quality, slight polyphony

Highest, loudest right before end

WINNOW
p. 540

Given at bottom of dives during display flight; sound created by air rushing through 2 outermost tail feathers. Very like Boreal Owl Song, especially at a distance. At close range, note slight polyphony due to 2 tail feathers vibrating in near-unison. Heard day or night, but especially at dawn and dusk.

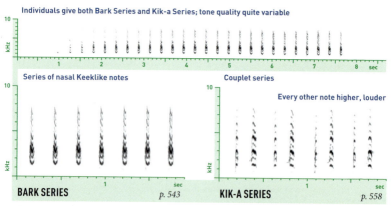

Individuals give both Bark Series and Kik-a Series; tone quality quite variable

Series of nasal Keeklike notes

Couplet series

Every other note higher, louder

BARK SERIES
p. 543

KIK-A SERIES
p. 558

Given by both sexes, usually from ground or low perch, but sometimes in flight. Function not entirely clear; couplet version reportedly used more often when mate present. Both versions used to communicate with chicks, especially in alarm. Low, noisy versions sometimes heard in alarm.

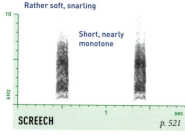

Rather soft, snarling

Short, nearly monotone

SCREECH
p. 521

By flushed birds; also during nocturnal migration and courtship pursuit flights. Only sound likely to be heard in migration and winter.

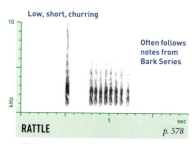

Low, short, churring

Often follows notes from Bark Series

RATTLE
p. 578

Little known. Possibly an alarm call to chicks; function needs more study.

AMERICAN WOODCOCK

Scolopax minor

Found on the ground in dense woods; males display in clearings at dawn and dusk. Probes for earthworms in moist soil with its long, flexible bill.

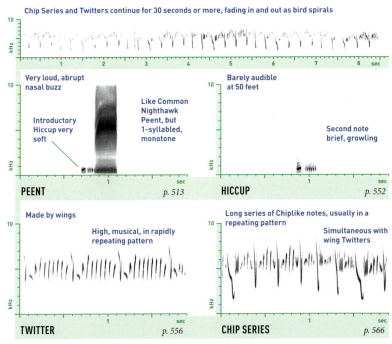

Chip Series and Twitters continue for 30 seconds or more, fading in and out as bird spirals

PEENT *p. 513*

Very loud, abrupt nasal buzz

Introductory Hiccup very soft

Like Common Nighthawk Peent, but 1-syllabled, monotone

HICCUP *p. 552*

Barely audible at 50 feet

Second note brief, growling

TWITTER *p. 556*

Made by wings

High, musical, in rapidly repeating pattern

CHIP SERIES *p. 566*

Long series of Chiplike notes, usually in a repeating pattern

Simultaneous with wing Twitters

Display sounds: by males, as early as Dec. in south, as late as June in north. In display, males sit on bare ground after sunset or before dawn, giving very soft Hiccups mixed with loud Peents. When bird takes wing, Hiccups and Peents replaced by Twitter, made by the 3 outer primary feathers. Chip Series, a vocal sound, added to Twitter at zig-zagging climax of flight display. Bird then drops silently back to same spot on ground to repeat performance. Twitter can also be heard when birds flush in daylight (at least males not in molt).

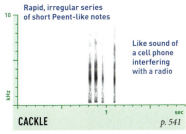

CACKLE *p. 541*

Rapid, irregular series of short Peent-like notes

Like sound of a cell phone interfering with a radio

Mostly Jan.–June, by territorial challengers flying low over Peenting rivals; also in flight chases and occasionally on ground.

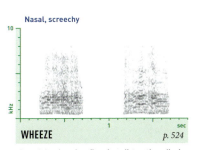

WHEEZE *p. 524*

Nasal, screechy

By adults in wing-flapping distraction display near nest or young. Plastic.

Spotted Sandpiper

Actitis macularius

Widespread along riverbanks and pond edges. Bobs tail often. Flies on stiff, shivering wings. Females larger; males tend nests and raise young.

Songs 2–10 seconds long, often in flight display; sometimes preceded by Pee notes

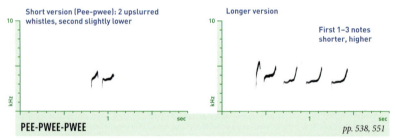

Rapid series of Weeplike notes

Slower, burry version

Each note 2-syllabled, with burry start

SONG
p. 558

Two main song patterns. Faster, simpler song (left) given by both sexes on ground or in flight, in territorial advertisement. Slower, more complex song (right) given on ground, mostly by female, in courtship. Intermediate songs also used in courtship.

Short version (Pee-pwee): 2 upslurred whistles, second slightly lower

Longer version

First 1–3 notes shorter, higher

PEE-PWEE-PWEE
pp. 538, 551

All year, in alarm and in flight, including nocturnal migration; 2-note version most common, especially from ground. Longer versions sometimes very long, sounding like (and perhaps grading into) Song. Lower and slower than Solitary Sandpiper Pwee-pwee-pwee, with series almost always downslurred.

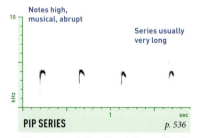

Notes high, musical, abrupt

Series usually very long

PIP SERIES
p. 536

Given in alarm by breeding adults, usually with chicks present. Loud, often incessant.

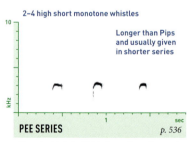

2–4 high short monotone whistles

Longer than Pips and usually given in shorter series

PEE SERIES
p. 536

Given by birds in flight at any time of year, including during sustained migration (day or night).

Solitary Sandpiper

Tringa solitaria

Aptly named. Breeds in the abandoned nests of other birds, in trees around muskeg bogs. At other seasons, favors wooded ponds or dense wetlands.

Songs usually embedded in plastic series of Klee notes

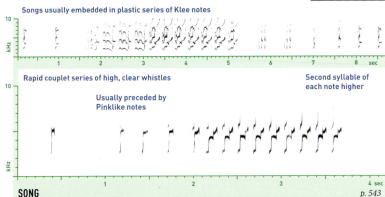

Rapid couplet series of high, clear whistles

Second syllable of each note higher

Usually preceded by Pinklike notes

SONG   *p. 543*

On breeding grounds, often in display flight, but also from ground or exposed perches. Variable; sometimes consists of 2 consecutive couplet series (top). Level and significance of variation needs more study. Associated with series of loud Klee notes, like Pink but longer, often broken.

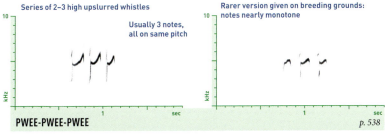

Series of 2–3 high upslurred whistles

Rarer version given on breeding grounds: notes nearly monotone

Usually 3 notes, all on same pitch

PWEE-PWEE-PWEE   *p. 538*

Commonly heard from migrants. Given in alarm or upon flushing; also during nocturnal flight. Almost never more than 3 notes. Infrequently heard version on breeding grounds (right) associated with Song; may represent intergrade with Pinklike notes.

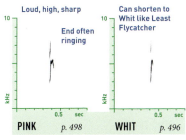

Loud, high, sharp

End often ringing

Can shorten to Whit like Least Flycatcher

PINK   *p. 498*   **WHIT**   *p. 496*

Variable, but calls from one bird usually identical, at least during migration. More plastic on breeding grounds.

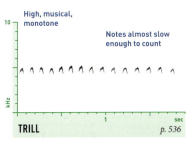

High, musical, monotone

Notes almost slow enough to count

TRILL   *p. 536*

Little known; apparently given by migrants in aggressive altercations. Sounds higher than most other vocalizations of the species.

GREATER YELLOWLEGS

Tringa melanoleuca

Common on mudflats in migration and winter; breeds in boreal bogs. Larger than Lesser Yellowlegs, with stouter, longer, slightly upturned bill.

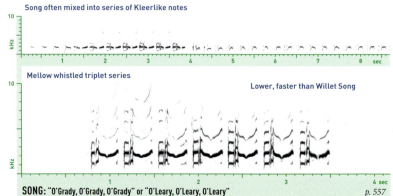

Song often mixed into series of Kleerlike notes

Mellow whistled triplet series

Lower, faster than Willet Song

SONG: "O'Grady, O'Grady, O'Grady" or "O'Leary, O'Leary, O'Leary"
p. 557

Mostly Apr.–June, by breeding males, but also occasionally by northbound migrants. Somewhat variable, but fairly stereotyped. Often given in flight display. Triplet pattern distinctive; tone quality may recall certain car alarms.

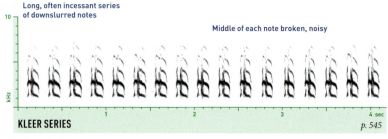

Long, often incessant series of downslurred notes

Middle of each note broken, noisy

KLEER SERIES
p. 545

All year, in high alarm. Variable and slightly plastic. Given on breeding grounds in defense of nest or chicks, but also by excited birds in a variety of situations in migration and winter. Grades into Klee Series.

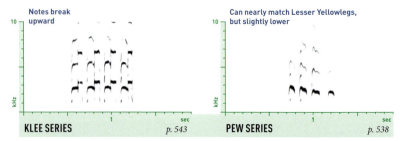

Notes break upward

Can nearly match Lesser Yellowlegs, but slightly lower

KLEE SERIES
p. 543

PEW SERIES
p. 538

All year; most common calls. Given in alarm, often when taking wing. Variable and plastic; most versions are series of at least 3 notes, but occasionally gives 1 or 2 notes alone. When notes are broken, higher second portion gives entire call the impression of high pitch.

LESSER YELLOWLEGS

Tringa flavipes

Common on mudflats in migration and winter; breeds in open boreal forest. Like Greater Yellowlegs, shows white rump and tail in flight.

Song often mixed into series of Peerlike notes

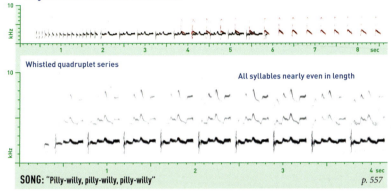

Whistled quadruplet series

All syllables nearly even in length

SONG: "Pilly-willy, pilly-willy, pilly-willy"

p. 557

Mostly Apr.–June, by breeding males, but also occasionally by northbound migrants. Females occasionally sing. Somewhat variable, but fairly stereotyped. Often given in flight display. Quadruplet pattern distinctive.

Rapid noisy series

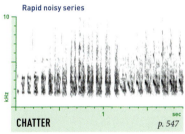

CHATTER

p. 547

All year, in altercations. Highly plastic; grades into other calls. Similar notes not known from Greater Yellowlegs.

Long series of downslurred whistles

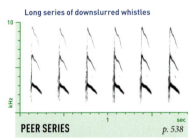

PEER SERIES

p. 538

All year, in high alarm. Variable and slightly plastic, but never with harsh or grating quality of Greater Yellowlegs' Kleer Series.

Often just 1 or 2 notes

More excited version

First note a longer Peep

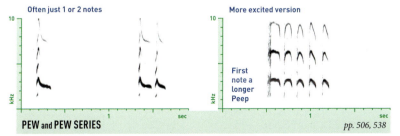

PEW and PEW SERIES

pp. 506, 538

All year; most common call. Given in alarm, often when taking wing. Variable and plastic; as excitement level rises, notes increase in number and are more likely to break upward, increasing potential for confusion with Greater Yellowlegs.

WILLET (WESTERN)

Tringa semipalmata inornata

Large; wing pattern striking. Breeds in freshwater marshes. Subtly larger and longer-billed than Eastern. Winters along coasts; rare north of New Jersey.

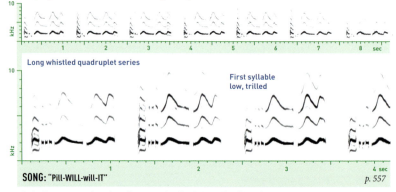

Slightly slower than Eastern Willet Song (8–12 phrases in 10 seconds)

Long whistled quadruplet series

First syllable low, trilled

SONG: "Pill-WILL-will-IT" *p. 557*

By both sexes, mostly on breeding grounds; often in long, high flight displays, sometimes at night. Females may join courting males in Song and on the wing. Female Song slightly lower than male's, with slightly less change in pitch. Eastern birds mostly unresponsive to playback of Western Song.

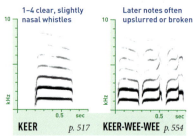

1–4 clear, slightly nasal whistles

Later notes often upslurred or broken

KEER *p. 517*   **KEER-WEE-WEE** *p. 554*

All year, in flight or in flock contact. Highly plastic; number of notes varies, and voice may break in any note.

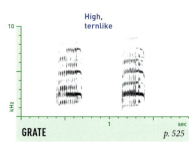

High, ternlike

GRATE *p. 525*

All year, in high alarm and when mobbing predators. Highly plastic; grades into both Keer and Klip. Often in series.

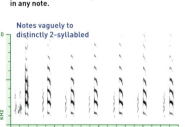

Notes vaguely to distinctly 2-syllabled

KEEK SERIES or **KA-LIP SERIES** *pp. 542, 558*

Given in long, rapid series, mostly during breeding season, in alarm and after Song. Variable, plastic, especially in speed.

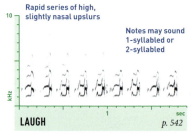

Rapid series of high, slightly nasal upslurs

Notes may sound 1-syllabled or 2-syllabled

LAUGH *p. 542*

All year, mostly in aggressive encounters between Willets; also sometimes upon landing. Rather soft and plastic.

WILLET (EASTERN)

Tringa semipalmata semipalmata

Large; wing pattern striking. Breeds in salt marshes; winters along coasts. Begins singing and displaying in March, well before wintering Westerns depart.

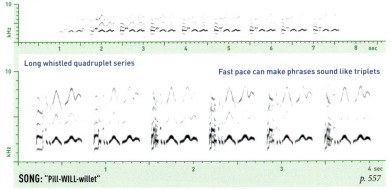

Like Western Willet Song, but slightly faster (13–15 phrases in 10 seconds)

Long whistled quadruplet series

Fast pace can make phrases sound like triplets

SONG: "Pill-WILL-willet" *p. 557*

By both sexes, mostly on breeding grounds; often in long, high flight displays, sometimes at night. Females may join courting males in Song and on the wing. Female Song slightly lower than male's, with slightly less change in pitch. Noticeably faster than Western Willet Song, and slightly higher.

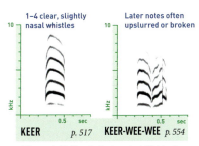

1–4 clear, slightly nasal whistles

Later notes often upslurred or broken

KEER *p. 517*   KEER-WEE-WEE *p. 554*

All year, in flight or in flock contact. Highly plastic; notes average barely higher and shorter than in Western, but much overlap.

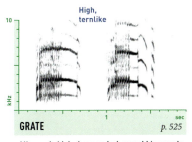

High, ternlike

GRATE *p. 525*

All year, in high alarm and when mobbing predators. Highly plastic; grades into Keer and Klip. Not known to be separable from Western's.

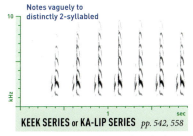

Notes vaguely to distinctly 2-syllabled

KEEK SERIES or KA-LIP SERIES *pp. 542, 558*

Given in long, rapid series, mostly during breeding season, in alarm and after Song. Variable, plastic, much like Western.

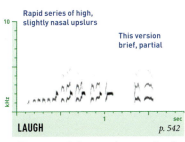

Rapid series of high, slightly nasal upslurs

This version brief, partial

LAUGH *p. 542*

All year, mostly in aggressive encounters between Willets; also sometimes upon landing. Rather soft and plastic.

Red-necked Phalarope

Phalaropus lobatus

Our smallest phalarope; bill short, very thin. Spends most of the year at sea, but some migrate through the center of the continent.

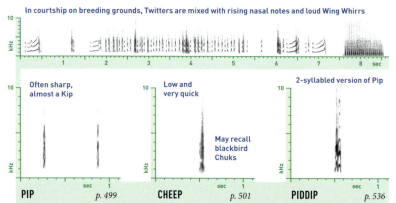

In courtship on breeding grounds, Twitters are mixed with rising nasal notes and loud Wing Whirrs

PIP — Often sharp, almost a Kip — *p. 499*

CHEEP — Low and very quick — May recall blackbird Chuks — *p. 501*

PIDDIP — 2-syllabled version of Pip — *p. 536*

Pip, Cheep, and Piddip are all variations on the most common call, given all year in contact and alarm. Highly plastic; all versions frequently heard in the constant pipping Twitter or Chitter of flocks. Other call types and Wing Whirrs not known away from breeding grounds.

Red Phalarope

Phalaropus fulicarius

Slightly larger than Red-necked; note thicker bill. Spends most of the year at sea; very rare inland. Fall and winter birds average paler than Red-necked.

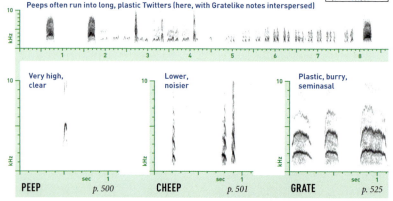

Peeps often run into long, plastic Twitters (here, with Gratelike notes interspersed)

PEEP — Very high, clear — *p. 500*

CHEEP — Lower, noisier — *p. 501*

GRATE — Plastic, burry, seminasal — *p. 525*

Peep given all year; most common call in migration and winter. Plastic. At least on breeding grounds, also gives lower, hoarser Cheeps more like Red-necked. Grates given all year, in altercations. In courtship on breeding grounds, gives Wing Whirrs like Red-necked plus brief, noisy rising notes.

Wilson's Phalarope

Phalaropus tricolor

Our largest phalarope. Breeds in wet meadows; never found offshore. Non-breeding adults like Lesser Yellowlegs, but paler, especially on face.

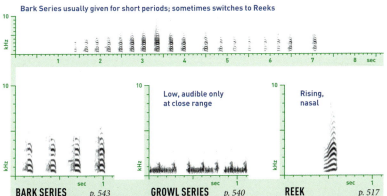

Bark Series usually given for short periods; sometimes switches to Reeks

BARK SERIES *p. 543*

Low, audible only at close range

GROWL SERIES *p. 540*

Rising, nasal

REEK *p. 517*

Bark Series apparently given mostly Apr.–July; may function in courtship. Growl Series given in close courtship. Reek appears to be most common call, given all year, by male in alarm, possibly by female in display. Barks and Reeks may recall short notes of Red-breasted Nuthatch.

JAEGERS
Genus *Stercorarius*

Related to gulls and terns, jaegers are darker in plumage; breeding adults have elongated central tail feathers. Three species breed in the North American Arctic, migrating mostly offshore and wintering primarily at sea. Their complex vocal repertoires are rather similar to those of gulls, except that jaegers are almost totally silent away from the breeding grounds. The few sounds they have been reported to make during migration and winter are described here briefly, but not included in the audio collection that accompanies this book.

Pomarine Jaeger *Stercorarius pomarinus*

Mostly silent at sea, but reportedly gives sharp Witch-yew and high Week when interacting with other Pomarine Jaegers at a food source or when attacking other species.

Parasitic Jaeger *Stercorarius parasiticus*

At sea, reportedly gives high Weet calls when interacting with other Parasitic Jaegers at a food source; less vocal than Pomarine and silent when pursuing other birds.

Long-tailed Jaeger *Stercorarius longicaudus*

No vocalizations reported at sea.

AUKS, PUFFINS, AND RELATIVES (Family Alcidae)

Known as "alcids," these mostly black-and-white seabirds are almost exclusively pelagic, coming ashore primarily for breeding. They dive in pursuit of fish and krill, propelling themselves underwater with their wings. All species in our area nest in seaside colonies, either on cliff ledges or in burrows. One species, the flightless, penguinlike Great Auk, was hunted to extinction by the middle of the nineteenth century.

The vocalizations of most alcid species consist of low grunts and groans, which accompany various types of visual displays to comprise a fairly complex system of communication. Vocal repertoire is most complex in the *Cepphus* guillemots, which give long series and phrases of high-pitched whistles, unlike other alcids.

BLACK GUILLEMOT

Cepphus grylle

Nests along rocky coasts, under rocks or in crevices. Unlike other alcids, stays mostly close to shore. Bright red mouth lining conspicuous when vocalizing.

Seet Series (p. 556): slower and longer than Song (sometimes 30 seconds), often rising and falling

Song: High, whistled phrase of 4–6 seconds

Starts with short series of Sreelike notes

Longest, loudest part of Song is a slightly rising couplet series

Sometimes ends with Sreelike notes

SONG
p. 566

Song and Seet Series are similar vocalizations that sometimes intergrade. Both given mostly May–July, near nest. Song often given by bird lying in nest, with bill vertical, wings sometimes drooped. Seet Series directed at other guillemots, likely in aggression, head tossed backward during each note.

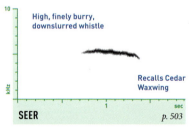

High, finely burry, downslurred whistle

Recalls Cedar Waxwing

SEER
p. 503

All year, in alarm; most common call. Plastic. Begging juveniles give coarsely trilled version, July–Oct., like calls of young gulls.

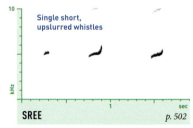

Single short, upslurred whistles

SREE
p. 502

Plastic; usually in irregular series. Version shown given by male in close courtship; similar calls reported all year.

Razorbill

Alca torda

Breeds in colonies on rocky islands and sea cliffs. Winters at sea. Resembles murres, but bill much deeper. In flight, tail extends beyond feet, unlike murres.

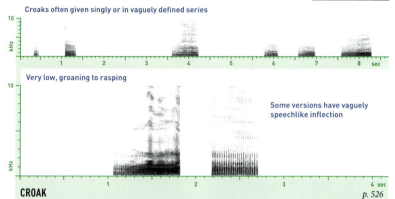

Croaks often given singly or in vaguely defined series

Very low, groaning to rasping

Some versions have vaguely speechlike inflection

CROAK

p. 526

Vocalizes mostly in breeding season, at nest and at sea nearby. Most sounds are highly plastic Croaks. Long Call described as accelerating series of Croaks culminating in a 2-second continuous rattling note with head vertical, then a few soft Croaks at end. Juvenile at sea may whistle like young murres.

Atlantic Puffin

Fratercula arctica

Well known and beloved for its colorful bill. Breeds in colonies on rocky islands and sea cliffs, nesting in burrows in soil or in rocky crevices. Winters at sea.

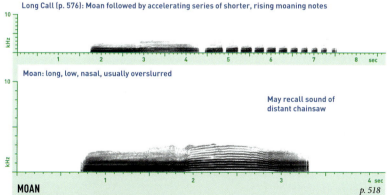

Long Call (p. 576): Moan followed by accelerating series of shorter, rising moaning notes

Moan: long, low, nasal, usually overslurred

May recall sound of distant chainsaw

MOAN

p. 518

Vocalizes mostly in breeding season, usually inside nesting burrow, rarely aboveground or on the water. Most sounds are long plastic Moans, singly or in series of 2–3; also soft Croaks audible at close range. Function of Long Call not well known. Young apparently beg with overslurred whistled notes.

Common Murre

Uria aalge

Breeds in colonies on rocks and sea cliffs. Winters at sea, sometimes near shore. Related Thick-billed Murre, rare in our area, has thicker bill, lower voice.

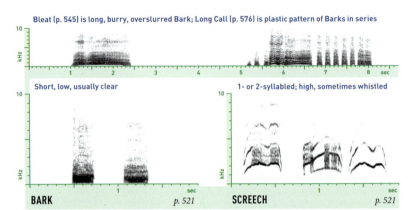

Bleat (p. 545) is long, burry, overslurred Bark; Long Call (p. 576) is plastic pattern of Barks in series

Short, low, usually clear

1- or 2-syllabled; high, sometimes whistled

BARK *p. 521*

SCREECH *p. 521*

Most vocal at nest, but adults sometimes give Barks and Bleats at sea. At nest, Barks given in alarm, often with bowing of head. Bleats and Long Call intergrade; both given in aggression as well as family contact. Also gives low Groan. Young birds beg with loud whistling Screeches in nest and at sea, June–Oct.

GULLS (Family Laridae, Subfamily Larinae)

Enduring symbols of the seashore, gulls are not restricted to it. Flocks gather across the continent in any open area where food is available, from landfills to agricultural fields to parking lots. Highly social, gulls have a complex system of visual displays often accompanied by loud vocalizations. All behaviors and sounds are likely innate.

Visual identification of gulls can be difficult, particularly before they acquire adult plumage, which can take up to four years. Species in the genus *Larus* are especially closely related, and some hybridize extensively.

Gull sounds tend to be highly variable and plastic. However, the general quality of the voice differs noticeably between species, and experienced observers can use it as an important supporting criterion in identification.

Common Calls and Displays of Large Gulls

The types of sounds described on the following pages are typical of most large gull species. Each sound is strongly associated with a particular visual display. Display postures tend to vary subtly between species, especially Long Call postures, but the postures may be perfunctory or absent in any given performance, especially if the bird is flying or swimming.

The vocal repertoires of small gulls tend to be less complex than those of large gulls, and display postures tend to differ.

Long Calls: Used in pair-bonding displays and aggressive encounters, often by multiple birds at once, Long Calls are given all year but are most frequent during the breeding season. Most have three distinct parts:

Introductory notes: Short, sometimes accelerating series of yelps or wails, often omitted; given in fairly neutral posture

High notes: The 1-3 highest and loudest notes, occasionally omitted; usually given with a deep bow of the head

Terminal series: A series of 1- or 2-syllabled notes, the least variable part of the call. Head sometimes tossed back at start; most notes given with head at 45 degree angle. Some species toss head with each note.

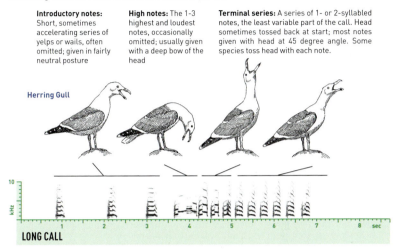

Herring Gull

LONG CALL

Wails ("Mew" display): Important in courtship and pair-bonding, Wails are also given all year by rivals in territorial encounters, typically in slow series, with the neck extended and the bill pointing downward. Two birds often Wail while walking side by side, sometimes with grass or other objects in the bill, before or after Long Calls.

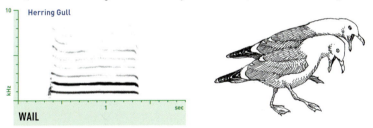

Herring Gull

WAIL

Begging calls (head-toss display): These calls are repeated at regular intervals to solicit the regurgitation of food. They are given by adults as well as juveniles, especially females begging from their mates during pair-bonding displays, with a characteristic quick toss of the head to a nearly vertical position.

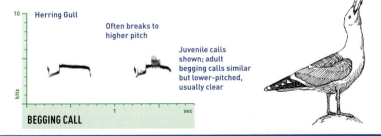

Herring Gull

Often breaks to higher pitch

Juvenile calls shown; adult begging calls similar but lower-pitched, usually clear

BEGGING CALL

Other Gull Sounds and Displays

Chuckles: Not strongly associated with any given posture, Chuckles are given in aggression, during fights, and in alarm.

Grunts ("choking" display): In courtship or high-intensity aggressive encounters, one or both birds may point the head downward and lower the tongue bone to change the facial expression, usually while giving a series of soft, low Grunts. Especially in courtship, a bird may actually settle on the ground as though preparing a nest.

Copulation calls: Intermediate between the Grunt and the Chuckle, this sound is made only by the male during copulation.

Single-note calls: Individual gulls give different versions of calls that more or less resemble single notes from the Long Call. Distinct versions may be given in nest defense, in contact, or in aggression. These calls are usually highly plastic, but at least some versions likely serve to identify individual birds.

Notes on the Voices of Young Gulls

Throughout much of their first year of life, most gulls have very high-pitched whistling or keening voices that can sound quite different from those of adults. In most species, not much is known about the transition between juvenile calls and adult calls, but it is apparently quite gradual. More study needed.

BLACK-LEGGED KITTIWAKE

Rissa tridactyla

Breeds in colonies on sea cliffs, often on offshore islands; occasionally on vertical manmade structures. Winters mostly at sea; rare inland.

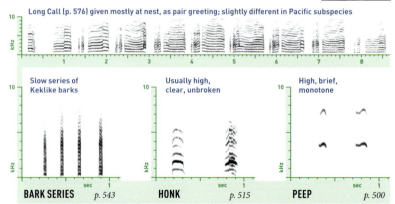

Long Call (p. 576) given mostly at nest, as pair greeting; slightly different in Pacific subspecies

Slow series of Keklike barks

Usually high, clear, unbroken

High, brief, monotone

BARK SERIES *p. 543*   **HONK** *p. 515*   **PEEP** *p. 500*

Most sounds apparently given all year, but especially at breeding colony, where kittiwakes can drown out all other species. Bark Series given in alarm; distinctively slow. Honks also often given in slow series. Peeps given by begging juveniles and adults, including at sea. Also gives long nasal Wails (p. 518).

Sabine's Gull

Xema sabini

An elegant species with a striking wing pattern. Breeds in the Arctic; migrates and winters mostly well offshore, but a few are found inland each fall.

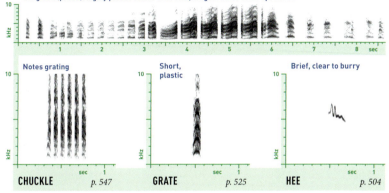

Long Call (p. 577): highly plastic series of Grates; longest notes often upslurred

Notes grating	Short, plastic	Brief, clear to burry
CHUCKLE *p. 547*	**GRATE** *p. 525*	**HEE** *p. 504*

Chuckle given in alarm, usually on the wing near nest. Grate is most common call of adults all year; highly plastic, sometimes given in rapid series. Hee is most common call of young birds, at least through fall; also apparently by adults at sea.

Little Gull

Hydrocoloeus minutus

Our smallest gull, barely larger than a Black Tern. Primarily a European species, rare on Northeast coast and Great Lakes in winter; a few nest in Canada.

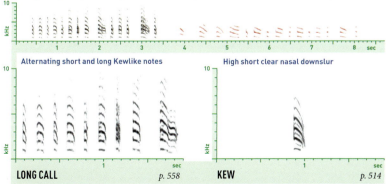

Long Call: plastic, but usually features a nasal couplet series

Alternating short and long Kewlike notes	High short clear nasal downslur
LONG CALL *p. 558*	**KEW** *p. 514*

Voice has distinctive quality: high, clear, nasal, like clear calls of Black Tern. Long Call given in flight during chases, or from ground with tail raised, head and body nearly horizontal. Often starts or ends with longer, single Kewlike notes. Kew is most common call; becomes very short in alarm.

Bonaparte's Gull

Chroicocephalus philadelphia

A small gull, rather like a tern in shape, habits, and voice. The only North American gull that regularly nests in trees. Breeds in boreal forests.

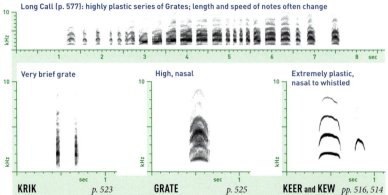

Long Call (p. 577): highly plastic series of Grates; length and speed of notes often change

Very brief grate	High, nasal	Extremely plastic, nasal to whistled
KRIK p. 523	**GRATE** p. 525	**KEER and KEW** pp. 516, 514

Krik given in agitation, especially near nest; almost never extended into a Chuckle. Most adult calls are versions of highly plastic Grate. Keers, Kews, and other whistled calls are most common sounds of first-winter birds; perhaps also given by feeding adults.

Black-headed Gull

Chroicocephalus ridibundus

A European species, rare in winter on Northeast coast; accidental elsewhere. Slightly larger than Bonaparte's, with red bill and black on underside of wing.

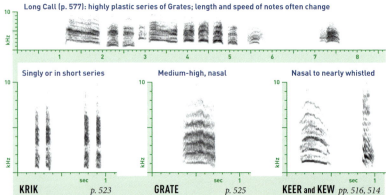

Long Call (p. 577): highly plastic series of Grates; length and speed of notes often change

Singly or in short series	Medium-high, nasal	Nasal to nearly whistled
KRIK p. 523	**GRATE** p. 525	**KEER and KEW** pp. 516, 514

Voice very similar to Bonaparte's Gull, but averages slightly lower and more nasal. Grate is most common call; exceedingly plastic. Krik plastic, but less grating than Bonaparte's, more noisy or nasal; sometimes extended into a Chuckle. Keers and Kews mostly by first-winter birds.

LAUGHING GULL

Leucophaeus atricilla

Common on coastal beaches, rare inland. The largest of our gulls with black heads, but still smaller than Ring-billed Gull. Social and vocal, even for a gull.

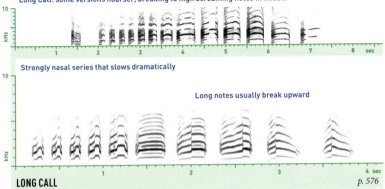

Long Call: some versions hoarser, breaking to high screaming notes in middle

Strongly nasal series that slows dramatically

Long notes usually break upward

LONG CALL *p. 576*

The namesake "laugh," often given in chorus. Variable and plastic, but nasal tone and slowing rhythm distinctive among gulls. Does not bow during call, but extends neck at 45 degree angle; tosses head back on 1–2 final notes. At least some high, hoarse, screeching versions (top) are from immatures.

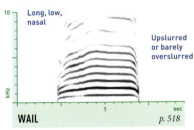

Long, low, nasal

Upslurred or barely overslurred

WAIL *p. 518*

All year, usually in slow series. May be accompanied by bowing of the head, at least during aggressive encounters.

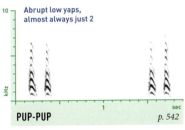

Abrupt low yaps, almost always just 2

PUP-PUP *p. 542*

All year, in alarm or aggression. Single notes sometimes heard. Long series of similar notes given during copulation and nest defense.

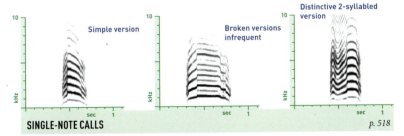

Simple version

Broken versions infrequent

Distinctive 2-syllabled version

SINGLE-NOTE CALLS *p. 518*

All year, in many situations. Medium-high pitch and clear nasal tone fairly distinctive. Squealing and gargling versions occasional, from excited birds. Characteristic 2-syllabled versions often given in flight; compare Marbled Godwit. Courting pairs give simple versions (left) with head-toss display.

FRANKLIN'S GULL

Leucophaeus pipixcan

Found mostly inland. Breeds on marshy ponds. Migrants often gather to hunt insects in plowed fields. Winters in southern South America.

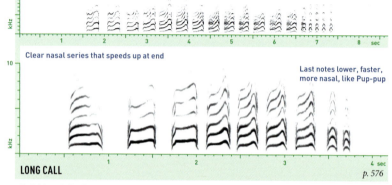

Long Call: long notes clear, upslurred, usually unbroken

Clear nasal series that speeds up at end

Last notes lower, faster, more nasal, like Pup-pup

LONG CALL
p. 576

Variable and plastic, but distinctive: starts with slow upslurs (2–3 notes/second), ends faster (4 notes/second), opposite of Laughing Gull's pattern. Often given on the wing. Does not bow during call, but extends head nearly horizontally; sometimes tosses head back on 1–2 final notes.

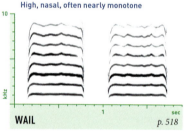

High, nasal, often nearly monotone

WAIL
p. 518

Quite plastic; similar to Laughing Gull's Wail but higher. Usually in series. Given in courtship, at nest, and in contact with chicks.

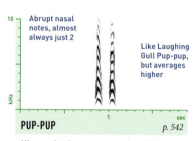

Abrupt nasal notes, almost always just 2

Like Laughing Gull Pup-pup, but averages higher

PUP-PUP
p. 542

All year, in alarm or aggression. Single notes sometimes heard. Long series of similar notes given during copulation and nest defense.

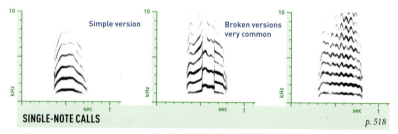

Simple version

Broken versions very common

SINGLE-NOTE CALLS
p. 518

All year, in many situations. Quite plastic. Much like Laughing Gull calls, but noticeably higher on average, and more often broken. Broken versions (center) may be given in courtship with head-toss display. Tremolo versions (right) given in high agitation.

RING-BILLED GULL

Larus delawarensis

The most common gull in much of North America, especially inland. Much smaller than Herring Gull. Acquires adult plumage in three years.

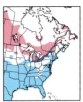

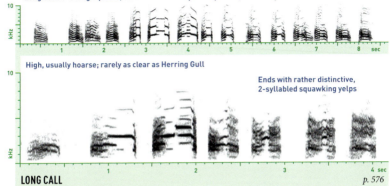

Long Call: 1–5 long Squeals; then a series of Yelps or Squawks, often 2-syllabled

High, usually hoarse; rarely as clear as Herring Gull

Ends with rather distinctive, 2-syllabled squawking yelps

LONG CALL *p. 576*

Variable and plastic. Final series relatively slow, usually 2–3 notes/second. Display postures distinctive: long squeals given with head down, sometimes nearly touching feet; most or all later notes delivered with bill vertical, head pumped up and down on each note.

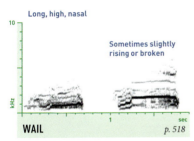

Long, high, nasal

Sometimes slightly rising or broken

WAIL *p. 518*

All year, in slow series. Apparently variable and plastic, but few recordings. Example above may be intergrade with Long Call.

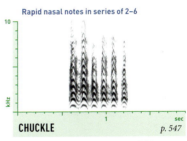

Rapid nasal notes in series of 2–6

CHUCKLE *p. 547*

All year, in alarm or aggression. Fairly stereotyped but variable. Averages faster and longer than Chuckles of larger gulls.

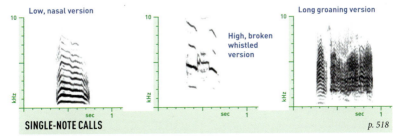

Low, nasal version

High, broken whistled version

Long groaning version

SINGLE-NOTE CALLS *p. 518*

Bewilderingly diverse. In addition to versions shown, also gives short Barks, high whistled Trills, and all manner of intermediates. High whistled versions heard more often from subadults; trilled and groaning versions more often in fights over food; other generalizations difficult.

CALIFORNIA GULL

Larus californicus

Mostly a western species, rather rare east of its breeding areas in the northern Great Plains. Intermediate in size between Ring-billed and Herring Gulls.

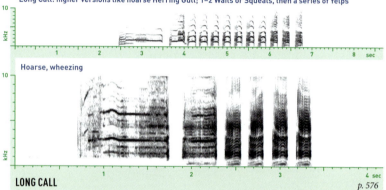

Long Call: higher versions like hoarse Herring Gull; 1–2 Wails or Squeals, then a series of Yelps

Hoarse, wheezing

LONG CALL *p. 576*

Highly variable and plastic. Often distinctively low, nasal, and hoarse, but many versions nearly as high and clear as Herring Gull Long Call. Final series almost always of single notes, not couplets; speed 3–5 notes/second. May slow slightly at end. Last few notes occasionally hoarser.

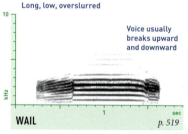

Long, low, overslurred

Voice usually breaks upward and downward

WAIL *p. 519*

All year, in slow series. Variable and plastic; low, groaning versions distinctive, but some are higher, more like Herring Gull.

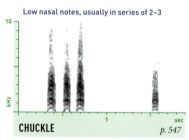

Low nasal notes, usually in series of 2–3

CHUCKLE *p. 547*

All year, in alarm or aggression. Fairly stereotyped but variable. Generally lower and more nasal than in most other gulls.

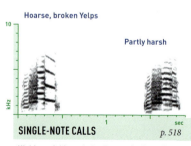

Hoarse, broken Yelps

Partly harsh

SINGLE-NOTE CALLS *p. 518*

Distinctive, low, braying

BLEAT *p. 545*

Highly variable and plastic sounds given all year in many situations. Most are Yelps and Squeals like those of Herring Gull, but noticeably hoarse. Distinctive Bleat given by birds of all ages in many situations. Bleating calls of other gull species are higher and rarer, restricted mostly to altercations.

HERRING GULL

Larus argentatus smithsonianus

The most common large gull in many areas, especially in winter. Like other large gulls, acquires adult plumage in 4 years; young birds highly variable.

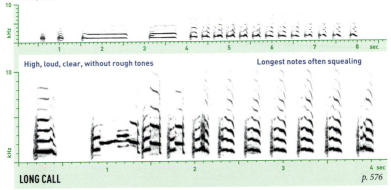

Long Call: 1–2 short Barks and/or 1–2 long Wails or Squeals; then a series of Yelps

High, loud, clear, without rough tones

Longest notes often squealing

LONG CALL *p. 576*

A quintessential sound of the seashore, given all year. Highly variable and plastic; may be as short as a single broken note followed by 4–5 Yelps. Final series may be 1- or 2-noted; speed 3–4 notes/second. Sometimes ends with 1–2 more widely spaced notes, or occasionally with Wails.

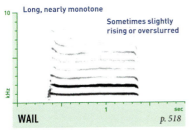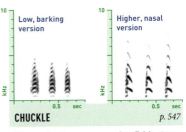

Long, nearly monotone

Sometimes slightly rising or overslurred

WAIL *p. 518*

All year, in slow series. Variable and plastic; some versions low and strongly nasal; some versions with 1–2 voice breaks.

Low, barking version

Higher, nasal version

CHUCKLE *p. 547*

All year, in alarm or aggression. Fairly stereotyped but highly variable; some versions rougher than shown.

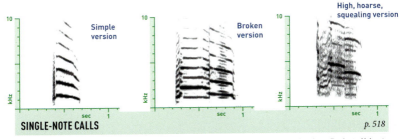

Simple version

Broken version

High, hoarse, squealing version

SINGLE-NOTE CALLS *p. 518*

A catch-all category for a wide variety of highly variable and plastic sounds ranging from Barks to Yelps to Squeals. Most common call; given all year in many situations. Usually high and clear, though low and rough versions exist. High trilled version (not shown) used against predators or in fights over food.

THAYER'S GULL

Larus thayeri

Breeds in colonies on cliffs; rare in East in winter. Smaller than Herring, with more rounded head, smaller bill, and more white in wingtip, especially below.

Long Call may not be safely separable from that of Herring Gull

At least some versions like slow, high versions of Herring Gull Long Call

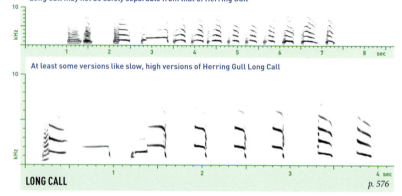

LONG CALL

p. 576

Voice poorly known. Long Call seems to resemble the higher-pitched Long Calls of Herring Gull. In the few available recordings, final series is of high, clear 1-syllabled notes; speed 2–2.5 notes/second. Other calls likely resemble those of Herring Gull; more study needed.

ICELAND GULL

Larus glaucoides kumlieni

Breeds in colonies on cliffs; most winter in the southern Arctic, a few in north-eastern North America. Very closely related to Thayer's Gull.

Here, Long Call preceded by Wails

This version extremely high, squealing

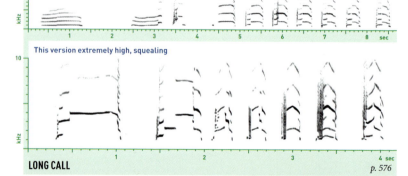

LONG CALL

p. 576

Voice poorly known. Long Call may average higher, more squealing than those of other large gulls. In the few available recordings, final series is of single notes or couplets, the notes high and often broken; speed 2–2.5 notes/second. Other calls may resemble those of Herring Gull; more study needed.

Glaucous Gull

Larus hyperboreus

One of our largest gulls. Rare but regular south of Canada in winter. The much smaller Iceland is the other gull that often shows pure white wingtips.

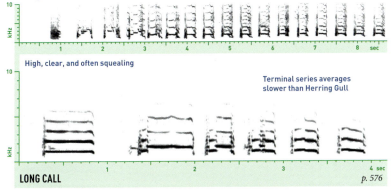

Long Call: 1–2 short Barks and/or 1–2 Wails or Squeals; then a series of Yelps

High, clear, and often squealing

Terminal series averages slower than Herring Gull

LONG CALL　　*p. 576*

Highly variable and plastic. Terminal series usually of 1-syllabled notes, around 2–2.5 notes/second. Sometimes ends with 1–2 more widely spaced notes, or occasionally with Wails.

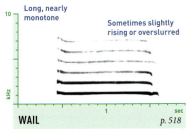

Long, nearly monotone

Sometimes slightly rising or overslurred

WAIL　　*p. 518*

All year, in slow series. Variable and plastic; some versions with 1–2 voice breaks.

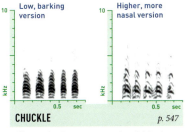

Low, barking version

Higher, more nasal version

CHUCKLE　　*p. 547*

All year, in alarm or aggression. Fairly stereotyped but highly variable; some versions harsher than shown.

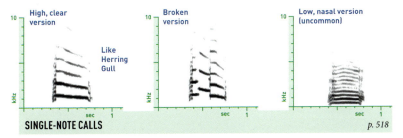

High, clear version

Like Herring Gull

Broken version

Low, nasal version (uncommon)

SINGLE-NOTE CALLS　　*p. 518*

Most common call; given all year in many situations. Usually high and clear, though low and rough versions exist. High trilled version (not shown) used against predators or in fights over food. Not known to be separable from similar calls of Herring Gull.

LESSER BLACK-BACKED GULL

Larus fuscus

A European species, uncommon but increasing in North America. Most are of the subspecies *graellsii*, from Western Europe; darker *fuscus* is very rare.

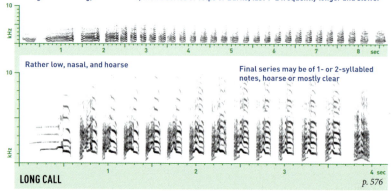

Long Call: 1–4 long, hoarse Wails; then a series of Yelps or Barks, last 1–2 frequently longer and slower

Rather low, nasal, and hoarse

Final series may be of 1- or 2-syllabled notes, hoarse or mostly clear

LONG CALL *p. 576*

All year. Highly variable and plastic. Usually distinctly lower, hoarser, and more nasal than Herring Gull Long Call, but some versions are difficult to separate. Final series rapid, 4–5 notes/second. Longest notes sometimes squealing. Call sometimes ends with Wails.

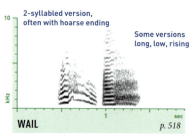

2-syllabled version, often with hoarse ending

Some versions long, low, rising

WAIL *p. 518*

All year, in slow series. Variable and plastic. Relatively high, 2-parted versions may be from females.

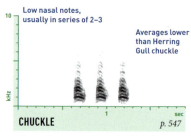

Low nasal notes, usually in series of 2–3

Averages lower than Herring Gull chuckle

CHUCKLE *p. 547*

All year, in alarm or aggression. Variable.

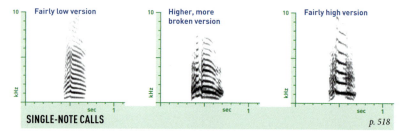

Fairly low version

Higher, more broken version

Fairly high version

SINGLE-NOTE CALLS *p. 518*

Highly variable and plastic sounds given all year in many situations. On average, lower, more nasal, and hoarser than corresponding sounds of Herring Gull, but much overlap.

GREAT BLACK-BACKED GULL

Larus marinus

Our largest gull, with the darkest back and the deepest voice. Fairly common on the north Atlantic coast and locally around the Great Lakes; rare inland.

Here, one bird gives Wails (red) while another starts a Long Call (black); background birds in cyan

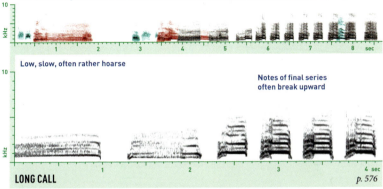

Low, slow, often rather hoarse

Notes of final series often break upward

LONG CALL   *p. 576*

All year; variable and plastic, but on average the lowest Long Call of any North American gull. Structure tends to be simple. Wails, when present, are quite low and groaning. Final series generally 3 notes/second or slower.

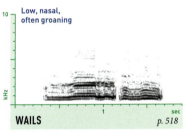

Low, nasal, often groaning

WAILS   *p. 518*

All year, in slow series. Variable and plastic, but distinctively low.

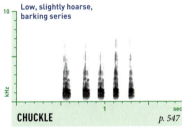

Low, slightly hoarse, barking series

CHUCKLE   *p. 547*

All year, in alarm or aggression. Fairly stereotyped but variable; averages lower than other gull Chuckles.

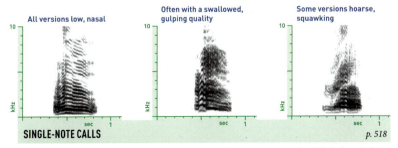

All versions low, nasal

Often with a swallowed, gulping quality

Some versions hoarse, squawking

SINGLE-NOTE CALLS   *p. 518*

Highly variable and plastic sounds given all year in many situations; some overlap with the lowest single-note calls of other gulls, but many may be identifiable by their low pitch alone. Calls of juveniles presumably higher, clearer.

SKIMMERS (Family Laridae, Subfamily Rynchopidae)

Represented by three species worldwide, the skimmers are highly distinctive ternlike birds that feed by flying along the surface of the water and dragging their elongated, razor-thin lower mandibles through it, snapping their bills shut upon contact with a fish. Sounds are likely innate, and vocal repertoires are quite simple.

TERNS (Family Laridae, Subfamily Sterninae)

Closely related to gulls, terns average smaller and more slender, often with longer tails. More closely tied to water than gulls, they feed primarily by plunge-diving, or by picking food from the water's surface in flight. Most species nest on the ground in dense, noisy colonies. Sounds are likely innate, and vocal repertoires are often highly complex. Many sounds accompany elaborate visual displays rather like those of gulls.

LOONS (Family Gaviidae)

Superficially similar to grebes, but not closely related; feet are webbed. Vocalizations are likely innate. Vocal repertoires are moderately complex; some species engage in unsynchronized duets and choruses, including at night.

BLACK SKIMMER

Rynchops niger

Unique in North America. Flight style distinctive even when not skimming surface of water for fish: wingbeats shallow and graceful, head held low.

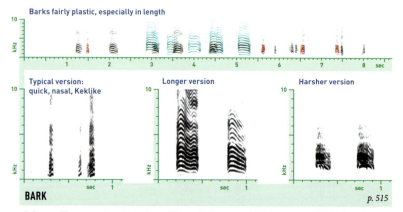

BARK

p. 515

Quite vocal in groups at any time of year. All vocalizations are variations on a nasal Bark. Short, low versions given in contact and mild alarm, becoming longer and somewhat higher in high alarm. Short harsh and burry versions sometimes heard during courtship and interactions near nest.

SOOTY TERN

Onychoprion fuscatus

Nests on the Dry Tortugas in Florida; widespread offshore in warm waters. Darker than Bridled Tern, especially on tail; heavier, more powerful in flight.

At colony, gives many variations on Kwiddy-wit, among other sounds.

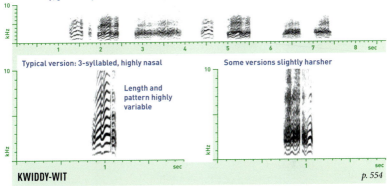

Typical version: 3-syllabled, highly nasal

Length and pattern highly variable

Some versions slightly harsher

KWIDDY-WIT — *p. 554*

Most common call. Possibly all year, but mostly heard Feb.–July, at breeding colony. Also given while foraging up to a few miles from colony. Variable; often much like Kee-erik of Gull-billed Tern, but averages slightly higher and longer. Female versions reportedly average higher than those of males.

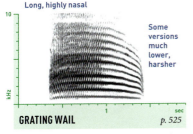

Long, highly nasal

Some versions much lower, harsher

GRATING WAIL — *p. 525*

Not well known; given infrequently at colony, possibly in aggression or agitation.

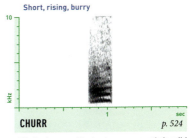

Short, rising, burry

CHURR — *p. 524*

Likely given mostly at colony, apparently in mild alarm and aggression.

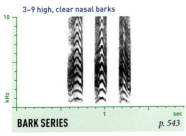

3–9 high, clear nasal barks

BARK SERIES — *p. 543*

Given at colony when other terns arrive nearby; sometimes given by several birds in unsynchronized chorus.

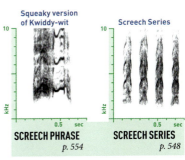

Squeaky version of Kwiddy-wit

Screech Series

SCREECH PHRASE — *p. 554*

SCREECH SERIES — *p. 548*

Both calls by juveniles; Screech Series mostly by younger birds still near nest, Screech Phrase up to several months after fledging.

BRIDLED TERN

Onychoprion anaethetus

A few breed in Florida; widespread off-shore in warm waters. Smaller and paler than Sooty Tern, especially on tail; often seen resting on floating debris.

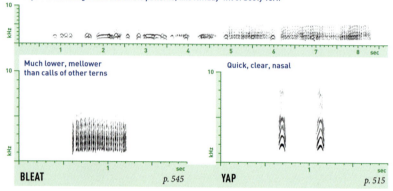

Yaps sometimes given in multinote patterns, like Kwiddy-wit of Sooty Tern

Much lower, mellower than calls of other terns

BLEAT *p. 545*

Quick, clear, nasal

YAP *p. 515*

Voice not well known; most commonly recorded sounds are shown. Bleats heard mostly near colony, but also from juveniles foraging at sea in fall. Yaps apparently given in alarm.

BROWN NODDY

Anous stolidus

Nests on the Dry Tortugas in Florida, where it is greatly outnumbered by Sooty Terns; rarely seen at sea north of south Florida.

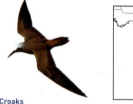

Possible Long Call (p. 548): series of short, harsh Croaks

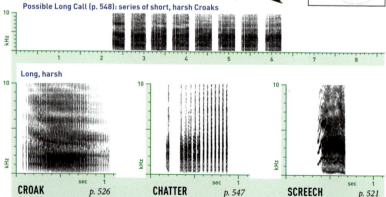

Long, harsh

CROAK *p. 526*

CHATTER *p. 547*

SCREECH *p. 521*

All sounds generally harsher, more croaking than those of other terns. Croak is most common call; highly plastic. Often slightly inflected like human speech. Chatter given at nest, sometimes continuing for many seconds. Screech given by begging juveniles, in addition to clear overslurred whistled notes.

Caspian Tern

Hydroprogne caspia

Our largest tern, as large as a Ring-billed Gull. Bill dark red-orange and quite thick; wingtips usually show more black below than above.

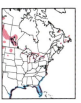

At nest colony, gives a variety of screechy Snarls and grunting notes

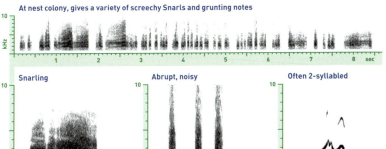

Snarling	Abrupt, noisy	Often 2-syllabled
KEE-KAREER *p. 554*	**GRUNT** *p. 521*	**WHISTLE** *p. 550*

Voice distinctive. Kee-kareer mostly Mar.–Aug., by adults, in contact and flight; noisy, but never grating. 1-syllabled versions given all year. Grunt given mostly near nest. Whistle by juveniles to at least Mar.; varies from very short to long, broken Squeal like that of young Red-tailed Hawk.

Gull-billed Tern

Gelochelidon nilotica

Medium-sized, with a very stout black bill. Found mostly along coasts; local inland. Nests on beaches, on barrier islands, and in marshes.

In high alarm, Yap Series sometimes extended into long, rapid Chuckle (p. 547)

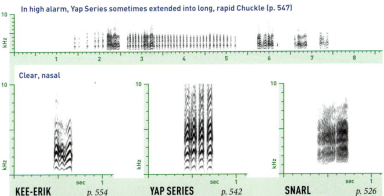

Clear, nasal		
KEE-ERIK *p. 554*	**YAP SERIES** *p. 542*	**SNARL** *p. 526*

Kee-erik is most common call of adults, given in family contact and in flight. Variable; often a 2-syllabled Kee-rik. Yap Series is given in alarm; usually 3- or 4-noted, but becomes longer, faster, and noisier as alarm increases. Snarl grades into short Wail (not shown), which may be associated with courtship.

ROYAL TERN

Thalasseus maximus

Found along sea coasts, foraging close to shore. Smaller than Caspian Tern, with orange bill; wingtips usually show more black above than below.

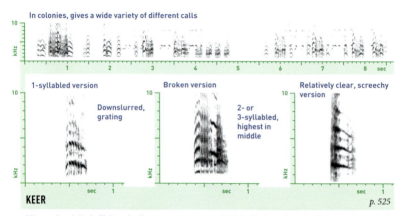

In colonies, gives a wide variety of different calls

1-syllabled version — Downslurred, grating

Broken version — 2- or 3-syllabled, highest in middle

Relatively clear, screechy version

KEER *p. 525*

All year, by adults in flight or in alarm; most common call. Typically downslurred and grating, but variable and plastic; some versions monotone. Different versions may serve different functions.

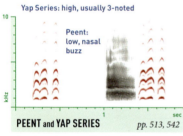

Yap Series: high, usually 3-noted

Peent: low, nasal buzz

PEENT and YAP SERIES *pp. 513, 542*

Apr.–June, in courtship, Yap Series given with bill pointed down, likely by females; Peent with wings drooped and crest erect, likely by males.

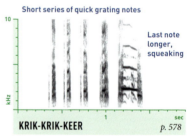

Short series of quick grating notes

Last note longer, squeaking

KRIK-KRIK-KEER *p. 578*

At least Apr.–June, by males in aggressive encounters; possibly also in other contexts. Rather variable and plastic.

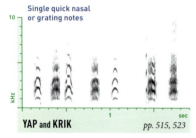

Single quick nasal or grating notes

YAP and KRIK *pp. 515, 523*

Not well known; often given in flight. Plastic; clear Yaps and grating Kriks may intergrade.

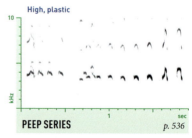

High, plastic

PEEP SERIES *p. 536*

Mostly July–Oct., by juveniles, often incessantly. Highly plastic; gradually grades into high, clear version of Grate as birds mature.

SANDWICH TERN

Thalasseus sandvicensis

Found along sea coasts, foraging close
to shore. Nests in colonies on islands,
like Royal. Yellow bill tip distinctive, but
can be hard to see at a distance.

In colonies, gives a wide variety of different calls

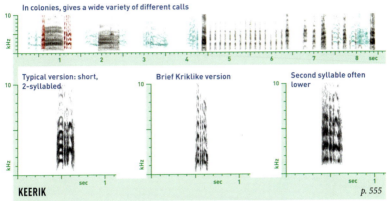

**Typical version: short,
2-syllabled**

Brief Kriklike version

**Second syllable often
lower**

KEERIK *p. 555*

All year, by adults in flight or in alarm; most common call. Somewhat less variable than similar calls of
Royal Tern; most versions are shorter, higher-pitched, and distinctly 2-syllabled.

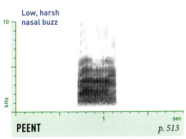

**Low, harsh
nasal buzz**

PEENT *p. 513*

Mostly Apr.–July, in courtship, given with wings
drooped and crest erect, likely by males. Females reportedly court with Peep Series.

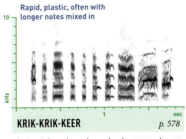

**Rapid, plastic, often with
longer notes mixed in**

KRIK-KRIK-KEER *p. 578*

At least Apr.–June, by males in aggressive encounters; possibly also in other contexts. Rather
variable and plastic.

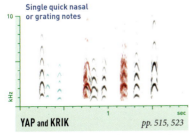

**Single quick nasal
or grating notes**

YAP and KRIK *pp. 515, 523*

Often given in flight, likely in contact or alarm.
Plastic; most versions clear and nasal, but some
are grating Kriks.

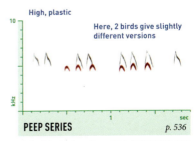

High, plastic

**Here, 2 birds give slightly
different versions**

PEEP SERIES *p. 536*

July–Oct., by juveniles, often incessantly; also
reportedly Apr.–July by courting females. Highly
plastic; grades into Grate as birds mature.

COMMON TERN

Sterna hirundo

Nests on rocky islands, in salt marshes, and on sandy barrier beaches. Similar to other *Sterna* terns; plumage of all species varies with age and season.

Attack sounds include rapid ticking Rattle followed by Snarl (p. 526), much like Arctic Tern

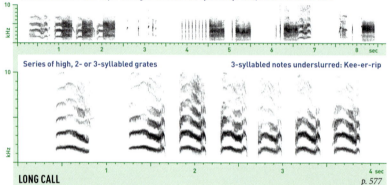

Series of high, 2- or 3-syllabled grates

3-syllabled notes underslurred: Kee-er-rip

LONG CALL *p. 577*

Mostly May–Aug. Somewhat variable and plastic; distinct versions likely given in courtship and aggression. Often given in flight chases. When given on ground, bill and head often lifted 45 degrees above horizontal, sometimes followed by head bows.

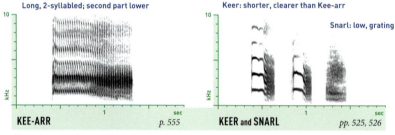

Long, 2-syllabled; second part lower

Keer: shorter, clearer than Kee-arr

Snarl: low, grating

KEE-ARR *p. 555*

KEER and SNARL *pp. 525, 526*

Mostly May–Aug. Kee-arr given when foraging, when approaching nest site with food, and in family contact. Typical version long, distinctly 2-syllabled, but grades into shorter, clearer Keers, given in high alarm and nest defense, often with long or short Snarls.

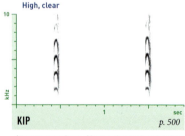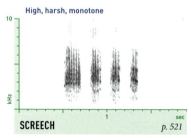

High, clear

High, harsh, monotone

KIP *p. 500*

SCREECH *p. 521*

In contact or alarm. Shorter, lower than Arctic's; sounds monotone. Some versions very brief, matching Roseate's.

Possibly July–Mar. Given by begging juveniles, in series or singly.

ARCTIC TERN

Sterna paradisaea

Nests in the Arctic and on offshore islands; at other seasons, found almost exclusively at sea. Rare inland. Bill and legs distinctively short.

In flight chases, Long Calls often mixed with Kews and Peeps

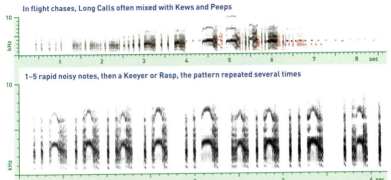

1–5 rapid noisy notes, then a Keeyer or Rasp, the pattern repeated several times

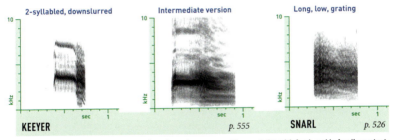

LONG CALL   *p. 577*

Mostly May–Aug. Distinct versions likely given in courtship and aggression. Often given in flight chases. When given on ground, bill and head often lifted 45 degrees above horizontal, sometimes followed by head bows. In attack, gives rapid ticking Rattle followed by Snarl (p. 578).

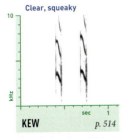

2-syllabled, downslurred   Intermediate version   Long, low, grating

KEEYER   *p. 555*   **SNARL**   *p. 526*

Mostly May–Aug. Keeyer given when foraging, when approaching nest site with food, and in family contact. Typical version distinctly 2-syllabled. Grades into Snarls, given in high alarm and nest defense. Intermediate versions can resemble Common Tern Kee-arr, but usually start higher.

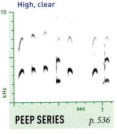
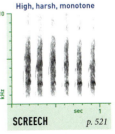

Clear, squeaky   High, clear   High, harsh, monotone

KEW   *p. 514*   **PEEP SERIES**   *p. 536*   **SCREECH**   *p. 521*

In contact or alarm. Higher, longer than Common's; monotone or downslurred.

Given in flight chases; possibly at other times. Usually in series.

Possibly July–Mar. Given by begging juveniles, in series or singly.

FORSTER'S TERN

Sterna forsteri

Breeds in freshwater and saltwater marshes; winters primarily along the coast. Winter head pattern distinctive: white with black patch behind eye.

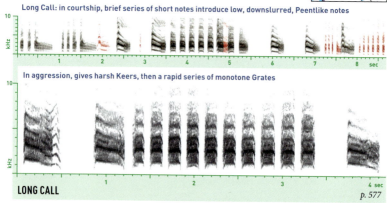

Long Call: in courtship, brief series of short notes introduce low, downslurred, Peentlike notes

In aggression, gives harsh Keers, then a rapid series of monotone Grates

LONG CALL *p. 577*

Mostly May–Aug. Distinct versions given in courtship and aggression, but these are highly plastic and intergrade. Often given during flight chases. When given on ground, courtship version often accompanied by bowed head, aggressive version by head thrown back to nearly vertical.

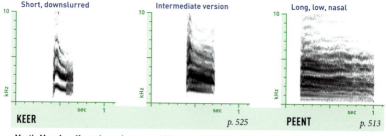

Short, downslurred

Intermediate version

Long, low, nasal

KEER   **KEER** *p. 525*   **PEENT** *p. 513*

Mostly May–Aug. Keer given when approaching nest site with food, in courtship flights with fish in bill, and in family contact. Typical version is distinctly brief and monosyllabic, but grades into Peents, given in high alarm and nest defense, often with Ticks and Kews. Peent lower, more nasal than Snarls of other *Sterna*.

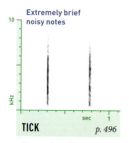

Extremely brief noisy notes

Clear, downslurred

High, harsh, monotone

TICK *p. 496*   **KEW** *p. 514*   **SCREECH** *p. 521*

All year, by both sexes, singly in contact, and in rapid series in aggression.

Mostly May–Aug., in nest defense, usually in series, often with Peents and Ticks.

July–Mar., by begging juveniles, in series or singly.

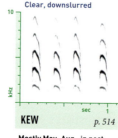

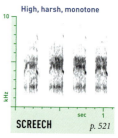

ROSEATE TERN

Sterna dougallii

A highly local breeder on offshore islands and sandbars in colonies with other terns. Bill usually entirely black, though base turns red in late summer.

Long Calls often given with Keers

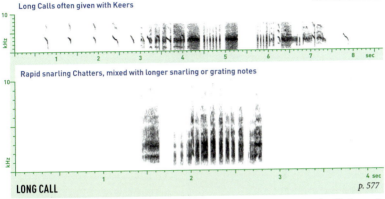

Rapid snarling Chatters, mixed with longer snarling or grating notes

LONG CALL   *p. 577*

Mostly May–Aug. Not well known. Distinct versions likely given in courtship and aggression. The few available recordings are from nesting colonies; may also be given during elaborate courtship display flights. At nest, harsh Chatter sometimes continues for many seconds, possibly in nest defense.

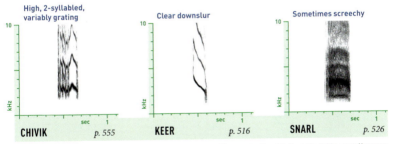

High, 2-syllabled, variably grating

Clear downslur

Sometimes screechy

CHIVIK   *p. 555*   **KEER**   *p. 516*   **SNARL**   *p. 526*

Mostly May–Aug. Chivik given in flight and when approaching nest site with food. Variable, usually more grating than Least Tern Ki-deek and higher than Sandwich Tern Keerik, but can be similar to both. Keer given in mild alarm, Snarl in high alarm, both usually near nest.

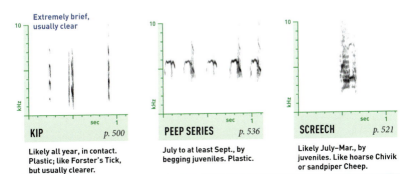

Extremely brief, usually clear

KIP   *p. 500*   **PEEP SERIES**   *p. 536*   **SCREECH**   *p. 521*

Likely all year, in contact. Plastic; like Forster's Tick, but usually clearer.

July to at least Sept., by begging juveniles. Plastic.

Likely July–Mar., by juveniles. Like hoarse Chivik or sandpiper Cheep.

LEAST TERN

Sternula antillarum

Our smallest tern. Yellow bill and white forehead distinctive. Nests on beaches and river sandbars. Some populations endangered.

Possible Long Call (p. 577): Screeches and Ki-deeks mixed in complex pattern; context little known

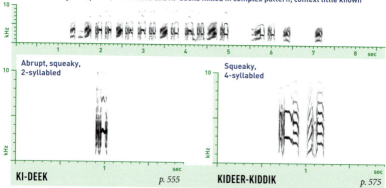

Abrupt, squeaky, 2-syllabled

KI-DEEK *p. 555*

Squeaky, 4-syllabled

KIDEER-KIDDIK *p. 575*

Most common and distinctive call. Variable, but fairly stereotyped within individuals; the two shown are the most typical. 4-syllabled version reportedly given by males carrying fish in courtship display flights; both versions given by both sexes when returning to nest, and in other contexts.

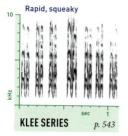

Rapid, squeaky

KLEE SERIES *p. 543*

Mostly Apr.–May, by males in close courtship, often in duet with female Tremolo.

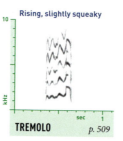

Rising, slightly squeaky

TREMOLO *p. 509*

Mostly Apr.–May, by female in close courtship, often in duet with male Klee Series.

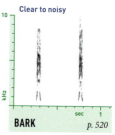

Clear to noisy

BARK *p. 520*

All year, in high alarm, usually in series, often with Screech.

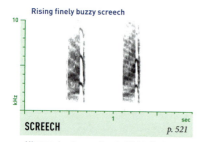

Rising finely buzzy screech

SCREECH *p. 521*

All year, in alarm, often in flight. Barks often mixed in as alarm level rises.

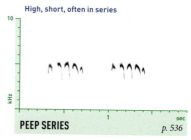

High, short, often in series

PEEP SERIES *p. 536*

By begging juveniles, at least well into fall. Highly variable and plastic; may resemble flight calls of some sandpipers.

BLACK TERN

Chlidonias niger

Quite small and short-tailed. Breeds in freshwater marshes; in winter, found largely offshore. Young birds and winter adults lack black body plumage.

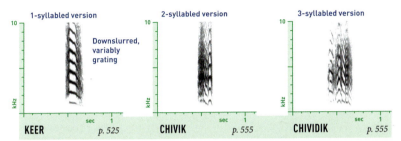

1-syllabled version

Downslurred, variably grating

KEER *p. 525*

2-syllabled version

CHIVIK *p. 555*

3-syllabled version

CHIVIDIK *p. 555*

Mostly May–Aug., by both sexes, in display flights and when approaching nest. Highly variable and fairly plastic; 1- and 2-syllabled versions are most common. Usually at least slightly grating. Averages somewhat lower than similar calls of most other terns.

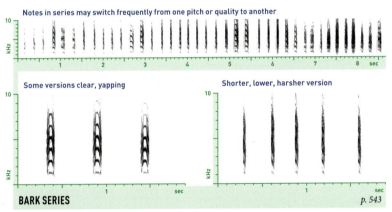

Notes in series may switch frequently from one pitch or quality to another

Some versions clear, yapping

Shorter, lower, harsher version

BARK SERIES *p. 543*

All year. A group of plastic, intergrading calls given in contact and alarm. In alarm, usually given in long series. Some versions can be quite similar to Bark Series of Wilson's Snipe or Keek Series of Willet, but usually more plastic in speed and quality.

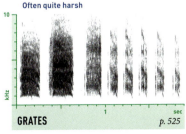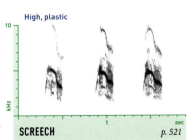

Often quite harsh

GRATES *p. 525*

Possibly all year, likely in aggression. Highly plastic; most versions are single notes. Multinote patterns like the one shown are rare.

High, plastic

SCREECH *p. 521*

Mostly June–Sept., by juveniles, singly or in series. Develops gradually into Keerlike calls.

COMMON LOON

Gavia immer

A symbol of the north woods, its haunting calls nearly synonymous with wilderness. Breeds on wooded lakes; winters on lakes and along coast.

Individual males have a single Song; geographic and individual variation are slight

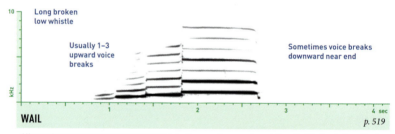

Rising 3-syllabled broken Wail, then slow whistled couplet series

SONG   *p. 568*

Mostly on breeding grounds, by males in aggressive conflicts or during nocturnal choruses. Can carry for miles. Individuals apparently recognize the Songs of neighbors and mates. Often given in "vulture" posture with body partway out of water, hunched back, extended neck, and partially spread wings.

Long broken low whistle

Usually 1–3 upward voice breaks

Sometimes voice breaks downward near end

WAIL   *p. 519*

All year, but mostly heard on breeding grounds. An evocative sound, reminiscent of the howl of a wolf, often given in nocturnal choruses. Number of voice breaks varies with motivation; simplest version, an unbroken whistle, used in mate contact. More breaks indicates higher level of excitement or alarm.

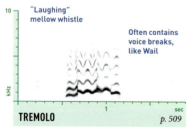

"Laughing" mellow whistle

Often contains voice breaks, like Wail

TREMOLO   *p. 509*

All year, especially on breeding grounds; in response to danger or territorial threat, in nocturnal choruses, and sometimes in flight.

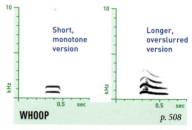

Short, monotone version

Longer, overslurred version

WHOOP   *p. 508*

All year, grading into one another and into Wail. Most common calls in migration and winter, mostly between mates.

RED-THROATED LOON

Gavia stellata

Our smallest loon, with a thin, upturned bill. Breeds in Arctic and winters on salt water; rare inland. Voice quite different from those of other loons.

Song: often a synchronized duet, with individuals alternating and overlapping phrases

2- or 3-syllabled burry phrases in series

Each phrase starts rough, ends higher, clearer

Here, another individual responds with Quacks

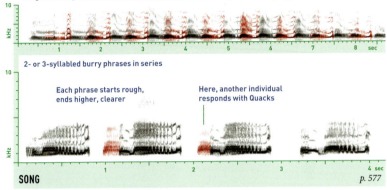

SONG *p. 577*

Mostly May–Aug., usually by pairs in duet or small groups in chorus, often with body raised 45 degrees out of water, neck extended, and head pointed downward. Often introduced by 1–2 Wails. Tone quality growling and slightly screeching. Male and female versions apparently differ slightly.

Long nasal Wail, rather gull-like

Often broken, partly hoarse or groaning

Loud, screechy

With loud splash made by feet

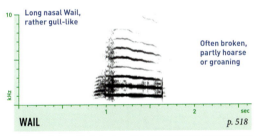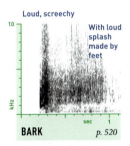

WAIL *p. 518*

BARK *p. 520*

All year, but mostly during breeding season. Plastic. By both sexes in territorial defense and in response to potential predators. Sometimes given in pair duet, usually unsynchronized.

Given with splashing dive when threatened by humans, other loons, etc.

Single long ducklike Quack

Varies from croaking to rather nasal

Short Quacklike notes accelerate, then end in longer note

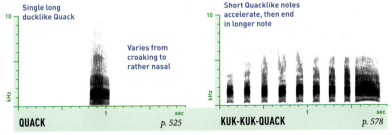

QUACK *p. 525*

KUK-KUK-QUACK *p. 578*

All year; the most common calls in winter. Variable and somewhat plastic; Quacks often in series, but also singly. Given in contact, in alarm, and frequently in flight. Mostly directed at other loons, but sometimes given in response to eagles or other predators.

PETRELS AND SHEARWATERS (Family Procellariidae)

These are highly pelagic birds, coming ashore only to nest, usually in inaccessible areas. Along with albatrosses and storm-petrels, they make up a group called "tubenoses" (order Procellariiformes). Of this group, only Northern Fulmar is treated in this book, because of the difficulty of hearing other species in our area.

STORKS (Family Ciconiidae)

Superficially similar to herons, storks are not closely related to them. Like vultures, they lack complex syringeal muscles and are therefore extremely limited in vocal ability. Their primary mode of long-distance communication is bill clattering. Only one species, Wood Stork, inhabits North America.

GANNETS AND BOOBIES (Family Sulidae)

These large seabirds hunt fish and squid by plunge-diving into the ocean from considerable height. They nest in large colonies. Only one species, Northern Gannet, is likely to be heard in our region.

NORTHERN FULMAR

Fulmarus glacialis

Nests in colonies on sea cliffs; spends the rest of its time at sea. Most likely to be heard near the nest, or when fighting over food at sea.

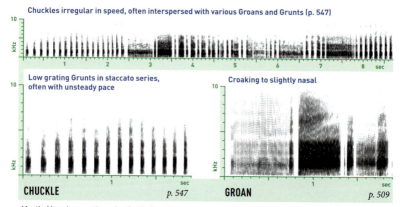

Chuckles irregular in speed, often interspersed with various Groans and Grunts (p. 547)

Low grating Grunts in staccato series, often with unsteady pace

Croaking to slightly nasal

CHUCKLE   *p. 547*

GROAN   *p. 509*

Mostly May–Aug.; rather plastic. Various Chuckle and Groan combinations given by pairs in courtship, by birds approaching the nest, and during territorial conflicts. Short Grunts (p. 521), like single note of Chuckle, given as contact call or during fights over food at sea; no recordings available.

Wood Stork

Mycteria americana

Our largest wading bird, local in and near swamps. When soaring, easily mistaken for a pelican. Adults nearly silent, but young birds can be vocal.

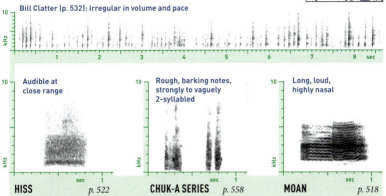

Bill Clatter (p. 532): Irregular in volume and pace

Audible at close range

HISS   *p. 522*

Rough, barking notes, strongly to vaguely 2-syllabled

CHUK-A SERIES   *p. 558*

Long, loud, highly nasal

MOAN   *p. 518*

Adults give Hiss and Bill Clatter mostly in displays at nest, Dec.–May; these and Groan occasionally when disturbed. Young birds are far more vocal, giving Chuk-a Series when distressed, and long, variable Moans when begging for food, mostly Apr.-Aug. Calls of young can carry for hundreds of meters.

Northern Gannet

Morus bassanus

Nests in colonies on sea cliffs; winters off coasts, often near shore. Hunts with spectacular plunge-dives into the sea, folding wings at last second.

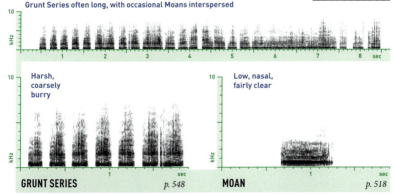

Grunt Series often long, with occasional Moans interspersed

Harsh, coarsely burry

GRUNT SERIES   *p. 548*

Low, nasal, fairly clear

MOAN   *p. 518*

Mostly vocal at breeding colonies. Grunt Series is most common call; varies slightly in different behavioral contexts: landing in the colony, displaying with mate, fishing in flocks, calling in alarm. Moans are less common, often given singly. At sea, may give single Grunts (p. 521); no recordings available.

PELICANS (Family Pelecanidae)

Well known for the large pouches below their bills which they use to catch fish, the pelicans are among our largest birds. Adults are nearly silent, giving only rare grunting sounds; juveniles beg loudly at the nest.

CORMORANTS (Family Phalacrocoracidae)

These long-necked, dark-plumaged birds catch fish by diving underwater. Their grunting and croaking vocalizations are given mostly near nests or roosts.

ANHINGA (Family Anhingidae)

The Anhinga is the only member of its family in North America. It resembles cormorants in shape and habits, but differs in bill structure, plumage, and soaring flight, as well as in the unique purring courtship display of the male.

AMERICAN WHITE PELICAN

Pelecanus erythrorhynchos

Very large and quite distinctive. Often soars. Flocks fly high in lines or V-formations. Feeds while swimming by dipping head below surface of water, often in cooperative groups.

Adults are usually silent, but do give brief, soft Croaks, mostly at nest site, in courtship or aggression. Begging juveniles give a range of plastic sounds, from Grunts (p. 521) to Groans.

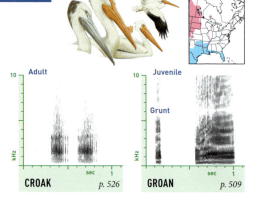

CROAK *p. 526*

GROAN *p. 509*

BROWN PELICAN

Pelecanus occidentalis

Smaller than American White Pelican. Almost exclusively coastal. Feeds by plunge-diving into the water from the air. Groups often glide low over the water in single file.

Adults are usually silent, except for soft Grunt or Groan given during courtship display; no recordings available. Begging juveniles give a range of plastic sounds, from Grunts to Groans to higher Screeches.

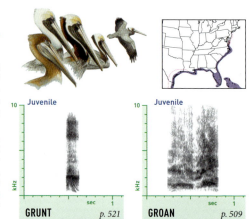

GRUNT *p. 521*

GROAN *p. 509*

DOUBLE-CRESTED CORMORANT

Phalacrocorax auritus

Our most widespread cormorant, occurring both inland and coastally. Like other cormorants, often swims with only neck protruding from water.

Colonies make a varied cacophony of Croaks, Grunts, and similar calls

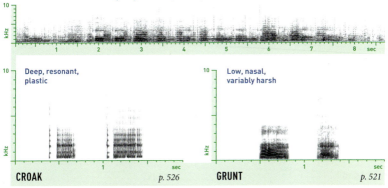

Deep, resonant, plastic

CROAK

p. 526

Low, nasal, variably harsh

GRUNT

p. 521

Calls often at colonies, rarely elsewhere, but may give Croaks when disturbed or in flight. Also gives Croaks to greet mate, with neck outstretched. In courtship, gives series of calls with head and tail vertical, wings lifted on each note; no recordings available. Also vocalizes during other displays. Young shriek in nest.

GREAT CORMORANT

Phalacrocorax carbo

Restricted to seacoasts. Some white on throat all year. Juveniles are palest on belly, while young Double-cresteds are palest on breast.

Colonies give various Bleats, Grunt Series, and other grunting calls

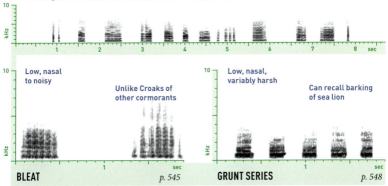

Low, nasal to noisy

Unlike Croaks of other cormorants

BLEAT

p. 545

Low, nasal, variably harsh

Can recall barking of sea lion

GRUNT SERIES

p. 548

Calls often at colonies, rarely elsewhere. Male's wing-waving courtship display is reportedly silent. Bleat given by both sexes to greet mate, with neck and head nearly touching back. Grunt Series given in various contexts with neck stretched, bill pointed down.

NEOTROPIC CORMORANT

Phalacrocorax brasilianus

Obviously smaller than Double-crested; tail proportionately longer. Found in a variety of freshwater, brackish, and saltwater habitats.

Colonies give near-constant Croaks, with the occasional Grunt

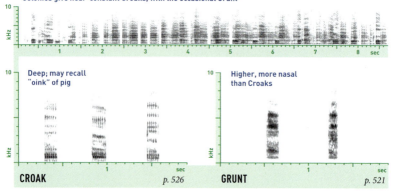

Deep; may recall "oink" of pig

Higher, more nasal than Croaks

CROAK *p. 526*

GRUNT *p. 521*

Calls often at colonies, rarely elsewhere, but may give Croaks when disturbed or in flight. Also gives Croaks to greet mate, with neck outstretched. In courtship, gives series of calls with head and tail vertical, wings lifted on each note; no recordings available. Also vocalizes during other displays.

ANHINGA

Anhinga anhinga

Locally common in freshwater swamps and along waterways. Often soars. Usually swims with only its snakelike neck above the water.

In displays and disputes, Croaks run into long series and/or become a Crackle (p. 523), like radio static

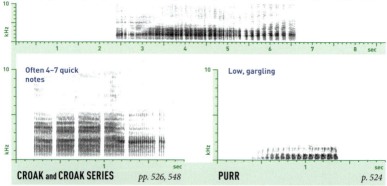

Often 4–7 quick notes

Low, gargling

CROAK and CROAK SERIES *pp. 526, 548*

PURR *p. 524*

Vocal at the nest; away from it, usually silent, but more vocal than cormorants. Highly plastic Croaks given in display at nest and all year when disturbed. In courtship, males flap wings alternately, then tip forward, spreading tail and giving soft Purr, uniform and stereotyped.

BITTERNS, HERONS, AND EGRETS (Family Ardeidae)

IBIS AND SPOONBILLS (Family Threskiornithidae)

These two families include medium-to-large wading birds with long legs, long necks, and long bills. Most species have limited vocal repertoires of grunting or croaking sounds, but bitterns give repeated songlike vocalizations.

Many species have a large number of visual displays performed in courtship or aggression, some accompanied by sounds. Common displays include an aggressive forward display with feathers/plumes raised and neck coiled as if to strike; a courtship neck-stretch display with the bill pointed upward, silently or with a call; and a greeting display between pairs when one arrives at the nest.

AMERICAN BITTERN

Botaurus lentiginosus

Stays well hidden in dense freshwater marshes. Like Least Bittern, sometimes responds to threats by freezing in place with bill and neck vertical.

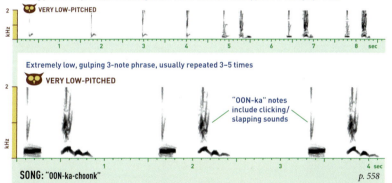

Song preceded by slow, accelerating series of soft clicks, as bird appears to take gulps of air

VERY LOW-PITCHED

Extremely low, gulping 3-note phrase, usually repeated 3–5 times

VERY LOW-PITCHED

"OON-ka" notes include clicking/slapping sounds

SONG: "OON-ka-choonk"   p. 558

Song is bizarre, unmistakable, and far-carrying; sounds like a huge rock being repeatedly plunged into water in a repeating 3-note pattern. Given mostly Mar.–July, especially from dusk to dawn. Not known whether both sexes sing. Sound appears to be made by expelling air from the inflated esophagus.

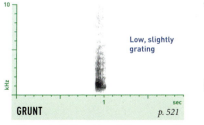

Low, slightly grating

GRUNT   p. 521

All year, in flight and in agitation, apparently only during the day. Infrequent and not very loud.

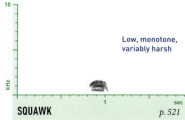

Low, monotone, variably harsh

SQUAWK   p. 521

All year, in nocturnal migration, usually singly. Has not been recorded during the day. Lower than Black-crowned Night-Heron Squawk.

LEAST BITTERN

Ixobrychus exilis

Our smallest heron, breeding in dense freshwater or brackish marshes. Often skulking and difficult to see, but can be quite confiding when in the open.

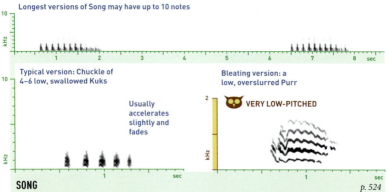

Longest versions of Song may have up to 10 notes

Typical version: Chuckle of 4–6 low, swallowed Kuks

Usually accelerates slightly and fades

Bleating version: a low, overslurred Purr

🦉 VERY LOW-PITCHED

SONG — p. 524

Mostly Mar.–Oct., reportedly only by males. Often given from deep cover. Quality varies from nasal to grunting. Some versions recall Song of Black-billed Cuckoo, but lower, less musical. Bleating version little known; may serve a slightly different function, or may represent individual variation.

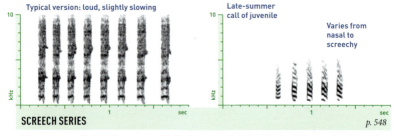

Typical version: loud, slightly slowing

Late-summer call of juvenile

Varies from nasal to screechy

SCREECH SERIES — p. 548

All year, apparently by both sexes, possibly in alarm or contact. Usually given infrequently, at long intervals, but variable, plastic versions may be given frequently in late summer, possibly by both adults and juveniles. Can be as short as 3 notes. Often mistaken for the call of a rail.

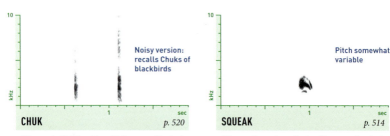

Noisy version: recalls Chuks of blackbirds

Pitch somewhat variable

CHUK — p. 520

All year, in flight and in agitation, usually singly. Like Keks of large rails, and sometimes given in response to them.

SQUEAK — p. 514

All year, in nocturnal migration, usually singly. Has not been recorded during the day. Generally lower and less plastic than Virginia Rail Squeak.

GREAT BLUE HERON

Ardea herodias

Our largest heron. South Florida form, "Great White Heron," resembles Great Egret, but larger and thicker-billed; hybridizes with blue forms.

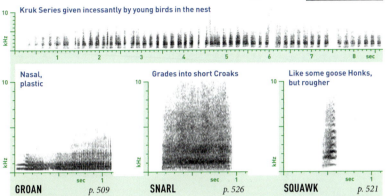

Kruk Series given incessantly by young birds in the nest

GROAN — Nasal, plastic — *p. 509*

SNARL — Grades into short Croaks — *p. 526*

SQUAWK — Like some goose Honks, but rougher — *p. 521*

All calls plastic. Croaking Groans given in early spring by males in neck-stretch display. Snarls, Croaks, and Squawks intergrade; all given in alarm or in flight. Kruk Series varies with age of young, but typically harsher than calls of young Great Egrets.

GREAT EGRET

Ardea alba

Locally common in wetland habitats. Much larger than Snowy Egret, with yellow bill and black legs. Forages deliberately, usually in an upright posture.

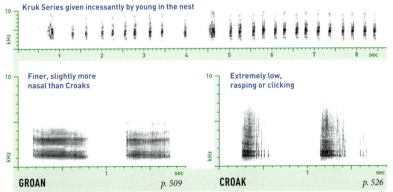

Kruk Series given incessantly by young in the nest

GROAN — Finer, slightly more nasal than Croaks — *p. 509*

CROAK — Extremely low, rasping or clicking — *p. 526*

Nearly silent in courtship; most displays are purely visual. All adult vocalizations seem to be plastic, on the continuum from Groans to Croaks, ranging from long to short; given in alarm and in interactions. Kruk Series of young in nest variable, but usually more nasal than Great Blue Heron's.

SNOWY EGRET

Egretta thula

Widespread in wetlands. Much smaller than Great Egret, with black bill and legs, yellow feet. Has recovered from ferocious hunting trade of the 1800s.

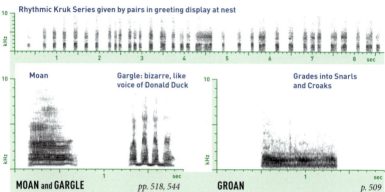

Rhythmic Kruk Series given by pairs in greeting display at nest

Moan

Gargle: bizarre, like voice of Donald Duck

Grades into Snarls and Croaks

MOAN and GARGLE *pp. 518, 544*

GROAN *p. 509*

Moan and Gargle given mostly Apr.–June, in courtship at nesting sites; Moan likely by both sexes, highly plastic Gargle reportedly only by males. Several other courtship vocalizations not shown. Groan, Snarls, and Croaks given all year, in alarm, in flight, and in aggressive interactions.

LITTLE BLUE HERON

Egretta caerulea

Found in freshwater and saltwater wetlands. First-year birds very similar to Snowy Egret; birds molting into adult plumage are variably patchy.

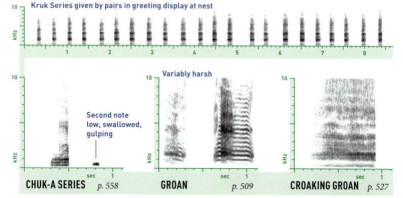

Kruk Series given by pairs in greeting display at nest

Second note low, swallowed, gulping

Variably harsh

CHUK-A SERIES *p. 558*

GROAN *p. 509*

CROAKING GROAN *p. 527*

Chuk-a given in series by courting male during neck-stretch display; descriptions vary, suggesting that some display sounds may be 1-noted. Highly plastic Groans not well known; may be given mostly at nest. Plastic Croaking Groans and Croaks given all year, in alarm and aggressive encounters.

TRICOLORED HERON

Egretta tricolor

Very slender and long-billed; about the size of Snowy Egret. Most common in saltwater habitats. Forages actively, sometimes running. No white morph.

Young birds beg with repeated short Screech Series (p. 548)

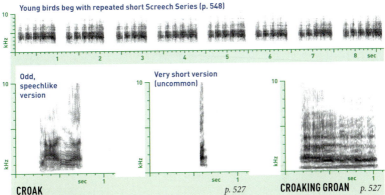

CROAK   Odd, speechlike version   Very short version (uncommon)   *p. 527*   **CROAKING GROAN**   *p. 527*

Display vocalizations poorly understood; no recordings available. Croaks and Croaking Groans intergrade, ranging from long to short; Croaking Groan is typical of most sounds given in alarm. Young beg with high Screeches; adults sometimes give Kruk Series (p. 548) from near nest.

REDDISH EGRET

Egretta rufescens

Forages on tidal flats and beaches, often running and flapping wings. Most U.S. birds are dark. White morph easily confused with other white egrets.

In twig-passing display, members of pair exchange a stick, giving Grunts and clapping bills

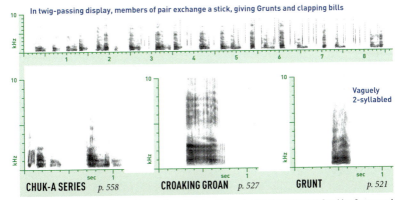

CHUK-A SERIES   *p. 558*   **CROAKING GROAN**   *p. 527*   **GRUNT**   *p. 521*   Vaguely 2-syllabled

In displays at nest, gives low, rather swallowed Chuk-a Series. In alarm, gives longer Croaking Groans and short Croaks like other herons. Typical call while foraging is a somewhat distinctive, slightly 2-syllabled Grunt, often fairly nasal. Calls of young in nest poorly known.

CATTLE EGRET

Bubulcus ibis

Originally an Old World species. Began colonizing North America in the 1950s; now widespread. Forages in fields, often in close association with cattle.

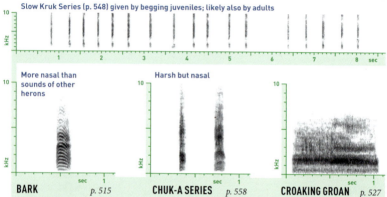

Slow Kruk Series (p. 548) given by begging juveniles; likely also by adults

More nasal than sounds of other herons

Harsh but nasal

BARK *p. 515*

CHUK-A SERIES *p. 558*

CROAKING GROAN *p. 527*

All sounds plastic. Barks given singly, grading into shorter, noisier Kruks in series, or into Chuk-a Series. All of these given at colony and possibly in alarm; single Chuk-a sounds sometimes given in flight. Croaking Groan rather harsh, almost snarling; given at nest, during altercations.

GREEN HERON

Butorides virescens

A small, squat, solitary heron of densely vegetated swamps, ponds, and wetlands. Often perches low above water, remaining motionless for long periods.

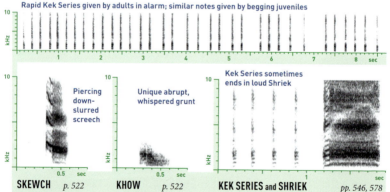

Rapid Kek Series given by adults in alarm; similar notes given by begging juveniles

Piercing down-slurred screech

Unique abrupt, whispered grunt

Kek Series sometimes ends in loud Shriek

SKEWCH *p. 522*

KHOW *p. 522*

KEK SERIES and SHRIEK *pp. 546, 578*

Skewch is most common call, given in alarm and flight; highly variable, grading into Keks. Khow often given in early breeding season from high in tree, possibly by males; rather quiet and difficult to locate. Kek Series given in alarm, grading into one or more plastic Shrieks in high alarm.

BLACK-CROWNED NIGHT-HERON

Nycticorax nycticorax

Found nearly worldwide in various wetland habitats. Active mostly between late evening and early morning; often overlooked during the day.

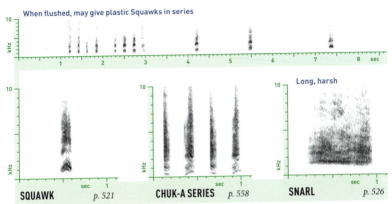

When flushed, may give plastic Squawks in series

SQUAWK *p. 521*

CHUK-A SERIES *p. 558*

Long, harsh

SNARL *p. 526*

Squawk is most common call, often given in flight; plastic but distinctive. Chuk-a Series given by young birds, which also give series of Clucks while still in the nest. Snarl given infrequently, during altercations. In display, gives a soft Pwut (p. 519), audible only at close range.

YELLOW-CROWNED NIGHT-HERON

Nyctanassa violacea

Slimmer, longer-necked than Black-crowned. Breeds in dense vegetation near water. Forages for crabs, its most important food, primarily at night.

When flushed, may give plastic Squawks in series

Typical version: harsher than Black-crowned

SQUAWK

Some versions clearer, more honking or screechy

p. 521

Nasal, grating version

QUACK *p. 525*

Squawks and Quacks given in alarm, often in flight. Versions range from nasal to screechy; some individuals give sounds that strongly resemble typical Squawk of Black-crowned. Young in nest give noisy series like other herons. Several other sounds reported in display.

GLOSSY IBIS

Plegadis falcinellus

Found primarily in freshwater marshes near the coast. Legs and facial skin bluish; eyes dark. North American range has expanded significantly.

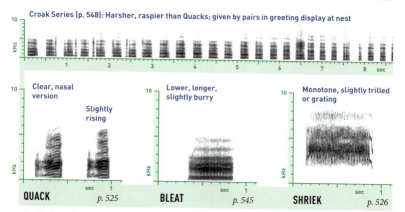

Croak Series (p. 548): Harsher, raspier than Quacks; given by pairs in greeting display at nest

Clear, nasal version — Slightly rising

QUACK — p. 525

Lower, longer, slightly burry

BLEAT — p. 545

Monotone, slightly trilled or grating

SHRIEK — p. 526

Quack given all year, in alarm, but most often near nest; most common call. Plastic; ranges from 1- to vaguely 2-syllabled, monotone to rising, clear and nasal to grunting or grating. Bleat given near nest, at least in spring; may function in courtship. Shrieks given by begging juveniles, including in flight.

WHITE-FACED IBIS

Plegadis chihi

Similar to Glossy Ibis; the two sometimes hybridize. Habitat similar. Legs, facial skin, and eye reddish. Only breeding adults have white facial border.

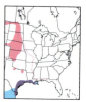

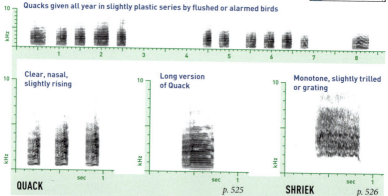

Quacks given all year in slightly plastic series by flushed or alarmed birds

Clear, nasal, slightly rising

QUACK

Long version of Quack

p. 525

Monotone, slightly trilled or grating

SHRIEK — p. 526

Quack similar to Quack of Glossy Ibis, and not usually separable, but averages shorter and clearer, with roughest versions heard mostly at nest. Likely gives a Bleat (p. 545) near nest and a Croak Series (p. 548) in greeting like Glossy Ibis; no recordings available. Shrieks given by begging juveniles, including in flight.

White Ibis

Eudocimus albus

Conspicuous in both freshwater and saltwater wetlands, often breeding in large colonies. Adults are bright white, juveniles mostly dark.

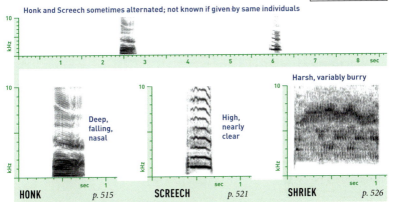

Honk and Screech sometimes alternated; not known if given by same individuals

Deep, falling, nasal

High, nearly clear

Harsh, variably burry

| HONK | p. 515 | SCREECH | p. 521 | SHRIEK | p. 526 |

Honk given all year, in mild alarm; some versions shorter, more monotone than shown, more like Quack of Glossy Ibis. Grades into Screech, which is poorly known; other ibis and night-herons may give similar calls. Shrieks given by begging juveniles, including in flight; some versions clearer, trilled.

Roseate Spoonbill

Platalea ajaja

Unique, with bright rosy plumage and spatulate bill. Local in swamps, mangroves, marshes, and other wetland habitats.

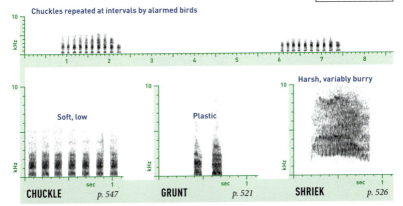

Chuckles repeated at intervals by alarmed birds

Soft, low

Plastic

Harsh, variably burry

| CHUCKLE | p. 547 | GRUNT | p. 521 | SHRIEK | p. 526 |

Not very vocal, even at nest. Chuckle given in alarm, likely all year, but soft and not often heard away from colony. Grunts are similar but more plastic, given by pairs in greeting and courtship at nest site. Shrieks and sometimes abrupt Cheeps (p. 501, not shown) given by begging juveniles.

VULTURES (Order Cathartiformes)

Adapted for scavenging carrion, vultures rarely take live prey. While foraging, they soar over open country or woodlands; when roosting, they often gather in large flocks in trees or on pylons. They do not build nests, but lay eggs on the ground in caves, rock crevices, hollow trees, or abandoned buildings.

The seven species of vultures in the Western Hemisphere are not closely related to the vultures of the Eastern Hemisphere. Once considered relatives of storks, American vultures are currently placed in the order Accipitriformes with the hawks, eagles, kites, and Osprey.

The vultures appear to lack the muscles that attach to the syrinx in most birds. As a result, they give only rough grunting and hissing sounds, when they vocalize at all. Most sounds do not carry far.

BLACK VULTURE

Coragyps atratus

Locates carrion primarily by sight. Often follows Turkey Vulture to carcasses and then displaces it. Soars on horizontal wings; wingbeats distinctively quick and snappy.

Distinctive short Grunts given by adults in interactions with other vultures at roosts or carcasses. Hiss given when birds feel threatened; juvenile version shown. Wings make a thrumming sound (p. 533) similar to wing sound of Mute Swan, audible at close range.

Quick, whispered

GRUNT *p. 521*

Roaring bass register

HISS *p. 522*

TURKEY VULTURE

Cathartes aura

Our most familiar vulture. Locates prey primarily by scent. Soars with wings in a shallow V, erratically tipping from side to side; wingbeats slow and labored. Juveniles have dark gray heads.

Heard even less often than Black Vulture. Juveniles and adults may give Hisses when threatened or in aggressive interactions near carcasses. Adults may also occasionally give shorter Grunts.

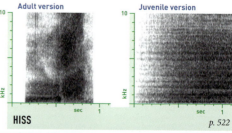

Adult version

Juvenile version

HISS *p. 522*

HAWKS, EAGLES, KITES, AND OSPREY
(Order Accipitriformes)

The sound of a raptor, in many people's minds, is the scream of a Red-tailed Hawk. Nevertheless, many birds in this family do not scream. In fact, many raptors actually sound much like gulls, both in tone quality and in pattern.

Raptor sounds are innate, as far as we know, but tend to be highly plastic. Hawks in the genus *Buteo* generally have simple repertoires: a series call used mostly in courtship, and a Scream or Wail given mostly in alarm. Accipiters give slower series in courtship and faster series in alarm. Kites, eagles, and harriers mostly give various whistled, nasal, or broken notes, often in series.

NORTHERN HARRIER

Circus cyaneus

Distinctively shaped, with long wings and tail. White rump often conspicuous in flight. Hunts low over marshes, with wings held in a shallow V.

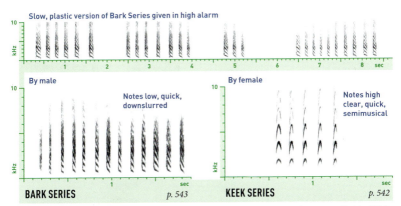

Slow, plastic version of Bark Series given in high alarm

By male — Notes low, quick, downslurred

By female — Notes high, clear, quick, semimusical

BARK SERIES *p. 543*

KEEK SERIES *p. 542*

All year, from alarmed or agitated birds in response to predators near the nest, harassment by blackbirds, etc. Male Bark Series and female Keek Series quite different in quality. In mild alarm, series are short; in high alarm, they become slower, longer, and more plastic.

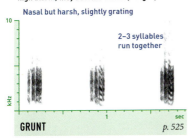

Nasal but harsh, slightly grating

2–3 syllables run together

GRUNT *p. 525*

Known only from males near the nest, but may occur in other situations; more study needed.

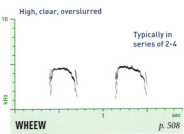

High, clear, overslurred

Typically in series of 2-4

WHEEW *p. 508*

Mostly Mar.–July, during acrobatic courtship flight, apparently by male; also reportedly in aggression and by female on the nest.

RAPTOR COPULATION SOUNDS

Most hawks and eagles have an excited vocalization, a Squeal Series (p. 545), that is given only during copulation. Often the female is most vocal, but both birds may participate. These sounds are given infrequently and only during the early breeding season, but they tend to be loud, far-carrying, and often characteristic of the species.

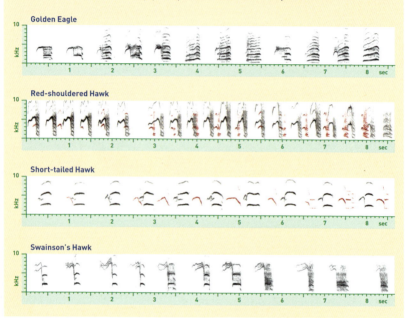

HOOK-BILLED KITE

Chondrohierax uncinatus

Rare in dense woods. Distinctive flight shape, with long tail and broad wings pinched at the base. Often seen in small family groups. Eats only tree snails.

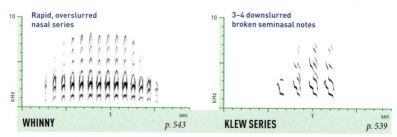

Whinny given all year; most common call. Quite similar to Pileated Woodpecker Whinny, but more overslurred. Klew Series given by subadults, possibly also adults. At a distance, can suggest titmouse song.

SWALLOW-TAILED KITE

Elanoides forficatus

A supremely graceful flier. Once ranged much farther north. Now rather scarce and local, nesting in tall trees adjacent to open country.

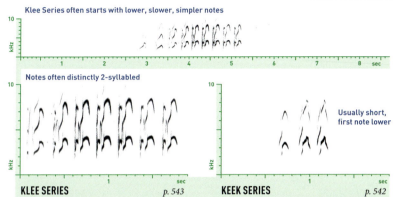

Klee Series often starts with lower, slower, simpler notes

Notes often distinctly 2-syllabled

KLEE SERIES *p. 543*

Usually short, first note lower

KEEK SERIES *p. 542*

All year. All sounds rather plastic, and can intergrade. Klee Series given mostly near the nest and in display flights; sometimes all or part of the series is 1-syllabled upslurs. Keek Series given in alarm or excitement, often by multiple birds on the wing. Begging juveniles give slow Pseep Series (p. 535).

WHITE-TAILED KITE

Elanus leucurus

Hunts on the wing over wetlands and savannas, hovering frequently, then dropping into the grass after small prey. Roosts communally in winter.

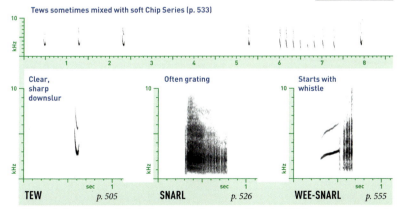

Tews sometimes mixed with soft Chip Series (p. 533)

Clear, sharp downslur

TEW *p. 505*

Often grating

SNARL *p. 526*

Starts with whistle

WEE-SNARL *p. 555*

All sounds given all year; all plastic. Tew given in pair contact and in territorial defense; most common call. Snarl given in territorial defense. Wee-snarl given in a variety of situations including courtship and alarm; whistle usually upslurred and snarl usually brief, but can be longer than 1 second.

SNAIL KITE

Rostrhamus sociabilis

Found in extensive wetlands, where it feeds exclusively on freshwater snails. Also found in the tropics, but Florida population may be just 100 to 200 birds.

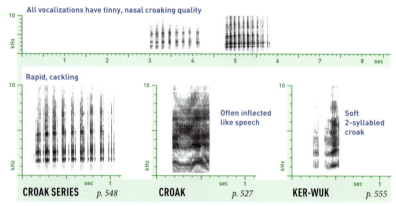

All vocalizations have tinny, nasal croaking quality

Rapid, cackling

Often inflected like speech

Soft 2-syllabled croak

CROAK SERIES *p. 548*

CROAK *p. 527*

KER-WUK *p. 555*

Croak series given all year by both sexes, in alarm, especially near nest; most common call. Single Croaks somewhat plastic; apparently used in courtship and territorial disputes. Ker-wuk given in many situations with other Snail Kites, such as at communal roosts.

MISSISSIPPI KITE

Ictinia mississippiensis

Locally common in forests and savannas, sometimes nesting in parks and old neighborhoods. Catches insects in graceful soaring flight.

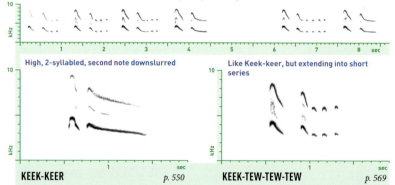

Keek-tew-tew-tew usually given in series beginning or ending with Keek-keer

High, 2-syllabled, second note downslurred

Like Keek-keer, but extending into short series

KEEK-KEER *p. 550*

KEEK-TEW-TEW-TEW *p. 569*

Keek-keer is most common call, given all year, but mostly near nest or in response to predator. Grades into Keek-tew-tew-tew, given mostly by excited birds near nest, in pair interactions or in reponse to nestlings. High, broken Squeal (p. 518) sometimes given, likely a variant of Keek-keer.

SHARP-SHINNED HAWK

Accipiter striatus

Our smallest accipiter, a fierce hunter of small birds. Smaller than Cooper's Hawk, with sharper tail corners, snappier wingbeats, different voice.

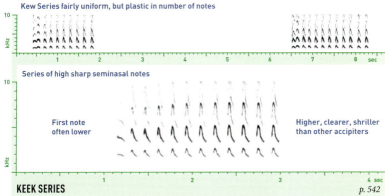

Kew Series fairly uniform, but plastic in number of notes

Series of high sharp seminasal notes

First note often lower

Higher, clearer, shriller than other accipiters

KEEK SERIES *p. 542*

Mostly Apr.–Sept., by adults near nest, especially in response to intruders; also frequently heard from fledglings in late summer. Series sometimes followed by a few Tews.

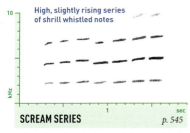

High, slightly rising series of shrill whistled notes

SCREAM SERIES *p. 545*

Likely Mar.–May, during copulation, possibly by female, but poorly known. Plastic. Notes and series both tend to be barely upslurred; tone clear.

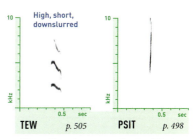

High, short, downslurred

TEW *p. 505*   **PSIT** *p. 498*

All year. Tews given near nest and on fall migration; grade into excited, sometimes twittering Psit. Also gives a clear Chip (p. 497; not shown).

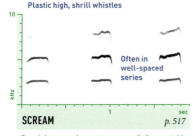

Plastic high, shrill whistles

Often in well-spaced series

SCREAM *p. 517*

By adults near the nest, apparently in courtship; possibly in other contexts also. Like Killdeer Deet, but higher.

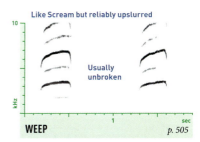

Like Scream but reliably upslurred

Usually unbroken

WEEP *p. 505*

July–Oct., by begging juveniles. Juvenile version of Scream, often given with Kew Series. Plastic.

COOPER'S HAWK

Accipiter cooperii

Intermediate in size between Sharp-shinned Hawk and Northern Goshawk. Found in a variety of habitats; in some areas, nests readily in towns.

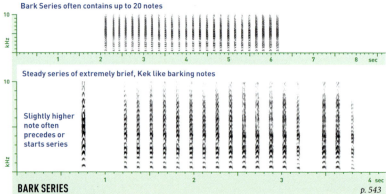

Bark Series often contains up to 20 notes

Steady series of extremely brief, Kek like barking notes

Slightly higher note often precedes or starts series

BARK SERIES

p. 543

Mostly Mar.–May, especially from females, and especially in response to intruders near the nest. Also heard from agitated birds in fall. Recently fledged juveniles in late summer give a higher-pitched version. Much lower and more nasal than series of other accipiters or Northern Flicker.

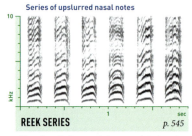

Series of upslurred nasal notes

REEK SERIES

p. 545

Mar.–May, during copulation, mostly by females; males sometimes join in near the end. Fairly plastic.

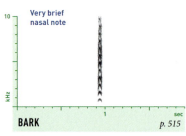

Very brief nasal note

BARK

p. 515

Mostly Mar.–May, especially by males, in contact and near nest. Most common call. Given regularly at dawn in breeding season.

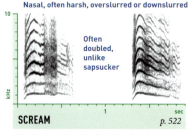

Nasal, often harsh, overslurred or downslurred

Often doubled, unlike sapsucker

SCREAM

p. 522

Mostly Jan.–June, by breeding females; males give lower, harsher versions. Plastic. Some versions are clear Mews (p. 517), like a sapsucker's.

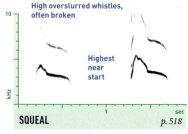

High overslurred whistles, often broken

Highest near start

SQUEAL

p. 518

Mostly July–Oct. Juvenile version of the Scream, given by begging fledglings and later by independent birds. Plastic and variable.

Northern Goshawk

Accipiter gentilis

Our largest accipiter. Breeds in mature forests, hunting squirrels, rabbits, and birds. Voice generally more like Sharp-shinned than like Cooper's.

Here, Keklike version of Screech Series introduced by Tew

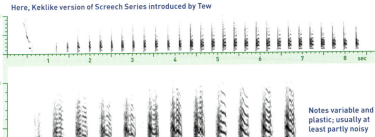

Notes variable and plastic; usually at least partly noisy

SCREECH SERIES *p. 548*

Mostly Mar.–June. Given in aggressive defense of nest or territory against all threats, especially by females. Some less excited versions have shorter notes, more like Cooper's Hawk Bark Series, but always higher-pitched and slower. Notes in series sometimes broken.

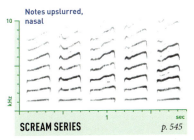

Notes upslurred, nasal

SCREAM SERIES *p. 545*

Mar.–May, by females during copulation, with males sometimes joining in near the end. Lower than Sharp-shinned Hawk Scream Series.

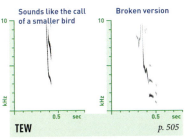

Sounds like the call of a smaller bird

Broken version

TEW *p. 505*

Apparently by both sexes, all year, likely in mild alarm. Variable and plastic, not very loud. Broken versions can recall Northern Flicker Kleer.

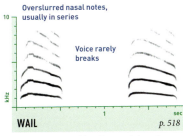

Overslurred nasal notes, usually in series

Voice rarely breaks

WAIL *p. 518*

Mostly Mar.–Aug. Given frequently by pairs near the nest; some Wails in each series may have a slight tremolo.

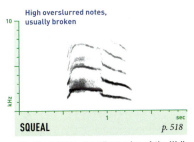

High overslurred notes, usually broken

SQUEAL *p. 518*

Plastic, variable. Juvenile version of the Wail, easily confused with Cooper's Hawk Squeal, but more often broken.

Red-tailed Hawk

Buteo jamaicensis

Our most common hawk, found almost everywhere. Variable in plumage, but vocalizations are the same throughout range and among subspecies.

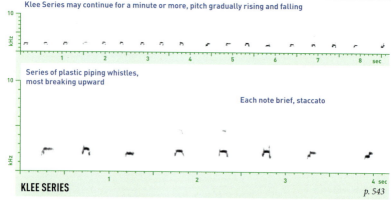

Klee Series may continue for a minute or more, pitch gradually rising and falling

Series of plastic piping whistles, most breaking upward

Each note brief, staccato

KLEE SERIES

p. 543

Mostly Mar.–June, by both sexes. Given in swooping, high-altitude courtship flight, sometimes by both members of a pair. Often sounds very faint because of great distance from the ground. Distinctive in most of the East; Swainson's Klee Series is similar but averages slower, with longer notes.

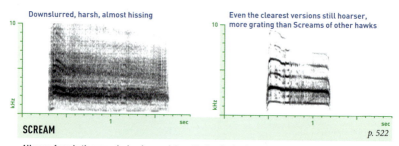

Downslurred, harsh, almost hissing

Even the clearest versions still hoarser, more grating than Screams of other hawks

SCREAM

p. 522

All year, from both sexes. An iconic sound, heard in hundreds of movies and commercials, often from the mouths of other species. Hissing, screechy quality distinctive; sounds louder than it actually is. Sexes may differ in pitch and tone quality; more study needed.

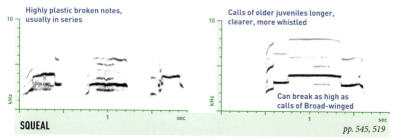

Highly plastic broken notes, usually in series

Calls of older juveniles longer, clearer, more whistled

Can break as high as calls of Broad-winged

SQUEAL

pp. 545, 519

Mostly July–Sept. Given by immature birds in their first summer and fall, including apparently independent birds. May transition gradually into adult Scream over course of first fall and winter; more study needed. Far more likely to cause identification confusion than adult Scream.

Harris's Hawk

Parabuteo unicinctus

Prefers mesquite and saguaro deserts; often breeds in small colonies, hunts in groups. Popular in falconry. Vocal repertoire may be larger than shown.

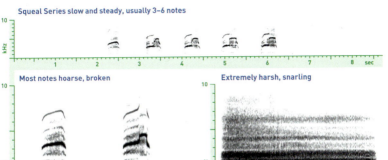

Squeal Series slow and steady, usually 3–6 notes

Most notes hoarse, broken

Extremely harsh, snarling

SQUEAL SERIES *p. 545*

SCREAM *p. 522*

Both sounds given all year. Squeal Series apparently given as a signal to mates or hunting partners; occasionally shortened to a single note. Scream given in several situations; most common call. Distinctively harsh, but fading at end. Juveniles give a clearer version.

Gray Hawk

Buteo nitidus

A small buteo of riparian areas and adjacent thorn scrub or savanna. Once rare in the U.S., but population seems to have increased in recent years.

Wail Series sometimes starts with soft chirping or squealing notes

Each note a long, low overslurred whistle, sometimes broken

Single nasal overslur with long, drawn-out end

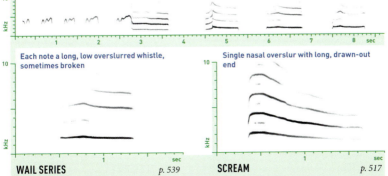

WAIL SERIES *p. 539*

SCREAM *p. 517*

Wail Series given mostly Mar.–Aug. Plastic; usually 3–5 notes, most smoothly overslurred, but some may break or sound 2-syllabled. Scream given all year; most common call. Sometimes noisy; juveniles beg with a higher, hoarser, often broken version. Version with Squeals given during copulation.

RED-SHOULDERED HAWK

Buteo lineatus

A small buteo of mature forests; hunts small prey mostly from perches. Highly vocal in spring, calling loudly in flight display just above treetops.

In courtship flight with dangling legs, gives Squeal Series (p. 545) of broken Kleerlike notes

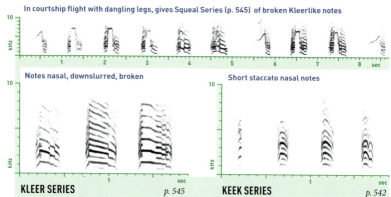

Notes nasal, downslurred, broken

Short staccato nasal notes

KLEER SERIES *p. 545*

KEEK SERIES *p. 542*

All sounds most common Feb.–Apr. Kleer Series is most common sound, often given in flight. Sometimes continues for long periods. Keek Series given in alarm or excitement; intergrades with Kleer Series. Occasional Keek given singly.

BROAD-WINGED HAWK

Buteo platypterus

Our smallest buteo, found in deciduous or mixed forests, where it hunts mostly from perches. Migrates in large flocks called kettles. Voice distinctively high.

Pee-tee Series typically 4–6 seconds long

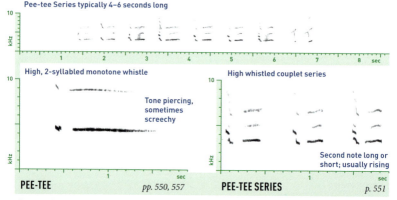

High, 2-syllabled monotone whistle

High whistled couplet series

Tone piercing, sometimes screechy

Second note long or short; usually rising

PEE-TEE *pp. 550, 557*

PEE-TEE SERIES *p. 551*

Pee-tee given all year in many situations; most common call. Rarely, second note breaks or is downslurred, or the 2 notes run together. Pee-tee Series not well known; highly plastic. Apparently used in territorial defense, possibly also in courtship.

Short-tailed Hawk

Buteo brachyurus

Widespread in the tropics. Fewer than 200 pairs breed in Florida, mostly dark-morph birds. Nests in mature, often wet forests; soars over other habitats.

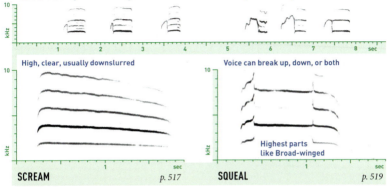

Screams and Squeals can intergrade over the course of a short Squeal Series (p. 545)

High, clear, usually downslurred

Voice can break up, down, or both

Highest parts like Broad-winged

SCREAM   *p. 517*   **SQUEAL**   *p. 519*

Scream and Squeal may be just variants of a single plastic call; both given all year. Distinctive in range, often nasal but never noisy; always clearer and higher than adult Red-tailed. Squeals sometimes given in slow series; function of series not known.

Swainson's Hawk

Buteo swainsonii

Common and widespread in the open country of the West. Migrates to South America in large flocks called kettles. Soars with wings held in a shallow V.

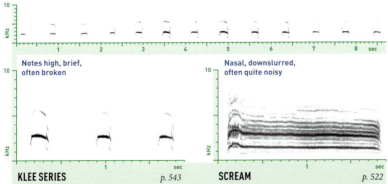

Klee Series much like Red-tailed's

Notes high, brief, often broken

Nasal, downslurred, often quite noisy

KLEE SERIES   *p. 543*   **SCREAM**   *p. 522*

Klee Series given mostly Mar.–June in swooping, high-altitude courtship flight, sometimes with Screams interspersed. Scream given all year; most common call. Usually clearer, more nasal than Red-tailed Scream. Female Scream is reportedly shorter and lower than male's; may also tend to be clearer.

White-tailed Hawk

Geranoaetus albicaudatus

Widespread but local in the tropics. In our area, found in coastal grasslands and savannas. Rather quiet and shy, but voice distinctive when heard.

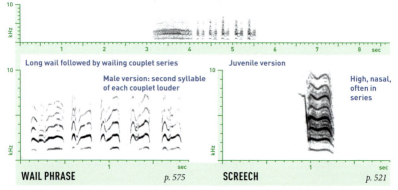

Female version of Wail Phrase: first syllable of each couplet louder

Long wail followed by wailing couplet series

Male version: second syllable of each couplet louder

Juvenile version

High, nasal, often in series

WAIL PHRASE   p. 575

SCREECH   p. 521

Wail Phrase given all year; may recall the voices of certain shorebirds. Male and female versions differ noticeably. Begging juveniles give harsh Screech. Adult version clearer, more like 1 note from end of Wail Phrase; function unknown.

Zone-tailed Hawk

Buteo albonotatus

Uncommon in western canyons. Almost exactly matches Turkey Vulture in flight style, shape, and coloration, perhaps to fool prey into allowing close approach.

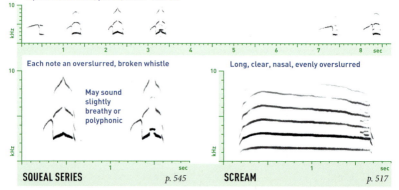

Squeal Series: slow, plastic, with 4–12 notes

Each note an overslurred, broken whistle

May sound slightly breathy or polyphonic

Long, clear, nasal, evenly overslurred

SQUEAL SERIES   p. 545

SCREAM   p. 517

Squeal Series given mostly Mar.–Aug., by females near nest, by begging juveniles, and by adults in high, swooping courtship flight. Scream given all year, in alarm and also in courtship; most common call. Longer and more monotone than Gray Hawk Scream.

Ferruginous Hawk

Buteo regalis

Our largest buteo, uncommon in short- to mid-grass prairies and sagebrush steppes. Soars with long, pointed wings held in a shallow V.

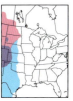

Variety of Screamlike sounds given in nest defense

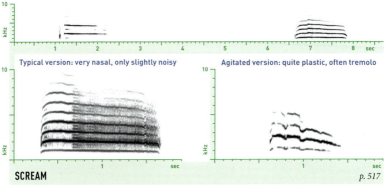

Typical version: very nasal, only slightly noisy

Agitated version: quite plastic, often tremolo

SCREAM

p. 517

Scream given all year, but infrequent except near nest. Scream lower than in most other hawks; some versions perhaps best called Wails (p. 518). Likely has a series call given in courtship, like other buteos, but no recordings available.

Rough-legged Hawk

Buteo lagopus

Breeds on Arctic cliffs; in winter, found in open areas. Plumage ranges from light to dark. Frequently hovers in place while hunting.

Variety of Screamlike sounds given in nest defense, including Squeals (p. 518)

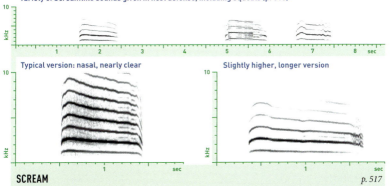

Typical version: nasal, nearly clear

Slightly higher, longer version

SCREAM

p. 517

Scream given all year, but infrequent except near nest. Scream lower than in most other hawks; some versions perhaps best called Wails (p. 518). Likely has a series call given in courtship, like other buteos, but no recordings available.

BALD EAGLE

Haliaeetus leucocephalus

Found mostly along rivers, lakeshores, and coasts; dives into water for live fish or scavenges on shore. Populations are rebounding after sharp declines.

Screams usually given in series of 2–4

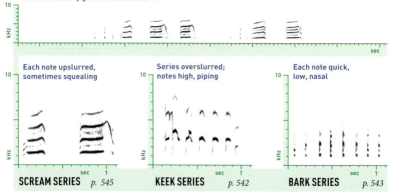

| SCREAM SERIES *p. 545* | KEEK SERIES *p. 542* | BARK SERIES *p. 543* |

Each note upslurred, sometimes squealing — Series overslurred; notes high, piping — Each note quick, low, nasal

All sounds given all year, and all quite plastic; function of most is poorly known. Scream Series plastic; some versions resemble Reek of female Wood Duck. Keek Series is most common call, often introduced by 4–5 well-spaced single Keeks. Bark Series given mostly near nest, likely in alarm.

GOLDEN EAGLE

Aquila chrysaetos

A large, powerful raptor of open areas and mountains, rare in the East. Not closely related to Bald Eagle. Hunts mostly rabbits and ground squirrels.

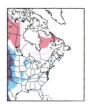

Klee Series can strongly resemble end of Herring Gull Long Call

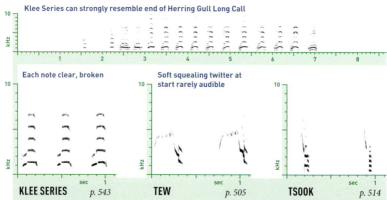

| KLEE SERIES *p. 543* | TEW *p. 505* | TSOOK *p. 514* |

Each note clear, broken — Soft squealing twitter at start rarely audible

Not very vocal. Klee Series given at least Jan.–July; may function in courtship. Tew and Tsook may be variants of one another; given by adults near the nest, and similar calls given by older begging juveniles. Tew lower than Osprey Tew; Tsook sometimes run into short rapid series.

OSPREY

Pandion haliaetus

A unique raptor; catches live fish by plunging feet-first into the water. Nests readily on artificial platforms, sometimes even in towns.

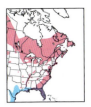

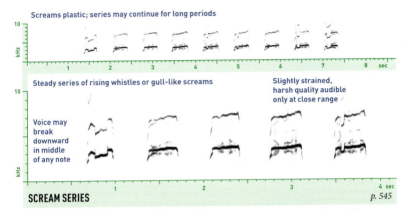

Screams plastic; series may continue for long periods

Steady series of rising whistles or gull-like screams

Slightly strained, harsh quality audible only at close range

Voice may break downward in middle of any note

SCREAM SERIES — *p. 545*

Mostly during breeding season: Nov.–Mar. in Florida, Mar.–June in New England. Given by both sexes, usually on the wing. Given continuously during male's fish flight display in courtship, along a shallowly undulating path high above the nest, legs typically dropped, sometimes carrying a fish.

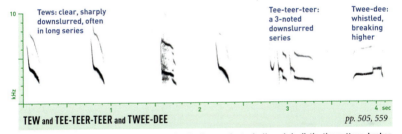

Tews: clear, sharply downslurred, often in long series

Tee-teer-teer: a 3-noted downslurred series

Twee-dee: whistled, breaking higher

TEW and TEE-TEER-TEER and TWEE-DEE — *pp. 505, 559*

All year. A few Tews often given alone in series, but long series typically ends in distinctive pattern: broken Tew, normal Tew, Tee-teer-teer, Twee-dee. Exact order varies, but typical performance is quite uniform rangewide. Function not well known; may generally signal alertness or excitement.

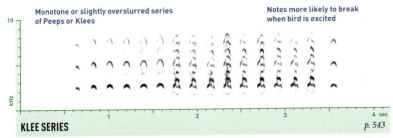

Monotone or slightly overslurred series of Peeps or Klees

Notes more likely to break when bird is excited

KLEE SERIES — *p. 543*

All year, by both sexes, but especially by females near nest. Highly plastic. Similar calls also used by females to beg for food from males in courtship.

OWLS (Order Strigiformes)

With large, forward-facing eyes adapted to hunting in low light, these birds of prey are primarily nocturnal, although some are also active in daylight. Owls can rotate their heads up to 270 degrees, and specially modified wing feathers make their wingbeats nearly inaudible. Most North American species are classified in the family Strigidae, except for the Barn Owl, which is in the family Tytonidae.

The songs of large owl species are low-pitched and hooting; smaller species give whistled or nasal songs. Although females are larger than males in body size, males tend to have lower-pitched voices. Many species give a variety of barking, wailing, or shrieking calls. All vocalizations are thought to be innate, and most are uniform throughout a species' range.

Barn Owl

Tyto alba

Widespread in open country, sometimes near human habitation. Nests in enclosed spaces ranging from barns to riverbank cavities. Highly nocturnal.

When cornered in nest or roost, gives long screeching Hiss in defensive posture (p. 522)

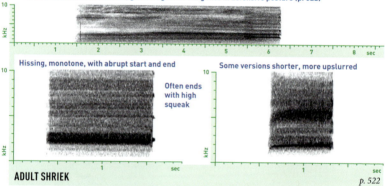

Hissing, monotone, with abrupt start and end

Often ends with high squeak

Some versions shorter, more upslurred

ADULT SHRIEK

p. 522

All year, especially during breeding season; most common vocalization. Often given repeatedly by adult males on the wing, sometimes with soft Wing Claps (p. 532). Shrieks of most other owl species usually given singly in flight or repeatedly from perch. Female version reportedly rarer, more broken.

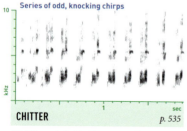

Series of odd, knocking chirps

CHITTER

p. 535

Variable and plastic. All year, mostly near nest, in food exchanges; also by young, by pairs in courtship, and occasionally in flight.

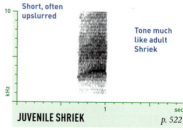

Short, often upslurred

Tone much like adult Shriek

JUVENILE SHRIEK

p. 522

By begging juveniles, usually every few seconds for long periods, often from a perch. Also by adult females. Variable and plastic.

GREAT HORNED OWL

Bubo virginianus

Our most familiar and widespread owl. Primarily nocturnal, but often active and vocal before full dark. Note large size and prominent ear tufts.

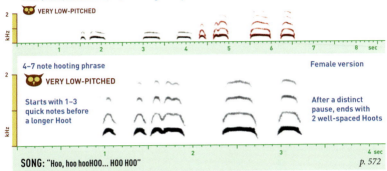

Male (black) and female (red) often sing in unsynchronized duets

🦉 VERY LOW-PITCHED

4–7 note hooting phrase

🦉 VERY LOW-PITCHED

Female version

Starts with 1–3 quick notes before a longer Hoot

After a distinct pause, ends with 2 well-spaced Hoots

SONG: "Hoo, hoo hooHOO... HOO HOO"   *p. 572*

All year, especially Dec.–Mar., by both sexes. Fairly uniform and stereotyped. Females average slightly higher-pitched, with more syllables per phrase. Number of syllables varies somewhat with agitation level, but individuals are often identifiable by subtle differences in song.

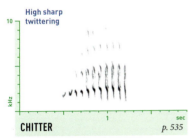

High sharp twittering

CHITTER   *p. 535*

Given infrequently, during copulation and around the nest. Quite plastic; some versions less staccato and more squealing.

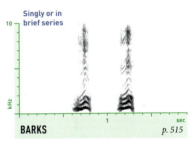

Singly or in brief series

BARKS   *p. 515*

Highly variable and plastic; given in agitation and close courtship. Sometimes Barks may introduce Song, or replace first few notes of Song.

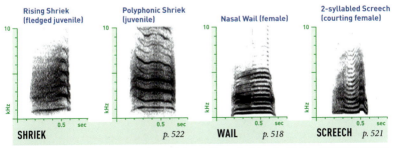

Rising Shriek (fledged juvenile)

Polyphonic Shriek (juvenile)

Nasal Wail (female)

2-syllabled Screech (courting female)

SHRIEK   *p. 522*   **WAIL**   *p. 518*   **SCREECH**   *p. 521*

A catch-all category of variable, plastic sounds heard all year. Food-begging Shrieks of juveniles begin in nest and continue for months after fledging; usually very noisy and upslurred. Adults, especially females, give Screeches and Wails that average more nasal, in close courtship and in agitation.

Snowy Owl

Bubo scandiacus

Breeds in Arctic; winters on prairies, beaches, and other open areas. Active both day and night. In some winters, irrupts far south of normal range.

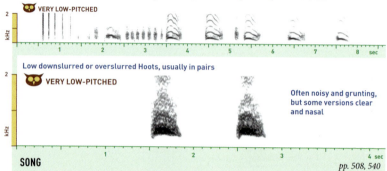

Hoots sometimes in steady series of up to 8; here, introduced by bill snaps and chuckles

🦉 VERY LOW-PITCHED

Low downslurred or overslurred Hoots, usually in pairs

🦉 VERY LOW-PITCHED

Often noisy and grunting, but some versions clear and nasal

SONG

pp. 508, 540

Mostly on breeding grounds, by males; females sing infrequently. Males sing with head lowered and tail raised, bowing with each note. Song given in territorial advertisement, often by multiple neighboring birds at once, as well as in response to threats.

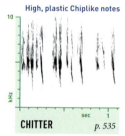

High, plastic Chiplike notes

CHITTER

p. 535

Mostly near nest, by juveniles and likely also adults.

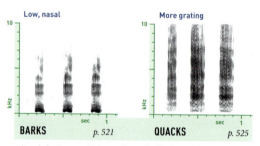

Low, nasal

BARKS

p. 521

More grating

QUACKS

p. 525

A varied category of intergrading calls. Barks and Quacks given all year as a threat, usually in slow series. Faster, softer Chuckles (p. 547) given in close courtship, among other situations.

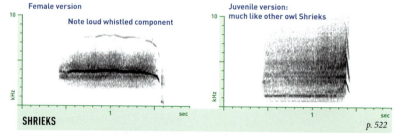

Female version

Note loud whistled component

Juvenile version: much like other owl Shrieks

SHRIEKS

p. 522

Variable and plastic. Most common call in winter, in clashes over territory; also given near nest, mostly by female. Some versions less noisy, almost a pure whistle. Shrieks given by juveniles for some time after fledging, often with Chitters.

Northern Hawk Owl

Surnia ulula

A large owl of boreal forests, active mostly during the day. Shape and behavior recall *Accipiter* hawks. In some winters, irrupts south of normal range.

Song can continue for 10 or more seconds without a pause

Long, slow trill on one pitch, slightly nasal

Fades in slowly at start

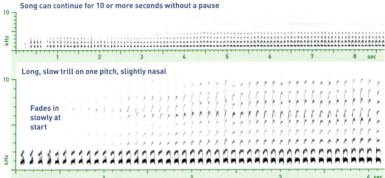

SONG *p. 540*

Mostly Feb.–Apr., by both sexes. Uniform. Male sings from perch or in circular display flight above territory that includes gliding and Wing Claps. Female song may average hoarser, shorter, and more unsteady in rhythm and pitch. Like Boreal Owl Long Song, but higher, faster, and more nasal.

Like soft, slow snippets of Song

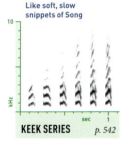

KEEK SERIES *p. 542*

Like flicker Keek Series, but short, high, and plastic. All year, in high alarm.

Occasionally gives single Screeches

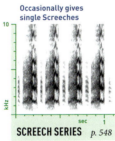

SCREECH SERIES *p. 548*

In alarm, often in flight, perhaps mostly near nest.

High, squeaky, chipping

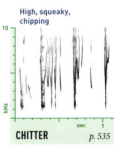

CHITTER *p. 535*

In alarm, perhaps mostly near nest. Extremely plastic.

Rapid nasal series or slow trills, variably screechy

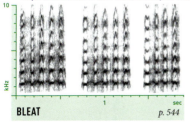

BLEAT *p. 544*

All year, by adults and juveniles in high alarm. Lower-intensity versions of this call reported from some observers in winter.

Long, slightly rising, ending in high Squeak

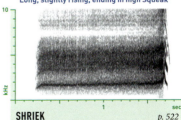

SHRIEK *p. 522*

By juveniles begging for food and by adults in pair interactions, in food exchanges, and in alarm near the nest.

BARRED OWL

Strix varia

A spooky voice of extensive forests and swamps. Primarily nocturnal, but occasionally active and vocal by day. Range has expanded north and west.

Mated pairs give maniacal duets with Songs, Series Songs, and cackling and bleating sounds (p. 572)

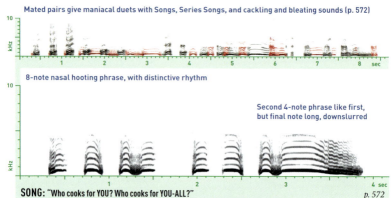

8-note nasal hooting phrase, with distinctive rhythm

Second 4-note phrase like first, but final note long, downslurred

SONG: "Who cooks for YOU? Who cooks for YOU-ALL?"   *p. 572*

All year, by both sexes, but especially Feb.–Apr. Uniform and fairly stereotyped, though occasionally truncated. "Who," "cooks," and "you" notes all on same pitch. Female voice averages higher, with burrier final note. Higher and louder than Great Horned Owl Song, with different rhythm.

6–9 Barks rising and crescendoing to a 2-note phrase

Quality like regular Song; rise in pitch is slight

SERIES SONG: "Who who who who WHO WHO for YOU?"   *p. 572*

All year, by both sexes. Reportedly given during confrontations between mated pairs; also given within pairs and by solo birds, often in association with Song. A key component of caterwauling duets.

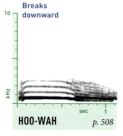

Breaks downward

HOO-WAH   *p. 508*

Like last note of Song. All year by both sexes, perhaps in pair contact.

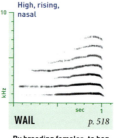

High, rising, nasal

WAIL   *p. 518*

By breeding females, to beg food from mate, and in contact.

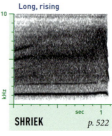

Long, rising

SHRIEK   *p. 522*

By begging juveniles, for months after fledging. Note length (1.5–2 seconds).

GREAT GRAY OWL

Strix nebulosa

Our largest owl, huge and distinctive. Rare in spruce bogs and boreal forest; active mostly at night. In some winters, irrupts south of normal range.

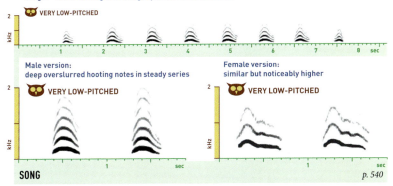

Song: 8–12 low hooting notes, slightly softer at start and end

🦉 VERY LOW-PITCHED

Male version:
deep overslurred hooting notes in steady series

🦉 VERY LOW-PITCHED

Female version:
similar but noticeably higher

🦉 VERY LOW-PITCHED

SONG   p. 540

Mostly Mar.–June, by both sexes, often near the nest. Usually given in full darkness. Slow, steady rhythm and low pitch are distinctive; notes are usually long enough to be distinctly overslurred.

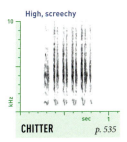

High, screechy

CHITTER   p. 535

Highly plastic. Given at nest during feedings of mate or young.

Rather high nasal overslur

BARK   p. 521

Little known; this example given at nest after male fed incubating female.

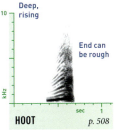

Deep, rising

End can be rough

HOOT   p. 508

Distinctive. Given by female asking to be fed by male; also in family contact.

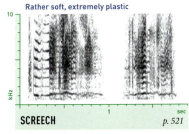

Rather soft, extremely plastic

SCREECH   p. 521

Extremely plastic. Similar sounds given in distraction display, near nest, and possibly also during copulation.

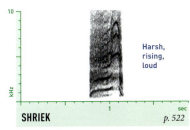

Harsh, rising, loud

SHRIEK   p. 522

Given by begging young, up to several months after fledging. Quite variable; can overlap with Shrieks of other owl species.

Long-eared Owl

Asio otus

Medium-sized. Nests and roosts in forests, conifers, or dense thickets; hunts in adjacent meadows, marshes, and other open areas. Highly nocturnal.

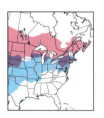

Song often given in flight display, along with well-spaced single Wing Claps (p. 532)

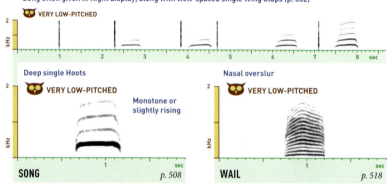

🦉 VERY LOW-PITCHED

Deep single Hoots

🦉 VERY LOW-PITCHED

Monotone or slightly rising

SONG *p. 508*

All year, by males, but mostly Feb.–June. Hoots given singly, usually 2–4 seconds apart, from perch, but also in zigzag display flight.

Nasal overslur

🦉 VERY LOW-PITCHED

WAIL *p. 518*

Mostly Feb.–Apr., by females, especially near nest. May be female version of Song. Can end with soft, low second syllable.

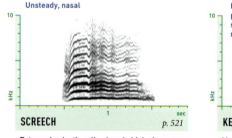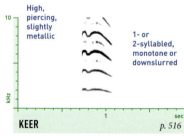

Male: low, nasal Barks

Female: higher, rising Wiks

Juvenile: high Wiks

BARK *p. 521* **WIK** *p. 513*

Highly variable, especially by age and sex, and somewhat plastic. Given all year in alarm or in response to disturbance, especially Mar.–Aug., near nest or dependent young. Grades into Wails.

Unsteady, nasal

SCREECH *p. 521*

Extremely plastic calls given in high alarm near nest, including during injury-feigning distraction display. Some are clear, wailing.

High, piercing, slightly metallic

1- or 2-syllabled, monotone or downslurred

KEER *p. 516*

Mostly May–Aug., by begging juveniles; often repeated every few seconds for long periods. Unlike any other owl sound.

Short-eared Owl

Asio flammeus

A medium-sized owl of open prairies and other open areas. Nests and usually roosts on the ground. Often active before sunset or after sunrise.

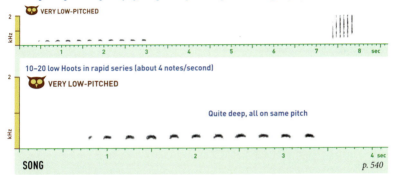

Song often given in flight display, along with rapid rattling series of Wing Claps (p. 532)

😚 VERY LOW-PITCHED

10–20 low Hoots in rapid series (about 4 notes/second)

😚 VERY LOW-PITCHED

Quite deep, all on same pitch

SONG *p. 540*

Mostly Feb.–Apr., by courting males during acrobatic "sky dancing" display flight. Sometimes also delivered from ground or prominent perch. Females apparently do not sing, but may respond to male song with barking or screeching calls.

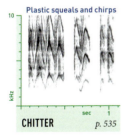

Plastic squeals and chirps

CHITTER *p. 535*

By adults in high alarm near the nest; similar calls given by juveniles.

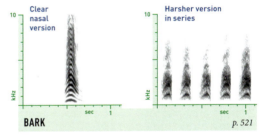

Clear nasal version

Harsher version in series

BARK *p. 521*

All year; given in alarm and during interactions, often in flight. Variable and somewhat plastic, ranging from clear and nasal to harsh and noisy; may be given singly or in series.

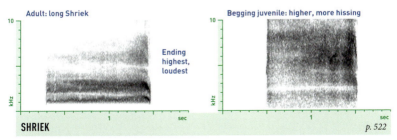

Adult: long Shriek

Ending highest, loudest

Begging juvenile: higher, more hissing

SHRIEK *p. 522*

Variable and plastic. Adults give Shrieks all year, in aggressive interactions. Juveniles give a variety of calls, including Shrieks, chittering Rattles, and more rarely a high whistled call rather like the Keer of Long-eared Owl, but generally noisier.

BOREAL OWL

Aegolius funereus

A little-seen bird of coniferous forests, where it nests in tree cavities. Highly nocturnal. In some winters, irrupts south of normal range.

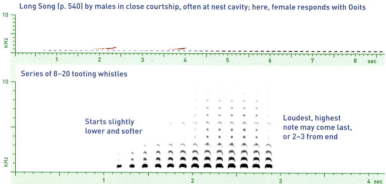

Long Song (p. 540) by males in close courtship, often at nest cavity; here, female responds with Ooits

Series of 8–20 tooting whistles

Starts slightly lower and softer

Loudest, highest note may come last, or 2–3 from end

SONG *p. 540*

All year, but mostly Feb.–June, by males. Female not known to sing. Recalls winnowing of Wilson's Snipe, but shorter, more whistled, and almost never given before full dark. Long Song, in presence of female, can continue for more than 10 seconds. A few soft songlike notes announce food deliveries at the nest.

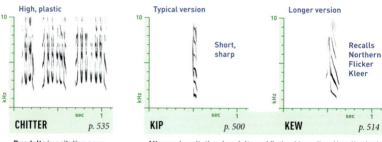

High, plastic

Typical version

Longer version

Short, sharp

Recalls Northern Flicker Kleer

CHITTER *p. 535*

KIP *p. 500*

KEW *p. 514*

By adults in agitation near nest; similar calls also by fledglings.

All year, in agitation, by adults and fledged juveniles. Usually single, but occasionally in short series. Usually higher and sharper than Northern Saw-whet Kews.

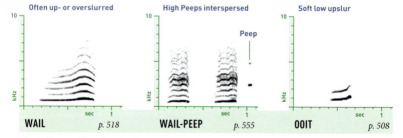

Often up- or overslurred

High Peeps interspersed

Soft low upslur

Peep

WAIL *p. 518*

WAIL-PEEP *p. 555*

OOIT *p. 508*

A set of intergrading, often plastic sounds, usually lower and shorter than Northern Saw-whet's. Wails and Wail-peeps have been recorded in early spring from agitated birds in response to playback of Song. Ooits are softer and appear to be a common call given in close-range contact.

Northern Saw-whet Owl

Aegolius acadicus

Breeds in coniferous and mixed forests, and occasionally in more open areas. In winter, roosts in dense vegetation, especially conifers. Highly nocturnal.

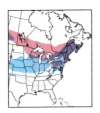

In excitement, Song may become irregular in rhythm

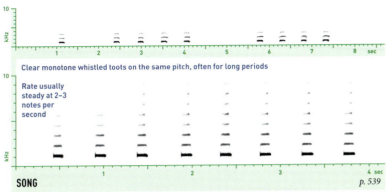

Clear monotone whistled toots on the same pitch, often for long periods

Rate usually steady at 2–3 notes per second

SONG

p. 539

Mostly Jan.–May, by males. Female Song infrequent, given only in courtship; usually softer, with less steady rhythm than male's. Songs of individual males vary slightly in pitch. In excitement, may speed up to nearly 5 notes/second for short periods. Song is easily imitated by human whistling.

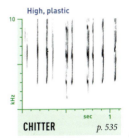

High, plastic

CHITTER   *p. 535*

By adults in agitation; similar calls also by fledglings.

High, downslurred, barking

KEW   *p. 514*

Often in brief, rapid series

KEEK SERIES   *p. 542*

All year, in agitation, by adults and fledged juveniles; also in response to playback of Northern Saw-whet or Boreal Owl Song. Highly variable and plastic.

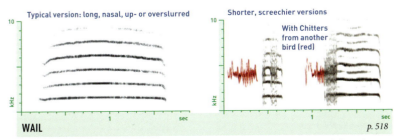

Typical version: long, nasal, up- or overslurred

Shorter, screechier versions

With Chitters from another bird (red)

WAIL

p. 518

All year, from agitated birds, often males, during interactions and in response to playback. Sometimes repeated every few seconds for extended periods. Variable and plastic, but most versions higher and longer than corresponding calls of Boreal Owl.

WESTERN SCREECH-OWL

Megascops kennicottii

Found in wooded habitats, especially deciduous riparian woods. Voice very different from Eastern Screech-Owl's; the two species rarely hybridize.

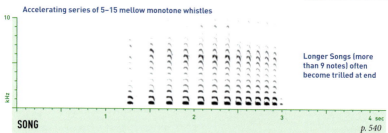

Accelerating series of 5–15 mellow monotone whistles

Longer Songs (more than 9 notes) often become trilled at end

SONG

p. 540

All year, by both sexes, but especially Dec.–Mar. by males. Heard mostly after dusk and before dawn; very rare during the day. Female version averages higher-pitched. Songs average longer in northern part of range, with faster ending, but length and speed vary in any given locale.

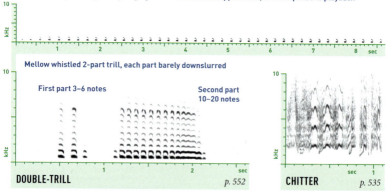

Bark Series (p. 543): given by singing males when female approaches, or in response to playback

Mellow whistled 2-part trill, each part barely downslurred

First part 3–6 notes

Second part 10–20 notes

DOUBLE-TRILL

p. 552

CHITTER

p. 535

All year, by both sexes, apparently in contact and to strengthen pair bond. Second part sometimes given alone. Pairs sometimes duet with one bird giving Song and the other Double-Trill.

In close interactions. Soft. Extremely plastic, ranging from rattling to squealing.

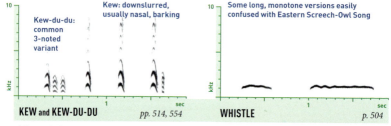

Kew-du-du: common 3-noted variant

Kew: downslurred, usually nasal, barking

KEW and KEW-DU-DU

pp. 514, 554

Some long, monotone versions easily confused with Eastern Screech-Owl Song

WHISTLE

p. 504

All year, by both sexes, in alarm and contact. All of these sounds are highly variable, plastic, and intergrading. In agitation, Barks tend to get higher and louder; they are often 2- or 3-noted, the first note highest. A whistled version of the Kew-du-du is a common female contact call. Also snaps bill in agitation.

EASTERN SCREECH-OWL

Megascops asio

Breeds in wooded habitats, sometimes including residential areas. Almost exclusively nocturnal. Most are gray, but some are bright rufous in color.

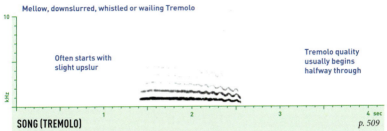

Mellow, downslurred, whistled or wailing Tremolo

Often starts with slight upslur

Tremolo quality usually begins halfway through

SONG (TREMOLO) *p. 509*

All year, by both sexes, but especially Mar.–June, mostly after dusk and before dawn; very rare during the day. Female version averages higher-pitched. More common than Trill later in breeding season; sometimes given along with Trill.

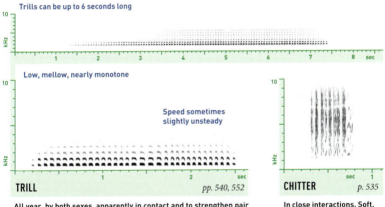

Trills can be up to 6 seconds long

Low, mellow, nearly monotone

Speed sometimes slightly unsteady

TRILL *pp. 540, 552*

CHITTER *p. 535*

All year, by both sexes, apparently in contact and to strengthen pair bond; also given by males advertising a potential nest site. More common than Song in early breeding season. Rare during day.

In close interactions. Soft. Extremely plastic, ranging from rattling to squealing.

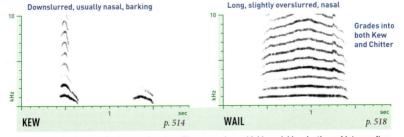

Downslurred, usually nasal, barking

Long, slightly overslurred, nasal

Grades into both Kew and Chitter

KEW *p. 514*

WAIL *p. 518*

All year, by both sexes, in alarm and contact. These sounds are highly variable, plastic, and intergrading. Most versions average longer and more nasal than corresponding sounds of Western Screech-Owl, but some versions are quite similar. Also snaps bill in agitation.

Elf Owl

Microthene whitneyi

The world's smallest owl, smaller than a starling. Breeds in old woodpecker holes in saguaros, thorn scrub, and riparian woods. Normally nocturnal.

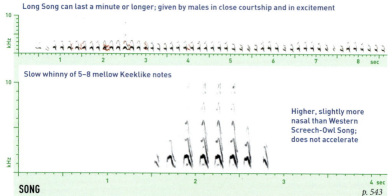

Long Song can last a minute or longer; given by males in close courtship and in excitement

Slow whinny of 5–8 mellow Keeklike notes

Higher, slightly more nasal than Western Screech-Owl Song; does not accelerate

SONG
p. 543

Mostly Mar.–June, by males; most frequent in the 1–2 hours after dusk and before dawn. Females not known to sing. In excitement, middle section of Song may rise in pitch; consecutive Songs may run together into Long Song. Song may grade into series of Kewlike or Keerlike calls.

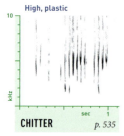
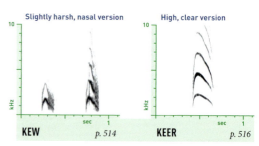

High, plastic

Slightly harsh, nasal version

High, clear version

CHITTER *p. 535*

During feedings; similar calls given in copulation and by juveniles.

KEW *p. 514*

KEER *p. 516*

Most common calls, often in series; variable and plastic. Different versions may serve different functions; soft whistled downslurs given in close contact, high shrill Keers in alarm.

"McCall's" Eastern Screech-Owl   *Megascops asio mccallii*

This subspecies of south Texas and northeast Mexico has a fairly distinctive voice. Song is infrequent and plastic; 2-parted effect in Trill can be subtle or absent.

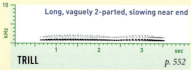
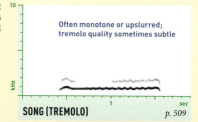

Often monotone or upslurred; tremolo quality sometimes subtle

Long, vaguely 2-parted, slowing near end

TRILL *p. 552*

SONG (TREMOLO) *p. 509*

BURROWING OWL

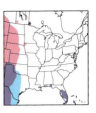

Athene cunicularia

Active both day and night. Most nest in prairie dog or ground squirrel burrows, but Florida birds can dig their own burrows. Population has declined.

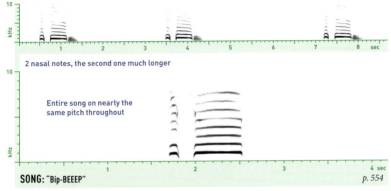

Song can include a third note that is low, nasal, and audible only at close range

2 nasal notes, the second one much longer

Entire song on nearly the same pitch throughout

SONG: "Bip-BEEEP"   *p. 554*

Mostly Mar.–May, by males, night or day. Uniform and fairly stereotyped, though first note is occasionally omitted, and soft, variable versions may be given in close courtship. Female may respond to male Song with "eep" notes similar to second note of Song, but softer, grading into Snarl.

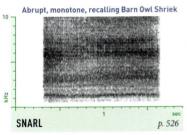

Abrupt, monotone, recalling Barn Owl Shriek

SNARL   *p. 526*

By juveniles and adults, begging for food and in high alarm. May startle predators by resembling rattlesnake rattles.

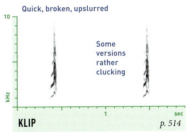

Quick, broken, upslurred

Some versions rather clucking

KLIP   *p. 514*

All year, in mild alarm. Like single note of Chuckle, which it grades into.

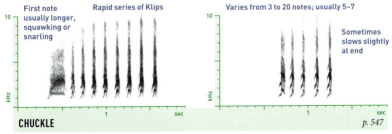

First note usually longer, squawking or snarling

Rapid series of Klips

Varies from 3 to 20 notes; usually 5–7

Sometimes slows slightly at end

CHUCKLE   *p. 547*

All year, in alarm; most common call. Slightly variable and quite plastic, becoming longer as agitation level rises. Notes usually slow enough to count. In high alarm, initial snarling note may be given by itself or repeated later in Chuckle.

Ferruginous Pygmy-Owl

Glaucidium brasilianum

Widespread in the tropics. In Texas, found in live oak and mesquite woodlands. Most active at dawn and dusk, but sometimes during the day.

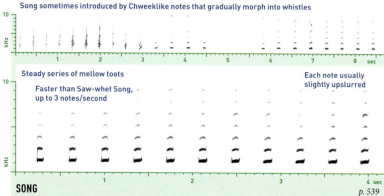

Song sometimes introduced by Chweeklike notes that gradually morph into whistles

Steady series of mellow toots

Faster than Saw-whet Song, up to 3 notes/second

Each note usually slightly upslurred

SONG *p. 539*

All year by both sexes, but especially Feb.–June, mostly at dawn and dusk; not often heard during the day or in full darkness. Easily imitated by human whistling. Female version averages slightly higher-pitched than male's; female may respond to male Song with Song of her own, or with Trill.

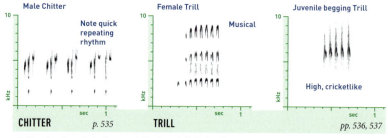

Male Chitter

Note quick repeating rhythm

Female Trill

Musical

Juvenile begging Trill

High, cricketlike

CHITTER *p. 535*　**TRILL** *pp. 536, 537*

Chitter given by male in close courtship. Female Trill given in courtship and during prey exchanges; also in response to male Song. Higher, more insectlike Trill of juvenile given for a month or so after fledging.

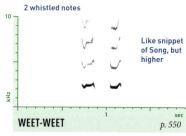
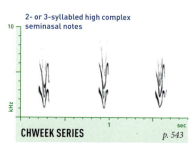

2 whistled notes

Like snippet of Song, but higher

WEET-WEET *p. 550*

Reportedly given by female near nest in response to potential predators, including humans.

2- or 3-syllabled high complex seminasal notes

CHWEEK SERIES *p. 543*

By both sexes, especially females, reportedly in aggression. Variable and plastic. Sometimes given by pairs in synchronized duet.

KINGFISHERS (Family Alcedinidae)

Our three species of kingfishers are rarely found far from water. Medium to large birds, with distinctively long, stout bills and ragged crests, they specialize in eating fish, which they capture by plunge-diving headfirst into slow-moving water. The Belted Kingfisher, and to a lesser extent the Ringed Kingfisher, frequently hovers prior to plunge-diving. Highly territorial, our kingfishers tend to be found singly or in pairs. Nests are in cavities, usually in holes excavated in earthen banks. Vocalizations are simple and apparently innate, consisting primarily of unmusical chatters, clicks, and clucks.

GREEN KINGFISHER

Chloroceryle americana

Our smallest kingfisher, about the size of a starling; rather quiet and inconspicuous. Found near clear freshwater streams and ponds.

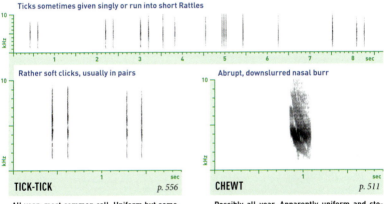

Ticks sometimes given singly or run into short Rattles

Rather soft clicks, usually in pairs

Abrupt, downslurred nasal burr

TICK-TICK   *p. 556*

All year; most common call. Uniform but somewhat plastic, especially in rhythm. Does not carry far.

CHEWT   *p. 511*

Possibly all year. Apparently uniform and stereotyped. Function unknown; given during interactions, and in flight, possibly in alarm.

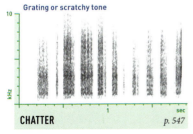

Grating or scratchy tone

CHATTER   *p. 547*

Possibly all year, by excited birds in interactions. Variable and plastic.

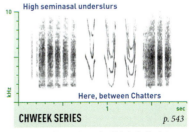

High seminasal underslurs

Here, between Chatters

CHWEEK SERIES   *p. 543*

Possibly all year, by excited birds during interactions, often with Chatters. Somewhat plastic.

BELTED KINGFISHER

Megaceryle alcyon

Found near water, almost anywhere suitable nest sites are available. Solitary, but often conspicuous and vocal, especially during defense of territory.

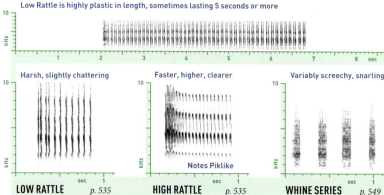

Low Rattle is highly plastic in length, sometimes lasting 5 seconds or more

Harsh, slightly chattering

Faster, higher, clearer

Variably screechy, snarling

Notes Piklike

LOW RATTLE *p. 535*   **HIGH RATTLE** *p. 535*   **WHINE SERIES** *p. 549*

All sounds variable and plastic. Most common call is Low Rattle, given all year in alarm and aggression. Semimusical High Rattle given by both sexes during and after confrontations, and as a pair greeting. Whine Series given mostly by male early in breeding season; short, often run into plastic Chitter.

RINGED KINGFISHER

Megaceryle torquata

Larger than Belted Kingfisher, with an even more massive bill. Range has expanded northward, reaching the U.S. in the mid-twentieth century.

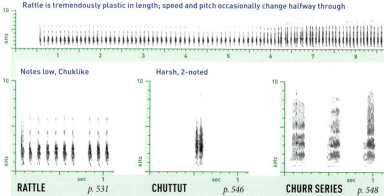

Rattle is tremendously plastic in length; speed and pitch occasionally change halfway through

Notes low, Chuklike

Harsh, 2-noted

RATTLE *p. 531*   **CHUTTUT** *p. 546*   **CHURR SERIES** *p. 548*

All sounds variable and plastic. Most common call is Rattle, given all year in alarm and aggression; often much lower and slower than in Belted Kingfisher, but some versions quite similar. Chuttut given in mild alarm or in long-distance flight. Churr Series given in interactions; rather soft and plastic.

WOODPECKERS (Family Picidae)

The woodpeckers are medium-sized to large birds well known for perching on the vertical surfaces of tree trunks and using their chisel-like bills to extract insect larvae from inside the wood. They are equally well known for drumming rapidly on resonant surfaces—dead limbs, telephone poles, houses, and even metal objects like chimney covers and transformer boxes—in order to communicate over long distances.

As far as is known, all sounds of woodpeckers are innate. Most vocal sounds are quite simple, with little individual or regional variation.

Identifying Woodpecker Species by Drum

Adult males and females of all North American woodpecker species drum, either in territorial advertisement, to strengthen the pair bond, or both. One species in the East, the Yellow-bellied Sapsucker, can be instantly recognized by its distinctive rhythm. Among other woodpeckers, the most important feature to listen to is the speed of the Drum.

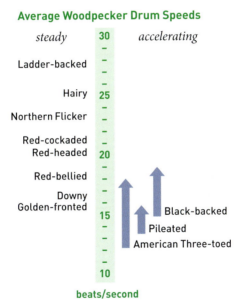

Average Woodpecker Drum Speeds

steady | 30 | *accelerating*

Ladder-backed

Hairy — 25

Northern Flicker

Red-cockaded
Red-headed — 20

Red-bellied

Downy
Golden-fronted — 15 — Black-backed
Pileated
American Three-toed

— 10

beats/second

Drum speed is variable, and there is much overlap between adjacent species in the table. However, it is often possible to separate the Drums of species that are far apart on the table, such as those of Hairy and Downy Woodpeckers.

Drumming vs. Tapping

In addition to drumming, most woodpeckers frequently engage in tapping, which is similar to drumming but generally softer, slower, and much more plastic. Tapping appears to be used mostly in short-range communication between members of a pair. The Red-cockaded Woodpecker occasionally drums its tongue on the surface of trees in mild excitement, and other species may do this as well; the resulting sound is said to resemble the rattle of a rattlesnake's tail.

GOLDEN-FRONTED WOODPECKER

Melanerpes aurifrons

Common and vocally conspicuous in mesquite scrub and open dry woodlands. Occasionally hybridizes with related Red-bellied Woodpecker where ranges overlap.

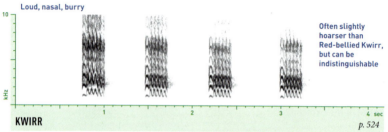

Loud, nasal, burry

Often slightly hoarser than Red-bellied Kwirr, but can be indistinguishable

KWIRR *p. 524*

All year, especially Feb.–May. Most common vocalization; often alternated with Drum as a territorial call. No equivalent of Red-bellied's Chuckle is known from Golden-fronted.

Kwirr and Drum often interspersed by territorial males

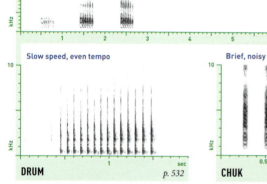

Slow speed, even tempo

DRUM *p. 532*

All year, especially Feb.–July. On average the slowest eastern woodpecker Drum, averaging about 15 notes/second.

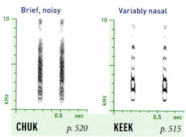

Brief, noisy

Variably nasal

CHUK *p. 520* **KEEK** *p. 515*

All year, in alarm or pair contact, often in long series. Chuk and Keek intergrade; each note usually shorter than Red-bellied Bark.

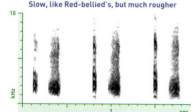

Slow, like Red-bellied's, but much rougher

CHUK-A SERIES *p. 558*

All year; in interactions with other Golden-fronteds. Function not well understood.

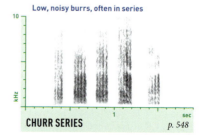

Low, noisy burrs, often in series

CHURR SERIES *p. 548*

All year; especially Feb.–May. Given in close contact, often by mated pairs. Averages finer than Red-bellied Woodpecker Churr.

Red-bellied Woodpecker

Melanerpes carolinus

Common in forests, particularly in the South, where in some places it is the most common woodpecker. Calls loudly and frequently.

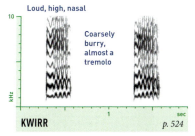

Loud, high, nasal

Coarsely burry, almost a tremolo

KWIRR   *p. 524*

All year, especially Feb.–July. Most common vocalization; often alternated with Drum as a territorial call.

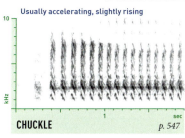

Usually accelerating, slightly rising

CHUCKLE   *p. 547*

All year, especially Feb.–July, in interactions. Infrequent. Faster, harsher, shorter than Northern Flicker Keek Series (2–3 seconds total).

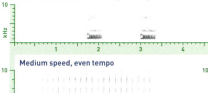

Kwirr and Drum often interspersed by territorial males

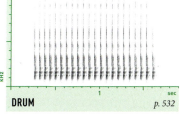

Medium speed, even tempo

DRUM   *p. 532*

All year, especially Feb.–July. Averages 18 notes/second. Often given in conjunction with Kwirr.

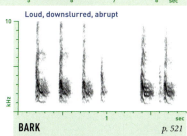

Loud, downslurred, abrupt

BARK   *p. 521*

All year, in pair contact and in alarm. Variable, but generally sharply downslurred, unlike Chuk of Golden-fronted, and rather screechy.

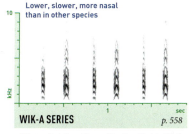

Lower, slower, more nasal than in other species

WIK-A SERIES   *p. 558*

All year, in interactions with other Red-bellieds. Function not well understood.

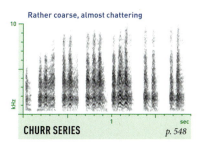

Rather coarse, almost chattering

CHURR SERIES   *p. 548*

All year; especially Feb.–May, in close contact, often by mated pairs.

RED-HEADED WOODPECKER

Melanerpes erythrocephalus

Locally common in forests and in open areas with a few mature trees. Often sits on snags or telephone poles, sallying out to catch insects on the wing.

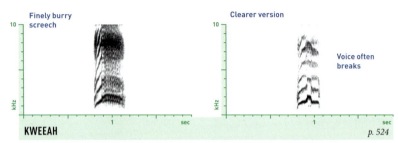

Finely burry screech

Clearer version

Voice often breaks

KWEEAH   p. 524

Mostly Feb.–July; often alternated with Drum. Variable and plastic. Higher and finer than Red-bellied Kwirr; often harsher.

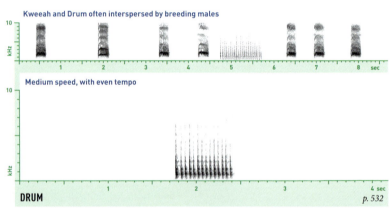

Kweeah and Drum often interspersed by breeding males

Medium speed, with even tempo

DRUM   p. 532

All year, but mostly Apr.–June; especially frequent during and immediately after intrusions by rivals into territory. Generally drums for shorter periods than other woodpeckers. Sometimes drums on manmade surfaces, including metal and ceramics.

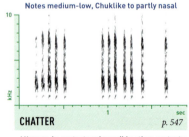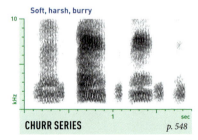

Notes medium-low, Chuklike to partly nasal

Soft, harsh, burry

CHATTER   p. 547

CHURR SERIES   p. 548

All year, in contact and possibly other contexts. Some frogs can sound very similar, especially Wood Frog and Gray Tree Frog.

All year, in aggressive interactions of various kinds.

Yellow-bellied Sapsucker

Sphyrapicus varius

Drills "wells" in trees, then returns to eat sap and insects. Spring wells are circular, in horizontal rows; summer wells are rectangular and vertically stacked.

Harsh, broken notes in slow series

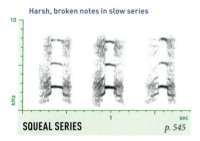

SQUEAL SERIES
p. 545

By both sexes, but especially males, Apr.–June; often with Drum. Like Red-headed Woodpecker Kweeah, but not burry; almost always in series.

Like Red-Bellied Woodpecker Chuckle, but lower, harsher, faster

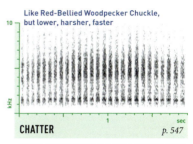

CHATTER
p. 547

Mostly early in breeding season, between members of a pair. Also in fluttering flight display and in response to intruders.

Drum often very long, later taps in couplets or Morse code–like pattern

Slow, irregular, decelerating

Taps often doubled so rapidly they sound single

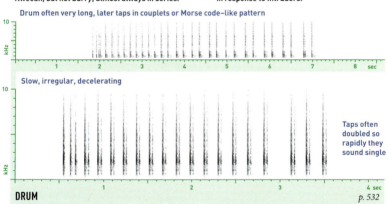

DRUM
p. 532

The most distinctive woodpecker Drum in the East, instantly recognizable by its extremely slow, decelerating rhythm, but length and exact pattern are highly plastic. By both sexes, but especially by unpaired males in spring.

Harsh Screeches in simple or couplet series

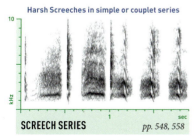

SCREECH SERIES
pp. 548, 558

Highly plastic, but usually harsh; soft or loud. Mostly in breeding season, in contact between mates and territorial squabbles.

Nasal downslur or overslur, sometimes broken

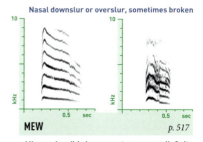

MEW
p. 517

All year, in mild alarm; most common call. Quite plastic; often noisy and/or broken, occasionally burry. Often in short series.

DOWNY WOODPECKER

Picoides pubescens

Widespread in woodlands and towns. Smaller and shorter-billed than Hairy Woodpecker; note differences in voice. Plumage varies regionally.

Rapid series of Pik notes, usually slightly downslurred

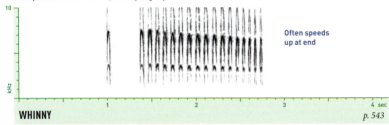

Often speeds up at end

WHINNY   *p. 543*

All year. Often follows Pik calls. Clearer than Hairy Woodpecker Rattle, and usually downslurred and accelerating, becoming slightly less musical at end. Beginning of Whinny can vary in speed; some versions begin with Kweek Series.

Drum relatively slow, but often repeated at short intervals (4 seconds or less)

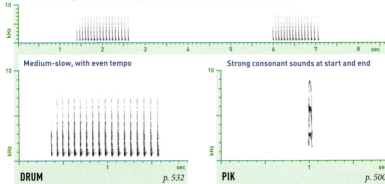

Medium-slow, with even tempo

Strong consonant sounds at start and end

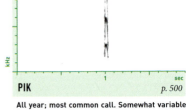

DRUM   *p. 532*

PIK   *p. 500*

All year, but especially winter and spring. Averages 16 notes/second; slow speed alone may be sufficient for identification in some regions.

All year; most common call. Somewhat variable in pitch, but fairly stereotyped. Averages lower than Hairy Peek, but sometimes similar.

Notes like Pik, but longer, often upslurred

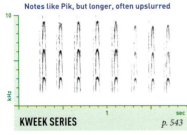

Soft and noisy

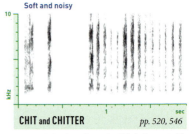

KWEEK SERIES   *p. 543*

CHIT and CHITTER   *pp. 520, 546*

All year, by excited birds, often in slow-flapping "butterfly" display flight. Varies from clear to noisy; can end in Whinny.

Mostly Feb.–June, by excited birds in courtship displays or disputes. Plastic, but fairly distinctive.

Hairy Woodpecker

Picoides villosus

Widespread, with a greater preference for dense forest and coniferous trees than Downy Woodpecker. Plumage varies regionally.

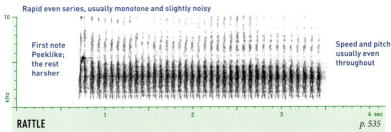

Rapid even series, usually monotone and slightly noisy

First note Peeklike; the rest harsher

Speed and pitch usually even throughout

RATTLE   *p. 535*

All year. Often follows Peek calls. Noisier, more monotone, and often longer than similar Downy Woodpecker Whinny. Agitated birds give a less noisy extended version, up to 10 seconds long, that may rise and fall in pitch, interspersed with fast series of Peeks.

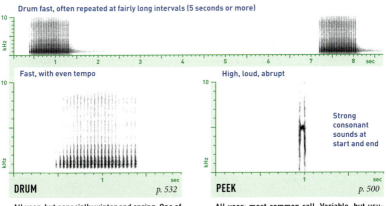

Drum fast, often repeated at fairly long intervals (5 seconds or more)

Fast, with even tempo

High, loud, abrupt

Strong consonant sounds at start and end

DRUM   *p. 532*

All year, but especially winter and spring. One of the fastest woodpecker Drums, averaging 25 notes/second.

PEEK   *p. 500*

All year; most common call. Variable, but usually distinctly higher than Downy Woodpecker Pik.

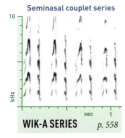
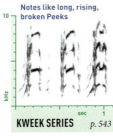
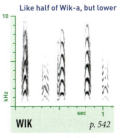

Seminasal couplet series

Notes like long, rising, broken Peeks

Like half of Wik-a, but lower

WIK-A SERIES   *p. 558*

Mostly Feb.–June, by excited birds in courtship displays or disputes.

KWEEK SERIES   *p. 543*

All year, from excited birds, often in slow-flapping display flight.

WIK   *p. 542*

All year. Variable; in pair greetings and low-grade disputes.

LADDER-BACKED WOODPECKER

Picoides scalaris

Common in deserts, thorn scrub, riparian woodlands, and areas dominated by cactus. Generally prefers drier habitats than Downy Woodpecker.

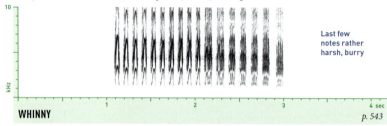

Rapid series of Piklike notes, becoming lower, slower, and rougher

Last few notes rather harsh, burry

WHINNY *p. 543*

All year; shorter version often given Mar.–May. Often given with Pik calls and Drums. Like Downy Woodpecker Whinny, but usually becomes distinctively rough in quality at end.

Drum fast, often repeated at medium to long intervals (5 seconds or more)

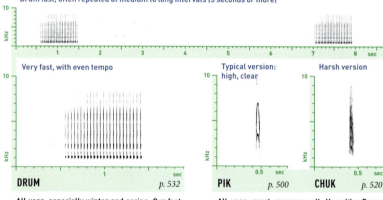

Very fast, with even tempo

Typical version: high, clear

Harsh version

DRUM *p. 532*

PIK *p. 500*

CHUK *p. 520*

All year, especially winter and spring. Our fastest woodpecker Drum on average, overlapping in speed with Hairy and Northern Flicker.

All year; most common call. Very like Downy Woodpecker Pik. Harsh versions apparently rare; function unknown.

Slightly squeaky series

Each note sounds upslurred, with consonant at start

Soft, low, noisy, short

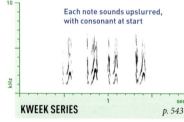

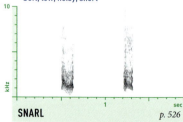

KWEEK SERIES *p. 543*

SNARL *p. 526*

Mostly Feb.–May, in aggressive encounters and in fluttering flight displays. Usually in short series; also occasionally single notes.

Mostly Feb.–May, in similar situations as Kweek, and often in conjunction with it. In some variants, notes are reportedly 2-syllabled.

Red-cockaded Woodpecker

Picoides borealis

Highly social, but rare and local in old-growth pine forests. Family groups breed cooperatively, drilling sap wells around the nest cavity to deter snakes.

Plastic, twittering series of Kriklike notes

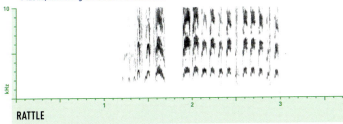

RATTLE   *p. 535*

All year, often by adults approaching the nest. Softer, less frequent, and more plastic than Rattle and Whinny calls of other woodpeckers; often irregular in rhythm and pitch, and sometimes also in quality, grading into peeping Twitters.

Rapid, even, usually soft

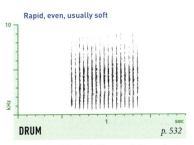

DRUM   *p. 532*

Mostly Mar.–May, by both sexes. Softer and more variable than other woodpecker Drums, used primarily for short-distance communication.

High, slightly nasal burr

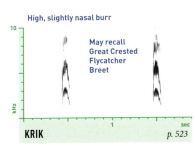

May recall Great Crested Flycatcher Breet

KRIK   *p. 523*

All year; most common call. Plastic; grades into Wik-a Series and Pip. Unique among North American woodpeckers.

Pips and Kriks often run into peeping Twitter (p. 536); may recall a flock of sandpipers

Squeaky couplet series

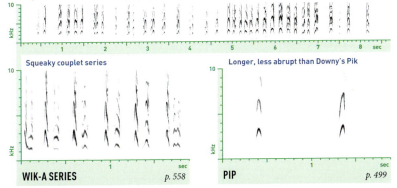

WIK-A SERIES   *p. 558*

By excited birds in interactions. High note sharper than in similar calls of other woodpeckers, more like "Pik-a."

Longer, less abrupt than Downy's Pik

PIP   *p. 499*

Plastic; grades into Krik. In group interactions, often runs into peeping Twitter, like Brown-headed Nuthatch's minus the nasal notes.

AMERICAN THREE-TOED WOODPECKER

Picoides dorsalis

Breeds in northern spruce forests; often forages for bark beetles by flaking off bark rather than drilling holes. Shares lack of fourth toe with Black-backed Woodpecker.

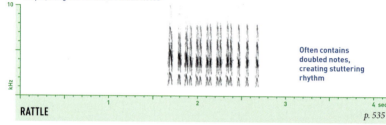

Rapid, irregular series of Piklike notes

Often contains doubled notes, creating stuttering rhythm

RATTLE — p. 535

All year, in threat and territorial displays; rather infrequent. Averages briefer and more plastic than Whinny and Rattle calls of other woodpeckers; stuttering rhythm distinctive when present. Last few notes may be lower, more churring.

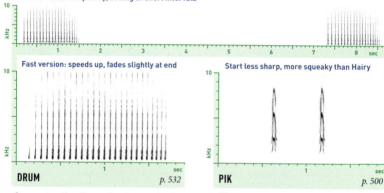

Drums rather frequently, at long or short intervals

Fast version: speeds up, fades slightly at end

DRUM — p. 532

Our most noticeably accelerating woodpecker Drum. Birds apparently give both fast and slow versions; some slow versions do not speed up.

Start less sharp, more squeaky than Hairy

PIK — p. 500

All year; most common call. Run into rapid series in high alarm, grading into Rattle.

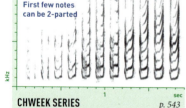

Fast series of nasal underslurs

First few notes can be 2-parted

CHWEEK SERIES — p. 543

Infrequent, mostly Feb.–May, in aggressive encounters with same or other species.

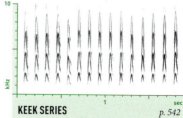

Notes like Pik but lower, more nasal

KEEK SERIES — p. 542

Possibly all year, by excited birds in a variety of situations, often between members of a mated pair.

BLACK-BACKED WOODPECKER

Picoides arcticus

Breeds in northern coniferous forests; specializes in wood-boring beetles. Like Three-toed, often irrupts into areas where fire or beetle outbreaks have killed many trees.

Downslurred, accelerating; quickly becomes harsh

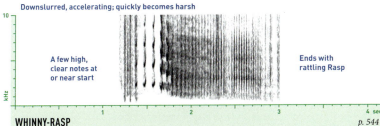

A few high, clear notes at or near start

Ends with rattling Rasp

WHINNY-RASP   *p. 544*

All year, by both sexes, mostly in aggressive encounters with same or other species; possibly also in establishing territories in spring. Plastic, but always distinctive. Abbreviated versions may include just the opening Whinny notes, or just the ending Rasp.

Drums rather frequently, at long or short intervals

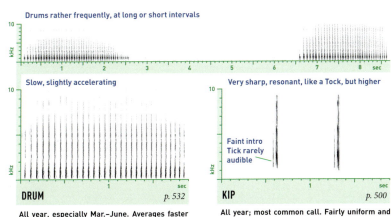

Slow, slightly accelerating

Very sharp, resonant, like a Tock, but higher

Faint intro Tick rarely audible

DRUM   *p. 532*

KIP   *p. 500*

All year, especially Mar.–June. Averages faster and more uniform than Three-toed Drum, with less acceleration, but difficult to separate.

All year; most common call. Fairly uniform and stereotyped; unique among woodpeckers.

Kip, then 1–4 quick Rasps

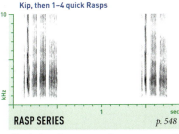

Notes harsh, Kiplike

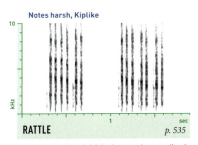

RASP SERIES   *p. 548*

RATTLE   *p. 535*

Mostly Apr.–June, apparently as a summons to mate. Distinctive, but compare Three-toed Rattle.

Apparently given in high alarm; sole recording is of captive bird. Rhythm irregular.

Northern Flicker

Colaptes auratus

Common and familiar. "Yellow-shafted" form of the East, "Red-shafted" form of the West mix freely in the Great Plains. All populations sound similar.

Long steady series on same pitch; starts slightly softer, lower, slower

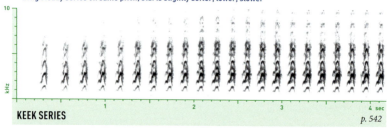

KEEK SERIES p. 542

All year, especially Mar.–June, mostly by unmated males, but also by females. Often long, up to 20 seconds. Usually given from a prominent perch along with Drum. Compare Pileated Woodpecker Keek Series.

Territorial males often alternate Keek Series with Drum, sometimes without pauses

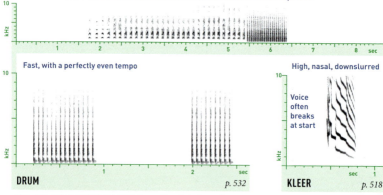

Fast, with a perfectly even tempo

DRUM p. 532

Mar.–Aug. Can be indistinguishable from other fast-drumming species; listen for Keek Series or other calls to confirm identification. Often given on metal to increase volume.

High, nasal, downslurred

Voice often breaks at start

KLEER p. 518

All year; most common call. Variable.

High nasal couplet series: Wik louder, upslurred; second note low, brief

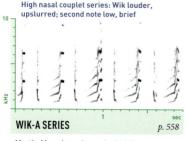

WIK-A SERIES p. 558

Mostly Mar.–June, by excited birds in courtship or in disputes. Variable; some versions omit second note, becoming Wik Series (p. 542).

Soft, low, somewhat musical

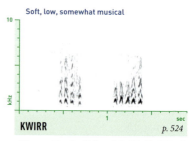

KWIRR p. 524

All year. Recalls Kwirr of Red-bellied Woodpecker, but usually given in flight and only audible at close range.

PILEATED WOODPECKER

Dryocopus pileatus

Our largest surviving woodpecker, the size of a crow, with distinctive plumage. Requires forests with large-diameter trees.

Keek Series can last 60 seconds or more

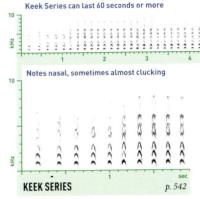

Notes nasal, sometimes almost clucking

KEEK SERIES
p. 542

All year, by both sexes, in alarm and in interactions, often in flight. Slower, often lower, and more plastic than Northern Flicker Keek Series.

Higher, faster, shorter than Keek Series

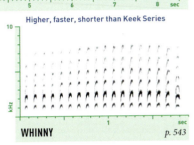

WHINNY
p. 543

Loud; rarely longer than 2 seconds. Last note typically lower. In spring, given repeatedly from high perch, suggesting territorial function.

Rather slow; slightly faster and softer at end

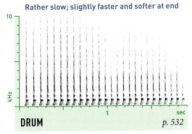

DRUM
p. 532

All year, but especially in early spring. Typically loud and far-carrying, acceleration subtle. Length variable, up to 3 full seconds.

Slow series of up- or underslurred nasal notes

Soft to medium-loud

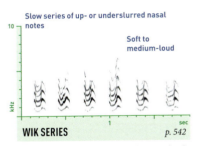

WIK SERIES
p. 542

Equivalent to Wik-a Series of other woodpeckers. Given in close pair interactions; louder versions may be given in disputes or in flight.

IVORY-BILLED WOODPECKER

Campephilus principalis

Extinct resident of southern virgin forests. Last undisputed U.S. sightings in 1938. Gave distinctive nasal Kent call like Red-breasted Nuthatch Yank, but louder and more abrupt. Drum doubled, bill striking tree only twice. Blue Jay Jeers filtered through trees can sound shockingly like distant Kent; other woodpeckers sometimes double-tap.

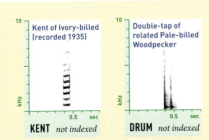

Kent of Ivory-billed (recorded 1935)

Double-tap of related Pale-billed Woodpecker

KENT *not indexed*

DRUM *not indexed*

FALCONS AND CARACARAS (Family Falconidae)

Long thought to be closely related to hawks and eagles, falcons and caracaras are now recognized as a very distinct family on the basis of genetic evidence. As far as is known, all vocalizations are innate.

Falcons (genus *Falco*) are built for aerial hunting. Their long, pointed wings and long tails give them great speed and agility, perfect for chasing down birds in flight, though they will also take a variety of other prey. For centuries, falconers have taken advantage of the hunting skills of captive birds. The North American species are all closely related and have very similar vocal repertoires.

Caracaras are more terrestrial than falcons, with longer legs and more rounded wings. Our sole species is a slow flier that feeds mostly on carrion.

PEREGRINE FALCON

Falco peregrinus

Legendary for aerial speed. Once rare and endangered, but now recovering. Many populations descend in part from released birds.

Scream Series ranges from nearly clear to quite screechy

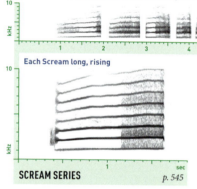

Each Scream long, rising

SCREAM SERIES   *p. 545*

All year, in a variety of contexts, but especially Mar.–June. Plastic; different forms may serve slightly different functions.

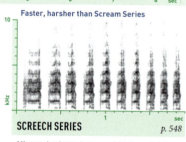

Faster, harsher than Scream Series

SCREECH SERIES   *p. 548*

All year, in alarm, but especially in nest defense. Rather plastic; may increase in pitch and speed in accordance with agitation level.

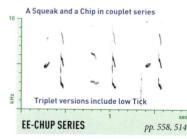

A Squeak and a Chip in couplet series

Triplet versions include low Tick

EE-CHUP SERIES   *pp. 558, 514*

Mostly Mar.–May, in courtship and nest defense. Quite plastic; in agitation, speeds up and syllables merge into Tsooklike notes.

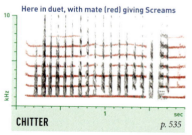

Here in duet, with mate (red) giving Screams

CHITTER   *p. 535*

Mostly Mar.–May, by males during copulation; also by either sex in high alarm. All our falcons give similar calls, but recordings are few.

PRAIRIE FALCON

Falco mexicanus

Breeds on mountain cliffs and prairie bluffs in the West; wanders farther east in winter. In flight, note dark feathers where the wing meets the body.

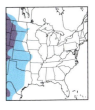

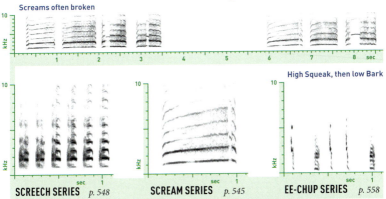

Screams often broken

High Squeak, then low Bark

SCREECH SERIES *p. 548*   **SCREAM SERIES** *p. 545*   **EE-CHUP SERIES** *p. 558*

Screech Series and Scream Series given all year, but primarily Mar.–May, near nest. Screech Series given mostly in alarm, Scream Series in courtship, Ee-chup Series in courtship display from prospective nest site on cliff ledge. All vocalizations plastic and similar to sounds of Peregrine Falcon.

GYRFALCON

Falco rusticolus

The largest falcon. Rare in winter in our area. Plumage ranges from nearly pure white (mostly Greenland breeders) to very dark; most birds are gray.

Scream Series slower than in other falcons

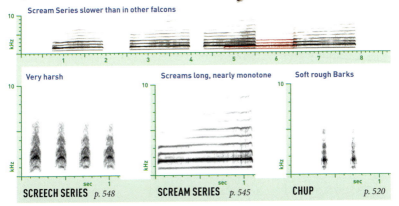

Very harsh

Screams long, nearly monotone

Soft rough Barks

SCREECH SERIES *p. 548*   **SCREAM SERIES** *p. 545*   **CHUP** *p. 520*

Vocalizes primarily on breeding grounds. Screech Series, given during aggressive encounters, is most likely call to be heard in winter. Scream Series and Chup associated mostly with courtship and nesting. Chup usually given in rapid plastic series, grading into louder, more regular Screech Series.

MERLIN

Falco columbarius

Slightly larger than American Kestrel, with white bands in tail. Breeds in open forests; in winter, found along woodland edges and in open areas.

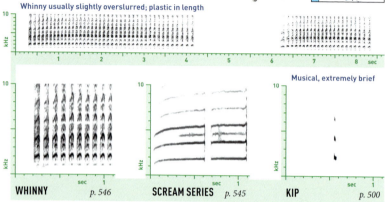

Whinny usually slightly overslurred; plastic in length

Musical, extremely brief

| WHINNY | p. 546 | SCREAM SERIES | p. 545 | KIP | p. 500 |

Whinny given all year by both sexes in courtship and aggression; most common call. Can resemble Pileated Woodpecker Whinny, but generally screechier. Scream Series given by juveniles and females begging for food. Kip given in courtship and pair contact; male version slightly higher.

APLOMADO FALCON

Falco femoralis

Once more widespread, but extirpated from the U.S. in the twentieth century. Small reintroduced population found in south Texas coastal savannah.

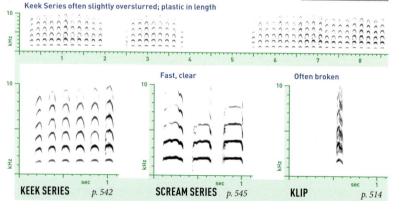

Keek Series often slightly overslurred; plastic in length

Fast, clear

Often broken

| KEEK SERIES | p. 542 | SCREAM SERIES | p. 545 | KLIP | p. 514 |

Keek Series given all year in aggression; distinctively clear, nasal. Only slightly faster than Scream Series, given by female near nest and by male in copulation. Klip given all year in pair contact and other interactions. Highly plastic; ranges from unbroken, nasal Kip to broken, squeaky Klee (p. 514).

AMERICAN KESTREL

Falco sparverius

Our smallest, most common, and most colorful falcon, found in various open habitats; often seen on roadside telephone wires. Nests in tree cavities.

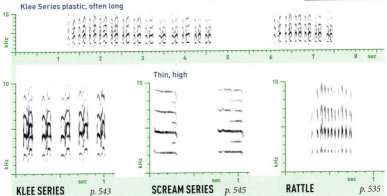

Klee Series plastic, often long

Thin, high

| KLEE SERIES | p. 543 | SCREAM SERIES | p. 545 | RATTLE | p. 535 |

Klee Series given all year, in alarm and in aerial displays; most common call by far. Notes can sound 1- or 2-syllabled. Scream Series given in courtship, in high alarm, and by begging fledglings; can be upslurred or monotone. Rattles heard mostly during pair interactions. All calls intergrade.

CRESTED CARACARA

Caracara cheriway

A unique raptor, local in open country. Eats carrion as well as live prey, which it sometimes steals from other birds. Soars infrequently.

In display, throws head back, giving Croak Phrase with odd speechlike inflection (pp. 527, 577)

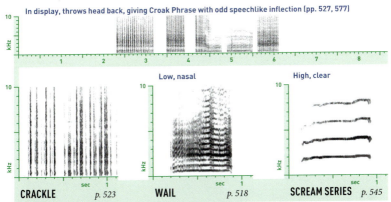

Low, nasal

High, clear

| CRACKLE | p. 523 | WAIL | p. 518 | SCREAM SERIES | p. 545 |

Mostly silent except during breeding season, or when groups gather to feed. Crackle given in agitation, Croak Phrase in aggressive displays, with head thrown back. Low, harsh Wails given rarely by adults in high alarm. High, keening Scream Series given frequently by young birds.

PARROTS AND PARAKEETS (Family Psittacidae)

Our only native parrot, the Carolina Parakeet, became extinct in 1918. Now, however, flocks of escaped or released parrots again roam parts of the region, especially south Florida, where dozens of species have been seen in the wild. Most nest in cavities, though Monk Parakeets build elaborate stick nests.

Parrots are among the most intelligent birds, with complex social interactions and vocal learning abilities. Although parrots are famous for imitation in captivity, wild parrots rarely imitate environmental sounds. Their natural voices are loud and varied, characterized by Yelps, Shrieks, Squeals, and Chatters, often in chorus. Identification by ear often requires experience; only a fraction of each species' repertoire is shown here.

BUDGERIGAR *Melopsittacus undulatus*

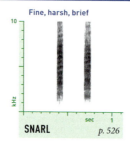

Fine, harsh, brief

SNARL *p. 526*

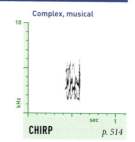

Complex, musical

CHIRP *p. 514*

Formerly in Florida; escapes elsewhere. Native to Australia. Calls frequent but rather soft. Chirps of flock merge into musical warbles.

MONK PARAKEET *Myiopsitta monachus*

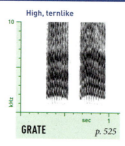

High, ternlike

GRATE *p. 525*

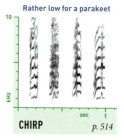

Rather low for a parakeet

CHIRP *p. 514*

Local in cities in the East. Native to S. America. Most sounds are versions of the Grate, but also gives Yelps and squealing Chirps.

GREEN PARAKEET *Psittacara holochlorus*

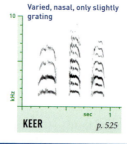

Varied, nasal, only slightly grating

KEER *p. 525*

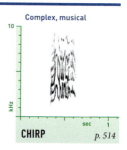

Complex, musical

CHIRP *p. 514*

S. Texas; some may be wild birds from native Mexico. Flocks give constant nasal notes, grading into excited squealing Chirps.

Mitred Parakeet *Psittacara mitratus*

Local in S. Florida. Most common sound is laughing series of Yanks, like loud Red-breasted Nuthatch.

Short nasal notes

YANK *p. 517*

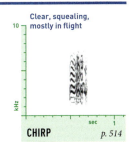

Clear, squealing, mostly in flight

CHIRP *p. 514*

Red-masked Parakeet *Psittacara erythrogenys*

Local in S. Florida. Native to Peru and Ecuador. All sounds like those of Mitred, but on average lower and slightly harsher.

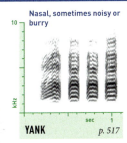

Nasal, sometimes noisy or burry

YANK *p. 517*

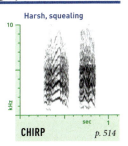

Harsh, squealing

CHIRP *p. 514*

White-eyed Parakeet *Psittacara leucophthalmus*

Local in S. Florida. Native to S. America. Calls average higher, harsher, screechier than those of our other parakeets.

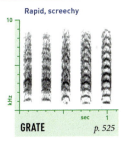

Rapid, screechy

GRATE *p. 525*

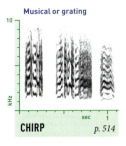

Musical or grating

CHIRP *p. 514*

Blue-crowned Parakeet *Thectocercus acuticaudatus*

Local in s. and cen. Florida. Native to S. America. Grates average long, high, and screechy. Chirps relatively rare.

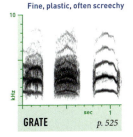

Fine, plastic, often screechy

GRATE *p. 525*

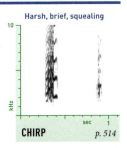

Harsh, brief, squealing

CHIRP *p. 514*

NANDAY PARAKEET *Aratinga nenday*

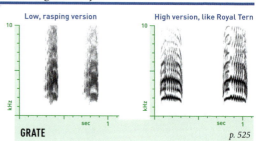

Low, rasping version

High version, like Royal Tern

Local in cen. Florida. Native to S. America. Voice distinctive; all sounds are ternlike Grates, without squealing or chirping.

GRATE

p. 525

WHITE-WINGED PARAKEET *Brotogeris versicolurus*

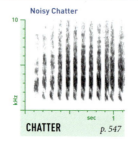

Noisy Chatter

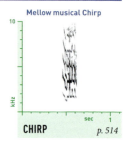

Mellow musical Chirp

Local in S. Florida. Native to S. America. This and the next species are our only parakeets that give noisy Chatters.

CHATTER   p. 547

CHIRP   p. 514

YELLOW-CHEVRONED PARAKEET *Brotogeris chiriri*

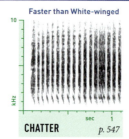

Faster than White-winged

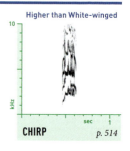

Higher than White-winged

Local in S. Florida. Native to S. America. Sounds from native range shown. In Florida, some sound like White-winged.

CHATTER   p. 547

CHIRP   p. 514

CHESTNUT-FRONTED MACAW *Ara severus*

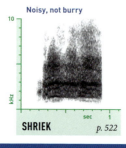

Noisy, not burry

Odd, mellow, nasal

Local in S. Florida. Native to S. America. Voice distinctive. Warbles sometimes repeated in complex series, or in chorus.

SHRIEK   p. 522

WARBLE   p. 572

Red-crowned Parrot *Amazona viridigenalis*

S. Texas; some may be wild birds from native Mexico. Voice hugely varied, as in all *Amazona*. Most distinctive calls are shown.

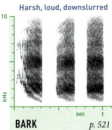

Harsh, loud, downslurred

BARK *p. 521*

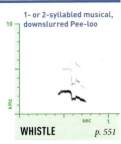

1- or 2-syllabled musical, downslurred Pee-loo

WHISTLE *p. 551*

Lilac-crowned Parrot *Amazona finschi*

Rare in S. Florida. Native to W. Mexico. Repertoire includes complex musical Chirps in flight, soft ravenlike Croaks on perch.

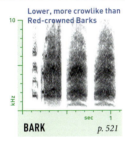

Lower, more crowlike than Red-crowned Barks

BARK *p. 521*

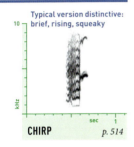

Typical version distinctive: brief, rising, squeaky

CHIRP *p. 514*

White-fronted Parrot *Amazona albifrons*

Rare in S. Florida. Native to Cen. America. Flock in chorus recalls screaming chimpanzees. Chak-Chak distinctive.

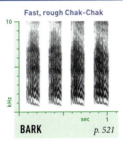

Fast, rough Chak-Chak

BARK *p. 521*

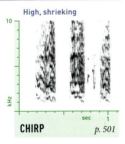

High, shrieking

CHIRP *p. 501*

Orange-winged Parrot *Amazona amazonica*

Local in S. Florida. Native to S. America. Voice averages higher and clearer than in other *Amazona*, with few harsh notes.

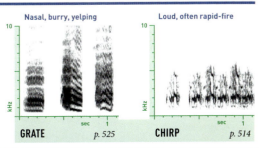

Nasal, burry, yelping

GRATE *p. 525*

Loud, often rapid-fire

CHIRP *p. 514*

TYRANT FLYCATCHERS (Order Passeriformes, Family Tyrannidae)

The tyrant flycatchers are small to medium-sized birds that sally out from perches to capture insects on the wing. "Tyrant" describes the highly aggressive behavior of certain species, especially kingbirds, which will readily defend territories against birds many times their size. In some species, even members of mated pairs may act aggressively toward one another. Because many flycatchers look very similar, vocalizations can be important in identifying them.

Flycatchers apparently do not learn any of their vocalizations; all sounds are innate. Sounds tend to be quite uniform and stereotyped, with little regional variation. Most species have extensive and complex vocal repertoires.

Dawn Songs and Day Songs

Flycatcher songs tend to be constructed of short phrases, often combined with one another according to strict syntactic rules that allow only certain combinations. For example, the Brown-crested Flycatcher's song phrases A, B, and C are never given alone, but only in combinations AC, ACC, BC, and BCC. Many species have a **Dawn Song** that males sing during the hour or so prior to sunrise during the breeding season. In some species the Dawn Song and the Day Song are different; in some species they are the same.

Flight Songs

Several flycatcher species—including the phoebes and Acadian, Willow, Yellow-bellied, and Least Flycatchers—have an additional type of complex song given very infrequently by breeding males in display flights. This **Flight Song** tends to be a precise, stereotyped recombination of calls and song phrases that are also given in other contexts. It may be given more often at dusk, or possibly upon detection of aerial predators. In some flycatchers, such as the kingbirds, the song given during flight displays is much like the typical Day Song or Dawn Song.

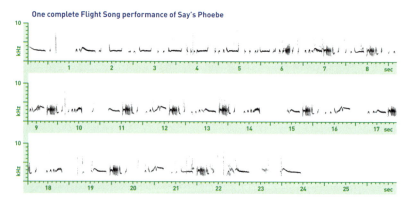

One complete Flight Song performance of Say's Phoebe

Other Sounds

Most flycatchers give several kinds of calls. Most kingbirds perform pair duets, albeit rather poorly sychronized ones; in some kingbirds, the male and the female duet calls are quite different. In some flycatchers, such as phoebes, males court females with unique call types given during "nest site showing" displays. All flycatchers snap their bills loudly when in pursuit of flying insects; many species also snap their bills in agitation.

Northern Beardless-Tyrannulet

Camptostoma imberbe

Nests in riparian and live oak woods, in clumps of Spanish moss and ball moss. Like a small *Empidonax*, but note dark eyeline and broken eye-ring.

Dawn Song (p. 560): "Pip... Peer... PEER-peeper"

3–9 slightly downslurred whistles

Series may be monotone or slightly downslurred

First notes often longest, most monotone

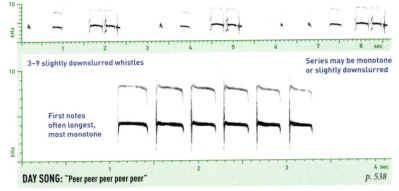

DAY SONG: "Peer peer peer peer peer"

p. 538

Mostly Mar.–Aug. Rather uniform and stereotyped, but some birds give more plastic versions in which notes break. Day Song may have territorial function; sometimes given by females. Dawn Song reportedly given only by males prior to sunrise in breeding season.

Notes usually burry, sometimes in series

INTERACTION CALLS

p. 574

Mostly May–Aug., in interactions. Highly plastic; all pitch patterns occur. Often mixed with various other calls by excited birds.

High semimusical trill, often downslurred

Here, with another bird giving Peers

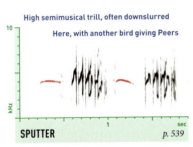

SPUTTER

p. 539

Mostly May–Aug., by both sexes, apparently in contact and some courtship situations. Plastic; some versions longer, more rattling.

Like Peer, but 2-syllabled, breaking downward at end

PEE-UK

p. 551

Simple downslurred whistle, like single note of Day Song

PEER

p. 506

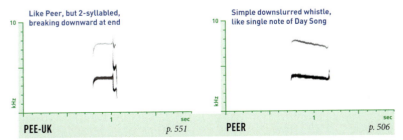

All year; most common calls. Pee-uk reportedly given only by female, while Peer is given by both sexes. Not known to what extent the 2 variants may differ in function. Both calls given at intervals while foraging; Peer also in response to playback of Dawn Song.

Western Wood-Pewee

Contopus sordidulus

Common in western pine forests and riparian woodlands. Nearly identical to Eastern Wood-Pewee, but voice differs; the two are not known to interbreed.

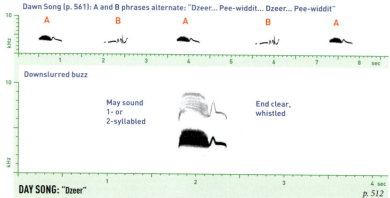

Dawn Song (p. 561): A and B phrases alternate: "Dzeer... Pee-widdit... Dzeer... Pee-widdit"

A   B   A   B   A

Downslurred buzz

May sound 1- or 2-syllabled

End clear, whistled

DAY SONG: "Dzeer"   *p. 512*

All year, but mostly May–Aug. Variable but stereotyped and highly distinctive. Presumed to be given mostly by males, in territorial advertisement and communication with mate. Dawn Song consists of Dzeer alternated with clear, whistled Pee-widdit phrase. Only Dzeer given during day; Dawn Song sometimes at dusk.

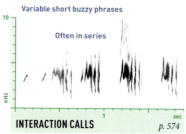

Variable short buzzy phrases

Often in series

INTERACTION CALLS   *p. 574*

Mostly May–Sept. A catch-all category of plastic sounds given in interactions, sometimes in flight; some versions rhythmic, repeating.

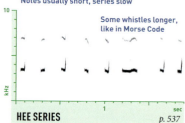

Notes usually short, series slow

Some whistles longer, like in Morse Code

HEE SERIES   *p. 537*

Mostly May–Aug. Function unclear, but apparently given in agitation; associated with Interaction Calls and Day Song.

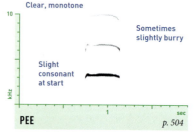

Clear, monotone

Sometimes slightly burry

Slight consonant at start

PEE   *p. 504*

All year; often given in migration. Plastic, grading into Dzree; occasionally slightly upslurred. Compare Eastern Wood-Pewee calls.

Upslurred or overslurred buzzy note

Whistled start

DZREE   *p. 512*

All year, in mate contact and nest defense; most common call. Slightly plastic. Similar to Day Song and often confused with it.

EASTERN WOOD-PEWEE

Contopus virens

Common in eastern deciduous forests, open pine forests, and other wooded habitats. Larger than an *Empidonax*, with longer wings and no eye-ring.

Dawn Song (p. 561): "C" phrase separates alternating "A" and "B" phrases

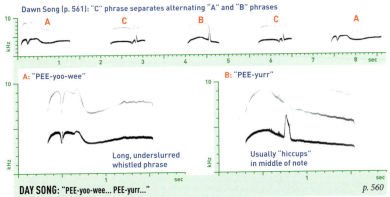

A: "PEE-yoo-wee"

Long, underslurred whistled phrase

B: "PEE-yurr"

Usually "hiccups" in middle of note

DAY SONG: "PEE-yoo-wee... PEE-yurr..."   *p. 560*

All year, but mostly May–Aug. Fairly uniform and stereotyped. Day Song consists of 2 phrase types (A and B) at long intervals; in most individuals, phrase A is 3 to 4 times more common than phrase B. Dawn Song, heard only prior to sunrise or at dusk, has a third phrase ("Pyur-piteet") and shorter pauses.

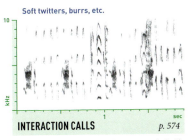

Soft twitters, burrs, etc.

INTERACTION CALLS   *p. 574*

Mostly May–Sept. A catch-all category of plastic sounds given in interactions, sometimes in flight; some versions rhythmic, repeating.

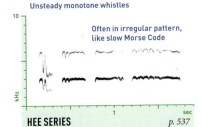

Unsteady monotone whistles

Often in irregular pattern, like slow Morse Code

HEE SERIES   *p. 537*

Mostly May–Aug. Function unclear, but apparently given in agitation; associated with Interaction Calls and Day Song.

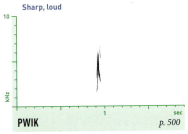

Sharp, loud

PWIK   *p. 500*

Rather infrequent; function unclear, but often accompanied by interaction calls. Compare Pip of Alder Flycatcher.

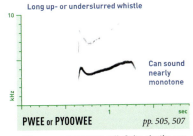

Long up- or underslurred whistle

Can sound nearly monotone

PWEE or PYOOWEE   *pp. 505, 507*

All year; most common call. Quite plastic; generally longer than Yellow-bellied Flycatcher Pwee, but some overlap.

OLIVE-SIDED FLYCATCHER

Contopus cooperi

Larger than wood-pewees, with larger head and shorter tail. Breeds in coniferous forests; frequently perches at the very top of tall dead snags.

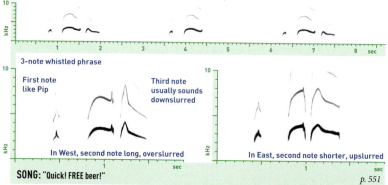

Possible Dawn Song from California: third note occasionally omitted

3-note whistled phrase

First note like Pip

Third note usually sounds downslurred

In West, second note long, overslurred

In East, second note shorter, upslurred

SONG: "Quick! FREE beer!"

p. 551

Subtle but consistent geographic variation; songs from Rockies and Pacific Coast slightly lower than those from Alaska east. Given all day; apparently no difference between Day Song and Dawn Song in many individuals, but in one dawn recording, third note is occasionally omitted, creating two different phrase types.

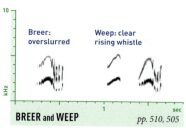

Breer: overslurred

Weep: clear rising whistle

BREER and WEEP

pp. 510, 505

Mostly May–July, in aggressive interactions. Breers and Weeps rather plastic, often mixed together in variable order.

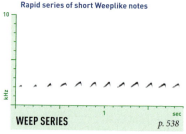

Rapid series of short Weeplike notes

WEEP SERIES

p. 538

Mostly May–July, in interactions, often mixed with other calls. Plastic.

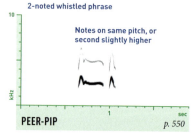

2-noted whistled phrase

Notes on same pitch, or second slightly higher

PEER-PIP

p. 550

Mostly May–July, repeated regularly for long intervals or mixed in with Pip Series, which it grades into. Function unclear.

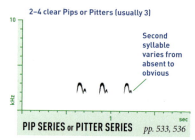

2–4 clear Pips or Pitters (usually 3)

Second syllable varies from absent to obvious

PIP SERIES or PITTER SERIES

pp. 533, 536

All year; most common call, in pair contact. Excited birds can give long series. Pitch somewhat variable; Pips more common than Pitters.

LEAST FLYCATCHER

Empidonax minimus

Breeds in deciduous woods. Like other members of the genus *Empidonax*, best identified by voice. A small, compact, and rather dull-plumaged "Empid."

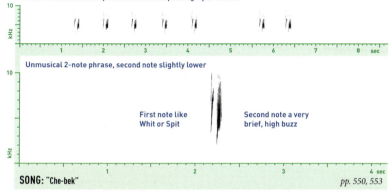

Che-bek sometimes repeated at rates surpassing 1 per second

Unmusical 2-note phrase, second note slightly lower

First note like Whit or Spit

Second note a very brief, high buzz

SONG: "Che-bek"   *pp. 550, 553*

Mostly Apr.–Aug., by males, but at least some singers are female. Highly uniform and stereotyped. Begins at dawn, but given all day. Higher and quicker than similar notes of other Empids, and often given at a far higher rate. Flight Song infrequent; includes rapid Che-beks with Pweew notes appended (p. 574).

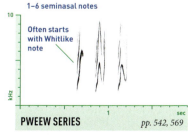

Quick downslurred burr

BURRT   *p. 510*

Often introduces Burrt

TWITTER   *p. 535*

A complex of rather plastic vocalizations, given mostly on breeding grounds, in interactions between males and females or adults and young. Also sometimes by adults upon landing after a brief flight. Rather soft; does not carry far.

1–6 seminasal notes

Often starts with Whitlike note

PWEEW SERIES   *pp. 542, 569*

Mostly in breeding season, during aggressive defense of nest or territory, often with bill snaps; also in lead-up to Flight Song.

Rapid, flicking upslur

WHIT   *p. 496*

All year; most common call. In mild alarm and pair contact. Less frequent on fall migration.

YELLOW-BELLIED FLYCATCHER

Empidonax flaviventris

Breeds in moist evergreen forests and muskeg bogs. Rather brightly colored for an Empid: greenish above and yellowish below, particularly on throat.

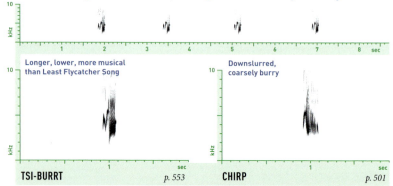

Tsi-burrt often repeated at short intervals, but averages slower than Least Flycatcher Song

Longer, lower, more musical than Least Flycatcher Song

Downslurred, coarsely burry

TSI-BURRT   *p. 553*

CHIRP   *p. 501*

Two different Song forms apparently serve different functions. 2-syllabled Tsi-burrt is highly uniform and stereotyped, given by breeding males mostly before pairing, including before dawn. 1-syllabled Chirp is more plastic, given more often by mated birds; may recall Henslow's Sparrow Song.

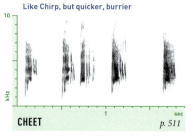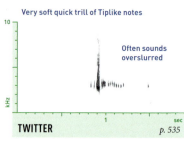

Like Chirp, but quicker, burrier

Very soft quick trill of Tiplike notes

Often sounds overslurred

CHEET   *p. 511*

TWITTER   *p. 535*

Rather plastic; usually sounds 1-syllabled. Given rapidly during aggressive encounters on breeding grounds.

Plastic; does not carry far. Sometimes given between Tsi-burrt Songs, or in interactions, but also occasionally by lone birds.

Overslurred

Downslurred

Upslurred

PSEEP   *p. 500*

PEER   *p. 506*

PWEE   *p. 505*

A confusing complex of contact calls, variable but fairly stereotyped. Pwee is most common call of breeders and migrants; usually shorter than Eastern Wood-Pewee Pwee. Pseep, Peer, and an underslurred Pewy (p. 511) are variants given mostly on wintering grounds; Pseep is very like Acadian Flycatcher's.

ACADIAN FLYCATCHER

Empidonax virescens

A rather large Empid of the interior of mature deciduous forests, with strong green tones above and relatively bright white underparts.

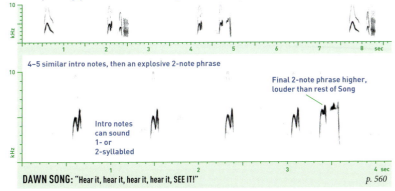

Complex Song (p. 561): Recombines virtually all calls and Song notes into several types of phrases

4–5 similar intro notes, then an explosive 2-note phrase

Final 2-note phrase higher, louder than rest of Song

Intro notes can sound 1- or 2-syllabled

DAWN SONG: "Hear it, hear it, hear it, hear it, SEE IT!"   *p. 560*

Mostly Apr.–Aug.; uniform and stereotyped. Dawn Song often starts before first light and continues almost without pause for up to an hour; very rare later in the day. Complex Song (top) given at any hour, mostly after aggressive territorial squabbles. Also has an even more complex Flight Song (p. 574).

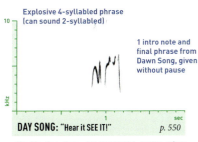

Explosive 4-syllabled phrase (can sound 2-syllabled)

1 intro note and final phrase from Dawn Song, given without pause

DAY SONG: "Hear it SEE IT!"   *p. 550*

Mostly Apr.–Aug., by territorial males after dawn, generally at fairly long intervals, often with soft Twitters in between. Unmistakable.

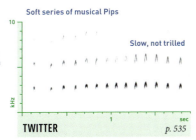

Soft series of musical Pips

Slow, not trilled

TWITTER   *p. 535*

Mostly Apr.–Aug., often in between Day Songs. Speed variable, sometimes irregular. May recall sound of Mourning Dove wings.

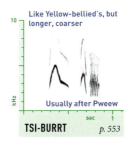

Like Yellow-bellied's, but longer, coarser

Usually after Pweew

TSI-BURRT   *p. 553*

Mostly Apr.–Aug., especially by female near nest, but also by males.

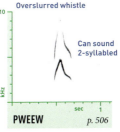

Overslurred whistle

Can sound 2-syllabled

PWEEW   *p. 506*

Apr.–Aug. Grades into Pseep, but usually distinct. Function little known.

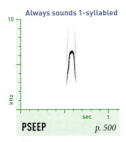

Always sounds 1-syllabled

PSEEP   *p. 500*

All year; most common call. Slightly variable, but generally distinctive.

ALDER FLYCATCHER

Empidonax alnorum

Nearly identical to Willow; the two were long considered one species, "Traill's Flycatcher." Breeds in northern willow and alder swamps and wet thickets.

Song: Single phrase (Free-BEE-ah) repeated over and over

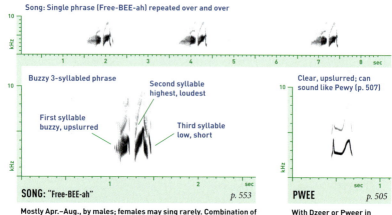

Buzzy 3-syllabled phrase

First syllable buzzy, upslurred

Second syllable highest, loudest

Third syllable low, short

SONG: "Free-BEE-ah" *p. 553*

Mostly Apr.–Aug., by males; females may sing rarely. Combination of buzzy, upslurred first syllable and 3-syllabled structure rule out all sounds of Willow Flycatcher, but final syllable can be faint.

Clear, upslurred; can sound like Pewy (p. 507)

PWEE *p. 505*

With Dzeer or Pweer in agitation, or with Song. Willow has rare similar call.

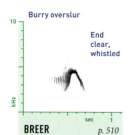

Burry overslur

End clear, whistled

BREER *p. 510*

Mostly May–Aug. Often repeated from a perch, like song. Function unclear.

Buzzy, downslurred

Can sound 2-syllabled (Ka-dzeer)

DZEER *p. 512*

Mostly Apr.–Aug., in aggression, often with Twitter or Pweer. Plastic.

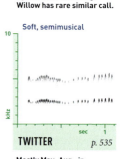

Soft, semimusical

TWITTER *p. 535*

Mostly May–Aug., in aggression, often introducing Dzeer. Plastic.

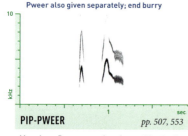

Pweer also given separately; end burry

PIP-PWEER *pp. 507, 553*

May–Aug. Pweer may be given without other calls, or with Dzeer or Twitter in agitation. Pip-pweer combination recalls Willow's Pete's-Beer.

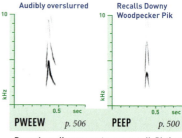

Audibly overslurred

PWEEW *p. 506*

Recalls Downy Woodpecker Pik

PEEP *p. 500*

Peep given all year; most common call. Distinctive, uniform, and stereotyped. Pweew a plastic, infrequent intergrade between Peep and Pweer.

WILLOW FLYCATCHER

Empidonax traillii

A brownish Empid with a faint eye-ring. Best separated from Alder Flycatcher by voice. Prefers drier habitats, but the two sometimes breed side by side.

Song: 3 phrases (A, B, C) in any order

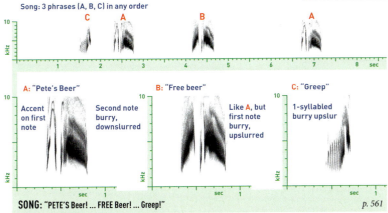

A: "Pete's Beer"
Accent on first note
Second note burry, downslurred

B: "Free beer"
Like A, but first note burry, upslurred

C: "Greep"
1-syllabled burry upslur

SONG: "PETE'S Beer! ... FREE Beer! ... Greep!"   *p. 561*

Mostly Apr.–Aug., by males, but females occasionally sing. Three-part syntax distinctive among eastern Empids; A, B, and C phrases rarely given outside context of full Song. Song begins at dawn, but given all day. "Petes' Beer" also transliterated as "Fitz-bew." Also has a complex Flight Song (p. 574).

Like B phrase of Song, but first note longer, coarser; second note shorter

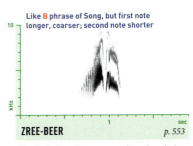

ZREE-BEER   *p. 553*

Mostly in breeding season, often given in long bouts from exposed perch. Easy to mistake for Alder Flycatcher Song, but 2-syllabled.

Soft, slightly grating

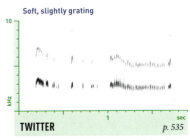

TWITTER   *p. 535*

Mostly on breeding grounds. Plastic. Given in aggressive encounters, often during Song or with Zree-beer.

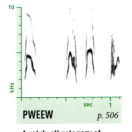

PWEEW   *p. 506*

A catch-all category of plastic, sometimes burry notes given in excitement.

Rapid overslur

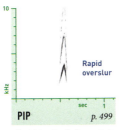

PIP   *p. 499*

Variable, plastic; by excited birds. Very like some Alder Flycatcher calls.

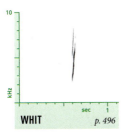

WHIT   *p. 496*

All year; most common call. Never given by Alder Flycatcher.

BLACK PHOEBE

Sayornis nigricans

Common near water, sometimes even beside backyard swimming pools. Dips tail frequently. Range expanding to the north and east.

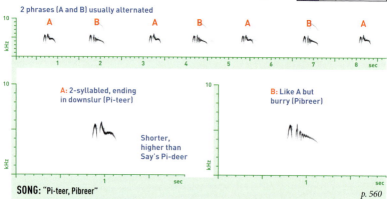

2 phrases (A and B) usually alternated

A: 2-syllabled, ending in downslur (Pi-teer)

Shorter, higher than Say's Pi-deer

B: Like A but burry (Pibreer)

SONG: "Pi-teer, Pibreer"

p. 560

Mostly Jan.–June. Most frequent at dawn but given all day, especially by unmated males. Pattern like Eastern Phoebe song but higher-pitched, less burry, and often faster. Sometimes incorporates Teer like a third phrase. Rarely, if ever, repeats A or B phrases without the other.

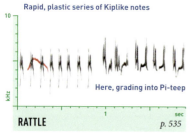

Rapid, plastic series of Kiplike notes

Here, grading into Pi-teep

RATTLE

p. 535

Highly plastic. Given near nest, when displaying potential nest site to mate, and in some aggressive situations.

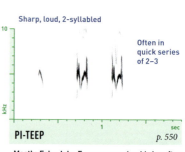

Sharp, loud, 2-syllabled

Often in quick series of 2–3

PI-TEEP

p. 550

Mostly Feb.–July. From aggressive birds; often used in flight display, along with song phrases. Extremely similar to Eastern's Pi-teep.

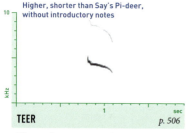

Higher, shorter than Say's Pi-deer, without introductory notes

TEER

p. 506

All year. Common at any time of day, often in long series, especially right before or after Song. Plastic; sometimes overslurred.

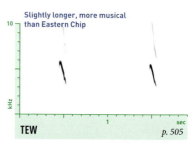

Slightly longer, more musical than Eastern Chip

TEW

p. 505

All year; most common call. Distinctively low and musical.

EASTERN PHOEBE

Sayornis phoebe

Found in open areas near water; nests on ledges under bridges or under eaves of buildings. Dips tail frequently. Range has expanded westward.

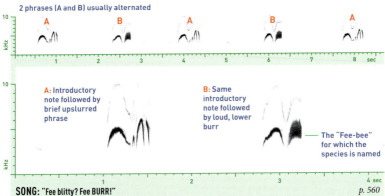

2 phrases (A and B) usually alternated

A: Introductory note followed by brief upslurred phrase

B: Same introductory note followed by loud, lower burr

The "Fee-bee" for which the species is named

SONG: "Fee blitty? Fee BURR!" *p. 560*

Mostly Mar.–July; Song is more plastic in fall. Most frequent at dawn but given all day, especially by unmated males, or in territorial altercations. Either phrase may be given repeatedly without the other: "Feeburr . . . Fee-burr" (p. 533) or "Fee-blitty? Fee-blitty?" (p. 553). Also gives complex Flight Song (p. 574).

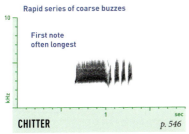

Rapid series of coarse buzzes

First note often longest

CHITTER *p. 546*

Highly plastic. Given near nest, when displaying potential nest site to mate, and in some aggressive situations.

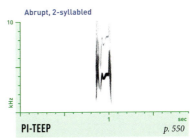

Abrupt, 2-syllabled

PI-TEEP *p. 550*

Rather stereotyped. Given infrequently, in aggressive encounters and in run-up to Flight Song.

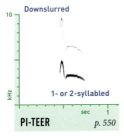

Downslurred

1- or 2-syllabled

PI-TEER *p. 550*

Probably all year, but infrequent; during or after aggressive encounters.

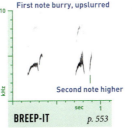

First note burry, upslurred

Second note higher

BREEP-IT *p. 553*

Mostly Mar.–July; variable. In interactions, usually right after short flights.

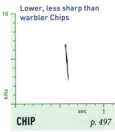

Lower, less sharp than warbler Chips

CHIP *p. 497*

All year. Most common call; given while foraging, in alarm, or near the nest.

SAY'S PHOEBE

Sayornis saya

Unlike other phoebes, prefers dry, open country, often far from water. Like other phoebes, readily nests under eaves of buildings, and migrates early in spring.

3 phrases (A, B, C); A by far most common; B and C almost never adjacent

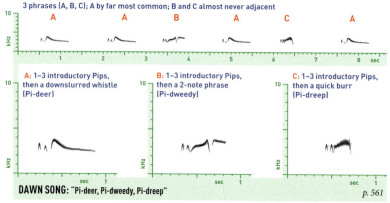

A: 1–3 introductory Pips, then a downslurred whistle (Pi-deer)

B: 1–3 introductory Pips, then a 2-note phrase (Pi-dweedy)

C: 1–3 introductory Pips, then a quick burr (Pi-dreep)

DAWN SONG: "Pi-deer, Pi-dweedy, Pi-dreep"   *p. 561*

Mostly Mar.–July. Phrase A predominates; B and C most frequent before dawn, becoming rarer throughout the morning; thus Dawn Song gradually transitions into Pi-deer calls, possibly a type of Day Song. One recording from Aug. includes a fourth phrase, a simple Pip; function unknown. See also Flight Song, p. 574.

Notes semimusical, slightly burry

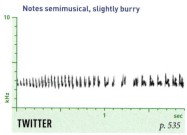

TWITTER   *p. 535*

Mostly Apr.–July, in family interactions near the nest, but apparently also sometimes in alarm. Highly plastic.

Rapid mix of burry phrases

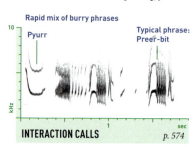

Pyurr

Typical phrase: Preer-bit

INTERACTION CALLS   *p. 574*

Mostly Apr.–July, in aggression. Highly plastic, except when repeated during Flight Song (p. 574).

1–3 introductory Pips, then a downslurred whistle

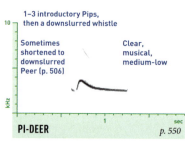

Sometimes shortened to downslurred Peer (p. 506)

Clear, musical, medium-low

PI-DEER   *p. 550*

All year. Like Song phrase A. Most common call; given in long strings during the day, near the nest, and in mild alarm.

Nearly monotone

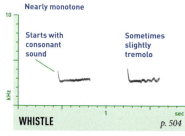

Starts with consonant sound

Sometimes slightly tremolo

WHISTLE   *p. 504*

All year, in response to aerial or other predators; also sometimes from apparently calm birds. Grades into Pi-deer.

Vermilion Flycatcher

Pyrocephalus rubinus

Strikingly plumaged. One of the few fly-catchers in which sexes differ. Found in open areas with low bushes and trees, often near water.

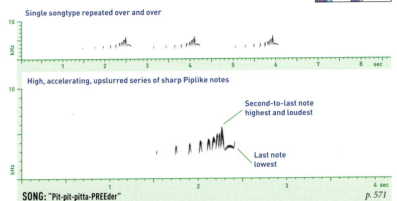

Single songtype repeated over and over

High, accelerating, upslurred series of sharp Piplike notes

Second-to-last note highest and loudest

Last note lowest

SONG: "Pit-pit-pitta-PREEder" *p. 571*

Mostly Feb.–July, by male. Quite uniform and stereotyped, but number of introductory Pips can vary. Given at any time of day, even at night, but especially at dawn. Songs may be repeated almost without pause, or with single bill snaps between Songs, especially during wing-fluttering flight display.

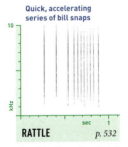

Quick, accelerating series of bill snaps

RATTLE *p. 532*

Possibly all year, by agitated birds in interactions. Also gives single bill snaps.

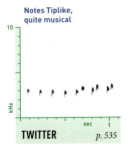

Notes Tiplike, quite musical

TWITTER *p. 535*

Given in pair interactions, perhaps courtship. Lower, softer than Pseep; plastic.

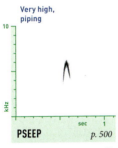

Very high, piping

PSEEP *p. 500*

All year; most common call. Somewhat plastic; clear to slightly burry.

HYBRID PHOEBES

Range expansions by Black and Eastern Phoebes have brought the species into contact in northern New Mexico and southern Colorado, where they now sometimes hybridize. First-generation hybrids are intermediate between the parent species in both plumage and voice. Some backcross hybrids may be best detected by voice.

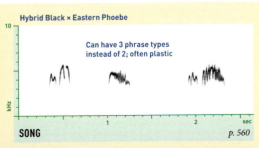

Hybrid Black × Eastern Phoebe

Can have 3 phrase types instead of 2; often plastic

SONG *p. 560*

Ash-throated Flycatcher

Myiarchus cinerascens

Found in dry country, from deserts to pinyon-juniper woodlands to riparian corridors. Paler than other *Myiarchus*, but best identified by voice.

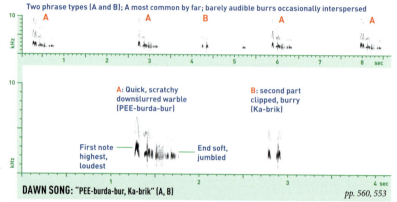

Two phrase types (A and B); A most common by far; barely audible burrs occasionally interspersed

A: Quick, scratchy downslurred warble (PEE-burda-bur)

First note highest, loudest

End soft, jumbled

B: second part clipped, burry (Ka-brik)

DAWN SONG: "PEE-burda-bur, Ka-brik" (A, B)

pp. 560, 553

Mostly Apr.–July, from predawn until just after sunrise; rarely if ever later in the day. Ending of phrase A sometimes doubled ("PEE-burda-burda-bur"). Rather similar to Brown-crested Flycatcher Dawn Song, but higher and scratchier, starting with a Piplike note rather than a whistle.

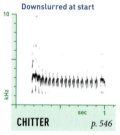

Downslurred at start

CHITTER *p. 546*

Mostly Apr.–July. Plastic; slightly screechy. First and last notes often accented.

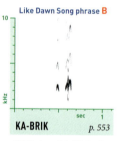

Like Dawn Song phrase **B**

KA-BRIK *p. 553*

Mostly Apr.–July. Distinctive. Second note barely longer and higher than first.

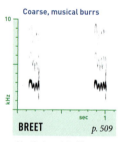

Coarse, musical burrs

BREET *p. 509*

Mostly Apr.–July. Short, monotone, and musical. Often in slow series.

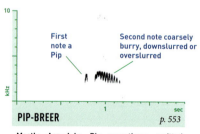

First note a Pip

Second note coarsely burry, downslurred or overslurred

PIP-BREER *p. 553*

Mostly Apr.–July. Pip sometimes omitted. Coarser than most similar flycatcher sounds.

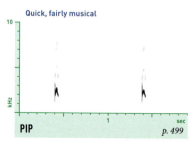

Quick, fairly musical

PIP *p. 499*

All year. Slightly higher and more abrupt than Brown-crested or Great Crested Wips. Sometimes run into brief, rapid Pip Series (p. 536).

GREAT CRESTED FLYCATCHER

Myiarchus crinitus

A large treetop flycatcher of deciduous woodland edges. This species, Brown-crested, and Ash-throated are the only cavity-nesting flycatchers in our region.

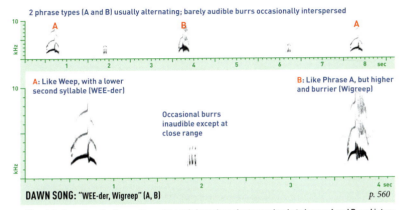

2 phrase types (A and B) usually alternating; barely audible burrs occasionally interspersed

A: Like Weep, with a lower second syllable (WEE-der)

B: Like Phrase A, but higher and burrier (Wigreep)

Occasional burrs inaudible except at close range

DAWN SONG: "WEE-der, Wigreep" (A, B)   *p. 560*

May–July. Dawn Song in above pattern given only by males prior to sunrise, but phrases A and B, and intergrades between them, also given separately during the day by both sexes, apparently for pair bonding.

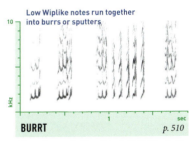

Low Wiplike notes run together into burrs or sputters

BURRT   *p. 510*

Given frequently May–July. Soft, highly plastic. Lower and shorter than Breet, given by calmer, often solo birds.

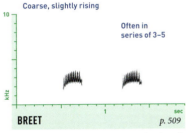

Coarse, slightly rising

Often in series of 3–5

BREET   *p. 509*

All year, in agitation or alarm. Loud and semi-musical. Can recall Evening Grosbeak Breet.

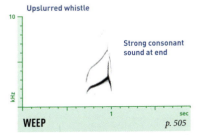

Upslurred whistle

Strong consonant sound at end

WEEP   *p. 505*

All year. Most common call. Distinctive, but variable; usually from solo birds. Some versions are burry, possibly given by female.

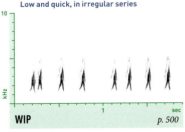

Low and quick, in irregular series

WIP   *p. 500*

May–July, in altercations. Plastic; can lack final downslur, sounding like Whit. Grades into Weep.

BROWN-CRESTED FLYCATCHER

Myiarchus tyrannulus

A large treetop flycatcher of riparian woods. Closely related to Great Crested, and very similar in plumage and habits, but voice differs.

Two compound phrases (AC, BC) in any order; endings often doubled (ACC, BCC)

ACC   BC   Barely audible burrs   ACC
                                      interspersed

ACC   BC

A: Upslurred   C: Overslurred   B: Downslurred
whistle   burr   whistle

DAWN SONG: "PLEASE put-it-HERE put-it-HERE... DEAR, put-it-HERE" (ACC, BC)

p. 560

Mostly Mar.–July, by males prior to sunrise. Uniform and stereotyped. Phrases combine only in patterns AC, ACC, BC, and BCC. Combinations starting with A are generally more common. Near nest, also gives long, soft Twitter (p. 535).

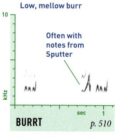

Low, mellow burr

Often with notes from Sputter

BURRT   p. 510

All year. Soft and plastic, from calm birds; grades into Sputter and Wee-burr.

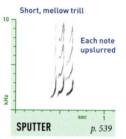

Short, mellow trill

Each note upslurred

SPUTTER   p. 539

All year, during interactions. Highly plastic; grades into Burrt and Wee-burr.

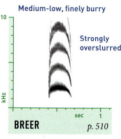

Medium-low, finely burry

Strongly overslurred

BREER   p. 510

All year, in agitation or alarm. Loud and frequent.

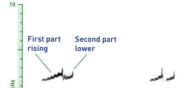

Low, 2-syllabled, variably burry phrase

First part   Second part
rising   lower

BREE-BURR   p. 553

All year, in excitement. Plastic; sometimes a single upslurred Weet, like Great Crested Weep but lower, with subtle burriness.

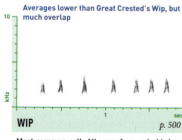

Averages lower than Great Crested's Wip, but much overlap

WIP   p. 500

Most common call. All year; from solo birds as well as in encounters. Some versions lack final downslur, sounding like Whit.

GREAT KISKADEE

Pitangus sulphuratus

Visually and vocally conspicuous in a variety of wooded habitats. Range has expanded northward. Diet more diverse than any other flycatcher's.

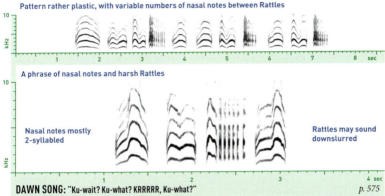

Pattern rather plastic, with variable numbers of nasal notes between Rattles

A phrase of nasal notes and harsh Rattles

Nasal notes mostly 2-syllabled

Rattles may sound downslurred

DAWN SONG: "Ku-wait? Ku-what? KRRRRR, Ku-what?" *p. 575*

Mostly Mar.–July. Not well known; infrequently heard, even at dawn, and apparently never given later in the day. Nasal notes much like daytime Reep calls in quality; rattling parts of Song may recall Low Rattle of Belted Kingfisher.

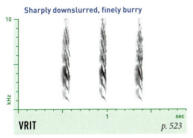

Sharply downslurred, finely burry

VRIT *p. 523*

All year; plastic. Given during interactions, often with other calls.

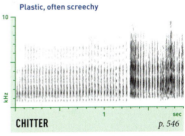

Plastic, often screechy

CHITTER *p. 546*

Mostly in breeding season, during high-intensity territorial chases and nest defense. Highly plastic.

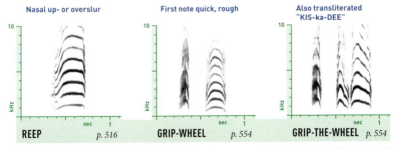

Nasal up- or overslur

REEP *p. 516*

First note quick, rough

GRIP-WHEEL *p. 554*

Also transliterated "KIS-ka-DEE"

GRIP-THE-WHEEL *p. 554*

All year; loud and frequent. Rather plastic; the 3 most common variants are shown, but Grips and nasal notes can mix in a variety of ways. Given by pairs maintaining contact, and in aggression; increasing number of notes seems to reflect increasing excitement.

TROPICAL KINGBIRD

Tyrannus melancholicus

Widespread in the American tropics, but rare and local in our area. Nearly identical to Couch's Kingbird, and best distinguished from it by voice.

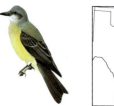

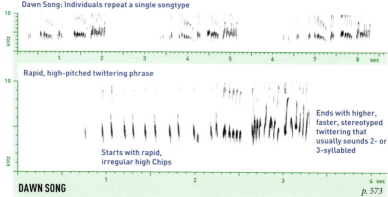

Dawn Song: Individuals repeat a single songtype

Rapid, high-pitched twittering phrase

Ends with higher, faster, stereotyped twittering that usually sounds 2- or 3-syllabled

Starts with rapid, irregular high Chips

DAWN SONG
p. 573

Mostly Mar.–July, apparently only by males and only prior to dawn. Somewhat variable but generally stereotyped. Tone semimusical, chittering, like Chimney Swift but with many notes slightly longer and more grating. Differs from Eastern and Gray Kingbird Dawn Songs in rhythm and lack of clear notes.

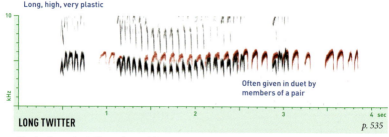

Long, high, very plastic

Often given in duet by members of a pair

LONG TWITTER
p. 535

Nearly all sounds of this species are high Twitters. Except for the Dawn Song, all vocalizations tend to be highly plastic and poorly differentiated from one another. Long Twitter tends to be given during pair interactions, often accompanied by fluttering wings.

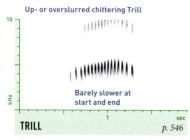

Up- or overslurred chittering Trill

Barely slower at start and end

TRILL
p. 546

Plastic, grading into Short Twitter. Apparently given in mild alarm and sometimes during greetings or in between foraging flights.

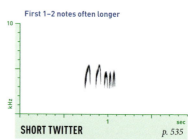

First 1–2 notes often longer

SHORT TWITTER
p. 535

Highly plastic, grading into Long Twitter. Given frequently in many contexts; some versions may function as a type of Day Song.

COUCH'S KINGBIRD

Tyrannus couchii

Vocally conspicuous in thorn scrub, riparian forest, overgrown fields, and other wooded areas. Much more common than Tropical Kingbird.

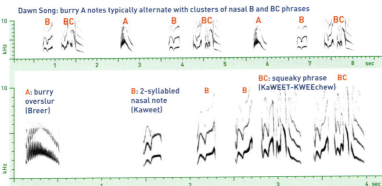

Dawn Song: burry A notes typically alternate with clusters of nasal B and BC phrases

A: burry overslur (Breer)

B: 2-syllabled nasal note (Kaweet)

BC: squeaky phrase (KaWEET-KWEEchew)

DAWN SONG: "Breer... kaweet, kaWEET-KWEEchew" (A, B, BC)   *pp. 510, 560*

Mostly Apr.–July, reportedly by males, up to an hour after sunrise. Typical pattern: A phrase, then a rising and accelerating series of B phrases, then 1–2 BC phrases. After sunrise, BC phrases become gradually less common, and Pips and other calls are interspersed. BC phrases may be given all day in aggression.

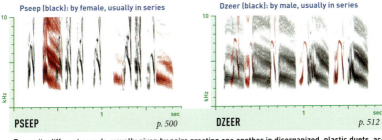

Pseep (black): by female, usually in series

Dzeer (black): by male, usually in series

PSEEP   *p. 500*

DZEER   *p. 512*

Two quite different sounds, usually given by pairs greeting one another in disorganized, plastic duets, accompanied by wing fluttering. Male gives downslurred buzzy Dzeer, female gives Pseep; both calls slightly plastic, each with a 2-syllabled version that starts with a harsh burr.

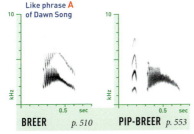

Like phrase A of Dawn Song

BREER   *p. 510*

PIP-BREER   *p. 553*

All year, especially by males guarding a territory, often with Pips. Breer finer, more overslurred than Ash-throated Flycatcher's.

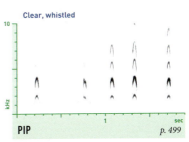

Clear, whistled

PIP   *p. 499*

All year, by both sexes, in interactions and in response to predators. Slightly plastic. No similar call known in Tropical Kingbird.

WESTERN KINGBIRD

Tyrannus verticalis

Conspicuous, pugnacious, and vocal. Uncommon in the East, but abundant in the West in a variety of open habitats. Range has expanded eastward.

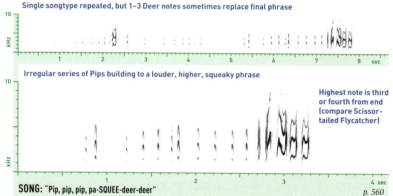

Single songtype repeated, but 1–3 Deer notes sometimes replace final phrase

Irregular series of Pips building to a louder, higher, squeaky phrase

Highest note is third or fourth from end (compare Scissor-tailed Flycatcher)

SONG: "Pip, pip, pip, pa-SQUEE-deer-deer"
p. 560

Mostly Apr.–July, reportedly only by males. Version shown above is given persistently prior to dawn, occasionally at other times during the day. In tumbling flight display, rises from perch to repeat "pa-SQUEE-deer-deer" phrase 3–6 times with 4–5 loud Wing Whirrs in between.

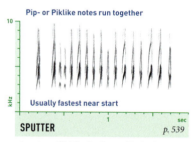

Pip- or Piklike notes run together

Usually fastest near start

SPUTTER
p. 539

Apr.–Aug. Slightly plastic; used by both sexes in pair greetings and by males patrolling territory; female version often shorter.

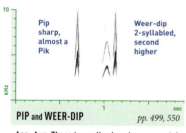

Pip sharp, almost a Pik

Weer-dip 2-syllabled, second higher

PIP and WEER-DIP
pp. 499, 550

Apr.–Aug. These two calls also given separately, but often in above pattern. Pip is like individual notes from Sputter.

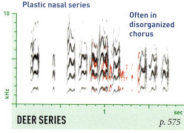

Plastic nasal series

Often in disorganized chorus

DEER SERIES
p. 575

Mostly Apr.–Aug. Extremely plastic; from agitated birds in interactions. Often in conjunction with Chitter.

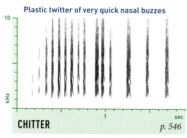

Plastic twitter of very quick nasal buzzes

CHITTER
p. 546

All year, in frequent aggressive encounters. Highly plastic. Usually in conjunction with other calls.

EASTERN KINGBIRD

Tyrannus tyrannus

Common and conspicuous in a variety of open and semi-open habitats. Well known for aggressively defending territory against all other birds.

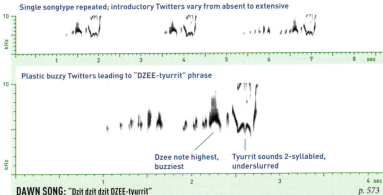

Single songtype repeated; introductory Twitters vary from absent to extensive

Plastic buzzy Twitters leading to "DZEE-tyurrit" phrase

Dzee note highest, buzziest

Tyurrit sounds 2-syllabled, underslurred

DAWN SONG: "Dzit dzit dzit DZEE-tyurrit"   *p. 573*

Mostly May–July. Given persistently in hour before dawn; almost never heard during the day. Unlike other kingbird Dawn Songs, not used in tumbling flight display. Introductory Twitters become more prolonged as singing bout continues.

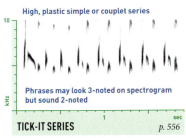

High, plastic simple or couplet series

Phrases may look 3-noted on spectrogram but sound 2-noted

TICK-IT SERIES   *p. 556*

Mostly Apr.–Aug., by both sexes in pair greetings and by males patrolling territory; often with wing flutter. Female version often shorter.

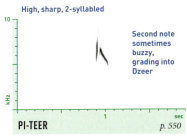

High, sharp, 2-syllabled

Second note sometimes buzzy, grading into Dzeer

PI-TEER   *p. 550*

Mostly Apr.–Aug., by both sexes; repeated at intervals in agitation or aggression.

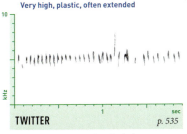

Very high, plastic, often extended

TWITTER   *p. 535*

Mostly Apr.–Aug., by female during nest construction and by male patrolling territory. Male versions usually end in Dzeer.

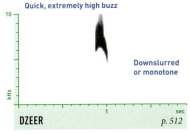

Quick, extremely high buzz

Downslurred or monotone

DZEER   *p. 512*

Apr.–Aug., mostly by males. Most common call, often given in aggression. Plastic; grades smoothly into Pi-teer.

SCISSOR-TAILED FLYCATCHER

Tyrannus forficatus

Beautiful, elegant, and conspicuous in summer in grasslands and savannas with scattered shrubs or trees. Eats mostly grasshoppers and beetles.

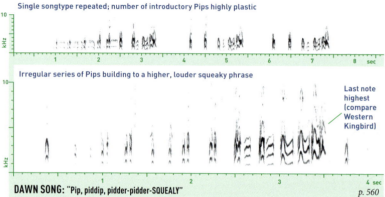

Single songtype repeated; number of introductory Pips highly plastic

Irregular series of Pips building to a higher, louder squeaky phrase

Last note highest (compare Western Kingbird)

DAWN SONG: "Pip, piddip, pidder-pidder-SQUEALY"
p. 560

Mostly Apr.–July, reportedly only by males. Version shown above is given persistently prior to dawn, occasionally at other times during the day. In tumbling flight display, rises from perch to repeat Pidder-pidder-SQUEALY phrase 3–6 times with 4–5 loud Wing Whirrs in between.

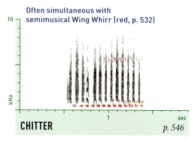

By male

Peer

Churr

PEER-CHURR
p. 575

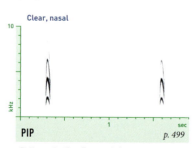

By female: highly plastic nasal burrs

DZEER
p. 512

Two quite different sounds, usually given by pairs greeting one another in disorganized, plastic duets, accompanied by wing fluttering. Male gives Peer and Churr, together or separately; female gives Dzeer notes, each usually followed by a softer snarling burst. All calls quite plastic.

Often simultaneous with semimusical Wing Whirr (red, p. 532)

CHITTER
p. 546

Mostly by males, often in Wing-Whirring flight display; females near the nest give slower, clearer versions, more like Pips in series.

Clear, nasal

PIP
p. 499

Rather plastic. Frequent in many contexts; higher and sharper in alarm. Averages lower than in Western Kingbird, but much overlap.

Gray Kingbird

Tyrannus dominicensis

A primarily Caribbean species. Breeds in coastal mangroves and locally in open pasturelands, rarely more than a few miles from the coast.

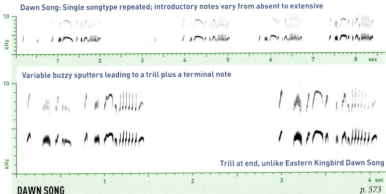

Dawn Song: Single songtype repeated; introductory notes vary from absent to extensive

Variable buzzy sputters leading to a trill plus a terminal note

Trill at end, unlike Eastern Kingbird Dawn Song

DAWN SONG *p. 573*

Mostly Apr.–July. Given by unpaired males at any time of day; given by paired males primarily in the hour prior to sunrise, but occasionally during tumbling flight displays or territorial disputes. Somewhat variable, but stereotyped. Generally quite similar to Eastern Kingbird Dawn Song, but pattern differs.

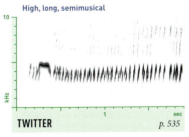

High, long, semimusical

TWITTER *p. 535*

Variable and plastic, grading into Short Twitter. Given by pairs in interactions; rather similar calls given in a variety of other contexts.

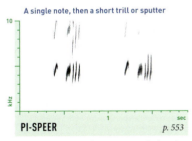

A single note, then a short trill or sputter

PI-SPEER *p. 553*

Variable; some versions stereotyped, given frequently by countercalling males, perhaps as a type of Day Song. Also sometimes by female.

SHRIKES (Family Laniidae)   *following pages*

The shrike family is primarily an Old World group. Only two species are found in North America. Both are almost entirely carnivorous, eating small birds and mammals as well as large insects such as grasshoppers. Shrikes are also known as "butcher birds" because of their habit of impaling prey on thorns or barbed wire fences, for reasons that are not entirely understood. Despite their hooked bills and carnivorous diet, shrikes are true songbirds, with learned and highly complex songs. In the two North American species, it is often difficult to distinguish between calls and songs; vocal communication in these birds needs more study.

Loggerhead Shrike

Lanius ludovicianus

Widespread in open habitats, but many populations have declined. Not often found in the same region at the same season as Northern Shrike.

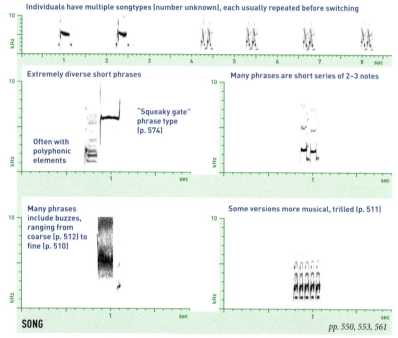

Individuals have multiple songtypes (number unknown), each usually repeated before switching

Extremely diverse short phrases

"Squeaky gate" phrase type (p. 574)

Often with polyphonic elements

Many phrases are short series of 2–3 notes

Many phrases include buzzes, ranging from coarse (p. 512) to fine (p. 510)

Some versions more musical, trilled (p. 511)

SONG

pp. 550, 553, 561

All year, by both sexes, but especially Feb.–June by male. Generally soft. Exceedingly variable, but usually stereotyped. Mostly consists of short, complex, rather musical phrases, often containing a brief trill or buzz, rarely totaling more than 2 syllables. Whines frequently mixed with Song; songs with more Whines reportedly serve a more territorial function. Immatures reported to give a continuous quiet Song, likely a form of subsong.

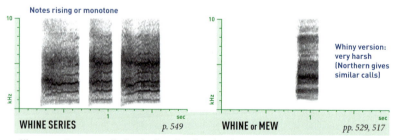

Notes rising or monotone

WHINE SERIES

p. 549

Whiny version: very harsh (Northern gives similar calls)

WHINE or MEW

pp. 529, 517

All year; most common call, given in alarm, but also mixed into song performances. Highly variable and plastic, ranging from metallic and buzzy to harsh and noisy. Different versions may serve different functions. Similar calls used in begging by both juveniles and adults.

Northern Shrike

Lanius excubitor

Slightly larger than Loggerhead Shrike, with thinner black face mask. Breeds in the Arctic, visiting our area only in winter. Uncommon in open habitats.

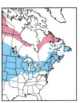

Individuals have multiple songtypes (number unknown), each usually repeated before switching

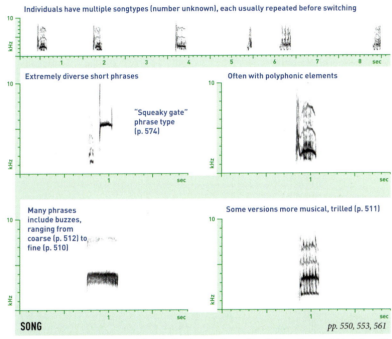

Extremely diverse short phrases

"Squeaky gate" phrase type (p. 574)

Often with polyphonic elements

Many phrases include buzzes, ranging from coarse (p. 512) to fine (p. 510)

Some versions more musical, trilled (p. 511)

SONG

pp. 550, 553, 561

All year, by both sexes, but especially Feb.–June by male. Generally soft. Exceedingly variable, but usually stereotyped. Mostly consists of short, complex, rather musical phrases, often containing a brief trill or buzz, rarely totaling more than 2 syllables. Whines frequently mixed with Song. More likely than Loggerhead to sing fast, complex Song in which consecutive phrases are different and separated by only short pauses, but most of the time, the two species are difficult to distinguish by voice.

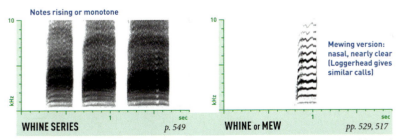

Notes rising or monotone

Mewing version: nasal, nearly clear (Loggerhead gives similar calls)

WHINE SERIES

p. 549

WHINE or MEW

pp. 529, 517

All year; most common call, given in alarm, but also mixed into song performances. Highly variable and plastic, ranging from metallic and buzzy to harsh and noisy. Some versions recall Black-billed Magpie. Different versions may serve different functions. Similar calls used in begging by both juveniles and adults.

VIREOS (Family Vireonidae)

Superficially similar to warblers, vireos have thicker, hook-tipped bills. They hunt caterpillars and other insects among leaves and branches with deliberate movements, pausing frequently. Most are subtly plumaged in shades of olive or gray.

The songs of vireos are complex and learned, usually consisting of short musical phrases. Many species give longer, more complex phrases in close courtship. One species, the White-eyed Vireo, regularly incorporates imitations of other bird species into its primary Song.

Call repertoires tend to be extensive, often including Whines, Chatters, Jits, and short musical Trills. All species snap their bills when highly agitated. Vireos are not known to call during nocturnal migration.

RED-EYED VIREO

Vireo olivaceus

One of the most abundant birds in eastern deciduous forests. Spends most of its time high in the canopy. Known for singing incessantly.

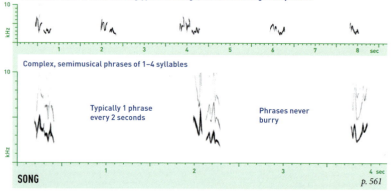

Individuals have 40 different songtypes on average; consecutive Songs always differ

Complex, semimusical phrases of 1–4 syllables

Typically 1 phrase every 2 seconds

Phrases never hurry

SONG *p. 561*

Mostly Apr.–Aug., by males. Females not known to sing. Song often given for long periods without a break, even on hot afternoons. Agitated birds sometimes sing with shorter pauses, and/or give soft Jits and other calls in between song phrases.

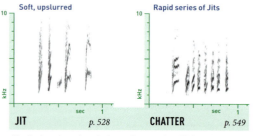 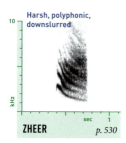

Soft, upslurred

Rapid series of Jits

Harsh, polyphonic, downslurred

JIT *p. 528*

CHATTER *p. 549*

ZHEER *p. 530*

Mostly Apr.–June, in agitation and aggression. Highly plastic. Also gives a short, musical, overslurred Trill, perhaps a version of Chatter.

All year, in alarm. Most common call. Distinctive. Some versions monotone.

Black-whiskered Vireo

Vireo altiloquus

Breeds in mangroves and tropical hard-wood hammocks. Very similar to Red-eyed Vireo; distinctive black "whisker" stripe can be difficult to see.

Males reportedly have 10–15 different songtypes; consecutive Songs always differ

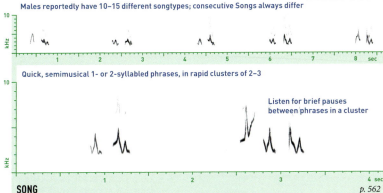

Quick, semimusical 1- or 2-syllabled phrases, in rapid clusters of 2–3

Listen for brief pauses between phrases in a cluster

SONG *p. 562*

Mostly Apr.–July, by males. Females not known to sing. Grouping of Song phrases into short clusters is unique among our vireos. Males reported to sing a faster, more complex version of Song in close courtship; no recordings available.

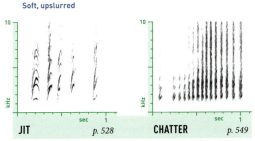

Soft, upslurred

JIT *p. 528*

CHATTER *p. 549*

Mostly Apr.–June, in agitation or aggression. Highly plastic; notes usually slightly polyphonic. Also gives a seminasal, whinnying Trill rather like Blue-headed Vireo's.

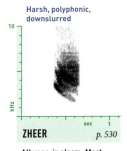

Harsh, polyphonic, downslurred

ZHEER *p. 530*

All year, in alarm. Most common call. Like Zheer of Red-eyed.

Yellow-green Vireo   *Vireo flavoviridis*

Rare and local summer resident in dense deciduous woodlands in south Texas. Resembles Red-eyed Vireo, but head markings less distinct, sides and flanks brighter yellow. Song faster than Red-eyed's, a phrase every 1–1.5 seconds; phrases average short and unmusical, some recalling House Sparrow Chirps. Chatter (p. 549) like Red-eyed; Zheer (p. 530) sometimes upslurred.

SONG *p. 561*

PHILADELPHIA VIREO

Vireo philadelphicus

Breeds mostly in young stands of birch, aspen, or alder. Voice like Red-eyed Vireo. Plumage like Warbling, but eye-stripe slightly darker, breast yellower.

Individuals probably have at least 15 different songtypes; consecutive Songs same or different

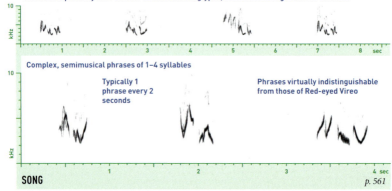

Complex, semimusical phrases of 1–4 syllables

Typically 1 phrase every 2 seconds

Phrases virtually indistinguishable from those of Red-eyed Vireo

SONG
p. 561

Mostly Apr.–Aug., by males. Females not known to sing. Agitated birds sometimes sing with shorter pauses, and/or give Zheers or other calls between Song phrases. Unlike Red-eyed, sometimes cycles through only 2–3 phrases for long periods, or even repeats a phrase over and over, before switching.

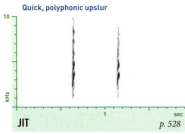

Quick, polyphonic upslur

JIT
p. 528

Possibly all year; infrequent and soft. Also gives soft noisy Chatters and a short, musical, overslurred Trill like Red-eyed's.

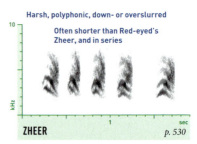

Harsh, polyphonic, down- or overslurred

Often shorter than Red-eyed's Zheer, and in series

ZHEER
p. 530

All year, in alarm, and between phrases of Song. Most common call, but still infrequent.

"WESTERN" WARBLING VIREO

Warbling Vireos breeding west of the Great Plains differ slightly in song, with high squeaky notes much more numerous and more evenly spread, creating a more herky-jerky rhythm and the impression of an overall higher pitch. Pattern like those of Blue Grosbeak and Varied Bunting; pitch intermediate between those 2 species. Excited Song of Western is very long, with many high squeaks and no pauses. Calls apparently do not differ.

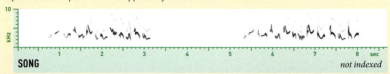

SONG
not indexed

WARBLING VIREO

Vireo gilvus

Breeds in mature deciduous woods; also in parks and residential areas. Yellowish coloring below, when present, brightest on sides and flanks.

Consecutive songtypes usually vary, sometimes in minor details

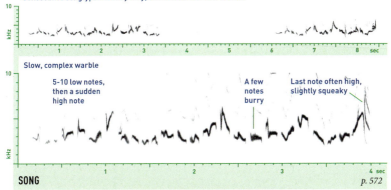

Slow, complex warble

5–10 low notes, then a sudden high note

A few notes burry

Last note often high, slightly squeaky

SONG
p. 572

At least Mar.–Sept., by males; females also reported to sing. Song can become quite long (10 seconds or more) when male is excited, but most Songs are 2–4 seconds in length. Distinctive pitch pattern: pitch of entire Song rises and falls 1–2 times per second. Repertoire size unknown.

Numerous Jits, squeaky notes, and whiny notes before and during Song

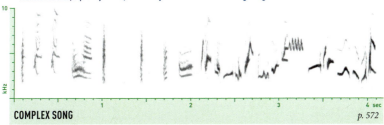

COMPLEX SONG
p. 572

Given by males in close courtship and during aggressive encounters. Both sexes occasionally give Jitlike and squeaky notes in agitation without Song. Females give rapid series of short whining notes in close courtship, somewhat like Trill calls of other vireos.

Quick, upslurred, polyphonic

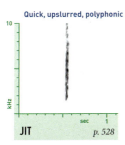

JIT
p. 528

By both sexes, in contact, in flight, and in mild alarm.

Often just a few Jits accelerating into a Whine

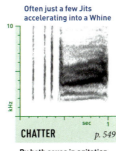

CHATTER
p. 549

By both sexes in agitation. Plastic; some versions are longer, harsher.

Harsh, slightly polyphonic, often upslurred

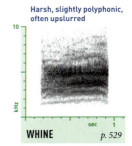

WHINE
p. 529

All year, in alarm; most common call. Plastic, but typically quite harsh.

White-eyed Vireo

Vireo griseus

Breeds in dense deciduous scrub; more often heard than seen. Nests are frequent targets of brood parasitism by Brown-headed Cowbirds.

Males have 9–16 songtypes, each usually repeated many times before switching

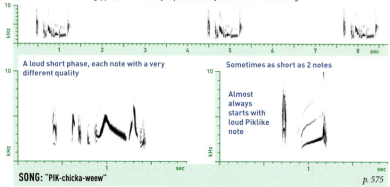

A loud short phase, each note with a very different quality

Sometimes as short as 2 notes

Almost always starts with loud Piklike note

SONG: "PIK-chicka-weew"
p. 575

Distinctive. Mostly Apr.–July; some Song in fall. Both sexes sing on wintering grounds, but only males reported to sing on breeding grounds. Most or all notes are excellent imitations of other bird species, but rapid delivery frequently makes the imitations difficult to recognize.

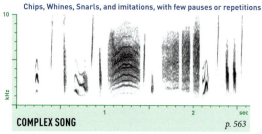

Chips, Whines, Snarls, and imitations, with few pauses or repetitions

COMPLEX SONG
p. 563

Given by adult males in several contexts, often during breeding season when a female is close by, as well as during male-male encounters and high-intensity chases.

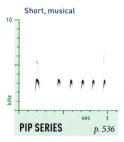

Short, musical

PIP SERIES
p. 536

Infrequent. By females in close courtship and males in aggressive encounters.

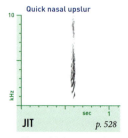

Quick nasal upslur

JIT
p. 528

Given softly by mated pairs as a contact note. Somewhat plastic.

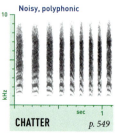

Noisy, polyphonic

CHATTER
p. 549

All year, by both sexes, in alarm and aggression; grades into Whine.

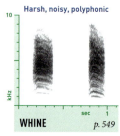

Harsh, noisy, polyphonic

WHINE
p. 549

Infrequent. All year, by both sexes, in alarm and aggression. Quite plastic.

BELL'S VIREO

Vireo bellii

Local in dense shrubby habitats, often near water, from riparian woods in the Great Plains to mesquite bosques in southwestern deserts.

Males average 10 songtypes; 2 or 3 alternated many times before switching to another set

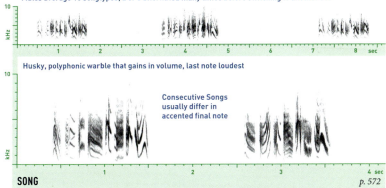

Husky, polyphonic warble that gains in volume, last note loudest

Consecutive Songs usually differ in accented final note

SONG *p. 572*

All year, but especially Apr.–Aug. Females sing occasionally. Alternating pattern of 2 songtypes sometimes suggests a "question-and-answer" pattern, with one songtype ending on a rising note, the other on a falling note. Mated males often intersperse Chatters of Jitlike notes between Songs.

Rapid soft warble with many polyphonic notes and squeaks; no pauses or repetitions

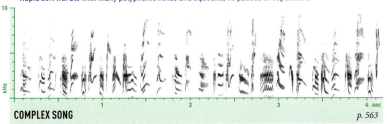

COMPLEX SONG *p. 563*

Reportedly only by males, mostly during breeding season, in a number of contexts, including close courtship and the aftermath of territorial disputes. Quality much like typical Song. Bell's apparently lacks an equivalent of the Trill call of other vireos.

Brief, usually polyphonic

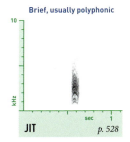

JIT *p. 528*

In quiet contact between mated pairs, or with Song. Variable, plastic.

Plastic, polyphonic

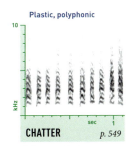

CHATTER *p. 549*

Infrequent, in alarm or agitation; grades into Whines.

Usually in series

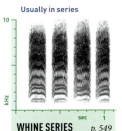

WHINE SERIES *p. 549*

All year, in alarm. Varies from polyphonic to rasping; usually in between.

Black-capped Vireo

Vireo atricapilla

Rare, local breeder in dense deciduous shrublands. Populations have seriously declined, primarily as a result of overgrazing and fire suppression.

Males may have hundreds of songtypes; consecutive Songs almost always different

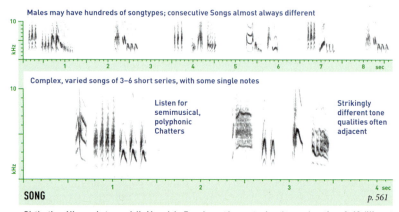

Complex, varied songs of 3–6 short series, with some single notes

Listen for semimusical, polyphonic Chatters

Strikingly different tone qualities often adjacent

SONG *p. 561*

Distinctive. All year, but especially Mar.–July. Females not known to sing. At any given time, 3–10 different songtypes may be given in any order, including alternating 2 songtypes or repeating 1 for a while, before gradually switching to a different set of 3–10 songtypes. A few notes may be imitations.

Like typical Song, but continuing much longer

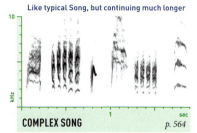

COMPLEX SONG *p. 564*

Possibly all year. Reportedly only by male, in close courtship and territorial conflicts.

Trill: musical

Whine: polyphonic

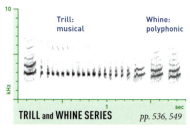

TRILL and WHINE SERIES *pp. 536, 549*

Both sounds plastic, intergrading. Functions little known; apparently during pair interactions, often with Song.

Slightly polyphonic

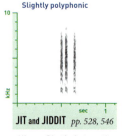

JIT and JIDDIT *pp. 528, 546*

All year. Plastic, but mostly 2-noted. Very like Ruby-crowned Kinglet Jiddit.

Polyphonic to harsh

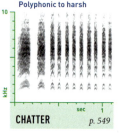

CHATTER *p. 549*

All year, in alarm. Quite plastic; grades into Jiddit and Rasp.

Usually burry, noisy

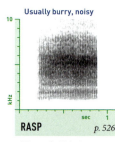

RASP *p. 526*

All year, in high alarm. Plastic, from polyphonic and burry to very harsh.

GRAY VIREO

Vireo vicinior

A nondescript resident of very arid open woodlands and brushland, often in areas dominated by junipers. Barely enters the East in west-central Texas.

Males have 20–50 songtypes; 4 or 5 given in the same order many times before switching to a new set

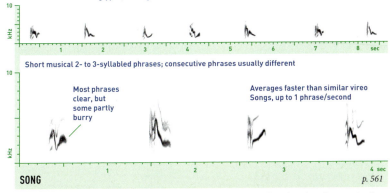

Short musical 2- to 3-syllabled phrases; consecutive phrases usually different

Most phrases clear, but some partly burry

Averages faster than similar vireo Songs, up to 1 phrase/second

SONG *p. 561*

All year, but especially Apr.–Aug. Females sing infrequently, generally only a few phrases at a time. Phrases generally shorter than in other vireos. Speed of Song varies with excitement level. Birds reportedly sometimes give the same songtype over and over for long periods, like Hutton's Vireo.

Faster than typical Song, with soft rising Squeaks between phrases

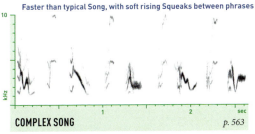

COMPLEX SONG *p. 563*

All year, reportedly only by male, in close courtship, after territorial conflicts, and in response to predators.

Musical, often downslurred

TRILL *p. 536*

All year, by both sexes, in many contexts involving pair contact. Quite plastic.

Soft, quick, noisy

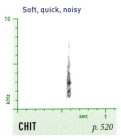

CHIT *p. 520*

All year, in pair contact, especially near the nest. Does not carry far.

Noisy, polyphonic

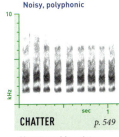

CHATTER *p. 549*

Given to scold predators, often near nest. Variable, plastic; grades into Whine.

Long, harsh version

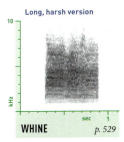

WHINE *p. 529*

Infrequent, in alarm; varies from House Finch–like Zree to harsh rasp.

YELLOW-THROATED VIREO

Vireo flavifrons

Breeds along edges and gaps in deciduous and mixed forests. Often sings from high in canopy, where it can be difficult to see.

Males have 2–8 song songtypes; consecutive Songs usually different

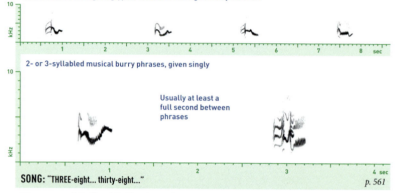

2- or 3-syllabled musical burry phrases, given singly

Usually at least a full second between phrases

SONG: "THREE-eight... thirty-eight..."

p. 561

All year, especially May–July. Females not known to sing. Almost all phases burry, but some can be clear. Rarely, one phrase may be repeated several times in a row. Most likely to be confused with Blue-headed Vireo song, but phrases usually lower, strongly burry; rare individuals may sing mostly clear phrases.

Varied jumble of Chatters and Tsweetlike notes

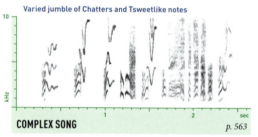

COMPLEX SONG   p. 563

Reportedly given by males in close courtship, as well as by mated males that have lost contact with mate. Soft and plastic.

Musical, often overslurred

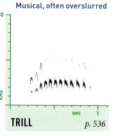

TRILL   p. 536

All year, in alarm and in pair contact. Some birds have a fast and a slow version.

Quick, upslurred

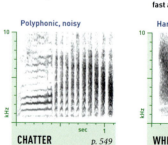

JIT   p. 528

All year, in contact. Highly plastic and quite soft.

Polyphonic, noisy

CHATTER   p. 549

All year, in agitation. Plastic. First note usually longest.

Harsh, burry

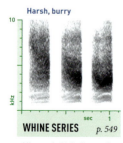

WHINE SERIES   p. 549

All year, in high alarm; often in series. Grades into Chatter.

Blue-headed Vireo

Vireo solitarius

Breeds in closed mature forests of several types. Formerly called "Solitary Vireo" and lumped with Cassin's and Plumbeous Vireos of the West.

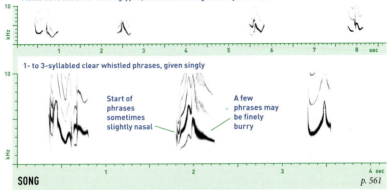

Males have about 10–20 songtypes; consecutive Songs usually different

1- to 3-syllabled clear whistled phrases, given singly

Start of phrases sometimes slightly nasal

A few phrases may be finely burry

SONG *p. 561*

All year, especially May–July; some singing in fall. Female not known to sing. Much like Red-eyed Vireo Song, but usually with longer pauses between phrases (averaging 1 song phrase every 2.5 seconds); rate varies somewhat with excitement level. Higher, clearer than Yellow-throated Vireo Song.

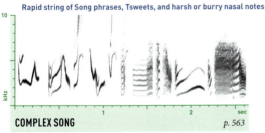

Rapid string of Song phrases, Tsweets, and harsh or burry nasal notes

COMPLEX SONG *p. 563*

Reportedly given by males in close courtship, as well as mated males that have lost contact with mate. Rather soft.

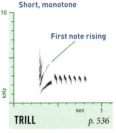

Short, monotone

First note rising

TRILL *p. 536*

Given in agitation in various contexts; volume varies with excitement level.

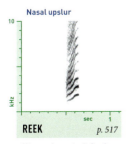

Nasal upslur

REEK *p. 517*

All year, in contact. Grades into Zer-wee; rough versions exist.

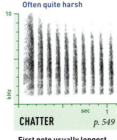

Often quite harsh

CHATTER *p. 549*

First note usually longest. Plastic; long harsh notes can be given singly.

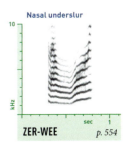

Nasal underslur

ZER-WEE *p. 554*

Highly distinctive; usually 2-syllabled. All year, in contact.

HUTTON'S VIREO

Vireo huttoni

Resident in mature mixed forests, usually near evergreen oaks. Beware the lookalike Ruby-crowned Kinglet. Range expanding in central Texas.

Males typically sing 1 songtype for long periods before switching; repertoire size unknown

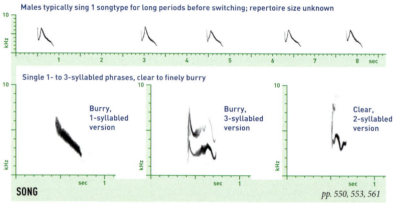

Single 1- to 3-syllabled phrases, clear to finely burry

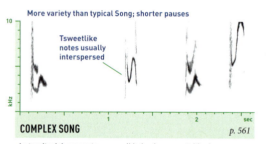

Burry, 1-syllabled version

Burry, 3-syllabled version

Clear, 2-syllabled version

SONG *pp. 550, 553, 561*

All year, especially Mar.–July. Unknown whether female sings. Phrases vary tremendously, sometimes recalling House Sparrow Song or calls of finches; incessant repetition of the same sound can be an excellent field mark, but singing occasionally deviates from this pattern.

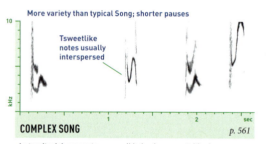

More variety than typical Song; shorter pauses

Tsweetlike notes usually interspersed

COMPLEX SONG *p. 561*

In territorial encounters; possibly in close courtship. In some versions, males cycle through songtypes like other vireo species.

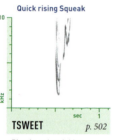

Quick rising Squeak

TSWEET *p. 502*

Plastic and variable; perhaps given in aggression.

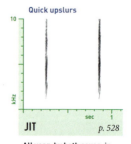

Quick upslurs

JIT *p. 528*

All year, by both sexes, in contact. Variable; some versions rather noisy.

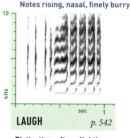

Notes rising, nasal, finely burry

LAUGH *p. 542*

Distinctive; often slightly whiny. First note usually longest.

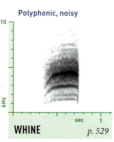

Polyphonic, noisy

WHINE *p. 529*

All year, in high alarm. Plastic; varies from fairly clear to quite harsh.

CROWS, RAVENS, JAYS, AND MAGPIES (Family Corvidae)

Members of this family, known as "corvids," are often considered to be among the most intelligent of all bird species, along with parrots. The crows and ravens are large and black; the jays and magpies are medium-sized, often with striking plumage. Some corvids cache enormous amounts of food, especially nuts and acorns, for retrieval during the colder months. Many species aggressively mob predators in noisy groups, sometimes making it easy to locate hawks and owls.

Although they belong to the oscine passerines, the group commonly known as "songbirds," the corvids rarely give "songs" in the traditional sense (though see p. 298). Many of their sounds appear harsh and simple to the human ear, but their vocal communication is apparently highly complex, as is their social behavior. Most corvid sounds likely have a learned component. At least in the Blue Jay, flockmates have been shown to match call types.

Rattle Calls of Female Corvids

All North American species of crows, ravens, and jays have a distinctive Rattle call that consists of a series of clicks, or sometimes two consecutive series of clicks, often delivered with a rapid up-and-down bobbing of the body. In the species that have been closely studied, these Rattle calls have been reported only from females. In ravens, they have been reported only from the dominant females in a group.

Rattles are apparently given in aggression toward other members of the species, in courtship, and sometimes in response to predators. They tend to be fairly quiet and are likely learned; in most species they vary geographically. No equivalent call has been reported in the Black-billed Magpie.

TAMAULIPAS CROW

Corvus imparatus

Smaller than American Crow; bill thin, plumage slightly glossy. Rare visitor from Mexico. Chihuahuan Raven is the only similar bird regular in south Texas.

Consecutive Croaks generally similar; some individual variation

Highly nasal, usually slightly rising

Sometimes inflected like human speech

Short, clicking

CROAK

pp. 526, 527

RATTLE

p. 531

Voice is highly distinctive: most common call is a variable but immediately recognizable nasal Croak. Rattle presumably given by females. Rattle likely variable; only available example shown.

AMERICAN CROW

Corvus brachyrhynchos

Common and widespread in various habitats, including residential areas. Often gathers in large flocks, especially outside the breeding season.

When mobbing potential predators, gives long, harsh, snarling versions of Caw

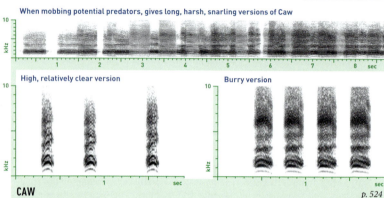

High, relatively clear version

Burry version

CAW *p. 524*

All year, by both sexes, in number of contexts; most common call. Often given in short series. Highly variable; consecutive calls from one bird usually the same, but individuals apparently have repertoires of different Caws, and occasionally switch types.

Very harsh, often rather snarling

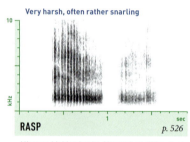

RASP *p. 526*

All year; highly plastic. Given when diving to attack potential predators, often with harsh Caws and rattling sounds.

Each note ticking or Chitlike

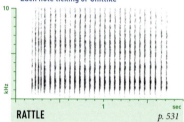

RATTLE *p. 531*

All year, especially July–Oct.; possibly by both sexes. Variable in quality, pitch, and speed; sometimes 2-parted. Function little known.

Soft cooing, warbling, or metallic phrase

Often includes Rattle

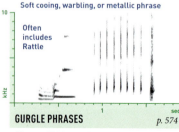

GURGLE PHRASES *p. 574*

All year. Highly variable and sometimes plastic; most versions very soft, but occasionally louder. Function little known.

Like Caw, but longer, more nasal, sometimes rougher

Can strongly resemble certain calls of Fish Crow

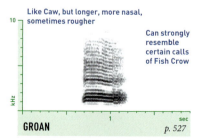

GROAN *p. 527*

Mostly June–Sept., by juveniles, but similar sounds given by females begging in courtship. Highly variable and plastic.

FISH CROW

Corvus ossifragus

Found along coastlines and large rivers; also locally in open inland habitats and urban areas. Virtually identical to American Crow, but voice is distinctive.

Several versions of Uh-oh and Aw calls often heard simultaneously from groups of crows

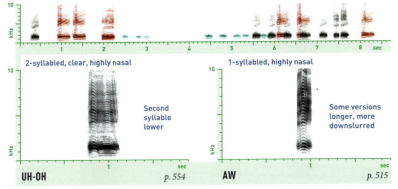

2-syllabled, clear, highly nasal

Second syllable lower

UH-OH p. 554

1-syllabled, highly nasal

Some versions longer, more downslurred

AW p. 515

All year, by both sexes; most common and distinctive calls. Female versions average slightly higher. Both likely function in contact; given frequently by crows in groups and on the wing. Highly variable; individuals may have several different versions of each.

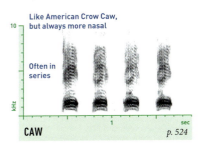

Like American Crow Caw, but always more nasal

Often in series

CAW p. 524

All year, by both sexes, mostly in territorial defense. Some versions rougher than shown. Note distinctive swallowed quality.

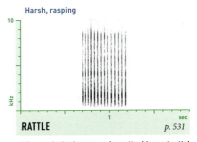

Harsh, rasping

RATTLE p. 531

All year, by both sexes, when attacking potential predators. May be the equivalent of American Crow's Rasp call rather than its Rattle.

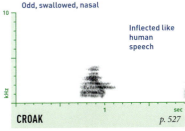

Odd, swallowed, nasal

Inflected like human speech

CROAK p. 527

Little known. Only recording is from a male bird near the nest. Fish Crow apparently lacks equivalent of American Crow's Gurgle Phrases.

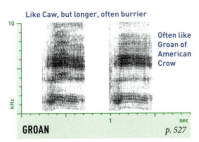

Like Caw, but longer, often burrier

Often like Groan of American Crow

GROAN p. 527

Mostly June–Sept., by juveniles, but similar sounds possibly given by females begging in courtship. Highly variable and plastic.

COMMON RAVEN

Corvus corax

Larger than crows; bill thicker, wings and tail longer. Found in mountains, forests, deserts, and some towns. Often soars; occasionally forms large flocks.

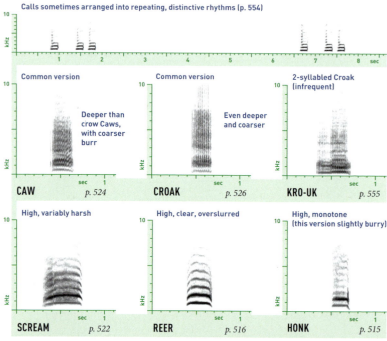

Calls sometimes arranged into repeating, distinctive rhythms (p. 554)

Common version

Deeper than crow Caws, with coarser burr

CAW *p. 524*

Common version

Even deeper and coarser

CROAK *p. 526*

2-syllabled Croak (infrequent)

KRO-UK *p. 555*

High, variably harsh

SCREAM *p. 522*

High, clear, overslurred

REER *p. 516*

High, monotone (this version slightly burry)

HONK *p. 515*

All year. These calls play a key role in the complex social lives of ravens, but variation is poorly understood. Individuals give many different types of croaklike calls, but consecutive calls from the same individual tend to be similar. Many versions briefer or longer than shown. Typical Caw resembles burrier versions of American Crow Caw, but deeper. Typical Croaks are distinctively deep, unlikely to be confused with any other bird sound. High Screams and Honks are quite diverse, ranging from coarsely burry to clear; a few versions approach Snow Goose Honks.

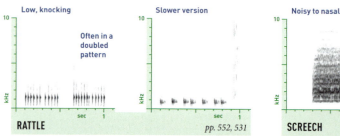

Low, knocking

Often in a doubled pattern

RATTLE *pp. 552, 531*

Slower version

Noisy to nasal

SCREECH *p. 521*

A highly variable set of sounds, given rather infrequently by females only, usually dominant females. Some versions more complex, gulping than shown. Often quite soft.

June–Sept., by juveniles. Variable, becoming clearer and lower with age.

CHIHUAHUAN RAVEN

Corvus cryptoleucus

Smaller than Common Raven; slightly more crowlike in shape. Found in open arid country, and in some towns. Often congregates in large flocks.

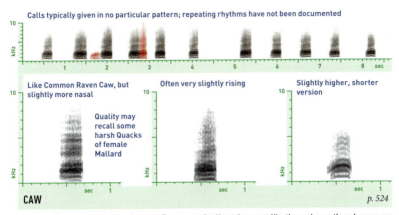

Calls typically given in no particular pattern; repeating rhythms have not been documented

Like Common Raven Caw, but slightly more nasal

Quality may recall some harsh Quacks of female Mallard

Often very slightly rising

Slightly higher, shorter version

CAW *p. 524*

All year. Far less variable than Common Raven; nearly all versions very like those shown, though some are higher-pitched. Common Raven Caw ranges from much lower than Chihuahuan's to much higher; can match Chihuahuan in pitch and pattern, but apparently rarely matches the nasal, slightly quacking quality.

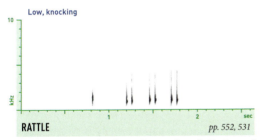

Low, knocking

RATTLE *pp. 552, 531*

Likely given in same situations as Common Raven Rattle. Range of variation poorly known; may have fast versions and gulping versions like Common Raven.

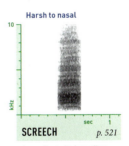

Harsh to nasal

SCREECH *p. 521*

June–Sept., by juveniles. Variable, becoming clearer and deeper with age.

IDENTIFICATION OF COMMON AND CHIHUAHUAN RAVENS

These two species present one of the most difficult identification problems among North American birds. Size differences are nearly impossible to judge except in the rare cases when the two species are seen side by side. Proportionally, Chihuahuan averages shorter-winged, shorter-tailed, and shorter-billed, with nasal bristles reported to extend more than halfway down the bill (vs. less than halfway on Common). Some birds approach or match this description but sound like Common Ravens. It is presently unknown whether these are small Commons, vocally atypical Chihuahuans, birds of mixed ancestry, or some combination of these. Vocally typical Chihuahuans may be rare in portions of their mapped range, such as southeast Colorado. More study needed.

BLUE JAY

Cyanocitta cristata

Common and vocal in woodlands, forest edges, and residential areas. Readily visits feeders. Often moves in loose groups, especially in fall and winter.

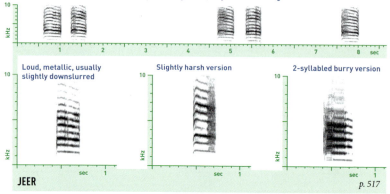

Jeer calls usually given singly or in pairs; in agitation, may be run into longer series

Loud, metallic, usually slightly downslurred

Slightly harsh version

2-syllabled burry version

JEER

p. 517

All year; most common call. Highly variable; individuals may have repertoires of multiple versions, serving slightly different functions. Flockmates apparently match call types. Jeers become harsher in alarm; 2-syllabled versions are common. Young birds beg with hoarse, highly plastic version.

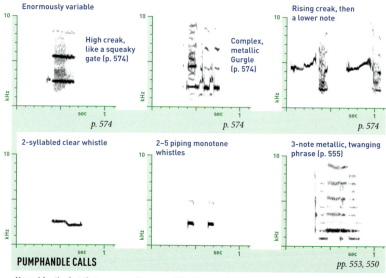

Enormously variable

High creak, like a squeaky gate (p. 574)

Complex, metallic Gurgle (p. 574)

Rising creak, then a lower note

p. 574

p. 574

p. 574

2-syllabled clear whistle

2–5 piping monotone whistles

3-note metallic, twanging phrase (p. 555)

PUMPHANDLE CALLS

pp. 553, 550

Named for the fact that some versions sound like the squeaky handle of an old-fashioned water pump. A hugely variable catch-all category of learned sounds. Individuals have multiple versions, but repertoire size unknown; 1 version usually repeated multiple times before switching. Given in a wide variety of situations, including mild alarm. Most versions are 2- to 3-syllabled, metallic, often slightly gurgling. May grade into the 2-syllable versions of the Jeer call.

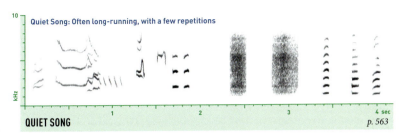

Quiet Song: Often long-running, with a few repetitions

QUIET SONG   *p. 563*

Likely all year, by adults and older juveniles. May consist of whistled, whiny, nasal, and/or Jeerlike notes; sometimes contains a few imitations of other bird species. Usually given by solo birds. Does not carry far.

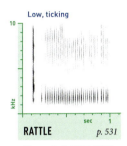

Low, ticking

RATTLE   *p. 531*

All year, by females. Often 2-parted, with 1 or more introductory clicks.

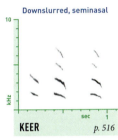

Downslurred, seminasal

KEER   *p. 516*

All year, sometimes after aggressive interactions. Not very loud.

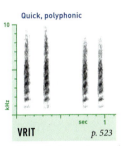

Quick, polyphonic

VRIT   *p. 523*

All year, in close contact. Often finely burry and upslurred. Not very loud.

JAY IMITATIONS

Blue Jays and Gray Jays are notorious for imitating the sounds of other birds. Blue Jays mostly imitate hawks, for reasons that are poorly understood. Gray Jays imitate a wider variety of species, often including shorebirds. In both species, imitations range from poor to nearly perfect.

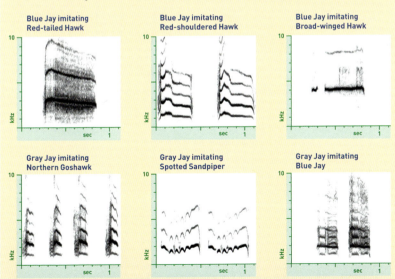

Blue Jay imitating
Red-tailed Hawk

Blue Jay imitating
Red-shouldered Hawk

Blue Jay imitating
Broad-winged Hawk

Gray Jay imitating
Northern Goshawk

Gray Jay imitating
Spotted Sandpiper

Gray Jay imitating
Blue Jay

Gray Jay

Perisoreus canadensis

Resident in boreal spruce forests. Often bold and curious around people, quick to seize unguarded food. Rather quiet, but vocal repertoire is very complex.

Quiet Song (p. 563): all year, by both sexes; usually without repetition, often with imitations

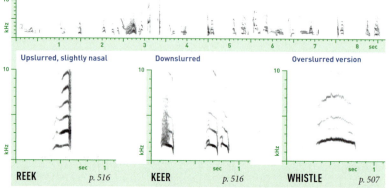

Upslurred, slightly nasal	Downslurred	Overslurred version
REEK p. 516	**KEER** p. 516	**WHISTLE** p. 507

A group of exceedingly variable sounds that seem to intergrade. Different versions likely serve different functions. Individuals may have several different Whistle types; one type usually given repeatedly, either singly or in series, before switching to another type. Some recall the Wail Series of Gray Hawk (p. 539).

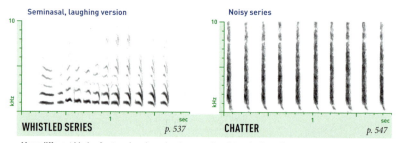

Seminasal, laughing version — **WHISTLED SERIES** p. 537

Noisy series — **CHATTER** p. 547

Many different kinds of notes given in series, in many situations; the 2 versions shown are regularly heard. Whistled Series may vary geographically. Chatters may be used in mobbing of predators, but are also given in other contexts.

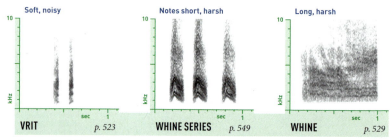

Soft, noisy	Notes short, harsh	Long, harsh
VRIT p. 523	**WHINE SERIES** p. 549	**WHINE** p. 529

A group of variable sounds, given singly or in short series. Soft Vritlike notes and short Whine Series typical in close contact. Long Whines typical of begging juveniles, but also given by adults in courtship, and occasionally directed at humans.

Green Jay

Cyanocorax yncas

Common and conspicuous in its limited U.S. range. Found in small, vocal family groups in open woodlands and brush. Difficult to confuse with any other bird.

Number of notes in series highly plastic, often as few as 2–3

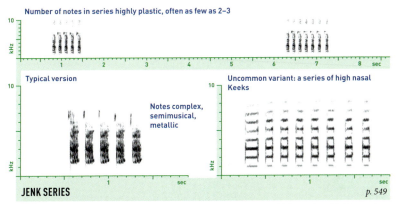

Typical version

Notes complex, semimusical, metallic

Uncommon variant: a series of high nasal Keeks

JENK SERIES
p. 549

All year, by both sexes. Highly variable; at least some individuals have multiple versions. Similar versions usually repeated at length, but 2 versions occasionally alternated. Repertoire size and precise function poorly known. Also gives complex Quiet Song (p. 563); reported to imitate hawks on occasion.

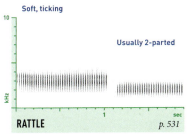

Soft, ticking

Usually 2-parted

RATTLE
p. 531

Mostly Apr.–June, by females. Usually accompanied by vigorous up-and-down bobbing of body. May serve courtship function.

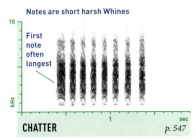

Notes are short harsh Whines

First note often longest

CHATTER
p. 547

All year, by both sexes, apparently in alarm and agitation. Uniform in tone; slight polyphony fairly distinctive. Plastic in length.

Brown Jay *Psilorhinus morio*

Rare and local along the Mexico border in south Texas. Much larger than other jays. Population in the U.S. has fluctuated, with few sightings in recent years. Vocal repertoire appears to be limited for a corvid; most common call is a nasal Keer, usually slightly downslurred, reminiscent of the individual notes of Red-shouldered Hawk, but usually shorter and slightly lower.

Singly or in plastic series

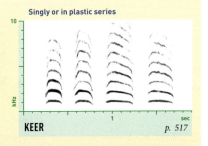

KEER
p. 517

Woodhouse's Scrub-Jay

Aphelocoma woodhouseii

Found in arid scrub, especially among oaks or junipers. Formerly lumped with California Scrub-Jay under the name Western Scrub-Jay.

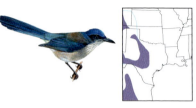

Quiet Song (p. 563): all year, by both sexes; usually without repetition, sometimes with imitations

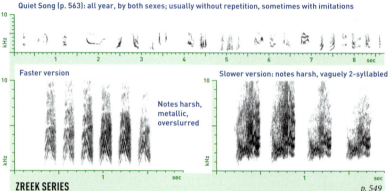

Faster version

Notes harsh, metallic, overslurred

Slower version: notes harsh, vaguely 2-syllabled

ZREEK SERIES *p. 549*

All year, by both sexes, usually in undulating display flight during territorial interactions and possibly courtship. Sometimes shortened to a single note. Males have two or more versions, each usually repeated before switching. Rivals often match each other's Zreek Series types, but mated pairs do not.

Rising, metallic, hoarse

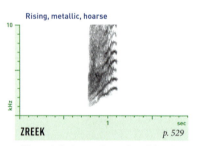

ZREEK *p. 529*

All year, by both sexes, in a variety of situations; most common call. Variable, but distinctive.

Rapid, often complex series of clicks

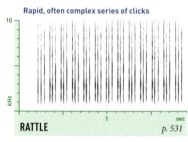

RATTLE *p. 531*

All year, by female, often while bobbing body, bill vertical. Usually in response to mate's Zreek Series, often overlapping it. Variable.

Quick, soft, noisy

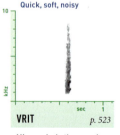

VRIT *p. 523*

All year, by both sexes, in close contact. Variable, plastic; sometimes whiny.

Harsh, noisy

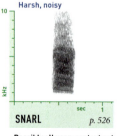

SNARL *p. 526*

Possibly all year; context not entirely clear. Not very loud.

Hoarse

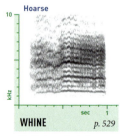

WHINE *p. 529*

Mostly June–Sept., by begging juveniles, but also by adult females. Plastic.

Florida Scrub-Jay

Aphelocoma coerulescens

Found only in native oak scrub forest of Florida; local, sedentary, and declining. Breeds in cooperative family groups, unlike most Western Scrub-Jays.

Quiet Song (p. 563): all year, by both sexes; usually without repetition, sometimes with imitations

Notes upslurred, burry, metallic, hoarse

Female often answers with Rattle (red)

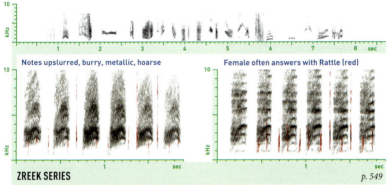

ZREEK SERIES
p. 549

All year, by both sexes, especially Jan.–Feb. and Aug.–Oct.; usually in undulating display flight during territorial interactions and possibly courtship. Sometimes shortened to a single note. Generally less variable than Zreek Series of Western Scrub-Jay; adults may have a single version each.

Rising, metallic, burry

Slow to fast, simple or couplet series of clicks

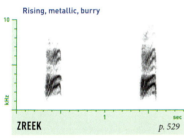

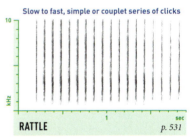

ZREEK
p. 529

All year, by both sexes, in a variety of situations; most common call. Variable, but distinctive.

RATTLE
p. 531

All year, by female, often while bobbing body. Usually in response to mate's Zreek Series, often overlapping it. Several local dialects exist.

Soft, low, harsh

Harsh, burry, polyphonic

Hoarse, polyphonic

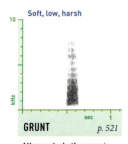

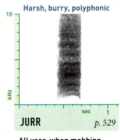

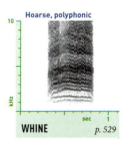

GRUNT
p. 521

All year, by both sexes, in close contact. Variable, plastic.

JURR
p. 529

All year, when mobbing predators and in territorial encounters.

WHINE
p. 529

Mostly June–Sept., by begging juveniles, but also by adult females. Plastic.

BLACK-BILLED MAGPIE

Pica hudsonia

A social and conspicuous bird of dry, open country, woodland edges, and sometimes residential areas. Nest is a large ball of sticks.

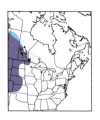

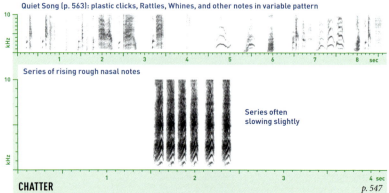

Quiet Song (p. 563): plastic clicks, Rattles, Whines, and other notes in variable pattern

Series of rising rough nasal notes

Series often slowing slightly

CHATTER
p. 547

All year, by both sexes, in alarm. Uniform but plastic; length and speed increase with agitation level. Female version reportedly averages higher than male's. Quiet Song reportedly given mostly by aggressive young males, but also by females; most frequent fall through early spring.

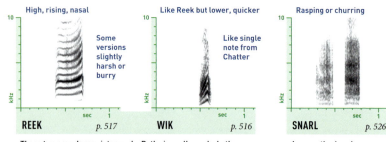

High, rising, nasal

Some versions slightly harsh or burry

REEK   *p. 517*

Like Reek but lower, quicker

Like single note from Chatter

WIK   *p. 516*

Rasping or churring

SNARL   *p. 526*

These two sounds may intergrade. Both given all year by both sexes, Wik in contact, Reek in various situations. Reek repeated loudly by females near nest at start of breeding season.

Apparently given in response to predators near nest site.

QUIET SONGS OF CORVIDS

All North American species of jay and magpie occasionally give highly complex songs, almost always at very low volume. Reported to function in close courtship, they are also given often by lone birds, at almost any time of year. In at least some species, both sexes sing, though males may sing more frequently. Many Quiet Songs are built primarily of notes de-

rived from the species' common calls, in addition to soft clicks, rattles, whines, and imitations of other bird species. Repetition tends to be fairly rare.

Crows and ravens apparently do not give complex Quiet Songs; in these species, the closest equivalent may be the Rattle, or in the case of American Crow, the Gurgle Phrase.

LARKS (Family Alaudidae)

This large family is represented in the New World by a single native species.

The Horned Lark is noteworthy for its apparently plastic calls and song notes, rarely repeated in precisely the same way. It may be that the Horned Lark sings in a completely innovative fashion — each note a unique, one-time utterance — or it may be that the repertoire of notes from which it constructs songs is almost unfathomably vast. Some breeding birds give variable Teer calls that seem to repeat over time, almost like a form of song. More study needed.

HORNED LARK

Eremophila alpestris

Common in deserts, prairies, tundra, and other open areas with sparse vegetation. Plumage varies geographically, but voice varies little.

Long Song (p. 561): Pitchitlike notes filling gaps between Short Songs

Short Song: accelerating, upslurred tinkling warble

Fastest and highest at end

SONG *p. 571*

Mostly Jan.–July. Short Song is more common. Long Song can last for several minutes and is more frequent later in the season and prior to dawn. Both may be given from the ground, from low perches, or in 300-foot-high flight display, though Short Songs predominate in flight.

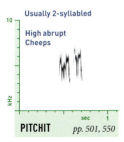

Usually 2-syllabled

High abrupt Cheeps

PITCHIT *pp. 501, 550*

All year, given frequently; highly plastic. Function not well known.

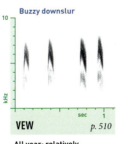

Buzzy downslur

VEW *p. 510*

All year; relatively stereotyped. Given as birds flush, likely in alarm.

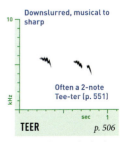

Downslurred, musical to sharp

Often a 2-note Tee-ter (p. 551)

TEER *p. 506*

All year. Extremely plastic; can be monotone or upslurred.

SWALLOWS (Family Hirundinidae)

Adapted for the aerial pursuit of flying insects, swallows spend much of their time on the wing. They have small bills but wide mouths, and pointed wings for elegant, swooping flight. Several species are colonial nesters. Migrating swallows often gather in huge flocks, often including multiple species, perching on wires or swarming over bodies of water.

Unlike swifts, swallows are songbirds (oscine passerines), and most species give highly complex Songs that are likely learned. Some species, such as Purple Martin, Barn Swallow, and Tree Swallow, have distinctive Dawn Songs given mostly on the wing prior to sunrise. In some swallow species, certain calls may be learned; more study needed.

CAVE SWALLOW

Petrochelidon fulva

Nests in caves, in culverts, and under bridges. Nest is an open cup like Barn Swallow's. Range has expanded. A few wander to New England in fall.

Typical Songs run 6 seconds or longer, but are often truncated

Complex gurgling phrases, separated by long, semimusical series

Includes whistled, noisy, and seminasal notes

Lacks the rasping, staticky quality of Cliff Swallow

SONG
p. 565

Mostly Mar.–Aug., presumably by male; not known whether female sings. Given often at nest site and on the wing nearby, including in response to disturbance; infrequent away from nest site. Variable and plastic; repertoire size unknown. Songs of Texas and Florida birds apparently similar.

Upslurred

Often slightly polyphonic

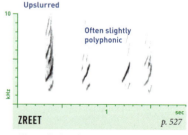

ZREET
p. 527

All year, by foraging birds. Plastic but fairly uniform. Often higher, clearer than Barn Swallow Jit. Cliff Swallow lacks similar notes.

Downslurred, polyphonic

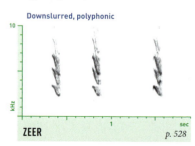

ZEER
p. 528

Likely all year, in alarm, especially near nest. Varible and plastic; much overlap with Cliff Swallow Zeer.

CLIFF SWALLOW

Petrochelidon pyrrhonota

Nests under bridges, in culverts, and on rock faces and canyon walls. Nest is an enclosed, gourd-shaped mud construction with a hole at the top.

Typical Songs run 6 seconds or longer, but can be truncated

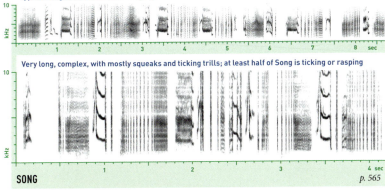

Very long, complex, with mostly squeaks and ticking trills; at least half of Song is ticking or rasping

SONG *p. 565*

Mostly Mar.–Aug., presumably by male; not known whether female sings. Given often at nest site and on the wing nearby, including in response to disturbance; infrequent away from nest site. Variable and plastic; repertoire size unknown. Usually not very loud.

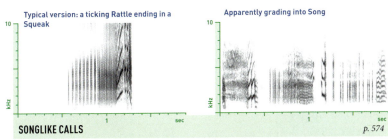

Typical version: a ticking Rattle ending in a Squeak

Apparently grading into Song

SONGLIKE CALLS *p. 574*

Mostly May–Aug., to recruit other members of the colony to a swarm of flying insects. Given mostly on chilly, overcast days when such swarms are infrequent. Variable and plastic; similar to brief snippets of Song. Unknown whether other swallow species have food recruitment calls; more study needed.

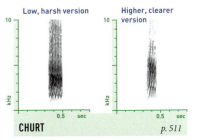

Low, harsh version **Higher, clearer version**

CHURT *p. 511*

All year, by birds in flocks, and in family contact near the nest. Slightly variable. Cave Swallow lacks similar notes.

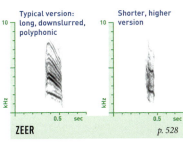

Typical version: long, downslurred, polyphonic **Shorter, higher version**

ZEER *p. 528*

Likely all year, in alarm, especially near nest. Variable, especially in pitch; rare versions are grating. Young beg with short, high Chirps.

BARN SWALLOW

Hirundo rustica

Common and widespread. Nests under bridges and the eaves of buildings, including houses in residential areas; forages over fields and wetlands.

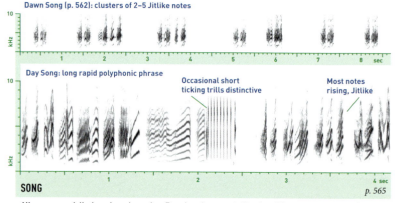

Dawn Song (p. 562): clusters of 2–5 Jitlike notes

Day Song: long rapid polyphonic phrase

Occasional short ticking trills distinctive

Most notes rising, Jitlike

SONG *p. 565*

All year, especially Apr.–Aug., by males. Females also reported to sing. Often preceded by or mixed with Jits. Repertoire size unknown; consecutive songs sound similar but vary greatly in details and in length. Dawn Song given prior to sunrise, often on the wing; grades gradually into Day Song.

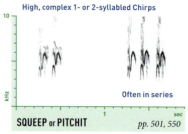

High, complex 1- or 2-syllabled Chirps

Often in series

SQUEEP or PITCHIT *pp. 501, 550*

All year, in mild alarm. Variable and slightly plastic, but distinctive. Similar calls given by adults when feeding fledglings.

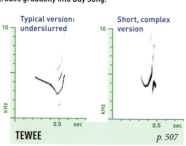

Typical version: underslurred

Short, complex version

TEWEE *p. 507*

Mostly Apr.–Aug., in high alarm. Highly variable and rather plastic, but most versions high, clear, and 2-syllabled.

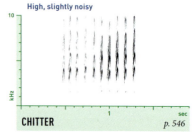

High, slightly noisy

CHITTER *p. 546*

Mostly Apr.–Aug., in alarm near the nest. Extremely plastic; notes often longer than shown, polyphonic, grading into Jits.

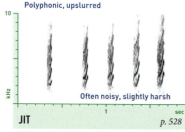

Polyphonic, upslurred

Often noisy, slightly harsh

JIT *p. 528*

All year; most common call. Highly plastic. Often in series. Generally rougher than House Finch Jirp.

Northern Rough-winged Swallow

Stelgidopteryx serripennis

Forages over rivers and fields near water; nests singly in riverbank burrows or rock crevices. Named for tiny barbs on the leading edge of the wing.

Song (p. 562): faint, rarely heard clusters of hoarse whistled and gurgling phrases

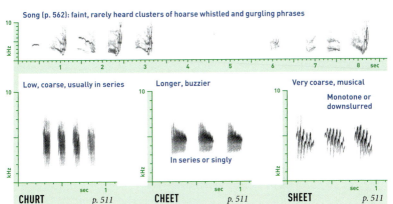

Low, coarse, usually in series

Longer, buzzier

Very coarse, musical

Monotone or downslurred

In series or singly

CHURT *p. 511*

CHEET *p. 511*

SHEET *p. 511*

Churt and Cheet given all year. Cheet higher than Churt, more grating, and less variable; both are higher and more musical than Bank Swallow Churt. Sheet rarely heard; function unknown. Recalls Tree Swallow Sheet and certain House Sparrow Songs. Sometimes gives Snarls near nest site.

Bank Swallow

Riparia riparia

Our smallest swallow. Nests in colonies of 10 to 1,000 pairs, in holes in vertical riverbanks or in piles of sand at gravel quarries.

Song (p. 541): An accelerating chatter of Churtlike notes, ending in soft, complex noisy gurgling

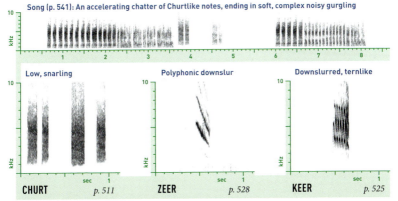

Low, snarling

Polyphonic downslur

Downslurred, ternlike

CHURT *p. 511*

ZEER *p. 528*

KEER *p. 525*

Churt given all year; most common call. Lower, harsher than Rough-winged Churt. Zeer and Keer given all year, in alarm. Zeer is given in warning by first bird to spot danger; flock responds with Keers. Song given Apr.–July, in flight, at nest hole, and to drive away rivals.

TREE SWALLOW

Tachycineta bicolor

Breeds in tree cavities or nest boxes, usually in semi-open areas near water. Arrives earlier in spring migration than other swallows.

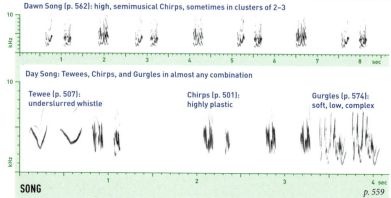

Dawn Song (p. 562): high, semimusical Chirps, sometimes in clusters of 2–3

Day Song: Tewees, Chirps, and Gurgles in almost any combination

Tewee (p. 507): underslurred whistle

Chirps (p. 501): highly plastic

Gurgles (p. 574): soft, low, complex

SONG p. 559

Mostly Apr.–July, by males. Females also reported to sing. Parts of Day Song given singly, sometimes with Sheets and other calls, or run together into long phrases. Dawn Song given prior to sunrise, often in flight; contains 1–7 (usually 2–3) different types of Chirps, often in repeating pattern.

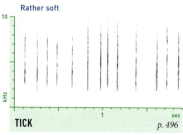

Rather soft

TICK p. 496

Mostly Apr.–June, by male swooping low over female in close courtship; also sometimes given when diving at predators.

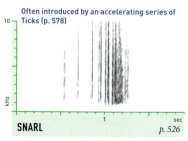

Often introduced by an accelerating series of Ticks (p. 578)

SNARL p. 526

Mostly Apr.–June. Above version given by male swooping low over female in close courtship; single Snarls or series directed at predators.

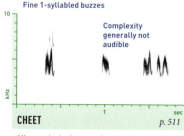

Fine 1-syllabled buzzes

Complexity generally not audible

CHEET p. 511

All year, by both sexes, in response to predator or competitor for nest site. Plastic.

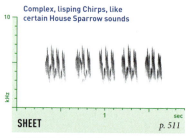

Complex, lisping Chirps, like certain House Sparrow sounds

SHEET p. 511

Mostly Apr.–July, by both sexes, near nest; may function in defense of nest site against competitors.

Purple Martin

Progne subis

Our largest swallow. In the East, nests almost exclusively in multi-compartment backyard birdhouses. Western populations still nest in tree cavities.

Dawn Song (p. 562): clusters of 1-syllabled burry, Veerlike notes; consecutive notes usually different

Complex musical, gurgling phrase

Usually contains Tews, and often Burrts

Most versions contain Ticks and ticking trills

DAY SONG *p. 565*

Mostly Apr.–Aug. Both sexes give versions of Day Song in courtship. Female version simpler, with mostly Tews and Burrts, lacking ticking trills. First-year males, in femalelike plumage, sing like adult males. Dawn Song given by males only prior to sunrise. Repertoire size unknown.

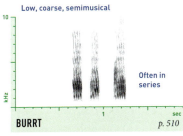

Low, coarse, semimusical

Often in series

BURRT *p. 510*

Likely all year, in a variety of contexts, possibly indicating mild excitement. Often given in stereotyped songlike series of 2–6.

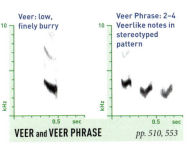

Veer: low, finely burry

Veer Phrase: 2–4 Veerlike notes in stereotyped pattern

VEER and **VEER PHRASE** *pp. 510, 553*

Likely all year, in high alarm or excitement. Individuals may have multiple versions. More frequent than Burrts.

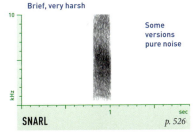

Brief, very harsh

Some versions pure noise

SNARL *p. 526*

Mostly Apr.–July, in high alarm and aggression, at bottom of dive while swooping at potential predators, or occasionally other bird species.

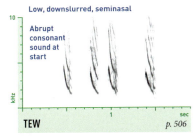

Low, downslurred, seminasal

Abrupt consonant sound at start

TEW *p. 506*

Likely all year, in a variety of contexts. Most common call. Somewhat variable and plastic; some versions partly noisy.

CHICKADEES AND TITMICE (Family Paridae)

Birds in this family, known as "parids," are small, active, highly social species that include some of our most common and familiar visitors to backyard bird feeders. Vocal repertoires and social structures are highly complex. Most species gather in winter flocks that have well-defined dominance hierarchies; each member dominates all lower-ranked birds and submits to all higher-ranked birds. In spring, flocks split up, with the highest-ranked pairs claiming the best breeding territories.

Three different kinds of parid vocalization can show a level of complexity or a communicative function usually associated with "songs":

Whistled Songs are simple, musical, and learned. They are given by all species of titmice but only some chickadees (in our region, only Black-capped and Carolina). They are strongly associated with breeding-season defense of territory by males, and may also help attract females. Male chickadees have repertoires of 2–3 (rarely 5) Whistled Songs, except in most Black-capped Chickadees, which create variety by shifting the pitch of a single songtype. Titmice have repertoires of 9–15 Whistled Songs.

Gurgle Songs are complex, musical, and learned, given by all chickadee species. They are strongly associated with aggression, although they can also be given in close-range courtship situations. Individual male Black-capped Chickadees have repertoires of 2–18 Gurgles (averaging 8), and other species are probably similar. The Tseet Songs of titmice may serve a similar function.

Chick-a-dee calls are the iconic sounds for which chickadees are named. Variable, versatile, and at least partly learned, they serve functions from maintaining flock contact to mobbing predators to giving the "all clear" after danger passes. In the species that have been well studied, Chick-a-dees consist of 3–6 different note types (usually four: A, B, C, D). All these may be repeated any number of times or left out altogether, but the order never varies: A always comes before B, B before C, etc. Different numbers and combinations of notes apparently send different messages. Titmice make similar, but simpler, sounds.

Boreal Chickadee

Poecile hudsonicus

A tame, fluffy bird of northern forests, usually associated with spruces. Unlike other eastern chickadees, has only one set of Songs (the Gurgle).

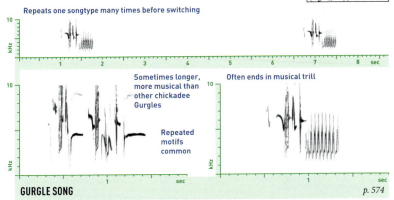

Repeats one songtype many times before switching

Sometimes longer, more musical than other chickadee Gurgles

Repeated motifs common

Often ends in musical trill

GURGLE SONG — *p. 574*

All year, especially during territory formation, Mar.–June. Apparently given only by males, usually in aggression, occasionally during copulation. Sometimes a complex series, with a short rapid phrase repeated 2–8 times. Repertoire size unknown, but probably each male has several Gurgles.

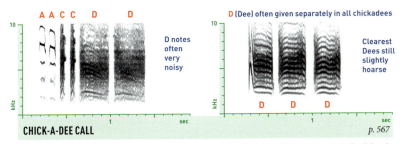

A A C C D D

D notes often very noisy

D (Dee) often given separately in all chickadees

Clearest Dees still slightly hoarse

D D D

CHICK-A-DEE CALL — *p. 567*

All year; plastic. Some noise and a hint of polyphony make the Dees sound hoarser, more strained than in Black-capped. Often abbreviated, with just 1–2 notes before a single Dee; such short versions rare in other chickadees. C notes frequently given in separate Chitters or singly, as Chit.

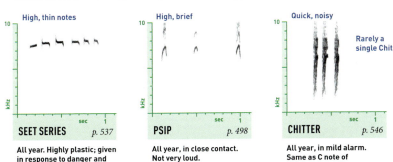

High, thin notes

High, brief

Quick, noisy

Rarely a single Chit

SEET SERIES — *p. 537*

All year. Highly plastic; given in response to danger and also in courtship.

PSIP — *p. 498*

All year, in close contact. Not very loud.

CHITTER — *p. 546*

All year, in mild alarm. Same as C note of Chick-a-dee call.

Black-capped Chickadee

Poecile atricapillus

Common and widespread in woodlands, parks, and towns; regular at bird feeders. Often flocks with other species, especially in winter.

"Fee bee" repeated 30–40 times at same pitch, then transposed higher or lower

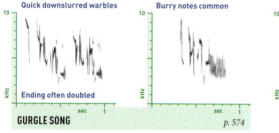

Typical version

Second note lower, often sounds 2-parted

Martha's Vineyard songtype

Notes in alternate songtypes often on the same pitch; rhythm variable

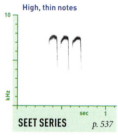

WHISTLED SONG: "Fee-bee" or "Hey, sweetie"
pp. 551, 567

Mostly Feb.–June. In spring, often heard in "song duels" between rival males. In most of North America, the "Fee-bee" (left) is the only songtype, and dueling males vary its pitch rather than its pattern. In some areas (such as Martha's Vineyard, right) males instead deploy 2–3 similar songtypes.

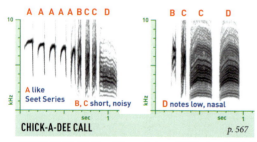

Quick downslurred warbles

Burry notes common

High, thin notes

Ending often doubled

GURGLE SONG
p. 574

SEET SERIES
p. 537

All year, especially in winter flocks. Brief musical gurgles, starting high and ending low; in this species given mostly by males, who average 8 Gurgle types apiece.

All year. Highly plastic; given in response to danger and also in courtship.

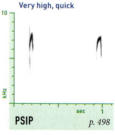

A A A A A B C C   D

B C   C   D

Very high, quick

A like Seet Series

B, C short, noisy

D notes low, nasal

CHICK-A-DEE CALL
p. 567

PSIP
p. 498

All year. Plastic; can have all four note types (ABCD, above left), but more often only AD, BC, AC, or BCD. Chitters of C notes (p. 546) and series of Dee notes common in agitation.

All year. The common contact call, given softly by calm birds.

CAROLINA CHICKADEE

Poecile carolinensis

Nearly identical to Black-capped; best identified by voice. Occasional hybrids with Black-capped may sound like either species, or intermediate.

Males average 2–3 songtypes; each repeated several times before switching

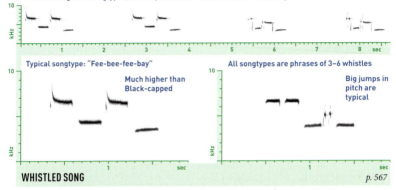

Typical songtype: "Fee-bee-fee-bay"

Much higher than Black-capped

All songtypes are phrases of 3–6 whistles

Big jumps in pitch are typical

WHISTLED SONG   *p. 567*

All year, but mostly Mar.–July. Males sing to defend territory and possibly to attract a mate; female Song evidently very rare. Much more variable than Black-capped Whistled Song; typically higher, with more notes. Alternating high and low notes are characteristic. Some songtypes are couplet series.

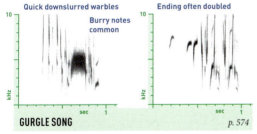

Quick downslurred warbles

Burry notes common

Ending often doubled

High, thin notes

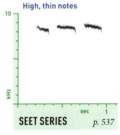

GURGLE SONG   *p. 574*

All year, by both sexes in aggression; also prior to copulation. Very similar to Black-capped Gurgle. Repertoire size unknown, but probably like Black-capped.

SEET SERIES   *p. 537*

All year. Highly plastic; given in response to danger and also in courtship.

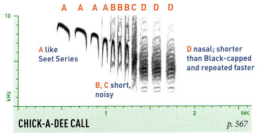

A   A   A   A B B B C D   D   D

A like Seet Series

D nasal; shorter than Black-capped and repeated faster

B, C short, noisy

CHICK-A-DEE CALL   *p. 567*

All year. Like Black-capped Chick-a-dee but distinctions between note types blurrier; Dee notes at end usually faster. More tendency to use C notes in separate Chitters (p. 546).

Very high, quick

PSIP   *p. 498*

All year. The common contact call, given softly by calm birds.

TUFTED TITMOUSE

Baeolophus bicolor

Common in woodlands and backyards; readily visits feeders. A loud and frequent vocalist, with an often bewildering variety of song variations and calls.

Males have 11–15 songtypes, typically repeated many times before switching

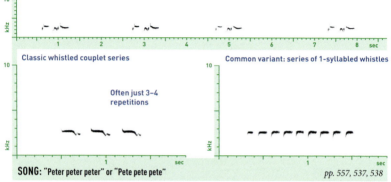

Classic whistled couplet series

Often just 3–4 repetitions

Common variant: series of 1-syllabled whistles

SONG: "Peter peter peter" or "Pete pete pete"   *pp. 557, 537, 538*

Almost all year, but especially Jan.–June. Familiar loud 2-noted "Peter peter" is most common, but some series are of single notes or triplets. Variants include whistled phrases (top, p. 550), sometimes with burry notes (p. 553). Always lower than Carolina Chickadee; much overlap with Black-crested Titmouse.

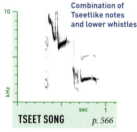

Combination of Tseetlike notes and lower whistles

TSEET SONG   *p. 566*

Little known, highly variable class of sounds, apparently given in aggression.

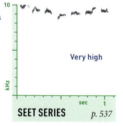

Very high

SEET SERIES   *p. 537*

Little known; given in many situations, often with Seer and Tsip.

Very high downslurred whistle; compare Seer calls of thrushes

SEER   *p. 503*

Possibly all year, in alarm and other situations. Sometimes in series.

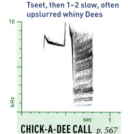

Tseet, then 1–2 slow, often upslurred whiny Dees

CHICK-A-DEE CALL *p. 567*

All year. Much shorter than similar Chick-a-dee calls, with long rising Dee notes.

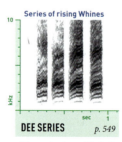

Series of rising Whines

DEE SERIES   *p. 549*

All year, in alarm. Like last note of Chick-a-dee call, and intergrades with it.

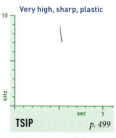

Very high, sharp, plastic

TSIP   *p. 499*

All year, apparently to maintain contact with flockmates.

Black-crested Titmouse

Baeolophus atricristatus

Closely related to Tufted Titmouse. Hybrids, which are frequent in a stable zone where the ranges meet, have intermediate crest and forehead color.

Males have 9–11 songtypes, typically repeated many times before switching

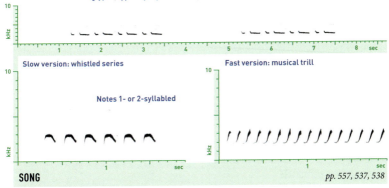

Slow version: whistled series

Notes 1- or 2-syllabled

Fast version: musical trill

SONG *pp. 557, 537, 538*

Almost all year, but especially Jan.–June. Highly variable. Series average longer and faster than in Tufted Titmouse; trilled songtypes are fairly distinctive, but where ranges meet, Songs are rarely separable. Songs from the Big Bend region differ from those elsewhere, often including syncopated series (top).

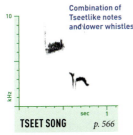

Combination of Tseetlike notes and lower whistles

TSEET SONG *p. 566*

Like Tufted Titmouse Tseet Song. At least in Big Bend, can start with Chatters.

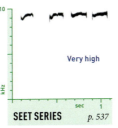

Very high

SEET SERIES *p. 537*

Little known; given in many situations, often with Seer and Tsip.

Very high downslurred whistle; compare Seer calls of thrushes

SEER *p. 503*

Possibly all year, in alarm and other situations. Sometimes in series.

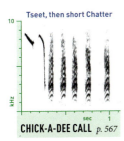

Tseet, then short Chatter

CHICK-A-DEE CALL *p. 567*

All year, usually with more and shorter Dee notes than Tufted Titmouse.

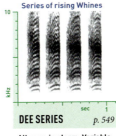

Series of rising Whines

DEE SERIES *p. 549*

All year, in alarm. Variable, often faster than Tufted; notes can be 2-syllabled.

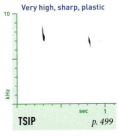

Very high, sharp, plastic

TSIP *p. 499*

All year, apparently to maintain contact with flockmates.

BUSHTIT (Family Aegithalidae)

The Bushtit is related to the long-tailed tits of Eurasia, which are only distantly related to chickadees and titmice. Vocalizations are simple and apparently innate.

VERDIN (Family Remizidae)

The Verdin is related to the penduline tits of Europe and Africa, in a sister group to chickadees and titmice. Song is likely learned.

NUTHATCHES (Family Sittidae)

Nuthatches cling to tree trunks like miniature woodpeckers, but are as likely to face downward as upward. Vocalizations are nasal or squeaky; songs may be learned.

CREEPERS (Family Certhiidae)

Creepers cling to tree trunks like nuthatches, but always facing skyward. Songs are complex, musical, and learned, with many local dialects.

BUSHTIT

Psaltriparus minimus

Forms large, active flocks, usually in dense bushes or low trees, sometimes in residential areas. Constant simple, quiet calls often announce its presence.

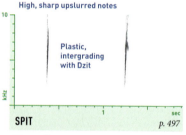

High, sharp upslurred notes

Plastic, intergrading with Dzit

SPIT   *p. 497*

All year. Most common contact call, given almost constantly by flocks.

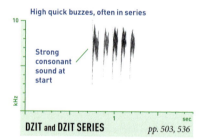

High quick buzzes, often in series

Strong consonant sound at start

DZIT and DZIT SERIES   *pp. 503, 536*

All year, in territorial encounters, in mobbing predators, or in response to humans. Plastic.

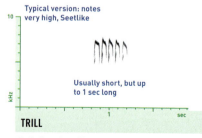

Typical version: notes very high, Seetlike

Usually short, but up to 1 sec long

TRILL

Hoarser, buzzier version

First note typically longest, loudest, highest

p. 537

Typical version of Trill given all year, by adult birds separated from the flock or warning of aerial predators. Some species of ground squirrel give very similar calls. Hoarse version given mostly May–Aug., by begging fledglings, but this or a similar call also given in various situations by highly agitated adults.

VERDIN

Auriparus flaviceps

Common and active, singly or in pairs, in deserts with thorny vegetation. Gives frequent simple, distinctive, rather loud vocalizations.

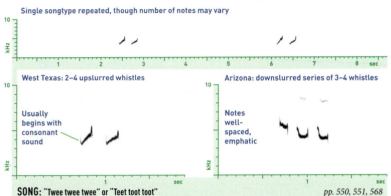

Single songtype repeated, though number of notes may vary

West Texas: 2–4 upslurred whistles

Usually begins with consonant sound

Arizona: downslurred series of 3–4 whistles

Notes well-spaced, emphatic

SONG: "Twee twee twee" or "Teet toot toot"   *pp. 550, 551, 568*

Mostly Feb.–July. Shows regional dialects, indicating that Song is likely learned; each male may have only a single songtype, but consecutive renditions can vary slightly. Males apparently use Tew to attract females; Song begins only after pairs form, so may be territorial. More study needed.

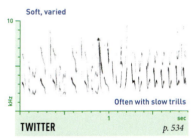

Soft, varied

Often with slow trills

TWITTER   *p. 534*

Feb.–July. Plastic; usually between mates, but sometimes by lone birds. May be learned; function unclear. Needs more study.

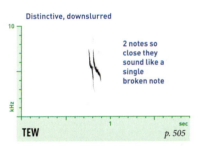

Distinctive, downslurred

2 notes so close they sound like a single broken note

TEW   *p. 505*

All year, but mostly Feb.–July, when males give it from high perches for long periods. Only slight geographic variation.

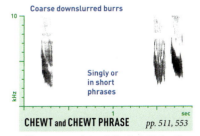

Coarse downslurred burrs

Singly or in short phrases

CHEWT and CHEWT PHRASE   *pp. 511, 553*

All year; somewhat variable. Probably functions as an alarm call, but more study needed.

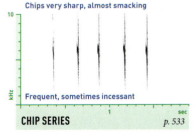

Chips very sharp, almost smacking

Frequent, sometimes incessant

CHIP SERIES   *p. 533*

All year. Most common call. Speed correlates with agitation; excited birds can accelerate into rapid chipping Twitters.

RED-BREASTED NUTHATCH

Sitta canadensis

Resident in northern coniferous forests; wanders irregularly south in winter. Often visits feeders. Usually excavates its own nesting cavities.

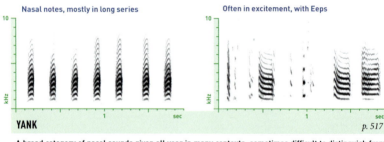

Slow Song continues at steady pace for long periods, occasionally a minute or more

Notes slow, slightly rising

Rapid series or slow trill

SLOW SONG *p. 545*

FAST SONG *p. 544*

All year, by both sexes, especially Mar.–Aug. Uniform. Individuals apparently have 2 distinct song patterns. Slow Song given in courtship, mostly before pairing; Fast Song apparently more territorial, predominating after pairing. Fast Song is 1–2 seconds long, repeated after a pause.

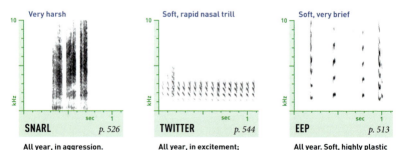

Nasal notes, mostly in long series

Often in excitement, with Eeps

YANK *p. 517*

A broad category of nasal sounds given all year in many contexts; sometimes difficult to distinguish from Song. Short versions almost always given in long series. Longer versions, like individual notes of Slow Song, given mostly in high agitation, with Eeps.

Very harsh

Soft, rapid nasal trill

Soft, very brief

SNARL *p. 526*

TWITTER *p. 544*

EEP *p. 513*

All year, in aggression. Plastic; often in series. Heard frequently at feeders.

All year, in excitement; made of Eeplike notes. Plastic.

All year. Soft, highly plastic contact calls, inaudible at a distance.

WHITE-BREASTED NUTHATCH

Sitta carolinensis

Most common in mature deciduous forests, including some residential areas. Rocky Mountain and West Coast forms differ strongly in vocalizations.

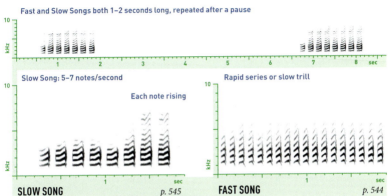

Fast and Slow Songs both 1–2 seconds long, repeated after a pause

Slow Song: 5–7 notes/second

Each note rising

Rapid series or slow trill

SLOW SONG p. 545

FAST SONG p. 544

All year, by males, but especially Jan.–May. Females not known to sing. Unlike in Red-breasted Nuthatch, Fast and Slow Songs are not well differentiated; given in similar contexts, apparently ends of a continuum of variation. Repertoire size unknown. Always higher than Red-breasted Nuthatch Song.

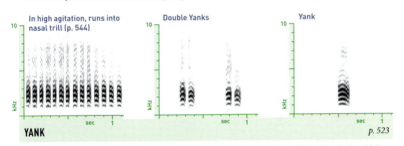

In high agitation, runs into nasal trill (p. 544)

Double Yanks

Yank

YANK p. 523

A broad category of nasal sounds given all year in many contexts; sometimes difficult to distinguish from Song. Trilled version given in high alarm and when mobbing predators. Double Yanks and Yanks given in mild alarm and in various interactions. Note distinctive slight burriness in most versions.

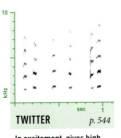
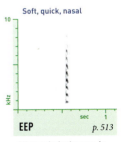

Soft seminasal downslur

Soft, quick, nasal

KEER p. 516

TWITTER p. 544

EEP p. 513

Soft, infrequent, by both sexes in courtship and high excitement.

In excitement, gives high Eeplike notes, but apparently not in trills.

All year, by both sexes, in close contact. Plastic. Audible only at close range.

Brown-headed Nuthatch

Sitta pusilla

The smallest nuthatch in the East. Found in vocal flocks, usually in pines. Often breeds in cooperative groups, several adults raising a single brood.

Skew-doo sometimes given for long periods from prominent perches, like a territorial song

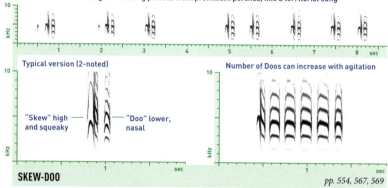

Typical version (2-noted)

"Skew" high and squeaky — "Doo" lower, nasal

Number of Doos can increase with agitation

SKEW-DOO

pp. 554, 567, 569

All year, by both sexes, in many situations; most common call. Often compared to the squeaking of a rubber dog toy. Slightly variable and highly plastic, especially in number of Doo notes. Skew notes also given separately.

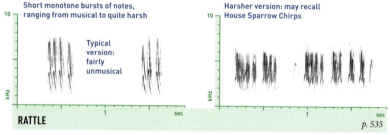

Short monotone bursts of notes, ranging from musical to quite harsh

Typical version: fairly unmusical

Harsher version: may recall House Sparrow Chirps

RATTLE

p. 535

All year, by both sexes, during interactions, especially in excitement. Highly plastic, varying widely in tone; different versions may serve different functions. Harshest versions take on a snarling, rasping quality; musical versions grade into Twitter.

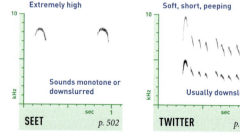

Extremely high

Sounds monotone or downslurred

SEET

p. 502

Associated with begging of food from mate; similar calls by begging juveniles.

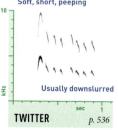

Soft, short, peeping

Usually downslurred

TWITTER

p. 536

All year. Function little known; may be a variant of Rattle.

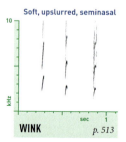

Soft, upslurred, seminasal

WINK

p. 513

All year, in contact; highly plastic, varying in pitch. Given constantly by flocks.

BROWN CREEPER

Certhia americana

Common in coniferous and mixed forests, but often inconspicuous. Slowly scales tree trunks like a woodpecker in upward spirals.

Dawn Song: stereotyped sequences of Tseews and Seets, often with complex Songs interspersed

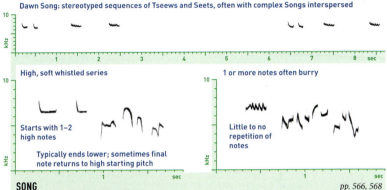

High, soft whistled series

Starts with 1–2 high notes

Typically ends lower; sometimes final note returns to high starting pitch

1 or more notes often burry

Little to no repetition of notes

SONG

pp. 566, 568

Apr.–July. Most males have only one songtype; some have two. Dawn Song begins with short repeated stereotyped phrases composed of Tseew- and Seetlike notes; as sun rises, these phrases become more variable and Song becomes more frequent. Dawn Song pattern may continue into late morning.

Rapid series of high Seetlike notes

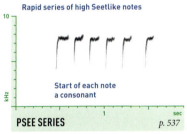

Start of each note a consonant

PSEE SERIES

p. 537

Mostly Apr.–July. Given infrequently, by males during territorial encounters; also perhaps in some other situations.

High, coarsely burry

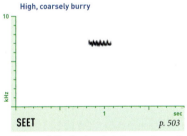

SEET

p. 503

All year. Evidently given by both sexes in a variety of situations. One of the loudest calls.

High, very brief

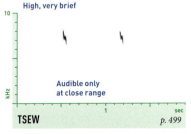

Audible only at close range

TSEW

p. 499

All year, in close contact. Also given in flight, but apparently not by night migrants.

Longer, louder than Tsew

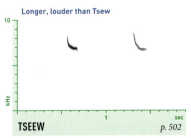

TSEEW

p. 502

All year, in alarm, during mobbing of predators, and during some territorial encounters.

WRENS (Family Troglodytidae)

The wrens are some of our most familiar songbirds, with one or more species common in almost every North American habitat. Most species are small and inconspicuously plumaged, but highly vocal, with loud, complex musical Songs. By contrast, the Cactus Wren of the desert Southwest is large and boldly patterned, with a monotonous, unmusical Song.

Subtle differences in vocalizations distinguish eastern and western populations of some species, including House Wren and Marsh Wren. Western populations of the Winter Wren were recently split into a separate species, Pacific Wren (*Troglodytes pacificus*), based in part on vocal differences. Field identification of these (sub-) species pairs can be difficult, sometimes requiring spectrographic analysis.

WINTER WREN

Troglodytes hiemalis

Tiny and shy, skulking in thick tangles and piles of downed wood. Breeds mostly in moist coniferous forests; winters in brushy woodlands.

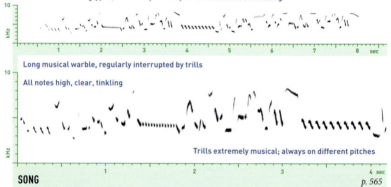

Males have 2–3 songtypes; each is repeated up to 40 times before switching

Long musical warble, regularly interrupted by trills

All notes high, clear, tinkling

Trills extremely musical; always on different pitches

SONG
p. 565

Mostly Mar.–Aug. One of the most remarkable bird Songs in North America, sweetly musical and extremely complex; rarely shorter than 4 seconds and sometimes longer than 30 seconds. Usually sings from a low exposed perch, but sometimes from high in trees; rarely in flight. Unknown whether female sings.

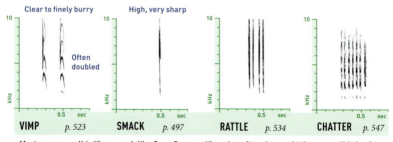

Clear to finely burry

Often doubled

High, very sharp

VIMP p. 523   **SMACK** p. 497   **RATTLE** p. 534   **CHATTER** p. 547

Most common call is Vimp, much like Song Sparrow Vimp, but often clearer. At times, possibly in alarm, Vimp grades into Smack, like call of Pacific Wren, which is sometimes run into rapid Rattle. In agitation, may also give a nasal Chatter with quality much like Vimp, as well as a soft Snarl (not shown).

House Wren

Troglodytes aedon

Common and widespread in brushy woodland edges, parks, and residential areas, where it nests in almost any type of small cavity.

Males estimated to sing 20–50 similar songtypes; consecutive songs usually same

(Eastern)

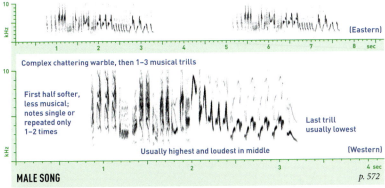

Complex chattering warble, then 1–3 musical trills

First half softer, less musical; notes single or repeated only 1–2 times

Last trill usually lowest

Usually highest and loudest in middle

(Western)

MALE SONG   *p. 572*

Mostly Mar.–July; sings occasionally in winter. Variable; songs tend to be shorter in West, with a Beertlike note near start. In close courtship or territorial disputes, can be given at rapid pace, one Song nearly running into the next. Subsong frequent fall through early spring; can suggest Winter Wren, but lower, more plastic.

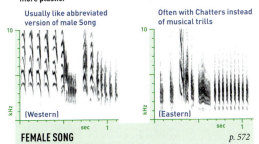

Usually like abbreviated version of male Song

Often with Chatters instead of musical trills

(Western)

(Eastern)

FEMALE SONG   *p. 572*

Infrequent, by females who have lost track of their mates, and in female-female aggression. Hugely variable and plastic, ranging from 1-note Squeals to versions matching male Song.

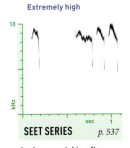

Extremely high

SEET SERIES   *p. 537*

In close courtship, often between Songs. Plastic; notes sometimes Chiplike.

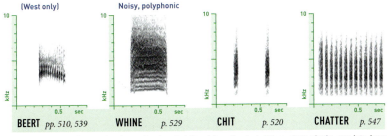

(West only)

Noisy, polyphonic

BEERT   *pp. 510, 539*   **WHINE**   *p. 529*   **CHIT**   *p. 520*   **CHATTER**   *p. 547*

All calls given all year, in alarm. Semimusical Beert heard only in West; given singly, run into long Sputters, or incorporated into Song. Plastic Whine recalls gnatcatchers or Gray Catbird. Chit varies from longer than shown to shorter, more like Dzik of Bewick's. Chatter noisy, variable in speed.

ROCK WREN

Salpinctes obsoletus

Found on open rocky slopes and eroded
bluffs. Nests in crevices; usually builds
a "pavement" of stones in front of the
nest, the purpose of which is unknown.

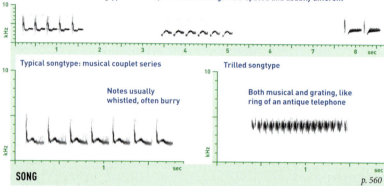

Individuals have 100 songtypes or more; consecutive Songs well-spaced and usually different

Typical songtype: musical couplet series

Notes usually
whistled, often burry

Trilled songtype

Both musical and grating, like
ring of an antique telephone

SONG
p. 560

Mostly Mar.–Aug., apparently by males. Most songtypes are simple series or couplet series; some are single
trills; a very small number are burry or buzzy phrases. Birds choose from the same 3–6 songtypes repeatedly for a few minutes, then switch to a new set.

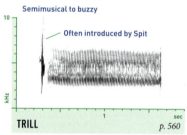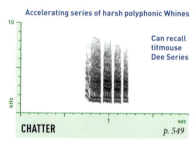

Semimusical to buzzy

Often introduced by Spit

TRILL
p. 560

Mostly May–Aug., in defense of nest or fledged
young. Highly plastic, especially in pitch; tone
may recall chirping cricket.

Accelerating series of harsh polyphonic Whines

Can recall
titmouse
Dee Series

CHATTER
p. 549

Mar.–May, at least, in agitation and during pair
interactions. Variable and plastic. Juveniles beg
with similar notes.

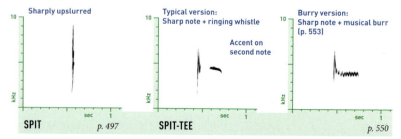

Sharply upslurred

SPIT
p. 497

Typical version:
Sharp note + ringing whistle

Accent on
second note

SPIT-TEE

Burry version:
Sharp note + musical burr
(p. 553)

p. 550

All year, by both sexes. Spit given in high agitation, often with Whine Series. Spit-tee given in a variety of
circumstances, the bird vigorously bobbing its body with each call. Variable and plastic; different versions
may serve slightly different functions.

CANYON WREN

Catherpes mexicanus

Almost never found away from vertical rock faces, usually in arid country. More often heard than seen, but often approachable when found.

Repeats 1 songtype many times before switching; excited birds may double Songs, as here

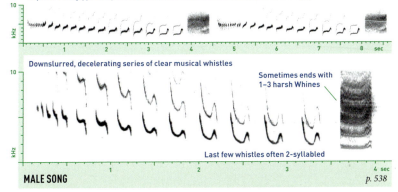

Downslurred, decelerating series of clear musical whistles

Sometimes ends with 1–3 harsh Whines

Last few whistles often 2-syllabled

MALE SONG *p. 538*

All year, but mostly Feb.–Aug. Variable, but distinctive for its musicality and decelerating, downslurred pattern. Individual notes may be upslurred, downslurred, or underslurred; speed at start varies greatly between songtypes. Repertoire size not fully known, but males appear to have at least 3 songtypes.

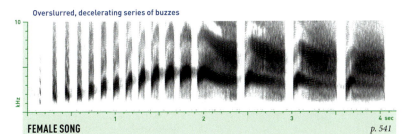

Overslurred, decelerating series of buzzes

FEMALE SONG *p. 541*

All year. Distinctive but infrequent; first notes like Veet, last ones slightly harsher. Apparently given only by females, either in response to male Song or in territorial conflicts with other females. When duetting with mate, female often begins singing halfway through male's Song.

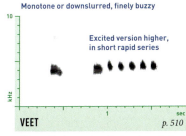

Monotone or downslurred, finely buzzy

Excited version higher, in short rapid series

VEET *p. 510*

All year. Most common call, given in contact and mild alarm.

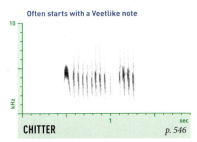

Often starts with a Veetlike note

CHITTER *p. 546*

Mostly from agitated adults near the nest or fledged young. Highly plastic; sometimes shortened to a single Chit.

CAROLINA WREN

Thryothorus ludovicianus

Vocally conspicuous in many wooded or brushy habitats, including near human habitation. Range has expanded slowly northward in recent decades.

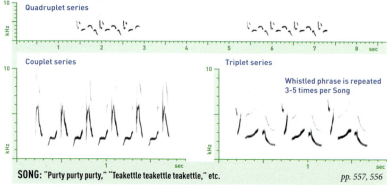

Males have 20–50 songtypes; consecutive songs are nearly always similar

Quadruplet series

Couplet series

Triplet series

Whistled phrase is repeated 3–5 times per Song

SONG: "Purty purty purty," "Teakettle teakettle teakettle," etc.   *pp. 557, 556*

All year, by males; females not known to sing. Triplet series are most common, then couplet series, then quadruplet series; quintuplet series are rare. Males match songtypes during song duels; if one lacks a songtype, he may substitute Beert or equivalent. Songs of S. Texas birds average lower, slower, simpler.

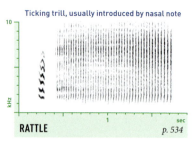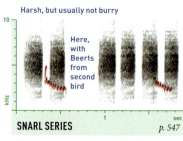

Ticking trill, usually introduced by nasal note

RATTLE   *p. 534*

All year, reportedly only by females, often in response to and overlapping male Song. Also used in female-female aggression.

Harsh, but usually not burry

Here, with Beerts from second bird

SNARL SERIES   *p. 547*

All year, in alarm; speed usually 2–3 notes/second. Agitated females may introduce series with 1–2 Pips; the result recalls chickadees.

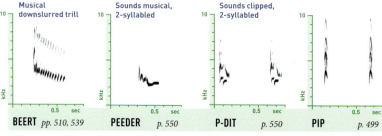

Musical downslurred trill

BEERT   *pp. 510, 539*

Sounds musical, 2-syllabled

PEEDER   *p. 550*

Sounds clipped, 2-syllabled

P-DIT   *p. 550*

PIP   *p. 499*

All year, in mild alarm and contact. First 3 are male versions; Beert is most widespread and distinctive, but replaced in some areas by Peeder, P-dit, or other versions, especially in southern part of range. Pip is female version, occasionally run into short trill; compare Vimp of Winter Wren.

BEWICK'S WREN

Thryomanes bewickii

A prodigious singer found in brushy areas, open woods, and junipers. Formerly common in the East, but now rare east of the Mississippi.

 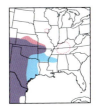

Males have 10–20 songtypes; consecutive Songs almost always similar

Long version: single notes, buzzes, and musical trills

Ends with 1–2 musical trills or series

Often starts with dissimilar single notes

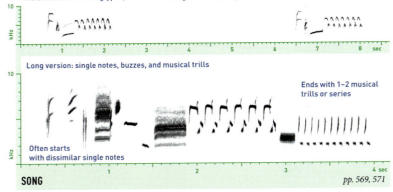

SONG pp. 569, 571

Nearly all year. Extremely varied, ranging from 1 to 4 seconds long, but always brightly musical. Long versions can be confused with Song Sparrow Song; short versions (top) have just 1–3 notes before a trill, recalling Black-throated Sparrow or Spotted Towhee. Female not known to sing.

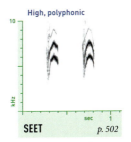

High, polyphonic

SEET p. 502

Given continuously between Songs by some males; also singly in other situations.

Notes often Dziklike

CHITTER p. 546

All year, in agitation. Highly individual; some versions rather rattling.

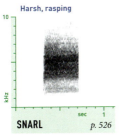

Harsh, rasping

SNARL p. 526

All year, in alarm. Sometimes given in series.

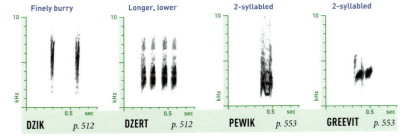

Finely burry

DZIK p. 512

Longer, lower

DZERT p. 512

2-syllabled

PEWIK p. 553

2-syllabled

GREEVIT p. 553

This species gives a bewildering array of 1- and 2-syllabled buzzy or noisy calls. Individuals usually give only one call type at a time, but one bird gave both Greevits and Dziks in mild agitation. Most common calls are 1-syllabled; some are intermediate between Dziks and Seets.

MARSH WREN (WESTERN)

Cistothorus palustris (*paludicola* group)

Locally common in freshwater cattail and bulrush marshes and coastal salt-marshes. Vocally conspicuous in breeding season; otherwise rather skulking.

In excitement, some Songs are extended into complex warbles with many noisy notes (p. 565)

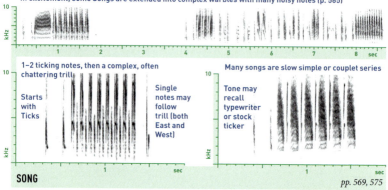

1–2 ticking notes, then a complex, often chattering trill

Starts with Ticks

Single notes may follow trill (both East and West)

Many songs are slow simple or couplet series

Tone may recall typewriter or stock ticker

SONG

pp. 569, 575

Mostly Mar.–July, by males only, sometimes at night. Subsong common Aug.–Oct. and Jan.–Apr. Males have 100–220 similar songtypes; consecutive songs almost always differ. In some Songs the Ticks are followed by a buzz before the trill (top left), or by two consecutive trills. Some Songs resemble Sedge Wren's.

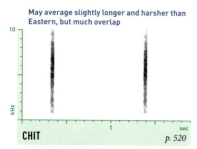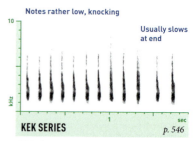

May average slightly longer and harsher than Eastern, but much overlap

CHIT

p. 520

All year, by both sexes, in contact and mild alarm; most common call. Fairly uniform. Juveniles give a sharp clear Chip (p. 497).

Notes rather low, knocking

Usually slows at end

KEK SERIES

p. 546

Little known; apparently given infrequently by territorial males. Western birds also give a Rasp like Eastern; no recordings available.

EASTERN AND WESTERN MARSH WRENS

Populations of Marsh Wren sort into two broad groups based on subtle differences in song forms and singing behavior. In addition to differences in introductory notes and the structure of terminal trills or series, Western males have larger repertoires and sing a greater variety of different-sounding songtypes. Call repertoires also apparently differ. In the zone of overlap between the two forms in the Great Plains, some interbreeding occurs, and some birds sing "mixed" songs with both Eastern and Western components. However, the majority of birds do not form mixed pairs. Outside the breeding season, subsinging birds may be identified with caution by general similarity of their songs to those of Eastern or Western breeding males.

Marsh Wren (Eastern)

Cistothorus palustris (*palustris* group)

Nearly identical to the western group of subspecies, though some coastal populations are distinctively colored. Habits and habitat are similar.

Soft Twitters and musical warbles may precede or follow Songs, especially in coastal birds (p. 565)

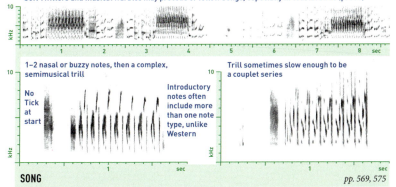

1–2 nasal or buzzy notes, then a complex, semimusical trill

No Tick at start

Introductory notes often include more than one note type, unlike Western

Trill sometimes slow enough to be a couplet series

SONG   *pp. 569, 575*

Mostly Mar.–July, by males only, sometimes at night. Subsong common Aug.–Oct. and Jan.–Apr. Males have 30–70 similar songtypes; consecutive songs usually differ, but 2 songtypes may alternate, or 1 repeat a few times before switching. Usually lacks harsh Chatters; doubled trills rarer than in West.

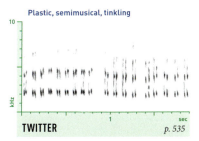

Plastic, semimusical, tinkling

TWITTER   *p. 535*

Little known; sometimes given in between songs, mostly by coastal birds. May recall distant Killdeer Trill.

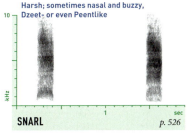

Harsh; sometimes nasal and buzzy, Dzeet- or even Peentlike

SNARL   *p. 526*

Similar to the notes that introduce Song; given separately by males during nestbuilding and sometimes between Songs.

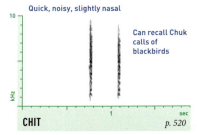

Quick, noisy, slightly nasal

Can recall Chuk calls of blackbirds

CHIT   *p. 520*

All year, by both sexes, in contact and mild alarm; most common call. Fairly uniform. Juveniles give a sharp clear Chip (p. 497).

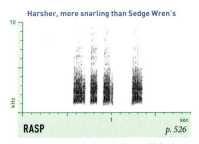

Harsher, more snarling than Sedge Wren's

RASP   *p. 526*

Given infrequently, in agitation or high alarm. Lower, harsher, and longer than Chit.

SEDGE WREN

Cistothorus platensis

A nomadic and opportunistic breeder in wet meadows of grass or sedges; may raise more than one brood per summer in widely separate locations.

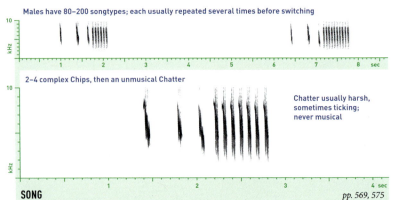

Males have 80–200 songtypes; each usually repeated several times before switching

2–4 complex Chips, then an unmusical Chatter

Chatter usually harsh, sometimes ticking; never musical

SONG *pp. 569, 575*

Mostly Feb.–July, by males. Females not known to sing. Unlike Marsh Wren, males usually repeat each songtype 10–20 times, but at dawn and in territorial conflicts, consecutive songs often differ. Many songtypes are very similar to the human ear.

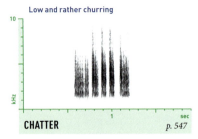

Low and rather churring

CHATTER *p. 547*

All year, by agitated birds, often with other calls. Highly plastic, but usually short (less than 2 seconds), unlike long Chatters of some other wrens.

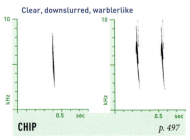

Clear, downslurred, warblerlike

CHIP *p. 497*

Apparently all year. Function and context little known. Generally higher than Chip of juvenile Marsh Wren.

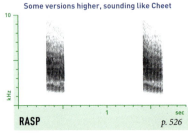

Some versions higher, sounding like Cheet

RASP *p. 526*

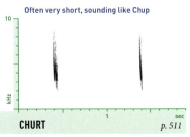

Often very short, sounding like Chup

CHURT *p. 511*

All year. Churt is the most common call; Rasp given in higher agitation. Both are plastic, intergrading with each other and variants. Compare to Chip and Churt of Common Yellowthroat (p. 511).

CACTUS WREN

Campylorhynchus brunneicapillus

By far our largest wren, conspicuous both visually and vocally in southwestern deserts and arid shrublands, including near human habitation.

Males can have 14 or more similar songtypes; each usually repeated several times before switching

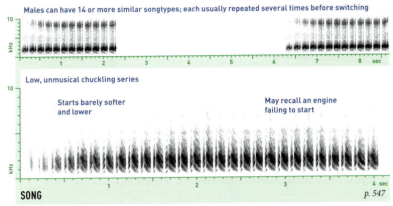

Low, unmusical chuckling series

Starts barely softer and lower

May recall an engine failing to start

SONG *p. 547*

All year, mostly by males. Female Song reportedly infrequent, higher and softer than male's. Rather uniform rangewide; usually 3–4 sec in length. Some versions fast enough to be called a slow trill. Some versions are couplet series.

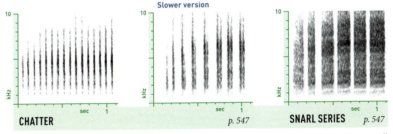

Slower version

CHATTER

SNARL SERIES *p. 547*

A set of plastic, intergrading calls given by both sexes in agitation or alarm. Chatters infrequent; can recall oriole Chatters, but usually more rasping and plastic. Series of noisy Keklike notes given frequently in mild alarm, grading into longer Snarl Series when mobbing potential predators.

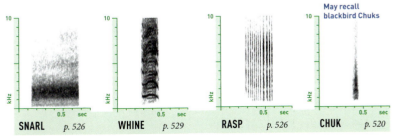

May recall blackbird Chuks

SNARL *p. 526*   **WHINE** *p. 529*   **RASP** *p. 526*   **CHUK** *p. 520*

A confusing array of calls, some of which grade into one another, but several of which likely serve different functions. Snarls and possibly Rasps given by mated pairs in greeting, Chuks in contact between adults and young. Whines may be given in high agitation.

GNATCATCHERS (Family Polioptilidae)

Some of our smallest and most active birds, gnatcatchers are slender, thin-billed, long-tailed, and subtly patterned. All species frequently flick their tails while gleaning insects from dense foliage.

Gnatcatcher vocal communication is highly complex and not well understood; this book only roughly outlines each species' repertoire. All vocalizations of a given species grade into one another, and nearly all occur in a variety of behavioral contexts. Many gnatcatchers, including Blue-gray but apparently not Black-tailed, include imitations of other bird species in their Complex Songs. All species snap their bills in alarm and aggression.

BLACK-TAILED GNATCATCHER

Polioptila melanura

A bird of desert thorn scrub, especially mesquite and creosote thickets. The most vocally distinctive gnatcatcher in North America.

Complex Song (p. 560): harsh, snarling notes with orderly series of Chips and Tinks interspersed

Series of harsh, noisy notes

Length of notes and series variable, but tempo usually even

Faster series associated with higher levels of excitement

SIMPLE SONG

p. 547

Variable; Simple and Complex Song grade smoothly into one another, and into fast scolding Chatter that can be indistinguishable from Chatter of House Wren. Even more complex versions of Song exist with less repetition, more like Blue-gray Complex Song, but always dominated by harsh notes.

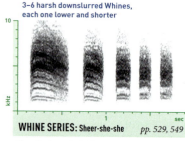

3–6 harsh downslurred Whines, each one lower and shorter

WHINE SERIES: Sheer-she-she *pp. 529, 549*

Given in alarm, including in response to humans. Whinier, less noisy than other Black-tailed sounds; easy to confuse with Blue-gray.

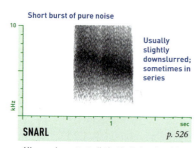

Short burst of pure noise

Usually slightly downslurred; sometimes in series

SNARL *p. 526*

All year, in contact; distinctively harsh, without any musicality. Like sound of letters "sh," but with abrupt start and end.

Blue-gray Gnatcatcher

Polioptila caerulea

Widespread in moist deciduous woods and woodland edges. Active and arborial, gleaning insects from leaves and occasionally catching them in midair.

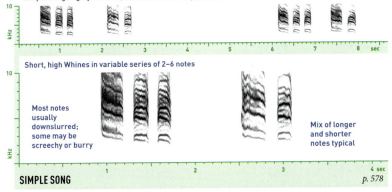

Simple Song: highly variable, but short series or parts of series sometimes repeat

Short, high Whines in variable series of 2–6 notes

Most notes usually downslurred; some may be screechy or burry

Mix of longer and shorter notes typical

SIMPLE SONG
p. 578

Given frequently, apparently serving various functions including territorial advertisement. Often mistaken for a series of call notes; best distinguished from calls by repeating patterns, syncopated rhythms, and variety of note types. Grades smoothly into Complex Song; intermediates common.

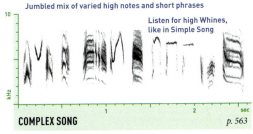

Jumbled mix of varied high notes and short phrases

Listen for high Whines, like in Simple Song

COMPLEX SONG
p. 563

Usually soft, not carrying far; by males in close courtship or in boundary disputes. May be continuous or given in short bursts; notes occasionally repeated. Often includes some mimicry.

High, downslurred Whine

Given singly; sometimes harsh

SHEER
p. 529

All year, in contact. Quality similar to notes of Simple Song.

"WESTERN" BLUE-GRAY GNATCATCHER

Blue-gray Gnatcatchers breeding west of central Texas and western Oklahoma prefer drier, shrubbier habitats than Eastern birds and differ slightly in vocalizations. Sheer calls average lower and harsher, but may not be consistently identifiable. Simple Songs are lower and less burry, with most notes distinctly overslurred.

BULBULS (Family Pycnonotidae)

following page

Two species from this Old World family have established feral populations in southern U.S. cities, in residential neighborhoods with mature exotic trees. Songs appear to be learned. At least in the Red-vented Bulbul, both sexes sing.

Red-vented Bulbul

Pycnonotus cafer

Native to South Asia; a small population has been feral in the urban Houston area since at least the 1990s. Sometimes visits feeders.

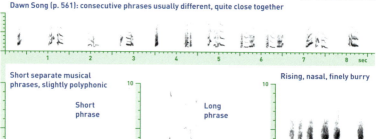

Dawn Song (p. 561): consecutive phrases usually different, quite close together

Short separate musical phrases, slightly polyphonic

Short phrase

Long phrase

Rising, nasal, finely burry

SONG

pp. 550, 568

VRIT

p. 523

Song mostly Jan.–Mar. in Texas, by both sexes. During day, Song phrases often 2–3 sec apart, and similar phrases often repeated before switching. Individuals apparently have many different phrases. Vrit is most common call, somewhat variable; often given in rapid series.

Red-whiskered Bulbul

Pycnonotus jocosus

Native to South Asia; since 1960, subspecies *emeria* established locally in urban Miami area, where it eats mostly fruit from ornamental trees.

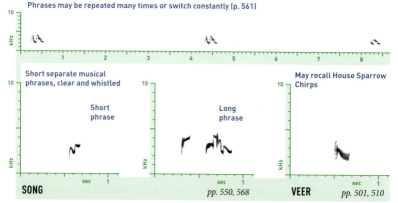

Phrases may be repeated many times or switch constantly (p. 561)

Short separate musical phrases, clear and whistled

Short phrase

Long phrase

May recall House Sparrow Chirps

SONG

pp. 550, 568

VEER

pp. 501, 510

Song mostly Jan.–Mar. in Florida, likely by both sexes. During day, Song phrases often 2–3 sec apart, similar phrases often repeated before switching. Individuals apparently have many different phrases. Dawn singing pattern not known. Most common calls sound like Veer or Veet; highly plastic.

KINGLETS (Family Regulidae)

Charismatic and active, these are the smallest birds in North America outside the hummingbird family. Despite their tiny size, they have large families, both occasionally laying clutches of more than 10 eggs. Both readily join mixed-species flocks in migration and winter.

Although similar in shape and plumage, our two kinglets are not particularly closely related, as their vocalizations attest. All Golden-crowned's vocalizations are rather soft and extremely high-pitched, much like Brown Creeper's, while Ruby-crowned has a loud, rich, rollicking song and a low, noisy call note.

GOLDEN-CROWNED KINGLET

Regulus satrapa

Found mostly in spruce trees year-round. Calls very like those of Brown Creeper, with which it often flocks outside the breeding season.

Sometimes repeats introductory Seet Series by itself (p. 537), much like Cape May Warbler song

Accelerating, overslurred series of Seetlike notes, ending in a lower, softer Chatter

Starts much like Ruby-crowned Kinglet Song

Ending reminiscent of Chick-a-dee calls

SONG *p. 567*

Mostly May–July. Slight geographic variation. Structure like the Chick-a-dee calls of parids (p. 567), with several different note types always in the same order, each repeated 0–6 times. Males appear to have only 1 songtype, but consecutive songs may vary considerably in number and type of notes.

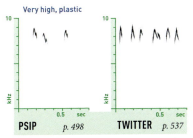

Very high, plastic

PSIP *p. 498*

TWITTER *p. 537*

Psip given all year, in close contact; can run into plastic Twitter. 2-noted Psittip (not shown) given constantly by fledglings in late summer.

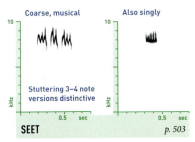

Coarse, musical

Also singly

Stuttering 3–4 note versions distinctive

SEET *p. 503*

All year. Single-note versions, often given in alarm, average higher, less musical than Brown Creeper's.

RUBY-CROWNED KINGLET

Regulus calendula

Breeds in coniferous and mixed forests; found in various habitats in migration and winter. Male's red crown inconspicuous unless flared in agitation.

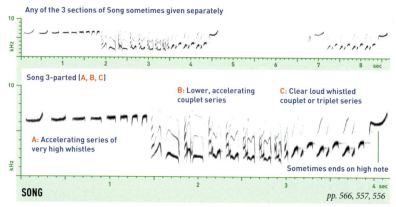

Any of the 3 sections of Song sometimes given separately

Song 3-parted (A, B, C)

A: Accelerating series of very high whistles

B: Lower, accelerating couplet series

C: Clear loud whistled couplet or triplet series

Sometimes ends on high note

SONG

pp. 566, 557, 556

Mostly Feb.–Aug.; versions on spring migration often plastic. Highly variable. Repertoire size unknown; individuals may have a single songtype. Female Song is shorter, apparently like first 2 parts of male Song; more study needed.

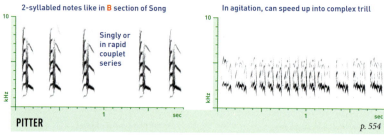

2-syllabled notes like in B section of Song

Singly or in rapid couplet series

In agitation, can speed up into complex trill

PITTER

p. 554

Mostly May–Aug. Given incessantly by alarmed or anxious birds. Highly variable; likely learned rather than innate. Second syllable may be higher or lower than first.

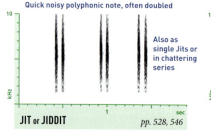

Quick noisy polyphonic note, often doubled

Also as single Jits or in chattering series

JIT or JIDDIT

pp. 528, 546

All year. Most common call, particularly in mixed-species foraging or mobbing flocks, but also from solo birds.

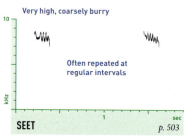

Very high, coarsely burry

Often repeated at regular intervals

SEET

p. 503

Mostly July–Aug., from begging juveniles; plastic. Very similar to burry Seets of Golden-crowned Kinglet and Brown Creeper.

THRUSHES (Family Turdidae)

The thrushes are among our most celebrated singers. They are remarkable aesthetically, because unlike many other birds, their fluting, polyphonic voices often contain harmonies similar to those in traditional Western music, resulting in a beauty easily appreciated by humans. But thrushes are also remarkable behaviorally, displaying highly developed vocal communication systems with multiple Song categories and large call repertoires. They are favorites of night migration enthusiasts, thanks to their loud, frequent, and rather distinctive nocturnal flight calls.

Excited Songs of Thrushes

Most North American thrushes have at least two distinct modes of singing: a loud, musical song used for long-distance communication, and a softer, more complex, often faster song used in close-range courtship and/or aggressive encounters. In the genus *Catharus* (Veery and Gray-cheeked, Bicknell's, Swainson's, and Hermit Thrushes), the Excited Song is much like the regular Song, but with additional components either in between the regular Song phrases, or appended to them. These additional components are often complex, polyphonic, and trilled.

Hermit Thrush: typical Song, with complex polyphonic phrases in between, or appended to end

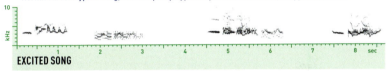

Veery: soft versions of typical Song with complex polyphonic trills appended at start

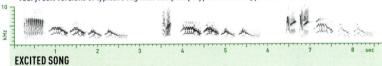

Swainson's Thrush: various calls and polyphonic trills at intervals, with typical Song mixed in

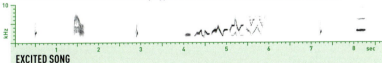

Call Repertoires of Veery and Bicknell's and Gray-cheeked Thrushes

These three closely related species have similar complex and poorly understood repertoires of burry call notes. Each adult has at least 5–10 different versions of these calls, often giving long series of identical calls before switching to a different type. Mates and neighbors have a strong tendency to match call types. Sometimes consecutive calls may vary, or one type may slowly grade into another over the course of several renditions. Similar calls are given by all three species during nocturnal migration; they may be less variable than those given during the day, but the relationship between daytime and nighttime calls is poorly understood. More study needed.

Townsend's Solitaire

Myadestes townsendi

Slender and long-tailed; often perches conspicuously on highest point in tree. Breeds in western mountains; in winter, usually found near junipers.

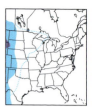

Complex musical warble, often lasting 30 seconds or more

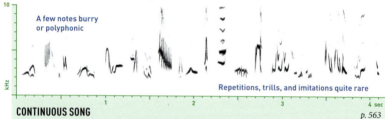

A few notes burry or polyphonic

Repetitions, trills, and imitations quite rare

CONTINUOUS SONG

p. 563

All year, by breeding males in high, slow, circling flight, and by both sexes to establish summer and winter territories. Extraordinarily long, complex, and beautiful. Pace rather slow; most notes are musical whistles, with a few chips, burrs, and creaks. Often soft, or fading slowly in or out.

In territorial disputes, may repeat the same phrase several times with only minor variation

Well-spaced musical phrases, variable in length; quality like Continuous Song

Consecutive phrases usually differ

DISJUNCT SONG

p. 561

All year; infrequent. Grades smoothly into Continuous Song. Shortest phrases resemble those of vireos, especially Red-eyed; slightly longer versions may recall a grosbeak or finch.

Bizarre 3-syllabled rising burry phrase

Burry parts inaudible at a distance

ZWEER-ZWEER-ZREE *p. 553*

Mostly Jul.–Sept., by begging fledglings. Unique, but not often heard.

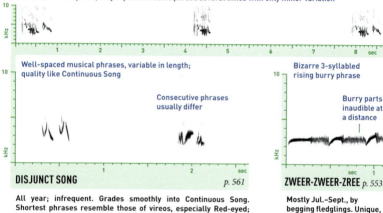

Lower, less musical than robin Cheep

CHEEP *p. 501*

Variably harsh

WHINE *p. 529*

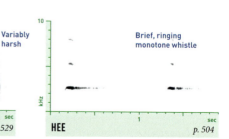

Brief, ringing monotone whistle

HEE *p. 504*

These calls given in defense of winter territory against waxwings and bluebirds (not robins or other solitaires). Fairly plastic.

All year; most common call. Distinctive; uniform and fairly stereotyped. Often repeated at 1- to 5-second intervals for long periods.

VEERY

Catharus fuscescens

Renowned for its ethereal, downward-spiraling song. In the East, breeds in dense deciduous forest understory; in the West, breeds in willow thickets.

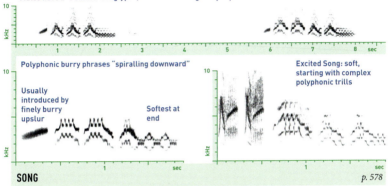

Males have 2–3 similar songtypes; consecutive songs may repeat or differ

Polyphonic burry phrases "spiralling downward"

Usually introduced by finely burry upslur

Softest at end

Excited Song: soft, starting with complex polyphonic trills

SONG *p. 578*

Mostly May–July. Distinctive and arrestingly beautiful. Overslurred phrases in downslurred series create "spiralling" sound; lower, more repetitive than Songs of Gray-cheeked or Bicknell's. Excited Song (p. 333) often heard in altercations; much softer than typical Song, audible only at close range.

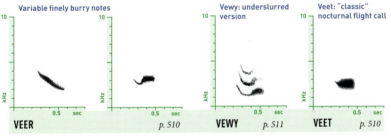

Variable finely burry notes

Vewy: underslurred version

Veet: "classic" nocturnal flight call

VEER *p. 510*   **VEWY** *p. 511*   **VEET** *p. 510*

All year, by both sexes, especially on breeding grounds during the day. Highly variable; adults have at least 5–10 different versions, some of which may differ in function. Lower than most calls of Gray-cheeked and Bicknell's; underslurred versions distinctive. Overslurred versions rare.

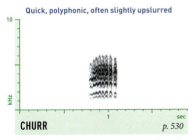

Quick, polyphonic, often slightly upslurred

CHURR *p. 530*

Apparently given by adults in aggression; also interspersed into bouts of Excited Song. Often introduced by soft, noisy Chuklike notes.

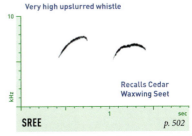

Very high upslurred whistle

Recalls Cedar Waxwing Seet

SREE *p. 502*

Usually upslurred, unlike similar calls of other thrushes, but sometimes nearly monotone. Given in altercations or in alarm.

GRAY-CHEEKED THRUSH

Catharus minimus

Our least-known thrush, breeding in tall, dense thickets and bogs. Secretive during migration, but northbound migrants sometimes sing or call.

Males have 3–6 similar songtypes; consecutive songs always differ

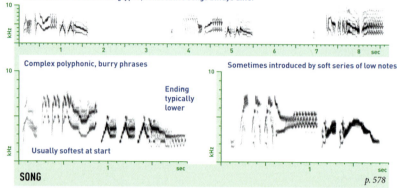

Complex polyphonic, burry phrases

Ending typically lower

Usually softest at start

Sometimes introduced by soft series of low notes

SONG

p. 578

Mostly late May–July. Remarkable Song like Veery's, often including overall downslurred trend, but higher and more complex, with irregular rhythm and little repetition of phrases. Some Gray-cheeked have final phrase identical to Bicknell's (top, third Song), but only in 1–2 songs of the repertoire.

Variable high, finely burry downslurs or overslurs

"Classic" nocturnal flight call

Sometimes burry only at end

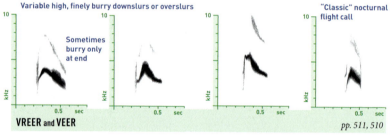

VREER and VEER

pp. 511, 510

All year, by both sexes, especially on breeding grounds during the day. Highly variable; adults have at least 5–10 different versions, some of which may differ in function. Usually higher than similar calls of Veery; never underslurred. Likely averages lower than Bicknell's, but much overlap.

Low, nasal

Slightly lower and less polyphonic than Veery Churr

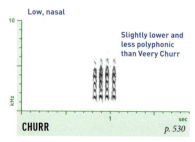

CHURR

p. 530

Given infrequently on breeding grounds, from agitated birds, including in response to playback. Plastic.

Very high, downslurred

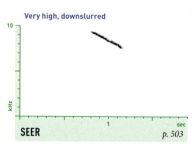

SEER

p. 503

Given infrequently on breeding grounds, possibly in alarm.

BICKNELL'S THRUSH

Catharus bicknelli

Split from Gray-cheeked Thrush due to differences in song and DNA. Slightly redder-tailed than Gray-cheeked, but probably only reliably identified by Song.

Males have 3–6 similar songtypes; consecutive songs always differ

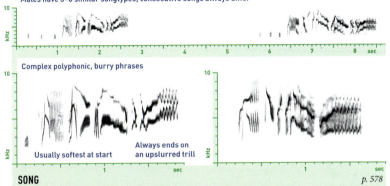

Complex polyphonic, burry phrases

Usually softest at start

Always ends on an upslurred trill

SONG

p. 578

Mostly late May–June. Remarkable song like Veery's, but higher and more complex, with irregular rhythm and little repetition of phrases. Upslurred final note is key distinction from Gray-cheeked Song. Mostly given before dawn and after dusk, for a few weeks at start of breeding season.

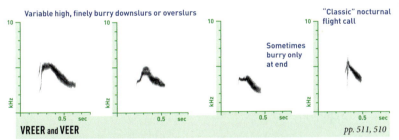

Variable high, finely burry downslurs or overslurs

Sometimes burry only at end

"Classic" nocturnal flight call

VREER and **VEER**

pp. 511, 510

All year, by both sexes, especially on breeding grounds during the day. Highly variable; adults have at least 5–10 different versions, some of which may differ in function. Usually higher than similar calls of Veery; never underslurred. Likely averages higher than Gray-cheeked, but much overlap.

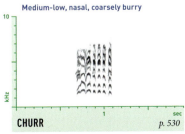

Medium-low, nasal, coarsely burry

CHURR

p. 530

Given infrequently on breeding grounds. Last part higher in only available recording; unknown whether this example is typical.

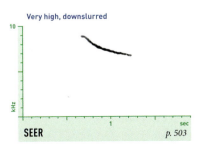

Very high, downslurred

SEER

p. 503

Given infrequently on breeding grounds, possibly in alarm.

SWAINSON'S THRUSH

Catharus ustulatus

Breeds in mixed and coniferous forests. Common in migration in many habitats. Eastern birds are olive; Pacific Northwest breeders redder, recalling Veery.

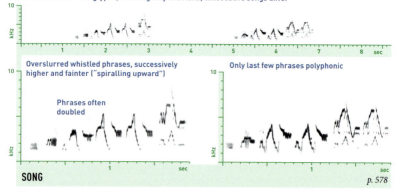

Males have 3–7 songtypes, differing only in details; consecutive songs differ

Overslurred whistled phrases, successively higher and fainter ("spiralling upward")

Phrases often doubled

Only last few phrases polyphonic

SONG
p. 578

Mostly Mar.–July. Often sings in spring migration, but for a rather short period on the breeding grounds, mostly in June in many areas, and mostly at dawn and dusk. No other thrush Song consistently rises in pitch. Excited Song adds trills at start of song phrases and/or call-like notes in between.

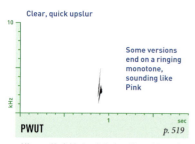

Clear, quick upslur

Some versions end on a ringing monotone, sounding like Pink

PWUT
p. 519

All year. Variable in pitch, length, and tone, but always at least vaguely upslurred.

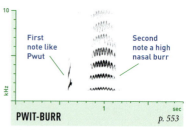

First note like Pwut

Second note a high nasal burr

PWIT-BURR
p. 553

Mostly May–Aug., with Pwuts in alarm near nest; also in territorial conflicts. Burr also given singly; like Veery Churr, but higher, clearer.

Clear, slightly upslurred whistle

Second half sometimes slightly burry

WEE
p. 505

All year. Given regularly on breeding grounds and in nocturnal migration. Much like the call of the Spring Peeper frog.

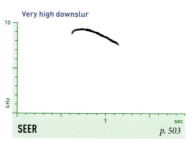

Very high downslur

SEER
p. 503

All year, from motionless, often hidden birds; warns of aerial or other predators. Like other downslurred thrush Seers.

HERMIT THRUSH

Catharus guttatus

Breeds in mixed and coniferous forests, where it gives extended Song performances at dawn and dusk. Like other thrushes, a vocal nocturnal migrant.

Males average 8–10 songtypes; consecutive songs always differ, often alternating high and low

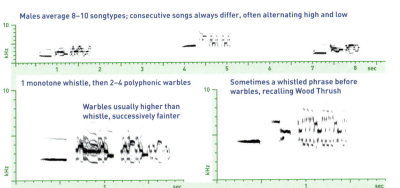

1 monotone whistle, then 2–4 polyphonic warbles

Warbles usually higher than whistle, successively fainter

Sometimes a whistled phrase before warbles, recalling Wood Thrush

SONG p. 578

Mostly May–Aug. A lovely, delicate song whose intricacy can be appreciated only at close range. Best distinguished from other thrushes by long monotone whistle at start of each song. Soft subsong and plastic song sometimes given in migration and winter. Excited Song (see p. 333) includes extra polyphonic warbles.

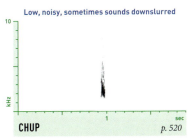

Low, noisy, sometimes sounds downslurred

CHUP p. 520

All year. Varies with level of agitation; higher-pitched versions and longer low, burry versions associated with extreme alarm.

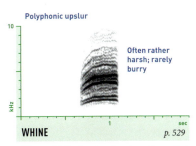

Polyphonic upslur

Often rather harsh; rarely burry

WHINE p. 529

All year. Distinctive; less harsh and burry than Spotted Towhee Wheeze. Often given with Chups; may indicate a lower level of alarm.

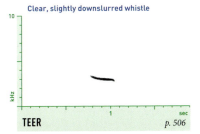

Clear, slightly downslurred whistle

TEER p. 506

All year. Given regularly by night migrants; also occasionally on breeding grounds, where finely burry versions have been recorded.

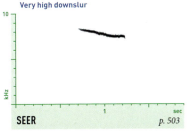

Very high downslur

SEER p. 503

All year, from motionless, often hidden birds; warns of aerial or other predators. Like other downslurred thrush Seers.

Wood Thrush

Hylocichla mustelina

A celebrated singer. Fairly common but declining in moist, mature deciduous and mixed forests with thick underbrush and leaf litter.

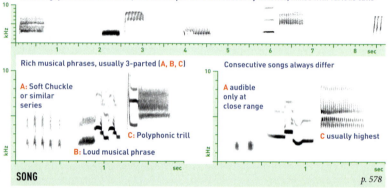

Trill Song (p. 560): musical trills, sometimes paired or introduced by whistles, mixed with various calls

Rich musical phrases, usually 3-parted (A, B, C)

A: Soft Chuckle or similar series

B: Loud musical phrase

C: Polyphonic trill

Consecutive songs always differ

A audible only at close range

C usually highest

SONG *p. 578*

Mostly Apr.–Aug.; sometimes sings just prior to and during spring migration. Individual males have 1–3 versions of part A, 2–8 of B, and 6–12 of C, which they combine in various ways to make 20 or so songtypes. Trill Song poorly understood; most common late in the breeding season.

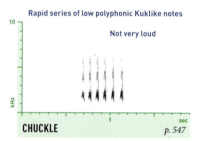

Rapid series of low polyphonic Kuklike notes

Not very loud

CHUCKLE *p. 547*

All year. Given in moderate alarm; grades into Weet Series as alarm level increases. Slower than other thrush Churrs.

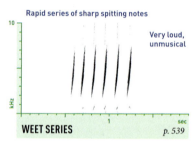

Rapid series of sharp spitting notes

Very loud, unmusical

WEET SERIES *p. 539*

All year, in high agitation. Slightly sharper than Curve-billed Thrasher Weet Series, and rarely heard in same range or habitat.

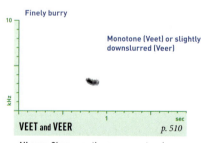

Finely burry

Monotone (Veet) or slightly downslurred (Veer)

VEET and VEER *p. 510*

All year. Given sometimes near nest and regularly during nocturnal migration. Like Hermit Thrush Teer, but strongly burry.

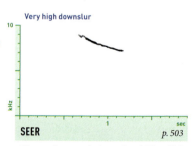

Very high downslur

SEER *p. 503*

All year. From motionless, often hidden birds; warns of aerial or other predators. Like other downslurred thrush Seers.

Eastern Bluebird

Sialia sialis

Familiar and beloved, found in meadows, orchards, and other open areas with scattered trees. Readily nests in boxes. An early spring migrant.

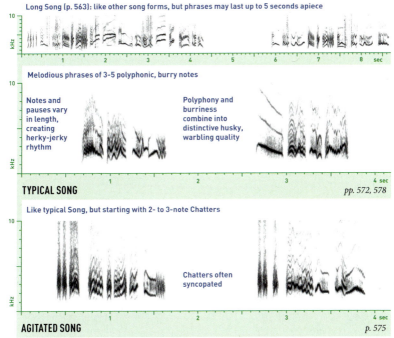

Long Song (p. 563): like other song forms, but phrases may last up to 5 seconds apiece

Melodious phrases of 3-5 polyphonic, burry notes

Notes and pauses vary in length, creating herky-jerky rhythm

Polyphony and burriness combine into distinctive husky, warbling quality

TYPICAL SONG *pp. 572, 578*

Like typical Song, but starting with 2- to 3-note Chatters

Chatters often syncopated

AGITATED SONG *p. 575*

Mostly Mar.–July, by both sexes. Typical Song given more often by males, loudly from prominent perch or very softly, often in proximity to mate. Individual birds apparently sing at least 3–16 stereotyped phrase types; consecutive songs almost always different. Grades into Agitated Song, given by both sexes in presence of potential predators, including humans, usually when mate is away from territory. Long Song little known; may be given in excitement, possibly in courtship.

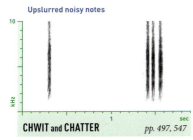

Upslurred noisy notes

CHWIT and CHATTER *pp. 497, 547*

All year, in agitation, often while flicking wings. Quite plastic in rhythm. Chatters more common than single Chwits.

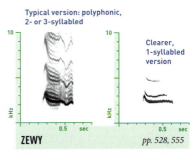

Typical version: polyphonic, 2- or 3-syllabled

Clearer, 1-syllabled version

ZEWY *pp. 528, 555*

All year; most common call, in contact. Often underslurred. Somewhat variable and plastic; male versions usually longer than female's.

AMERICAN ROBIN

Turdus migratorius

One of our most familiar birds, found in almost any habitat. Forms large flocks in migration and winter; famously pulls earthworms from lawns in summer.

After dawn, often lacks "hissely" phrases at end of each series

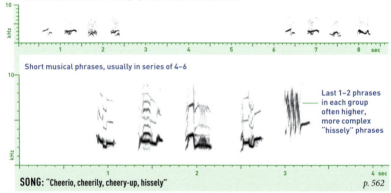

Short musical phrases, usually in series of 4–6

Last 1–2 phrases in each group often higher, more complex "hissely" phrases

SONG: "Cheerio, cheerily, cheery-up, hissely"   *p. 562*

Well known. All year, but especially Mar.–June. Both sexes reported to sing. Individuals have 6–20 "cheerio" phrases, which are medium-low, usually of 2–3 syllables, rarely burry or polyphonic. In a series, phrases usually switch constantly, but are sometimes repeated up to several times in a row.

A continuous string of "hissely" phrases, usually audible only at close range

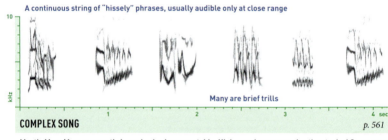

Many are brief trills

COMPLEX SONG   *p. 561*

Mostly Mar.–May, apparently by males in close courtship. Higher and more complex than typical Song, composed mostly of "hissely" phrases, of which an individual may have 50–100. Some louder Complex Songs mix "cheerio" phrases, "hissely" phrases, and multiple versions of Whinny; function unknown.

Version with Cheeplike notes · Version with Kuklike notes · Version with Squeeplike notes

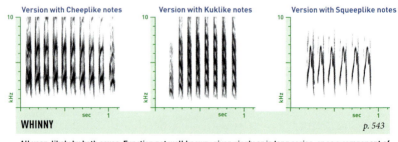

WHINNY   *p. 543*

All year, likely by both sexes. Function not well known; given singly or in long series, or as a component of Complex Song. Individuals have several fairly stereotyped versions, each usually repeated before switching, but some birds switch frequently. First and last note often lower.

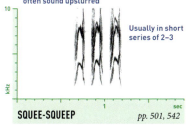

Notes higher, slightly longer than Cheep; often sound upslurred

Usually in short series of 2–3

SQUEE-SQUEEP   *pp. 501, 542*

All year, by both sexes; function not well understood. May grade into both Whinny and Cheep. Often followed by a series of soft Kuks.

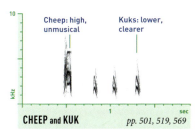

Cheep: high, unmusical

Kuks: lower, clearer

CHEEP and KUK   *pp. 501, 519, 569*

All year, by both sexes, in alarm. Often in above pattern, with 1–2 Cheeps followed by 3–4 Kuks (p. 569), but both sounds also given separately.

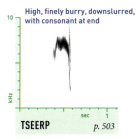

High, finely burry, downslurred, with consonant at end

TSEERP   *p. 503*

All year. Sometimes in flight, including by night migrants. Can introduce Kuks.

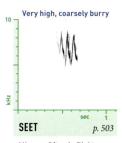

Very high, coarsely burry

SEET   *p. 503*

All year. Often in flight, including by night migrants. Can introduce Kuks.

Very high, downslurred

SEER   *p. 503*

From motionless, often hidden birds; warns of aerial or other predators.

CATEGORIES OF SONG

Many bird species have multiple categories of song, given in different situations and often composed from separate repertoires of notes or phrases.

Complex Song is the term usually used in this book for categories of song that are more complex than a species' typical Song. Such songs can be used in courtship, but their function varies between species. Some authors refer to Complex Song as "alternate song," "excited song," or "flight song."

Whisper Song describes Complex Songs (or occasionally typical Songs) given very softly, in intense, close-range courtship or territorial encounters. It differs from subsong (p. 27) in being highly stereotyped and given by very excited adults, usually with other adults nearby.

Dawn Song, a complex song given only prior to sunrise, is distinguished from Day Song in some species. Dawn songs are best known and most distinct in certain flycatchers, but can also be heard from various other species, including American Robin, *Piranga* tanagers, and some swallows. Although many bird species deliver their typical songs more often and with more variety at dawn (contributing to the widespread burst of birdsong known as the **dawn chorus**), this book reserves the term "Dawn Song" for songs that differ significantly from typical daytime songs in form or pattern.

First-category and **second-category song** are special terms used to describe the two different ways in which some warbler species change song patterns in different situations (see p. 378).

CATBIRDS, THRASHERS, AND MOCKINGBIRDS
(Family Mimidae)

Members of this family, known as "mimids," are medium-sized, long-billed, long-tailed birds, generally plumaged in grays and browns, that tend to skulk in deep cover. Songs are learned, and famous for including vocal imitations of other bird species. All North American mimids incorporate at least some imitations into their Songs, and some, such as the Northern Mockingbird, even occasionally imitate non-bird sounds such as phone ringtones and car alarms. In addition to Songs, at least some call types may be learned.

Mimids tend to have huge Song repertoires. The Brown Thrasher is estimated to have the largest repertoire of any bird species yet studied, with individuals likely capable of singing 1,500–2,000 different phrase types, and possibly more. Northern Mockingbirds can learn new sounds throughout their lives, and it has been suggested that some mimids may innovate new Song phrases continually. In these species, the patterns in which the sounds are arranged are more important for species identification than the types of sounds that are sung.

Regional Variation in Curve-billed Thrashers

In the U.S., eastern and western populations of Curve-billed Thrashers differ subtly in plumage and in some vocalizations. Eastern birds of the nominate subspecies, found from southeast Arizona eastward, tend to give Weet Series in which all the notes have equal emphasis (p. 539); western birds of the subspecies *palmeri* (sometimes called "Palmer's Thrasher") tend to give a distinctive "wit-WEET" or "wit-WEET-WEET" in which the first note is softer and ends on a lower pitch than the later notes. However, individuals in some areas give calls that do not closely match either pattern, suggesting that the whistled calls may be learned rather than innate.

Western
Curve-billed Thrasher

WIT-WEET   p. 539

Curve-billed Thrashers also give a large number of other calls. Regional variation in these sounds is difficult to assess, because so few recordings are available. As of this writing, recordings of the Churr and Chup are available only from western birds, and recordings of the Chatter and Ratchet are available only from eastern birds, but it is likely that all populations give versions of these calls.

GRAY CATBIRD

Dumetella carolinensis

Common, but more often heard than seen. Skulks in dense deciduous undergrowth, especially along woodland edges. Named for its catlike Mew call.

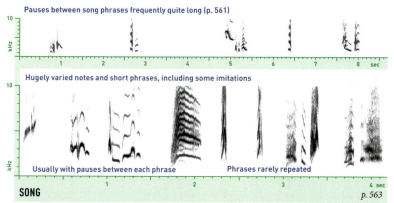

Pauses between song phrases frequently quite long (p. 561)

Hugely varied notes and short phrases, including some imitations

Usually with pauses between each phrase   Phrases rarely repeated

SONG   *p. 563*

Mostly Apr.–Aug., by male; female sings rarely and softly. Males may have 100 or more song phrases. Soft subsong often heard in fall. Length of pauses between phrases varies greatly, but fairly constant within a singing bout. Occasionally repeats phrases several times, suggesting a mockingbird.

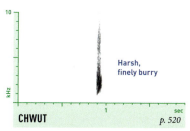

Harsh, finely burry

CHWUT   *p. 520*

All year, in mild alarm. Like Hermit Thrush Chup, but rougher, slightly rising.

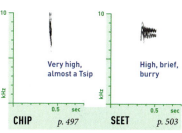

Very high, almost a Tsip

High, brief, burry

CHIP   *p. 497*   **SEET**   *p. 503*

Many types of high, brief calls given by begging juveniles, including a rising Chwit (p. 497). Adults sometimes give some of these calls.

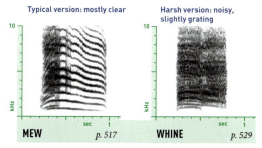

Typical version: mostly clear

Harsh version: noisy, slightly grating

MEW   *p. 517*   **WHINE**   *p. 529*

All year; Mew is most common and distinctive call. Highly variable and plastic, becoming harsh Whine in excitement or alarm. Always at least partly polyphonic.

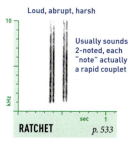

Loud, abrupt, harsh

Usually sounds 2-noted, each "note" actually a rapid couplet

RATCHET   *p. 533*

Given at dawn, in escape flight, or during chases. Plastic, but always short.

Brown Thrasher

Toxostoma rufum

Found in dense deciduous brush and woodland edges. Named "thrasher" for its habit of foraging by tossing up leaf litter on the ground.

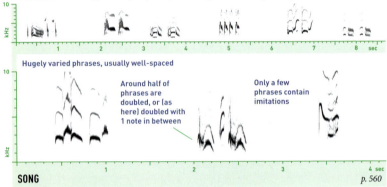

Pauses between song phrases quite plastic, but usually at least as long as the phrases themselves

Hugely varied phrases, usually well-spaced

Around half of phrases are doubled, or (as here) doubled with 1 note in between

Only a few phrases contain imitations

SONG *p. 560*

Mostly Mar.–July, by males, usually from conspicuous perch; females not known to sing. Individual males estimated to have more than 1,000 song phrases. Very soft versions of Song sometimes heard in close courtship; soft subsong also given in fall and winter.

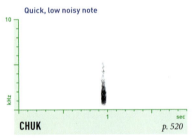

Quick, low noisy note

CHUK *p. 520*

All year, likely in alarm. Variable, but all versions low, brief, and fairly soft.

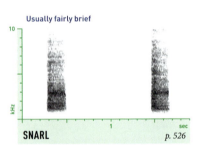

Usually fairly brief

SNARL *p. 526*

All year, in agitation or alarm.

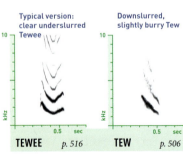

Typical version: clear underslurred Tewee

Downslurred, slightly burry Tew

TEWEE *p. 516* **TEW** *p. 506*

All year, possibly in alarm or contact. Quite variable and plastic; can resemble certain calls of Veery, but usually clearer.

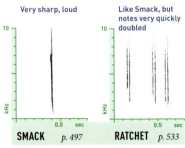

Very sharp, loud

Like Smack, but notes very quickly doubled

SMACK *p. 497* **RATCHET** *p. 533*

All year. Smack is most common call, apparently given in alarm. Ratchet rare; notes may be more smacking than in Gray Catbird.

LONG-BILLED THRASHER

Toxostoma longirostre

Much like Brown Thrasher, but slightly darker brown. Generally prefers drier habitats than Brown Thrasher, and denser thickets than Curve-billed.

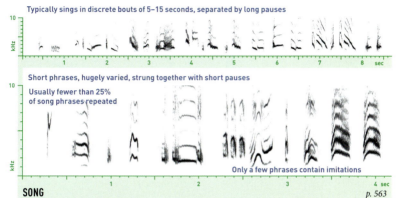

Typically sings in discrete bouts of 5–15 seconds, separated by long pauses

Short phrases, hugely varied, strung together with short pauses

Usually fewer than 25% of song phrases repeated

Only a few phrases contain imitations

SONG *p. 563*

Nearly all year, but especially Mar.–July, by males, usually from conspicuous perch. Females not known to sing. Individual repertoire size unknown, but clearly very large. Very difficult to separate from Curve-billed Thrasher song; may average slightly fewer phrases/second and fewer repeated phrases.

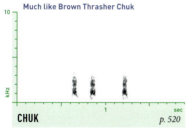

Much like Brown Thrasher Chuk

CHUK *p. 520*

All year, likely in alarm. Variable, but all versions low, brief, and fairly soft.

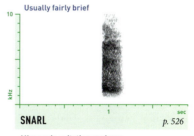

Usually fairly brief

SNARL *p. 526*

All year, in agitation or alarm.

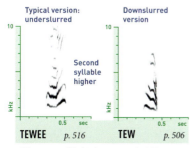

Typical version: underslurred

Downslurred version

Second syllable higher

TEWEE *p. 516*   **TEW** *p. 506*

All year, possibly in alarm or contact. Even more variable and plastic than in Brown Thrasher; upslurred and multinote versions exist.

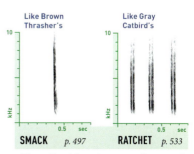

Like Brown Thrasher's

Like Gray Catbird's

SMACK *p. 497*   **RATCHET** *p. 533*

All year. Smack is most common call, apparently given in alarm. Ratchet less common; similar to Gray Catbird's.

Curve-billed Thrasher

Toxostoma curvirostre

Breeds in open arid habitats, including sparsely vegetated residential areas. Eastern birds spotted below, with faint white wingbars and tail corners.

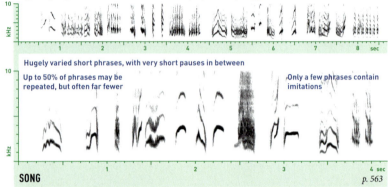

Typically sings in discrete bouts of 5–15 seconds, separated by long pauses

Hugely varied short phrases, with very short pauses in between

Up to 50% of phrases may be repeated, but often far fewer

Only a few phrases contain imitations

SONG *p. 563*

Nearly all year, but especially Mar.–July, by males, usually from conspicuous perch. Not known whether females sing. Individual repertoire size unknown, but clearly very large. Very difficult to separate from Long-billed Thrasher Song; may average slightly more phrases/second and more repeated phrases.

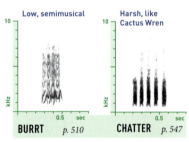

Low, semimusical

Harsh, like Cactus Wren

BURRT *p. 510* **CHATTER** *p. 547*

Only recording of Burrt is from an agitated bird in early spring. Chatter given in alarm near nest and in altercations.

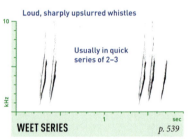

Loud, sharply upslurred whistles

Usually in quick series of 2–3

WEET SERIES *p. 539*

All year; most common and distinctive call of the eastern subspecies (see p. 344). Series becomes longer in agitation.

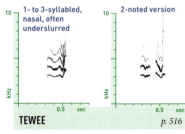

1- to 3-syllabled, nasal, often underslurred

2-noted version

TEWEE *p. 516*

Not well known; apparently given in courtship and possibly during or after territorial altercations. Highly variable and plastic.

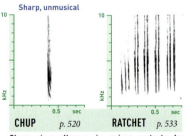

Sharp, unmusical

CHUP *p. 520* **RATCHET** *p. 533*

Chups given all year, in various contexts, by adults and juveniles. Only recording of Ratchet is from two birds in a flight chase.

Northern Mockingbird

Mimus polyglottos

Visually and vocally conspicuous in open areas with scattered trees, including residential areas. In flight, shows white patches in the wings and tail.

Typical Song has significant pauses between series

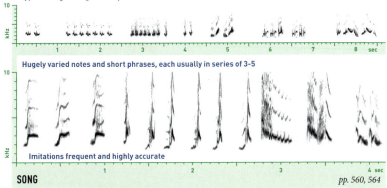

Hugely varied notes and short phrases, each usually in series of 3–5

Imitations frequent and highly accurate

SONG — pp. 560, 564

Nearly all year. Both sexes sing, females much less often. Unmated males may sing all night long. Individual males have 50–200 song phrases, and continue to learn new phrases throughout their lives. Imitates many sounds, including nonbird sounds. Song best recognized by pattern of repetition.

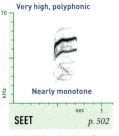

Very high, polyphonic

Nearly monotone

SEET — *p. 502*

Given by begging juveniles; similar calls near nest by adults.

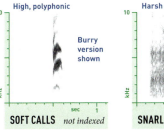

High, polyphonic

Burry version shown

SOFT CALLS — *not indexed*

A varied set of highly plastic calls given by both sexes near the nest.

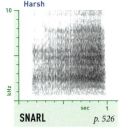

Harsh

SNARL — *p. 526*

All year, in high agitation, especially in response to predators.

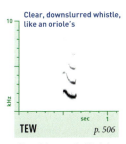

Clear, downslurred whistle, like an oriole's

TEW — *p. 506*

Given infrequently, likely by both sexes, perhaps in alarm.

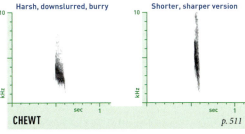

Harsh, downslurred, burry

Shorter, sharper version

CHEWT — *p. 511*

Most common call, given by both sexes in mild alarm. Quite variable. Most versions quite abrupt and harsh. Often given in rapid series; rarely too fast to count.

STARLINGS AND MYNAS (Family Sturnidae)

This large and diverse family is native to the Old World; our species are introduced. Vocalizations are extremely complex and diverse. Songs are learned, often incorporating accurate imitations of other birds and environmental sounds.

EUROPEAN STARLING

Sturnus vulgaris

Released in New York City in 1890; now one of the most abundant birds in North America. A remarkable vocalist. Competes with native species for nest sites.

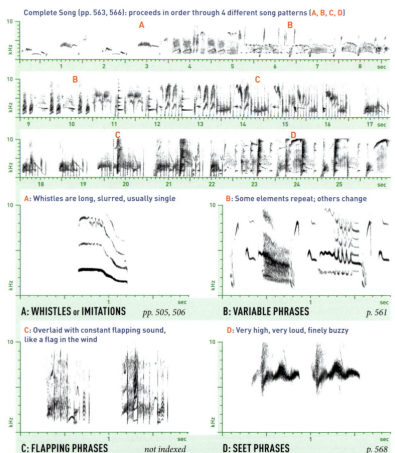

Complete Song (pp. 563, 566): proceeds in order through 4 different song patterns (A, B, C, D)

A: Whistles are long, slurred, usually single

A: WHISTLES or IMITATIONS *pp. 505, 506*

B: Some elements repeat; others change

B: VARIABLE PHRASES *p. 561*

C: Overlaid with constant flapping sound, like a flag in the wind

C: FLAPPING PHRASES *not indexed*

D: Very high, very loud, finely buzzy

D: SEET PHRASES *p. 568*

All year, but especially Apr.–June. Four distinct patterns make up the complete Song, up to 40 seconds total, but any pattern may be omitted or given by itself. Patterns A–C usually quite soft; Pattern D is far louder. Each male has 2–12 Whistles (A), 10–35 Variable Phrases (B), 2–14 Flapping Phrases (C), and up to 6 Seet Phrases (D), for a total of up to 70 phrase types per bird. On average, 18 of these phrases are imitations of other species, in patterns A or B (rarely C). Females also sing, but lack Seet Phrases (D).

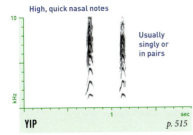

High, quick nasal notes

Usually singly or in pairs

10

kHz

1 sec

YIP *p. 515*

All year, in response to hawks, in nest defense, and in chases of other starlings. Fairly uniform; slightly polyphonic or screechy.

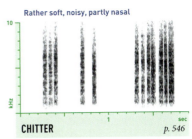

Rather soft, noisy, partly nasal

10

kHz

1 sec

CHITTER *p. 546*

All year, when individuals join groups and in mild aggressive encounters. Fairly uniform. Sometimes shortened to a single Chit.

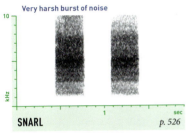

Very harsh burst of noise

10

kHz

1 sec

SNARL *p. 526*

All year, but especially in alarm at nest, often with Yips. Varies in length; sometimes in series.

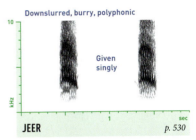

Downslurred, burry, polyphonic

Given singly

10

kHz

1 sec

JEER *p. 530*

All year, by solo birds in flight and by small flocks. Juveniles beg with a similar call.

HILL MYNA

Gracula religiosa

Native to southeast Asia; subspecies *intermedia* now local in south Florida, where it can be quite conspicuous both visually and vocally.

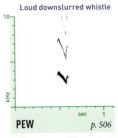

Song: single notes, short phrases, and occasional trills, all widely spaced; consecutive songs differ

10

kHz

1 2 3 4 5 6 7 8 sec

Phrases of 1–4 syllables; tone qualities vary tremendously

Listen for distinctive ringing, harmonica-like notes

10

kHz

1 2 sec

SONG *pp. 555, 561*

Immensely varied. Repertoire size unknown, but probably vast. Some sounds, especially whistles, are quite starlinglike; others have unique, almost electronic twanging quality.

Loud downslurred whistle

10

kHz

sec 1

PEW *p. 506*

All year; one of the species' most common calls. Highly variable.

COMMON MYNA

Acridotheres tristis

Native to Asia; now common in parts of Florida, frequenting parking lots and other manmade habitats. Likely competes with native birds for nest cavities.

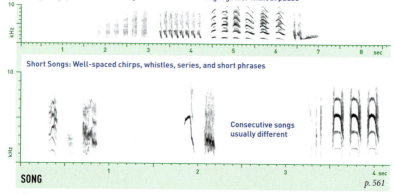

Long Song (p. 564): several very different series strung together without pause

Short Songs: Well-spaced chirps, whistles, series, and short phrases

Consecutive songs usually different

SONG
p. 561

All year. Poorly studied, but astonishingly varied. Songs may be long, short, or some combination; differences in function, if any, unknown. Tone varies from musical to harsh; sometimes contains imitations. Flocks may sing all night at roosts. Overall, recalls Great-tailed Grackle.

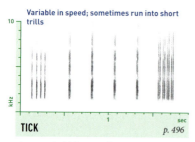

Variable in speed; sometimes run into short trills

TICK
p. 496

Incorporated into some Songs; may also occur independently as a type of call, but function unknown.

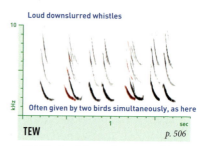

Loud downslurred whistles

Often given by two birds simultaneously, as here

TEW
p. 506

All year; one of the most common call types. Variable and slightly plastic. Also incorporated into Songs, usually in series.

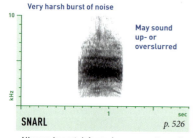

Very harsh burst of noise

May sound up- or overslurred

SNARL
p. 526

All year, in nest defense, in response to predators, and in other high-stress situations.

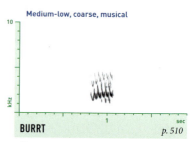

Medium-low, coarse, musical

BURRT
p. 510

All year; one of the species' most common calls. Function unclear, but may serve in pair contact, or as a greeting at the nest.

WAXWINGS (Family Bombycillidae)

The waxwings are handsome, medium-sized birds with crests and black masks, named for the red waxy tips on their secondary feathers. They eat mainly fruit, along with some insects, and are highly social and nomadic outside the breeding season, gathering in flocks that wander widely in search of fruiting trees. In flight silhouette they resemble starlings, with short tails and triangular wings. All their vocalizations are simple and apparently innate. Both species' vocal repertoires are traditionally considered to consist entirely of calls, lacking Songs altogether.

PIPITS (Family Motacillidae)

Pipits are small birds of open country with generally cryptic, streaked plumage. They eat primarily insects and nest on the ground. They resemble larks and long-spurs in their elongated hind claws and walking (instead of hopping) gait, and they sometimes flock with larks and longspurs in migration and winter, but the three families are not closely related.

Two species of pipit are regularly found in North America outside Alaska. Both perform long flight displays on the breeding grounds, and both show white outer tail feathers in flight. The bolder and more gregarious American Pipit wags its tail frequently, while the shyer and more solitary Sprague's Pipit does so rarely if at all. Songs in both species are apparently learned. Calls are probably innate, and call repertoires are quite simple.

Cedar Waxwing

Bombycilla cedrorum

A very social bird that spends much of the year in wandering flocks in search of fruiting trees. Vocalizations simple and very high-pitched.

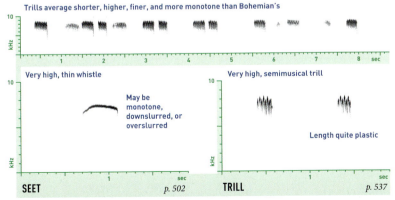

Trills average shorter, higher, finer, and more monotone than Bohemian's

Very high, thin whistle

May be monotone, downslurred, or overslurred

SEET p. 502

Very high, semimusical trill

Length quite plastic

TRILL p. 537

Both calls given all year. Seet given frequently, especially during or prior to flight and in response to predators. Trill maintains flock contact, often in flight. Very short, soft versions given while feeding and in close courtship. Begging juveniles give longer, more Bohemian-like version.

BOHEMIAN WAXWING

Bombycilla garrulus

An irregular winter visitor from the far north; invades towns and woodlands in flocks in search of fruiting trees, sometimes with Cedar Waxwings.

Trills lower than Cedar Waxwing's; longer, more downslurred on average

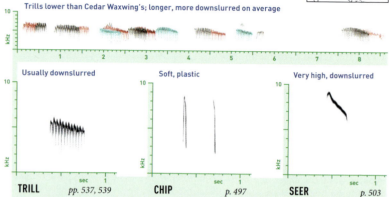

Usually downslurred	Soft, plastic	Very high, downslurred
TRILL pp. 537, 539	**CHIP** p. 497	**SEER** p. 503

Trill is most common call all year; given frequently by winter flocks. Very short, soft versions given while feeding. Chips given all year, possibly in alarm, resembling single note of Trill, and grading into it. Seer given in response to predator near nest or young; given rarely, if ever, in winter.

SPRAGUE'S PIPIT

Anthus spragueii

A local and declining bird of native prairies with a remarkable flight display. When flushed, rises steeply and flies far, then plummets back into the grass.

Songs sometimes introduced by a short series of Squeeplike notes

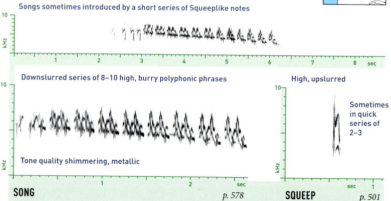

Downslurred series of 8–10 high, burry polyphonic phrases

High, upslurred

Sometimes in quick series of 2–3

Tone quality shimmering, metallic

SONG p. 578	**SQUEEP** p. 501

Song May–Aug. Recalls Veery Song, but longer, higher, and likely to be heard only from far overhead in native prairies. Display flights can last up to 3 hours. Repertoire size unknown, but consecutive songs similar. Squeep given all year, by flushed birds. Quite uniform; no other calls known from the species.

AMERICAN PIPIT

Anthus rubescens

Breeds on Arctic tundra and atop a few
New England mountains. In migration
and winter, found in flocks in fields and
along shorelines. Bobs tail frequently.

Series short to extremely long (30 seconds or more); consecutive series may repeat or differ

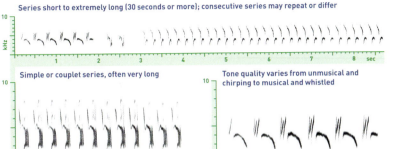

Simple or couplet series, often very long

Tone quality varies from unmusical and
chirping to musical and whistled

SONG: "Chew chew chew chew" or "Teer teer teer teer"   *pp. 537, 557*

Mostly May–Aug. Given in long, slow display flight or from ground. Repertoire size unknown; territorial
birds may give many series consecutively without pauses. When excited, may switch series often, like
American Goldfinch Song. Some series faster, like slow trills; some are brief, just a few repetitions.

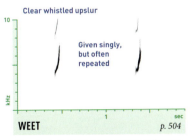

Clear whistled upslur

Given singly,
but often
repeated

WEET   *p. 504*

May–Sept., in alarm near nest or fledglings; apparently rare away from breeding grounds.
Compare Cheep of Sprague's Pipit.

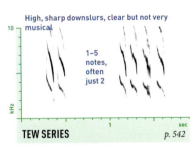

High, sharp downslurs, clear but not very
musical

1–5
notes,
often
just 2

TEW SERIES   *p. 542*

All year; most common call. Plastic, but consecutive calls fairly consistent in pitch and pattern,
unlike similar calls of Horned Lark.

OLD WORLD SPARROWS (Family Passeridae)

The House Sparrow and Eurasian Tree Sparrow are not closely related to New World sparrows. The songs of both species are likely learned.

WAXBILLS, MANNIKINS, & MUNIAS (Family Estrildidae)

This diverse family of finchlike birds is closely related to the Old World Sparrows.

WEAVERS (Family Ploceidae)

Birds in this family, including the Northern Red Bishop, weave elaborate nests.

HOUSE SPARROW

Passer domesticus

Common near human habitation of all kinds. Has spread nearly everywhere on the continent since its introduction from Europe to New York in 1851.

Fast Song: males cycle rapidly through many different Chirplike notes

Complex, 1- to 3-syllabled, semimusical Chirps

Often downslurred

3-syllabled version

Can have shimmering quality, like Tree Swallow Sheet call

SONG (CHIRP)

pp. 501, 559

All year, by males in a variety of situations; rarely by females. Males reportedly have 4–12 different Chirp types; usually one is repeated for long periods, or two types are alternated, before switching to a new type. Fast Song may be associated with courtship, but also given by lone birds.

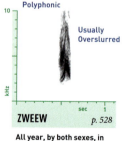

Polyphonic

Usually Overslurred

ZWEEW *p. 528*

All year, by both sexes, in mild alarm. Somewhat plastic. Recalls House Finch.

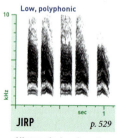

Low, polyphonic

JIRP *p. 529*

All year, singly or in series, in high alarm. Generally plastic.

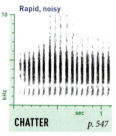

Rapid, noisy

CHATTER *p. 547*

All year, mostly by females, in interactions. Highly variable and plastic.

EURASIAN TREE SPARROW

Passer montanus

Introduced to St. Louis from Europe in 1870; local and uncommon in residential areas and farmland. Often found with House Sparrows.

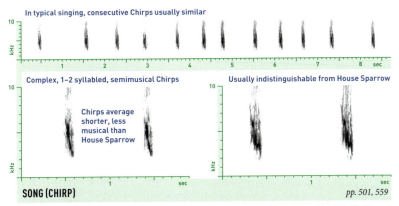

In typical singing, consecutive Chirps usually similar

Complex, 1–2 syllabled, semimusical Chirps

Chirps average shorter, less musical than House Sparrow

Usually indistinguishable from House Sparrow

SONG (CHIRP) *pp. 501, 559*

All year, by males in a variety of situations; rarely by females. Males likely have at least 5–10 different chirps. Usually one is repeated for long periods, or two types are alternated, before switching to a new type. Also reportedly gives a Fast Song, in which songtypes switch constantly.

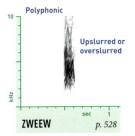

Polyphonic

Upslurred or overslurred

ZWEEW *p. 528*

All year, by both sexes, in mild alarm. Somewhat plastic.

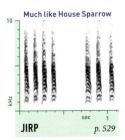

Much like House Sparrow

JIRP *p. 529*

All year, singly or in series, in high alarm. Generally plastic.

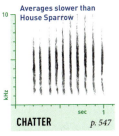

Averages slower than House Sparrow

CHATTER *p. 547*

All year, by both sexes, in agitation. Plastic; often shortened to 1–2 notes.

BRONZE MANNIKIN

Spermestes cucullatus

A popular cage bird, native to Africa; small numbers now feral in parts of Houston, Texas, and southern California. Adults resemble Scaly-breasted Munia (next page) but darker above, with central belly pure white. Young birds pale buffy-brown. Breet is most common call of adults; Chatter seems to be associated with begging of juveniles. In close courtship, male may give a slightly more complex Song.

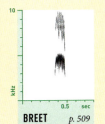

BREET *p. 509*

High, short; can recall Evening Grosbeak

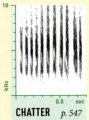

CHATTER *p. 547*

Plastic, often long, much like House Sparrow Chatter

SCALY-BREASTED MUNIA

Lonchura punctulata

Also called Nutmeg Mannikin or Spice Finch. Native to Asia; feral birds found mostly in parks and residential areas. Adult shown; juveniles are plain brown.

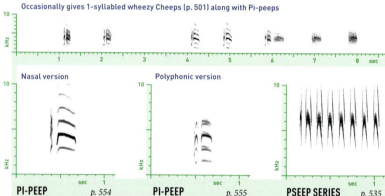

Occasionally gives 1-syllabled wheezy Cheeps (p. 501) along with Pi-peeps

Nasal version

Polyphonic version

PI-PEEP *p. 554*   **PI-PEEP** *p. 555*   **PSEEP SERIES** *p. 535*

Often rather quiet, but sometimes vocal in flocks. Most common call is a distinctive 2-noted Pi-peep, usually nasal or polyphonic. Clear whistled versions exist, and first note occasionally missing. Pseep Series given loudly by begging juveniles.

NORTHERN RED BISHOP

Euplectes franciscanus

Also known as Orange Bishop. Native to Africa. Feral in parts of Texas and California; has also been seen in Florida and Arizona. Favors weedy floodplains.

Song (p. 556): a couplet series alternating Chwits and Tseets, often introduced by Tseets

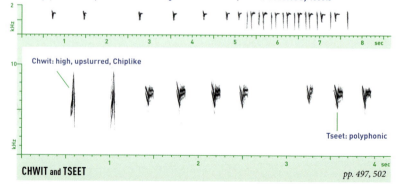

Chwit: high, upslurred, Chiplike

Tseet: polyphonic

CHWIT and TSEET *pp. 497, 502*

Most common calls appear to be Chwit and Tseet, given at least by males in many situations, often mixed together as above. Chwit also given separately in mild alarm. Couplet series of Chwit- and Tseetlike notes is apparent Song; also gives plastic Chirps and faint Whines, possibly a form of subsong.

FINCHES (Family Fringillidae, Subfamily Carduelinae)

This group includes some of our most vocally and behaviorally complex birds. Songs and at least some calls are learned; songs tend to be long and complex, made up of large repertoires of notes. Many finches have more than one type of song. In some species, such as Lesser Goldfinch, Pine Siskin, and Pine Grosbeak, songs can incorporate excellent imitations of other bird species.

Most finches give several different types of calls, many of which are polyphonic, resulting in a characteristic "finchy" whining quality common to many species. The begging calls of many species are distinctively 2-noted, giving rise to transcriptions like "tea-cup," "chit-too," and "teer-weet."

Call Matching in Finches

Laboratory studies have shown that in many species of cardueline finches, mated birds change their calls to match their mate's. In most such cases the calls change only in minor details, but call matching can occur even among members of different finch species when they are kept in the same cage. Recordings suggest that, in the redpolls, members of winter flocks may match calls with each other; more study is needed. It is possible that finches in the wild may in rare cases match the call notes of a different species.

Red Crossbill Types

Specially adapted to extracting seeds from closed conifer cones, the nomadic Red Crossbill (p. 362) is exceedingly complex in biology and behavior. At least 10 populations in North America, called "types," differ in vocalizations, bill structure, size, distribution, and ecology. Some have proposed that the types are best recognized as separate species.

Differences in size and bill structure apparently optimize different Red Crossbill types for feeding on different species of conifers, and different types usually do not interbreed where they overlap in range. However, field identification of the types depends on vocalizations that are learned, including the flight calls and the excitement calls. In rare cases, individuals have been shown to give calls of 2 different types. In one documented instance, a female that paired with a male of a different type changed her flight call type to match her mate's.

When identifying Red Crossbill types in the field, it is very helpful and sometimes necessary to make an audio recording. Observers with recording equipment have the capability to expand our knowledge of the status, distribution, behavior, and identification of these fascinating populations.

LONGSPURS AND SNOW BUNTINGS (Family Calcariidae)

Formerly considered close relatives of the North American sparrows, longspurs and the Snow Bunting are now classified in their own family. Most species are found in open areas such as tundra, grasslands, agricultural fields, and beaches. Songs are complex and apparently learned; call repertoires are well developed.

In some species, individual birds have repertoires of several short whistled calls which appear to be learned. These calls show marked geographic variation. In the breeding season, in alarm near the nest or young, adults often cycle continuously through several or all of their whistled calls. Flocks also give them on the wintering grounds. The functions and patterns of variation of these calls need further study.

PURPLE FINCH

Haemorhous purpureus

Breeds in a variety of forest types and woodland edges. In winter, found in small flocks in many different habitats. Sometimes visits feeders.

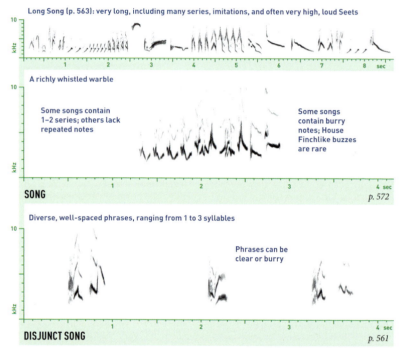

Long Song (p. 563): very long, including many series, imitations, and often very high, loud Seets

A richly whistled warble

Some songs contain 1–2 series; others lack repeated notes

Some songs contain burry notes; House Finchlike buzzes are rare

SONG
p. 572

Diverse, well-spaced phrases, ranging from 1 to 3 syllables

Phrases can be clear or burry

DISJUNCT SONG
p. 561

Nearly all year. Both sexes reported to sing, though many brown-plumaged singers are likely first-spring males. Typical Song is longer and more plastic in late winter, becoming shorter and more stereotyped by early summer; rarely but regularly includes imitations of other bird species. Individual repertoire size unknown; consecutive songs tend to start similarly but ends and middles often vary. Long Song directed at female, with crest raised and wings fluttering; often rather soft.

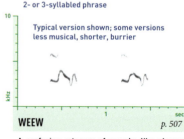

2- or 3-syllabled phrase

Typical version shown; some versions less musical, shorter, burrier

WEEW
p. 507

A confusing category of sounds, like phrases from the Disjunct Song, but given over and over, by both sexes and all ages. More study needed.

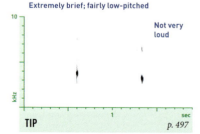

Extremely brief; fairly low-pitched

Not very loud

TIP
p. 497

All year; most common call. Distinctive, like a Tink call, but much lower.

HOUSE FINCH

Haemorhous mexicanus

Common in semi-open habitats; often visits feeders. Originally a bird of the arid Southwest, introduced to the East in 1939; has expanded continent-wide.

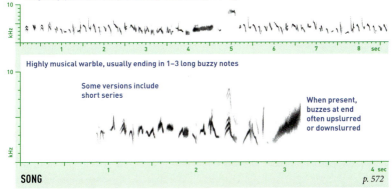

Long Song (p. 563): courting males sing continuously, with high loud Seets mixed in

Highly musical warble, usually ending in 1–3 long buzzy notes

Some versions include short series

When present, buzzes at end often upslurred or downslurred

SONG
p. 572

All year, but mostly Mar.–Aug., by males. Females sing occasionally, mostly in spring. Individual males have 2–10 songtypes; consecutive songs may be same or different. Songs frequently truncated, lacking buzzy notes at end. Usually lacks repeated notes. Very rarely includes imitations of other species.

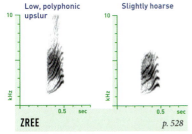

Low, polyphonic upslur

Slightly hoarse

ZREE
p. 528

All year, by perched birds. Slightly to highly plastic. One of the species' most common and characteristic calls.

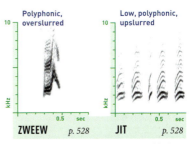

Polyphonic, overslurred

Low, polyphonic, upslurred

ZWEEW
p. 528

JIT
p. 528

Zweew given by females in close courtship; juveniles beg with similar notes. Jit poorly known.

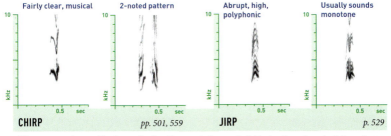

Fairly clear, musical

2-noted pattern

Abrupt, high, polyphonic

Usually sounds monotone

CHIRP
pp. 501, 559

JIRP
p. 529

All year, sometimes with other calls. Function poorly known. Individuals often alternate 2–3 versions; a few versions are 2-noted. Often highly plastic, but some versions fairly stereotyped. Several versions are easily confused with sounds of House Sparrow.

RED CROSSBILL

Loxia curvirostra

Ten distinct groups of Red Crossbills, called "types," breed in North America (see pp. 362–364). Six are regular in the East. Field identification requires experienced listening and often spectrographic analysis. All types are nomadic; range maps courtesy of Matt Young.

Flight calls

These are the most common calls, variable but stereotyped, usually given in short series, from a perch or in flight. They are diagnostic for each type.

Excitement calls

These are variable but stereotyped, given in agitation or alarm.

Type 2 (other types similar)

CHIT-TOO *p. 555*

Songs

Red Crossbills may have two categories of song. Short Songs usually consist of a couplet or triplet series, whistled or burry, with a few single notes at the start or end (pp. 559, 564). They grade into Long Songs, which are more plastic, often including strings of notes much like the flight or excitement calls of the same or a different Red Crossbill type.

Song repertoire size is unknown, but likely large. The spectrograms of song shown in this book represent a fraction of the variation in each type. Song differences between types need more study.

Other calls

All crossbills give very soft, plastic calls when feeding. All juvenile crossbills beg with a distinctive Chit-too or Chit-too-too.

Type 1 Red Crossbill

One of the most genetically distinct crossbill types, found mostly in the Appalachians. Medium bill and body size. May prefer to feed on red spruce, eastern white pine, eastern hemlock, and various hard-coned pine species in the Southeast.

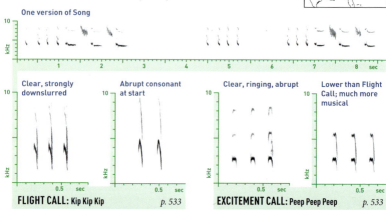

One version of Song

Clear, strongly downslurred

Abrupt consonant at start

Clear, ringing, abrupt

Lower than Flight Call; much more musical

FLIGHT CALL: Kip Kip Kip *p. 533*

Averages sharper than Type 2 Flight Call, but identification by ear is difficult; spectrogram shows downward hook at start (often subtle).

EXCITEMENT CALL: Peep Peep Peep *p. 533*

Averages slightly higher than Type 2 Excitement Call, but difficult to separate from it in the field. Notes on spectrogram show flatter shape.

Type 2 Red Crossbill

The most widespread crossbill type in North America, especially common in the West. One of our largest crossbills in body and bill size. Often feeds on large-seeded, hard-coned pine species such as ponderosa pine, but uses a wide variety of conifer species.

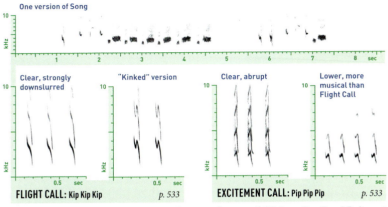

One version of Song

Clear, strongly downslurred

"Kinked" version

FLIGHT CALL: Kip Kip Kip   p. 533

Highly variable. Kinked versions, common in West, sound only slightly different from nonkinked versions.

Clear, abrupt

Lower, more musical than Flight Call

EXCITEMENT CALL: Pip Pip Pip   p. 533

Averages slightly lower than Type 1 Excitement Call; notes on spectrogram usually show more pronounced backward-N shape.

Type 3 Red Crossbill

Most common in the Pacific Northwest, but regular in the Great Lakes region, and wanders farther east; large numbers sometimes irrupt as far south as Maryland. Often found with Types 4 and 10, especially in winter. Small in bill and body size; often feeds on soft-coned conifers such as spruces and hemlocks.

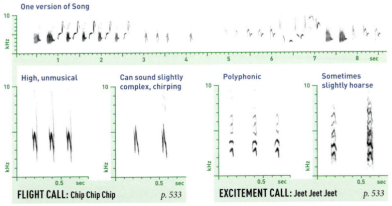

One version of Song

High, unmusical

Can sound slightly complex, chirping

FLIGHT CALL: Chip Chip Chip   p. 533

Quite uniform. Averages higher, less musical than Types 1 and 2; usually sounds monotone. On spectrogram, note high peak, above 5 kHz.

Polyphonic

Sometimes slightly hoarse

EXCITEMENT CALL: Jeet Jeet Jeet   p. 533

Distinctive among eastern Red Crossbills; may recall Jit of White-winged Crossbill. Abrupt; sounds monotone. Only slightly musical.

Type 4 Red Crossbill

Most common in the Pacific Northwest, but rarely irrupts eastward as far as Maine and Maryland. Medium bill and body size. May feed preferentially on cones of Douglas-fir, but also forages on spruces and pines.

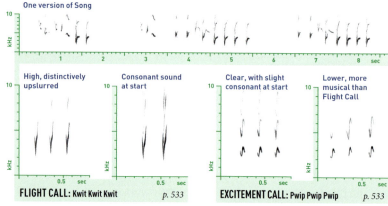

One version of Song

High, distinctively upslurred

Consonant sound at start

FLIGHT CALL: Kwit Kwit Kwit   *p. 533*

Highly uniform. When heard well, can be confused only with Type 10; averages less musical, with initial consonant sound.

Clear, with slight consonant at start

Lower, more musical than Flight Call

EXCITEMENT CALL: Pwip Pwip Pwip   *p. 533*

Fairly variable. Like Type 2 Excitement Call, but usually higher and slightly less musical. Often sounds slightly upslurred.

Type 10 Red Crossbill

Usually the most common Red Crossbill type in New England. Small in bill and body size. Sometimes flocks with Types 3 and 4. Associated with Sitka spruce along the Pacific Coast; in the East, likely feeds preferentially on soft-coned conifers such as spruces and hemlocks.

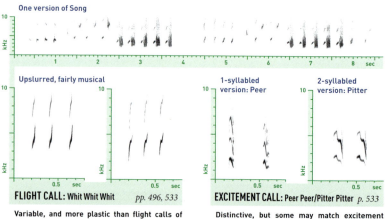

One version of Song

Upslurred, fairly musical

FLIGHT CALL: Whit Whit Whit   *pp. 496, 533*

Variable, and more plastic than flight calls of other types. Recalls Whit of Empidonax flycatchers, but usually more musical.

1-syllabled version: Peer

2-syllabled version: Pitter

EXCITEMENT CALL: Peer Peer/Pitter Pitter *p. 533*

Distinctive, but some may match excitement calls of Types 4 and 2. Pitter may indicate high alarm; recalls Pitter of Ruby-crowned Kinglet.

WHITE-WINGED CROSSBILL

Loxia leucoptera

Breeds in northern coniferous forests; feeds primarily on cones of spruce and tamarack. In some winters, large numbers wander south of breeding range.

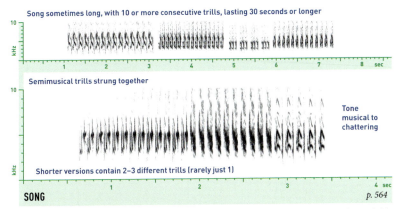

Song sometimes long, with 10 or more consecutive trills, lasting 30 seconds or longer

Semimusical trills strung together

Tone musical to chattering

Shorter versions contain 2–3 different trills (rarely just 1)

SONG *p. 564*

All year, by both sexes, but mostly Jan.–July, by males. Distinctive. Repertoire size unknown, but probably large; consecutive songs often differ in number and type of trills. Recalls Continuous Songs of redpolls, but trills longer, and single notes very rare.

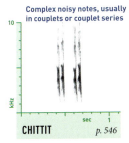

Complex noisy notes, usually in couplets or couplet series

CHITTIT *p. 546*

All year. Slightly variable; like redpoll Flight Calls, but averages harsher.

Rising, polyphonic

ZREET *p. 527*

All year; usually given singly. Variable, but shorter than goldfinch Zrees.

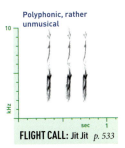

Polyphonic, rather unmusical

FLIGHT CALL: Jit Jit *p. 533*

All year; most common call. Variable; often upslurred.

Type 8 Red Crossbill

Endemic to Newfoundland, where other Red Crossbill types are rare. Medium bill and body size. Calls are quite variable but usually distinctive. Flight Calls (p. 533) usually sound downslurred, often clear, like Types 1 and 2, but some versions more complex. Excitement Calls (p. 533) low, complex, recalling Type 2 Excitement Calls but averaging less musical.

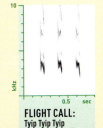

FLIGHT CALL:
Tyip Tyip Tyip

EXCITEMENT CALL:
Trit Trit Trit

COMMON REDPOLL

Acanthis flammea

A finch of far-northern forest edges; winters in the northern part of our area, in some years irrupting farther south in numbers. Often visits feeders.

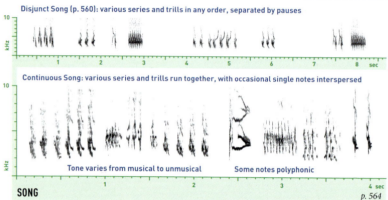

Disjunct Song (p. 560): various series and trills in any order, separated by pauses

Continuous Song: various series and trills run together, with occasional single notes interspersed

Tone varies from musical to unmusical   Some notes polyphonic

SONG   *p. 564*

Mostly in breeding season. Continuous and Disjunct versions of Song intergrade, but Disjunct Song seems more common. Each bird uses several types each of series and trills; unclear how/whether these differ from Flight Calls and Buzzes. Not known to be distinguishable from Hoary Redpoll Song.

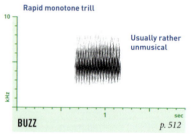

Rapid monotone trill

Usually rather unmusical

BUZZ   *p. 512*

All year. Variable; birds may have multiple versions. Often used in Song, but also given by winter flocks.

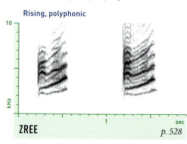

Rising, polyphonic

ZREE   *p. 528*

All year; rather plastic. Apparently serves to call groups together, and to signal excitement or alarm.

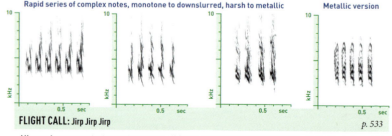

Rapid series of complex notes, monotone to downslurred, harsh to metallic   Metallic version

FLIGHT CALL: Jirp Jirp Jirp   *p. 533*

All year, in many contexts; most common call from winter flocks. Hugely variable; generally similar within flocks, but can differ greatly between them. Unknown whether individuals might have multiple call types or change call types over time. More study needed.

HOARY REDPOLL

Acanthis hornemanni

Very closely related to Common Red-poll, and difficult to distinguish from it. Breeds farther north, in tundra; rare in winter flocks of Commons.

Disjunct Song (p. 560): various series and trills in any order, separated by pauses

Continuous Song: various series and trills run together, with occasional single notes interspersed

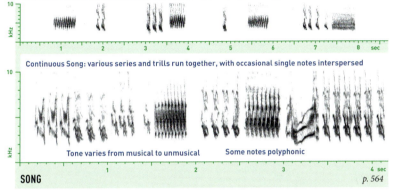

Tone varies from musical to unmusical Some notes polyphonic

SONG *p. 564*

Mostly in breeding season. Continuous and Disjunct versions of Song intergrade, but Disjunct Song seems more common. Each bird uses several types each of series and trills; unclear how/whether these differ from Flight Calls and Buzzes. Not known to be distinguishable from Common Redpoll Song.

Rapid monotone trill or buzz   Rising, polyphonic

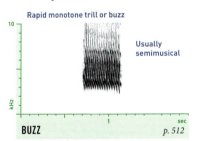

Usually semimusical

BUZZ *p. 512*

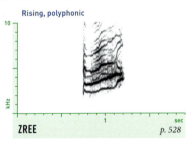

ZREE *p. 528*

All year. Variable; birds may have multiple versions. Often used in Song, but also given by winter flocks.

All year; rather plastic. Apparently serves to call groups together, and to signal excitement or alarm.

Rapid series of complex notes, monotone to downslurred, harsh to metallic   Metallic version

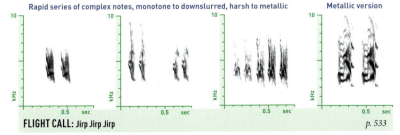

FLIGHT CALL: Jirp Jirp Jirp *p. 533*

All year, in many contexts; most common call from winter flocks. Hugely variable; generally similar within flocks, but can differ greatly between them. Unknown whether individuals might have multiple call types. Hoaries in flocks of Commons may change calls to match flockmates.

AMERICAN GOLDFINCH

Spinus tristis

Common and widespread in weedy fields, second growth, and backyards; often visits feeders. Yellow plumage of male is present only in summer.

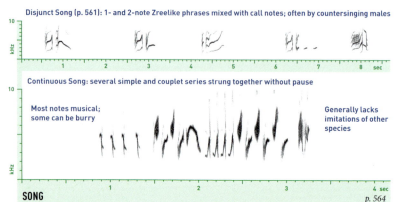

Disjunct Song (p. 561): 1- and 2-note Zreelike phrases mixed with call notes; often by countersinging males

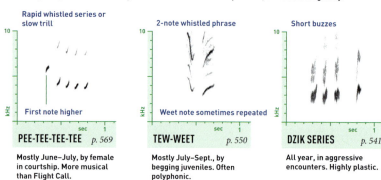

Continuous Song: several simple and couplet series strung together without pause

Most notes musical; some can be burry

Generally lacks imitations of other species

SONG *p. 564*

Mostly Apr.–Aug., from high perch or in circling flight. Short Songs of 2–4 series may be repeated at intervals, recalling Songs of Indigo Bunting or various warblers; at other times, consecutive songs may differ greatly in content and length. Most series contain 3–4 repetitions, but this varies greatly.

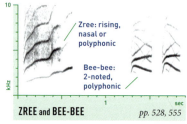

Rapid whistled series or slow trill

First note higher

PEE-TEE-TEE-TEE *p. 569*

Mostly June–July, by female in courtship. More musical than Flight Call.

2-note whistled phrase

Weet note sometimes repeated

TEW-WEET *p. 550*

Mostly July–Sept., by begging juveniles. Often polyphonic.

Short buzzes

DZIK SERIES *p. 541*

All year, in aggressive encounters. Highly plastic.

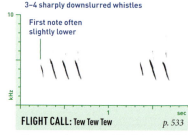

Zree: rising, nasal or polyphonic

Bee-bee: 2-noted, polyphonic

ZREE and BEE-BEE *pp. 528, 555*

All year, in alarm. Two main call types; Zree is more common, Bee-bee typically given in higher alarm, especially near nest.

3–4 sharply downslurred whistles

First note often slightly lower

FLIGHT CALL: Tew Tew Tew *p. 533*

All year; most common call. Plastic in length, but otherwise uniform rangewide. Often given in flight. Sharp but fairly musical.

LESSER GOLDFINCH

Spinus psaltria

Common in a variety of habitats, from oak woods to pine forests to backyards; often visits feeders. Prefers more arid areas than American Goldfinch.

Disjunct Song (p. 561): various 1- and 2-note whistled and polyphonic phrases delivered at intervals

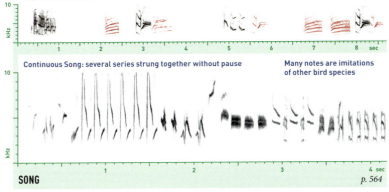

Continuous Song: several series strung together without pause

Many notes are imitations of other bird species

SONG   *p. 564*

Mostly Feb.–July, from high perch or in circling flight. Continuous Song grades into Disjunct Song. Consecutive songs may differ greatly in content and length; most series contain 3–4 repetitions, but this varies. Imitations of other species are excellent, but can go unnoticed inside complex jumble of Song.

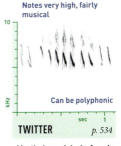

Notes very high, fairly musical

Can be polyphonic

TWITTER   *p. 534*

Mostly June–July, by female in courtship. Highly plastic.

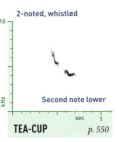

2-noted, whistled

Second note lower

TEA-CUP   *p. 550*

Mostly July–Sept., by begging juveniles. Notes can be given separately.

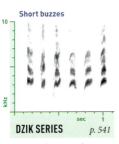

Short buzzes

DZIK SERIES   *p. 541*

All year, in aggressive encounters. Highly plastic.

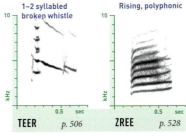

1–2 syllabled broken whistle

Rising, polyphonic

TEER   *p. 506*   **ZREE**   *p. 528*

All year. Teer is most common and distinctive call. Zree lower, burrier than other finches'. Individuals likely have multiple versions of both.

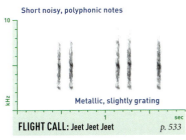

Short noisy, polyphonic notes

Metallic, slightly grating

FLIGHT CALL: Jeet Jeet Jeet   *p. 533*

All year, in contact and flight. Somewhat variable, but distinctive. Higher, more metallic than Pine Siskin Flight Call.

PINE SISKIN

Spinus pinus

Breeds in coniferous forests; wanders widely in winter, sometimes remaining to breed south of normal range. Often visits feeders.

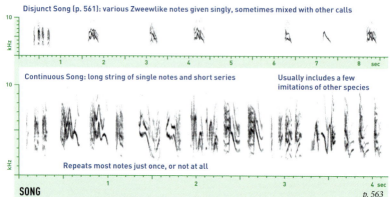

Disjunct Song (p. 561): various Zweewlike notes given singly, sometimes mixed with other calls

Continuous Song: long string of single notes and short series

Usually includes a few imitations of other species

Repeats most notes just once, or not at all

SONG p. 563

Mostly Jan.–June, during breeding activity, reportedly only by males. Many notes resemble Zweew call; listen for inclusion of occasional diagnostic Zhree calls. Song often begins on high perch and continues during long circling flight; may conclude back on the perch. Continuous Song grades into Disjunct Song.

Long, rising polyphonic burr

Often followed by Zweew and Jits

ZHREE p. 530

All year; unmistakable. Given during Song or singly, usually while flashing yellow patches in wings and tail, perhaps in aggression.

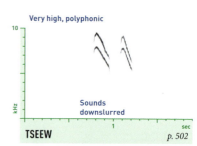

Very high, polyphonic

Sounds downslurred

TSEEW p. 502

Given near nest, and by individuals while foraging; function unclear. Fairly uniform; usually given in 1s and 2s.

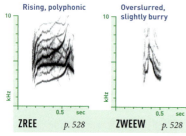

Rising, polyphonic

Overslurred, slightly burry

ZREE p. 528

ZWEEW p. 528

All year. Zree possibly in alarm; Zweew, possibly a variant of Zree, often with Flight Call. Phrases in Disjunct Song may resemble both calls.

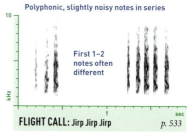

Polyphonic, slightly noisy notes in series

First 1–2 notes often different

FLIGHT CALL: Jirp Jirp Jirp p. 533

All year, in contact and flight. Only slightly variable. Lower, less musical than Lesser Goldfinch Flight Call.

EVENING GROSBEAK

Coccothraustes vespertinus

A social, somewhat nomadic finch of coniferous and mixed forests. In some winters, wanders far south of breeding range. Sometimes visits feeders.

To distinguish apparent Song from apparent Breet calls, listen for steady rhythm and doubling

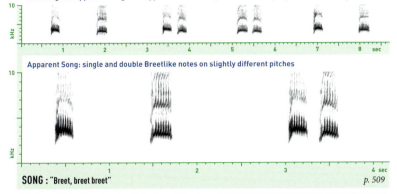

Apparent Song: single and double Breetlike notes on slightly different pitches

SONG : "Breet, breet breet"   *p. 509*

What constitutes Song is not entirely clear. At least some males have repertoires of 2–6 different stereotyped Breetlike notes, which they give in 1s and 2s from a high perch for long periods, in the manner of a song. Pzeers may be incorporated; perhaps some songs contain only Pzeers. More study needed.

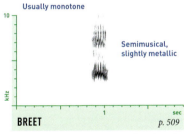

Usually monotone

Semimusical, slightly metallic

BREET   *p. 509*

All year; frequently heard. Most often given singly or with Pzeers, in a random rhythm more typical of calls than songs.

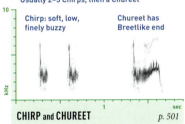

Usually 2–3 Chirps, then a Chureet

Chirp: soft, low, finely buzzy

Chureet has Breetlike end

CHIRP and CHUREET   *p. 501*

Poorly known; soft vocalizations given mostly by foraging flocks. Lower, softer, and less musical than Pzeer.

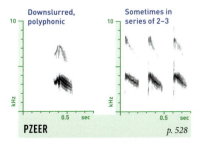

Downslurred, polyphonic

Sometimes in series of 2–3

PZEER   *p. 528*

All year; most common call. Often recalls certain sounds of House Sparrow. Both versions shown above correspond to "Type 3" (see right).

GEOGRAPHIC VARIATION IN EVENING GROSBEAK CALLS

Five "types" have been identified among the Pzeerlike calls of Evening Grosbeaks. These seem to represent regional variation, unlike Red Crossbill call types, which represent ecologically specialized populations that overlap in range. Shown here is "Type 3" Evening Grosbeak, the only type known from the East.

PINE GROSBEAK

Pinicola enucleator

A large, tame, but often inconspicuous northern finch of open coniferous forests. In some winters, wanders well south of breeding range.

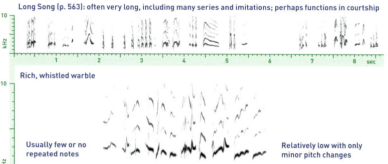

Long Song (p. 563): often very long, including many series and imitations; perhaps functions in courtship

Rich, whistled warble

Usually few or no repeated notes

Relatively low with only minor pitch changes

SONG

p. 572

Almost all year, mostly by males. Females apparently sing rarely; singers without red plumage may be first-spring males. Song grades into Long Song; some versions are short, but contain imitations. Repertoire size unknown; consecutive songs may start similarly, but then vary.

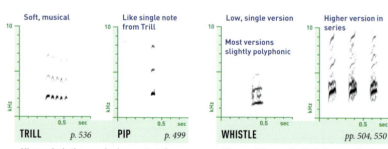

Soft, musical

Like single note from Trill

Low, single version

Most versions slightly polyphonic

Higher version in series

TRILL *p. 536*   **PIP** *p. 499*   **WHISTLE** *pp. 504, 550*

All year, by both sexes, in close contact. Generally audible only at close range, often mixed with other, louder calls. Plastic.

All year, infrequently, by both sexes, in alarm. Highly geographically variable and fairly plastic; most versions are nearly monotone.

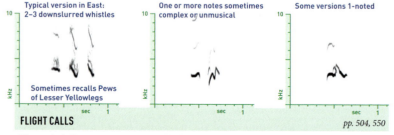

Typical version in East: 2–3 downslurred whistles

One or more notes sometimes complex or unmusical

Some versions 1-noted

Sometimes recalls Pews of Lesser Yellowlegs

FLIGHT CALLS *pp. 504, 550*

All year, by both sexes; most common call. Given frequently by perched birds. Extremely geographically variable and quite plastic; as in redpolls and Red Crossbills, flockmates appear to share a call type, and flocks with dissimilar call types reportedly do not mix. More study needed.

LAPLAND LONGSPUR

Calcarius lapponicus

Breeds on Arctic tundra; winters in open areas, often in flocks with Horned Larks, Snow Buntings, and/or other longspur species.

Males typically have a single songtype, repeated with only minor truncations

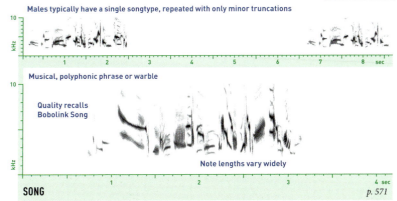

Musical, polyphonic phrase or warble

Quality recalls Bobolink Song

Note lengths vary widely

SONG   *p. 571*

Almost exclusively on breeding grounds. Geographically variable, but males on adjacent territories tend to sing the same songtype. Usually about 2 seconds in length, but 2 or more songs sometimes strung together without pause, especially in courtship.

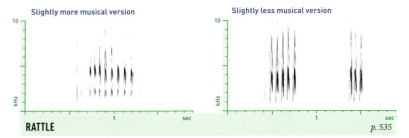

Slightly more musical version   **Slightly less musical version**

RATTLE   *p. 535*

All year; frequently heard in winter. Somewhat plastic, especially in number of notes; most versions are 4–5 notes long, but shorter and longer versions, including single Piks, are not uncommon. Averages slightly lower and more musical than Smith's Rattle, longer and faster than McCown's, but overlaps both.

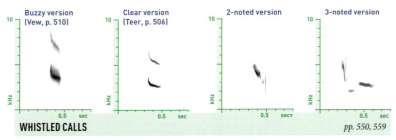

Buzzy version   **Clear version**   **2-noted version**   **3-noted version**
(Vew, p. 510)   **(Teer, p. 506)**

WHISTLED CALLS   *pp. 550, 559*

All year, in alarm, flock contact, and possibly other contexts. Extremely variable; each bird has a repertoire of perhaps 6 versions. Tone may be clear, buzzy, or polyphonic; most versions 1-noted and downslurred, but some are monotone, and multinoted versions are fairly common.

Smith's Longspur

Calcarius pictus

Breeds in Arctic; winters in mowed fields with three-awn grass (*Aristida*). Unusual breeding biology: males mate with multiple females and vice versa.

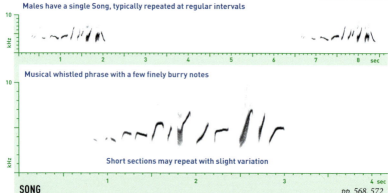

Males have a single Song, typically repeated at regular intervals

Musical whistled phrase with a few finely burry notes

Short sections may repeat with slight variation

SONG

pp. 568, 572

Mostly on breeding grounds, but reported to be given occasionally on spring migration. Most similar to Chestnut-collared Longspur Song; lacks polyphonic quality of Lapland Longspur. Lacks trills; some versions lack burry notes. Compare American Tree Sparrow Song.

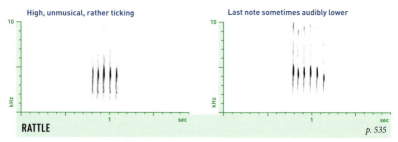

High, unmusical, rather ticking

Last note sometimes audibly lower

RATTLE

p. 535

All year; most common call. Somewhat more uniform than other longspur Rattles; plastic in length, but consistently unmusical. Notes rather ticking, recalling Rattle of Summer Tanager, but higher. Lower final note rather distinctive, but not always present.

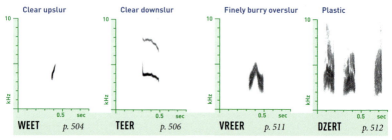

Clear upslur

Clear downslur

Finely burry overslur

Plastic

WEET *p. 504* **TEER** *p. 506* **VREER** *p. 511* **DZERT** *p. 512*

All year, in chases on breeding grounds and in winter flocks. Weets may predominate in winter. Dzerts become plastic in high agitation. Unlike in Lapland and McCown's, apparently not given in songlike sequences, and quite uniform rangewide; only calls like the 4 shown have been recorded.

Chestnut-collared Longspur

Calcarius ornatus

Breeds in native midgrass prairies, especially those that have been moderately mowed or grazed; winters in more arid grasslands.

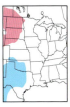

Males appear to have a single songtype, usually repeated at steady intervals

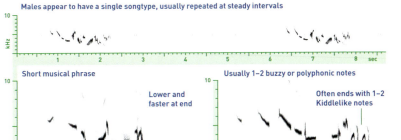

SONG   *p. 571*

Short musical phrase
Lower and faster at end
Starts with 1–3 high clear whistles

Usually 1–2 buzzy or polyphonic notes
Often ends with 1–2 Kiddlelike notes

Mostly Apr.–July. Given from ground or low perch, or in flight display as male glides to ground on spread wings and tail, sometimes with slow flaps. Pattern and quality can strongly resemble Western Meadowlark Song, but includes buzzy, polyphonic, and/or Kiddle-like notes.

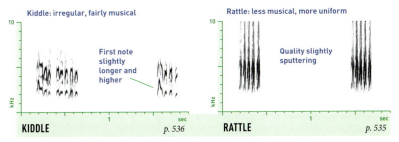

KIDDLE   *p. 536*   **RATTLE**   *p. 535*

Kiddle: irregular, fairly musical
First note slightly longer and higher

Rattle: less musical, more uniform
Quality slightly sputtering

All year. Kiddle is most common and distinctive call, given in contact and mild alarm. Highly plastic. Rattle is high, rapid, and rather unmusical; associated with male Song, responses to playback, and aggressive encounters. Both call types intergrade extensively.

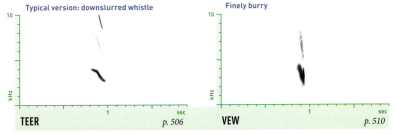

TEER   *p. 506*   **VEW**   *p. 510*

Typical version: downslurred whistle

Finely burry

All year. Given in alarm near nest, usually mixed with Kiddles or Rattles (p. 559); also given in winter flocks. Individuals may have multiple versions, but overall, less variable than Whistled Calls of Lapland and Mc-Cown's; the vast majority are clear Teerlike downslurs.

McCown's Longspur

Rhynchophanes mccownii

Breeds and winters in semiarid short-grass prairie. Generally prefers shorter grass than Chestnut-collared Longspur, but the two often found together.

Long Song (p. 563): phrases run together for up to 5 seconds; short sections may be repeated

Disjunct Song: high tinkling musical phrases, separated by distinct pauses

Quality recalls Horned Lark, but phrases longer, more complex, and more musical

SONG
p. 562

Mostly Apr.–June. Each male apparently has at least 10 phrase types. Disjunct Song often given from ground or low perch, Long Song in flight display as male glides to ground on spread wings and tail. The 2 song forms intergrade; pauses in Disjunct Song tend to become shorter as Song progresses.

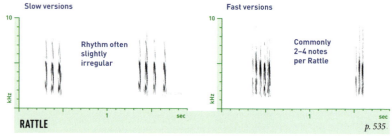

Slow versions

Rhythm often slightly irregular

Fast versions

Commonly 2–4 notes per Rattle

RATTLE
p. 535

All year; most common call. Quite plastic in length, but averages shorter than other longspur Rattles, usually 4 or fewer notes. Slow, irregular rhythm fairly distinctive when present.

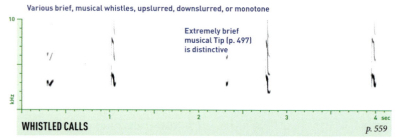

Various brief, musical whistles, upslurred, downslurred, or monotone

Extremely brief musical Tip (p. 497) is distinctive

WHISTLED CALLS
p. 559

All year. Individuals have perhaps 4 versions. In alarm on breeding grounds, birds tend to cycle through several versions continuously, mixing them with short Rattles. Also heard from flocks in migration and winter. Almost always clear and 1-syllabled, but buzzy versions exist.

Snow Bunting

Plectrophenax nivalis

Breeds in rock cavities on Arctic tundra, often near human habitation; winters in open areas, often in flocks with Horned Larks and/or longspurs.

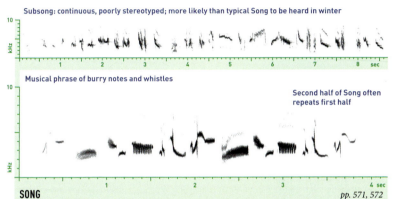

Subsong: continuous, poorly stereotyped; more likely than typical Song to be heard in winter

Musical phrase of burry notes and whistles

Second half of Song often repeats first half

SONG *pp. 571, 572*

Mostly on breeding grounds, but subsong can be heard in late winter. Males have a single songtype, often truncated. Quality may recall Blue Grosbeak; generally lacks polyphony of Lapland Longspur song, and more likely to be constructed of repeated phrases, especially in flight display.

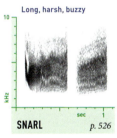

Long, harsh, buzzy

SNARL *p. 526*

All year, in aggressive interactions. Highly plastic.

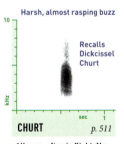

Harsh, almost rasping buzz

Recalls Dickcissel Churt

CHURT *p. 511*

All year, often in flight. No longspur has a similar call.

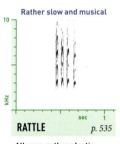

Rather slow and musical

RATTLE *p. 535*

All year; rather plastic. Averages slightly more musical than longspur Rattles.

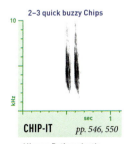

2–3 quick buzzy Chips

CHIP-IT *pp. 546, 550*

All year. Rather plastic; often replaces first 1–2 notes of Rattle.

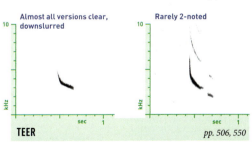

Almost all versions clear, downslurred

TEER *pp. 506, 550*

Rarely 2-noted

All year; in alarm near nest, and when flocking in winter. At least some birds have repertoires of a few versions, but most versions quite similar.

WARBLERS (Family Parulidae)

Warblers are small, highly active, and often brightly colored. Their beauty and diversity make them very popular with birders. Most are highly migratory.

Warblers learn their songs. They sort into 3 broad groups by repertoire type:

Warblers with Primary and Complex Songs
(*Prothonotary, Worm-eating, Swainson's, Ovenbird, waterthrushes, all* Geothlypis)

In this group, vocal repertoires are very similar to those of many sparrows. Males tend to have a single short **primary song** as well as a single longer **complex song**, which is often given during a flight display and introduced by a series of call notes. Complex songs are given infrequently.

Warblers with "First-Category" and "Second-Category" Songs
(*Blue-winged, Golden-winged, Black-and-white, Canada, most* Setophaga)

Many species in this group also have two different types of singing behaviors. **First-category singing** is done mostly during the day by unpaired males, often directed at females. **Second-category singing** is done mostly at dawn by unpaired males or during the day by paired males, often directed at other males. Second-category singing tends to be characterized by long series of call-like notes in between songs, by shorter pauses between songs, and/or by a greater tendency for consecutive song-types to differ.

In species such as the Black-throated Green Warbler, first- and second-category songs are easy to distinguish in the field, because they differ consistently in form. In other species, songs in the two categories may be difficult to distinguish. In fact, in species such as the Yellow Warbler, the very same songtype may be used in first-category singing by one individual and in second-category singing by another individual (or even the same individual). In these cases it is not the form of the songs that determines their category, but the pattern in which they are sung.

Warblers with No Clear Distinction Between Song Categories
(*Wilson's, Palm, Bay-breasted, Blackpoll, all* Oreothlypis)

These warblers have not been documented to sing two different categories of songs, but more study may reveal that the categories do in fact exist.

Warbler Calls

The call repertoires of most warblers include a Seetlike or Dzitlike call given in contact and in flight (including by night migrants); a high brief Tink given in high alarm; and a Chiplike note given in mild alarm. In addition, many species give Twitters or Rattles in alarm or aggression (see p. 535).

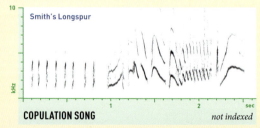

COPULATION SONGS OF LONGSPURS

Smith's Longspur has been recorded giving a complex, stereotyped song during copulation that is quite different from the typical song that males use to advertise for a mate. Other species in the longspur family may also have specialized copulation songs; more study is needed.

Smith's Longspur

COPULATION SONG   *not indexed*

OVENBIRD

Seiurus aurocapilla

A very distinctive, thrushlike warbler. Breeds in mature deciduous or mixed woods, foraging mostly in leaf litter; builds dome-shaped nest on ground.

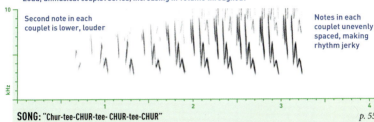

Loud, unmusical couplet series, increasing in volume throughout

Second note in each couplet is lower, louder

Notes in each couplet unevenly spaced, making rhythm jerky

kHz

10

1 2 3 4 sec

SONG: "Chur-tee-CHUR-tee- CHUR-tee-CHUR" *p. 557*

Mostly Apr.–Aug., by males, sometimes from high perches; females not known to sing. Somewhat variable; individuals have a single version. Rare versions are triplet series, or 2 consecutive couplet series. Often transliterated as "Teacher Teacher," but the accent is always on the lower "CHUR" notes.

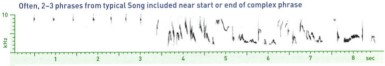

Often, 2–3 phrases from typical Song included near start or end of complex phrase

10

kHz

1 2 3 4 5 6 7 8 sec

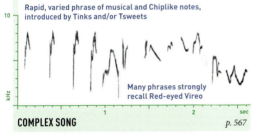

Rapid, varied phrase of musical and Chiplike notes, introduced by Tinks and/or Tsweets

10

kHz

Many phrases strongly recall Red-eyed Vireo

1 2 sec

COMPLEX SONG *p. 567*

Mostly Apr.–Aug., by males, sometimes in flight display, but also on ground; often at dusk. Infrequent, but not rare; partial flight songs common in courtship and aggressive interactions.

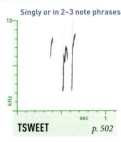

Singly or in 2–3 note phrases

10

kHz

sec 1

TSWEET *p. 502*

Highly variable; 1 typical version shown. By males, in interactions.

Rising, mostly clear

10

kHz

sec 1

SREET *p. 501*

All year, often by night migrants. Some versions grade into Tsweet.

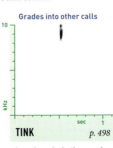

Grades into other calls

10

kHz

sec 1

TINK *p. 498*

Apr.–Aug., by both sexes in interactions. Sometimes run into rapid Twitter.

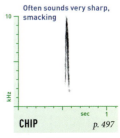

Often sounds very sharp, smacking

10

kHz

sec 1

CHIP *p. 497*

All year, in alarm. Highly plastic, especially in pitch; becomes lower in alarm.

NORTHERN WATERTHRUSH

Parkesia noveboracensis

Large, stocky, and short-tailed. Breeds in boggy thickets and along wooded streams. Walks along the water's edge, frequently bobbing tail up and down.

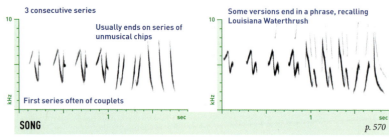

3 consecutive series

Usually ends on series of unmusical chips

First series often of couplets

Some versions end in a phrase, recalling Louisiana Waterthrush

SONG   *p. 570*

Mostly Apr.–July, reportedly only by males. Individuals apparently have a single songtype. Pitch distinctively low, tone emphatic. A few Songs have 4 sections, or only 2. Consecutive series tend to increase slightly in speed and descend slightly in pitch.

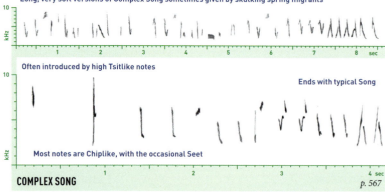

Long, very soft versions of Complex Song sometimes given by skulking spring migrants

Often introduced by high Tsitlike notes

Ends with typical Song

Most notes are Chiplike, with the occasional Seet

COMPLEX SONG   *p. 567*

Given infrequently, mostly on breeding grounds, often in display flight. Repertoire size unknown. Becomes louder and more recognizable toward end, like Complex Song of Ovenbird, but unlike that of Louisiana Waterthrush.

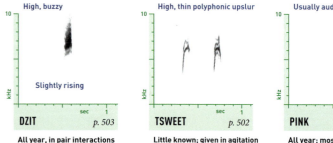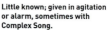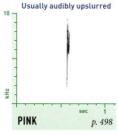

High, buzzy

Slightly rising

DZIT   *p. 503*

All year, in pair interactions and in flight, including by night migrants.

High, thin polyphonic upslur

TSWEET   *p. 502*

Little known; given in agitation or alarm, sometimes with Complex Song.

Usually audibly upslurred

PINK   *p. 498*

All year; most common call. Fairly distinctive; sometimes run into a Sputter.

LOUISIANA WATERTHRUSH

Parkesia motacilla

Breeds along rushing forest streams. Like Northern Waterthrush in shape and behavior, but differs in vocalizations and details of plumage.

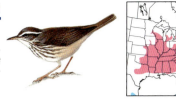

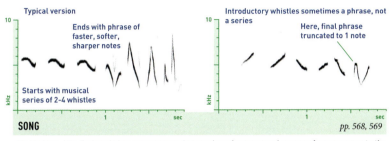

Typical version

Ends with phrase of faster, softer, sharper notes

Starts with musical series of 2-4 whistles

Introductory whistles sometimes a phrase, not a series

Here, final phrase truncated to 1 note

10 kHz | 1 | sec

SONG   pp. 568, 569

Mostly Mar.–July. Each male has one songtype, but number of notes at end can vary from one song to the next, grading smoothly into the Complex Song; typical Song can even be thought of as a truncation of the Complex Song. Short versions can recall Swainson's or Hooded Warbler Songs.

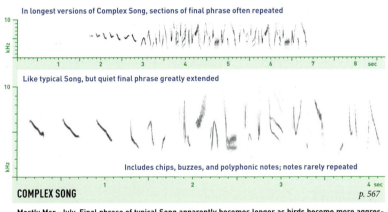

In longest versions of Complex Song, sections of final phrase often repeated

10 kHz | 1 2 3 4 5 6 7 8 sec

Like typical Song, but quiet final phrase greatly extended

Includes chips, buzzes, and polyphonic notes; notes rarely repeated

10 kHz | 1 2 3 4 sec

COMPLEX SONG   p. 567

Mostly Mar.–July. Final phrase of typical Song apparently becomes longer as birds become more aggressive or territorial; Complex Song given more often when countersinging with rival males. Also given in slow-flapping flight display, introduced by long series of call notes.

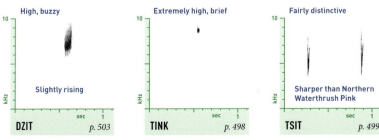

High, buzzy

Slightly rising

DZIT   p. 503

All year, in pair interactions and in flight, including by night migrants.

Extremely high, brief

TINK   p. 498

All year, in high alarm. Plastic; grades into Tsit.

Fairly distinctive

Sharper than Northern Waterthrush Pink

TSIT   p. 499

All year. In agitation, becomes lower, noisier, and runs into rapid Chatter.

Worm-eating Warbler

Helmitheros vermivorum

Breeds in forests with thick understory, usually on slopes. The "worms" it eats are caterpillars, which it seeks among dead leaves. Nests on the ground.

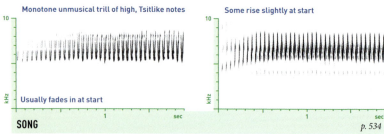

Monotone unmusical trill of high, Tsitlike notes

Some rise slightly at start

Usually fades in at start

SONG *p. 534*

Apr.–July. Males have a single songtype; pitch and speed of trill vary from male to male. Usually 2 seconds long; generally faster, higher, and more insectlike than Chipping Sparrow Song, with a greater tendency to fade in at the start, but some may not be separable by ear.

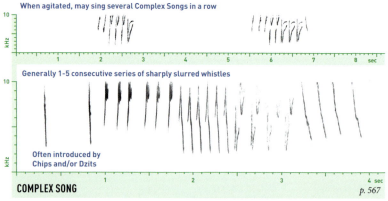

When agitated, may sing several Complex Songs in a row

Generally 1-5 consecutive series of sharply slurred whistles

Often introduced by Chips and/or Dzits

COMPLEX SONG *p. 567*

Apr.–July. Infrequently heard, from males in aggressive encounters with other males, usually in flight displays through the forest midstory. Variable; consecutive flight songs from a bird may differ slightly, most being truncations of the complete pattern. Some versions include buzzes.

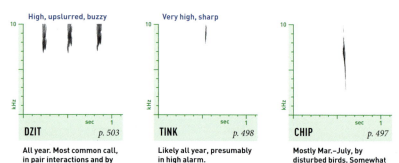

High, upslurred, buzzy

Very high, sharp

DZIT *p. 503*

TINK *p. 498*

CHIP *p. 497*

All year. Most common call, in pair interactions and by night migrants.

Likely all year, presumably in high alarm.

Mostly Mar.–July, by disturbed birds. Somewhat plastic.

PROTHONOTARY WARBLER

Protonotaria citrea

Breeds in flooded forests and bayous; once aptly known as "Golden Swamp Warbler." The only eastern warbler that nests in cavities.

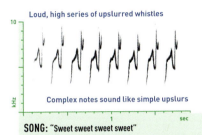

Loud, high series of upslurred whistles

Complex notes sound like simple upslurs

SONG: "Sweet sweet sweet sweet"

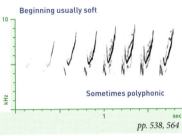

Beginning usually soft

Sometimes polyphonic

pp. 538, 564

Mar.–Aug. Males apparently have only a single songtype, which varies only in number of notes. In some rare versions, a second series of slightly different notes is appended to the first. Generally uniform; nearby males may have different songtypes, but similar songtypes are heard rangewide.

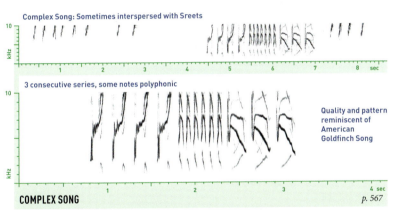

Complex Song: Sometimes interspersed with Sreets

3 consecutive series, some notes polyphonic

Quality and pattern reminiscent of American Goldfinch Song

COMPLEX SONG

p. 567

Likely Mar.–Aug. Poorly known; apparently given by males in close courtship of females, sometimes in flight display. No information on geographic variation. In the sole available recording, 1 songtype is repeated many times with little variation.

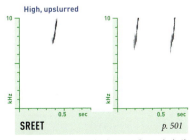

High, upslurred

SREET

p. 501

Likely all year, by males between Songs, by both sexes in pair interactions, and by night migrants. Slight burriness hard to hear.

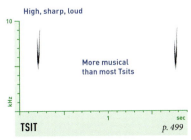

High, sharp, loud

More musical than most Tsits

TSIT

p. 499

All year. Most common call. Somewhat variable; like Louisiana Waterthrush Tsit but higher. Compare Hooded Warbler Chip.

SWAINSON'S WARBLER

Limnothlypis swainsonii

An uncommon, local, and skulking bird of swampy lowland canebrakes, as well as forested upland ravines with an understory of rhododendron.

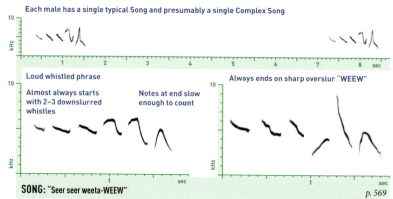

Each male has a single typical Song and presumably a single Complex Song

Loud whistled phrase

Almost always starts with 2–3 downslurred whistles

Notes at end slow enough to count

Always ends on sharp overslur "WEEW"

SONG: "Seer seer weeta-WEEW" p. 569

Mostly Apr.–July. Moderate variation between individuals. Very like Louisiana Waterthrush Song, but ending slower and simpler; last note always lowest. Easily confused with variant of Hooded Warbler Song that starts with downslurs, but typically slower.

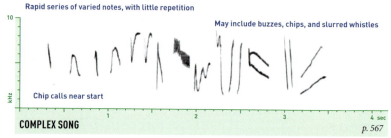

Rapid series of varied notes, with little repetition

May include buzzes, chips, and slurred whistles

Chip calls near start

COMPLEX SONG p. 567

Soft and infrequently given; reportedly sometimes given in display flight. In the sole available recording, 1 stereotyped songtype is repeated several times. Rhythm rather jerky and disjointed, with less repetition than most other warbler Complex Songs.

 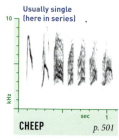 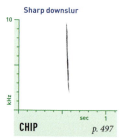

High, clear, upslurred

SREET p. 501

All year, sometimes in flight, and presumably by night migrants.

Usually single (here in series)

CHEEP p. 501

Mostly Apr.–Aug., in high alarm. Highly plastic, varying from Chirps to Whines.

Sharp downslur

CHIP p. 497

All year. Plastic; grades into Cheeps, especially in agitation.

BLACK-AND-WHITE WARBLER

Mniotilta varia

Found in mature forests and wood-land edges. Unlike other warblers, creeps along tree trunks and large branches, much like a nuthatch.

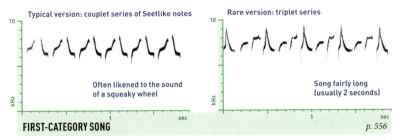

Typical version: couplet series of Seetlike notes

Rare version: triplet series

Often likened to the sound of a squeaky wheel

Song fairly long (usually 2 seconds)

FIRST-CATEGORY SONG

p. 556

Mostly Apr.–Aug. Little studied, but likely functions like First-Category Songs of other warblers. Repertoire size unknown, but males may have a single version. Consecutive songs usually similar, but males some-times switch frequently between a First-Category and a Second-Category songtype.

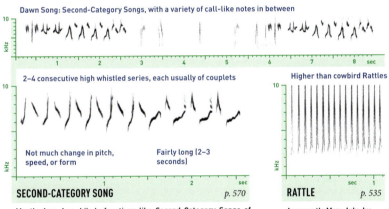

Dawn Song: Second-Category Songs, with a variety of call-like notes in between

2–4 consecutive high whistled series, each usually of couplets

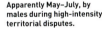

Higher than cowbird Rattles

Not much change in pitch, speed, or form

Fairly long (2–3 seconds)

SECOND-CATEGORY SONG   *p. 570*

RATTLE   *p. 535*

Mostly Apr.–Aug. Likely functions like Second-Category Songs of other warblers, but more study needed. Individuals may have a sin-gle version; consecutive songs usually similar.

Apparently May–July, by males during high-intensity territorial disputes.

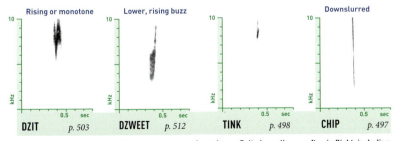

Rising or monotone

Lower, rising buzz

Downslurred

DZIT   *p. 503*

DZWEET   *p. 512*

TINK   *p. 498*

CHIP   *p. 497*

Gives a variety of plastic calls; most common versions shown. Dzit given all year, often in flight, including by night migrants. Dzweet apparently given in aggressive encounters. Tink given in high alarm. Chip given all year in mild alarm; sometimes has initial upslur like single note from Rattle.

BLUE-WINGED WARBLER

Vermivora cyanoptera

Common in deciduous second growth, forest edges, and shrublands. Appears to be expanding northward, displacing Golden-winged Warbler in many areas.

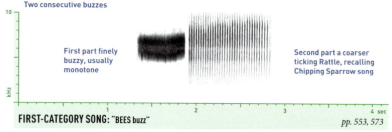

Two consecutive buzzes

First part finely buzzy, usually monotone

Second part a coarser ticking Rattle, recalling Chipping Sparrow song

FIRST-CATEGORY SONG: "BEES buzz" *pp. 553, 573*

Mostly May–July, by male only. Fairly uniform. Second buzz usually sounds lower than first, but sometimes higher; rarely, order of buzzes reversed. Some males sing like typical Golden-winged Warblers; rarely, males will alternate typical Blue-winged and Golden-winged Songs.

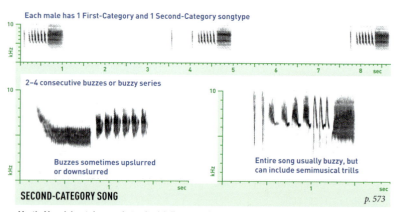

Each male has 1 First-Category and 1 Second-Category songtype

2–4 consecutive buzzes or buzzy series

Buzzes sometimes upslurred or downslurred

Entire song usually buzzy, but can include semimusical trills

SECOND-CATEGORY SONG *p. 573*

Mostly May–July, at dawn or in territorial disputes, often with strings of Chips or Sreetlike notes before each song. Stereotyped, but highly variable. Occasionally given in flight display with a few additional notes at start. Not consistently separable from Golden-winged's Second-Category Song.

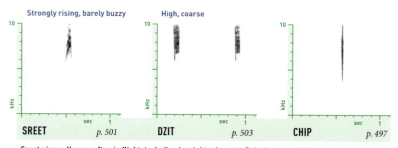

Strongly rising, barely buzzy

High, coarse

SREET *p. 501*   **DZIT** *p. 503*   **CHIP** *p. 497*

Sreet given all year, often in flight, including by night migrants. Dzit given mostly May–Aug., by adults and juveniles in family groups. Highly plastic in length; grades into Chip. Chip is most common call, given all year in alarm. Likely also gives Tink in high alarm (p. 498); no recordings available.

GOLDEN-WINGED WARBLER

Vermivora chrysoptera

Uncommon and declining in deciduous second growth. Range contracting with reforestation, urban development, and expansion of Blue-winged Warbler.

A buzz, then 2–5 slightly lower buzzes

Buzzes generally similar in quality, shorter than in Blue-winged Warbler Song

FIRST-CATEGORY SONG: "Bees, buzz buzz buzz"   *pp. 553, 573*

Mostly May–July, by males only. Fairly uniform, but after pairing, males sing fewer buzzes at end. Rare 2-noted versions resemble Blue-winged Song, but note buzz length and quality. Some males sing like typical Blue-winged Warblers.

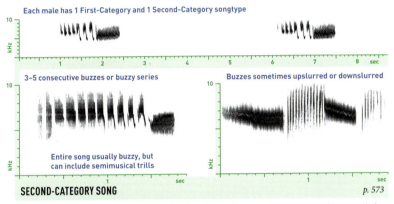

Each male has 1 First-Category and 1 Second-Category songtype

3–5 consecutive buzzes or buzzy series

Buzzes sometimes upslurred or downslurred

Entire song usually buzzy, but can include semimusical trills

SECOND-CATEGORY SONG   *p. 573*

Mostly May–July, at dawn or in territorial disputes, often with strings of Chips or Sreetlike notes before each song. Highly variable from male to male. Occasionally given in flight display with a few additional notes at start. Not consistently separable from Blue-winged's Second-Category Song.

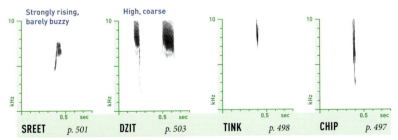

Strongly rising, barely buzzy

High, coarse

SREET *p. 501*   **DZIT** *p. 503*   **TINK** *p. 498*   **CHIP** *p. 497*

Sreet given all year, often in flight, including by night migrants. Somewhat plastic; not known to be separable from Blue-winged Sreet. Dzit given mostly May–Aug., by adults and juveniles in family groups. Highly plastic in length; grades into Chip. Tink given in high alarm. Chip given all year in alarm.

Tennessee Warbler

Oreothlypis peregrina

An abundant breeder in boreal forests, especially during outbreaks of spruce budworm. Like Orange-crowned, but undertail always whiter than breast.

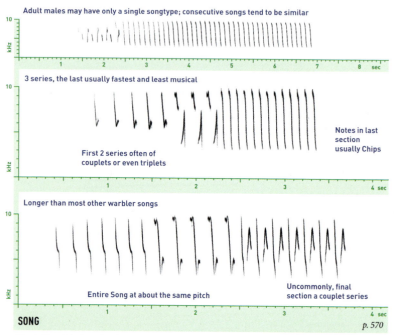

Adult males may have only a single songtype; consecutive songs tend to be similar

3 series, the last usually fastest and least musical

First 2 series often of couplets or even triplets

Notes in last section usually Chips

Longer than most other warbler songs

Entire Song at about the same pitch

Uncommonly, final section a couplet series

SONG

p. 570

Mostly May–July, by males; females not known to sing. Variable, but usually fairly distinctive. Can be confused with Northern Waterthrush Song; note higher pitch and less musical quality. Some males sing 2-parted Song that can be confused with Nashville Warbler. Multiple songtypes and Complex Songs have not been documented.

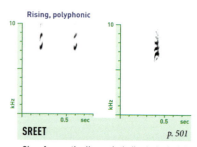

Rising, polyphonic

SREET

p. 501

Given frequently all year, including by both night and daytime migrants. Somewhat plastic; resembles other warbler Sreets.

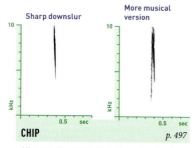

Sharp downslur

More musical version

CHIP

p. 497

All year, in alarm. Typical version is sharp, downslurred; more musical overslurred version may serve a different function.

ORANGE-CROWNED WARBLER

Oreothlypis celata

Common, especially in the West, nesting in open shrubby woods. Western populations are greener, eastern populations grayer; undertail coverts always yellowish.

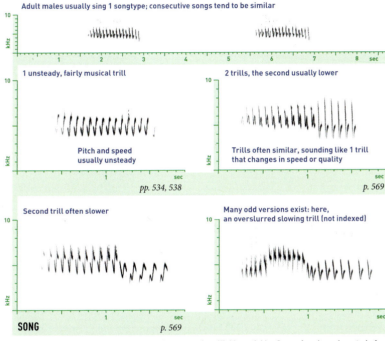

Adult males usually sing 1 songtype; consecutive songs tend to be similar

1 unsteady, fairly musical trill

Pitch and speed usually unsteady

pp. 534, 538

2 trills, the second usually lower

Trills often similar, sounding like 1 trill that changes in speed or quality

p. 569

Second trill often slower

Many odd versions exist: here, an overslurred slowing trill (not indexed)

SONG   *p. 569*

Mostly Mar.–July, by males; females not known to sing. Highly variable. Song of spring migrants is frequently plastic; becomes stereotyped shortly after arrival on breeding grounds. Most males apparently have a single songtype; 1 male documented singing 2 songtypes. Complex Songs have not been documented. Typically more musical and faster than Wilson's, but some songs are very similar. Trills reportedly average slightly faster in West (10–25 notes/second) than in East (7–18 notes/second).

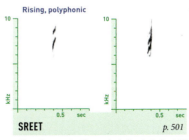

Rising, polyphonic

SREET   *p. 501*

Given frequently all year, including by both night and daytime migrants. Somewhat plastic; resembles other warbler Sreets.

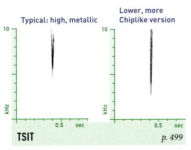

Typical: high, metallic

Lower, more Chiplike version

TSIT   *p. 499*

All year, in alarm. Usually fairly distinctive, but Nashville can give similar calls.

NASHVILLE WARBLER

Oreothlypis ruficapilla

Breeds in spruce bogs and second growth. Eastern population sometimes pumps tail; western population, called "Calaveras Warbler," pumps tail often.

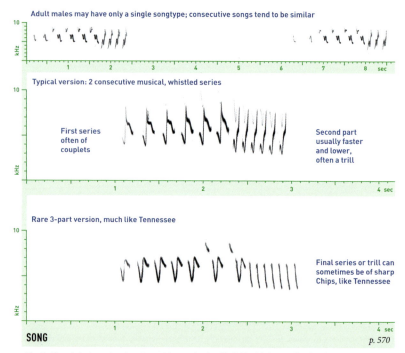

Adult males may have only a single songtype; consecutive songs tend to be similar

Typical version: 2 consecutive musical, whistled series

First series often of couplets

Second part usually faster and lower, often a trill

Rare 3-part version, much like Tennessee

Final series or trill can sometimes be of sharp Chips, like Tennessee

SONG

p. 570

Mostly May–July, by males; females not known to sing. Variable; birds occasionally sing only first series, sometimes for up to a few minutes at a time. Some versions difficult to separate from certain songs of Wilson's and Tennessee Warblers. Multiple songtypes and Complex Songs have not been documented.

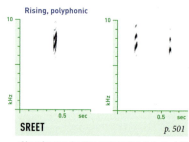

Rising, polyphonic

SREET

p. 501

Given frequently all year, including by both night and daytime migrants. Somewhat plastic; resembles other warbler Sreets.

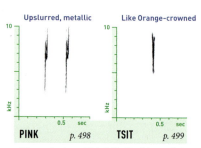

Upslurred, metallic

Like Orange-crowned

PINK *p. 498*   **TSIT** *p. 499*

All year, in alarm. Pink distinctive, intermediate between Tsit and Sreet. Not shown is a sharp, downslurred Chip (p. 497).

Common Yellowthroat

Geothlypis trichas

Common and widespread in wetlands and dense undergrowth near water. Plumage varies geographically, especially in amount of yellow on belly.

Males have 1 typical Song, given repeatedly, and 1 Complex Song, given rarely

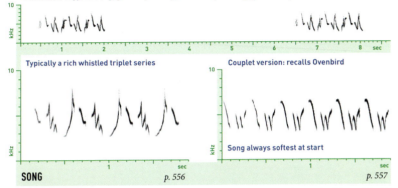

Typically a rich whistled triplet series

Couplet version: recalls Ovenbird

Song always softest at start

SONG   *p. 556*   *p. 557*

Mostly Apr.–July. Distinctive, but highly variable; western birds more likely to sing quadruplet or quintuplet series. About 3 percent of males sing a 2-part Song of consecutive complex series. Details of each male's Song are unique, but broadly similar songtypes may occur across entire regions.

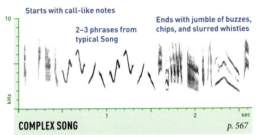

Starts with call-like notes

2–3 phrases from typical Song

Ends with jumble of buzzes, chips, and slurred whistles

COMPLEX SONG   *p. 567*

Given infrequently by males on breeding grounds, usually in flight display. May serve a territorial function, but possibly also given in response to potential predators.

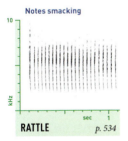

Notes smacking

RATTLE   *p. 534*

Frequent from aggressive breeding males. Highly plastic in length.

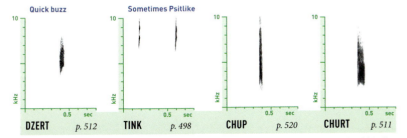

Quick buzz

Sometimes Psitlike

DZERT   *p. 512*   **TINK**   *p. 498*   **CHUP**   *p. 520*   **CHURT**   *p. 511*

Gives a variety of calls. Dzert given all year, including by night migrants; somewhat plastic, but fairly distinctive. Most common call is Chup, given all year in alarm. Tink given in high alarm; Churt is longer, burrier version of Chup. Both Tink and Churt highly plastic, grading into Chup.

Kentucky Warbler

Geothlypis formosa

Breeds in extensive mature forest with dense undergrowth; usually skulks on or near ground, but will sing from high perches, often well concealed.

Males have a single songtype, repeated at intervals

Typical version: musical couplet series

Second note of each phrase often higher

Rare version: 2 consecutive series (p. 570)

Notes often semicomplex or slightly burry

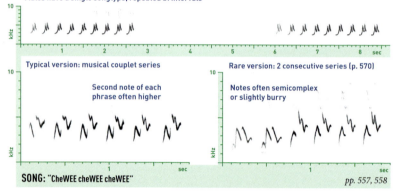

SONG: "cheWEE cheWEE cheWEE"

pp. 557, 558

Mostly Apr.–Aug., reportedly only by males. Variable but stereotyped; some series only vaguely 2-noted. Usually higher and clearer than 1-series versions of Mourning Warbler Song; longer and less varied than couplet versions of Carolina Wren Song; more musical than Ovenbird Song.

Soft warble of Chiplike notes and short buzzes

Often introduced by accelerating series of Chiplike notes

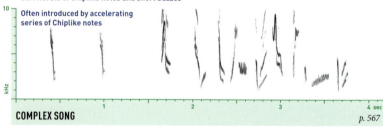

COMPLEX SONG

p. 567

Mostly Apr.–Aug., apparently by males only. Infrequent; reportedly sometimes given in flight display, but also softly by birds on the ground, sometimes for extended periods. Not well known; individuals likely have a single version.

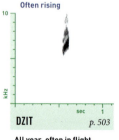

Often rising

DZIT *p. 503*

All year, often in flight, including by night migrants.

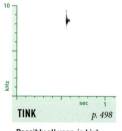

TINK *p. 498*

Possibly all year, in high alarm. Grades into Chip; often interspersed with it.

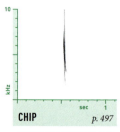

CHIP *p. 497*

All year; most common call, given in alarm.

MOURNING WARBLER

Geothlypis philadelphia

Breeds in dense brush, second growth, and regenerating clearcuts. Skulking. Females and immatures have narrow, broken white eye-rings.

Males apparently have a single songtype, repeated at intervals

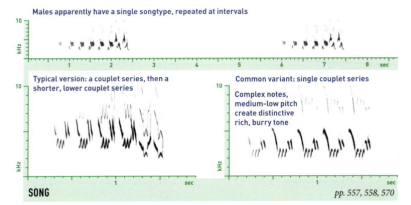

Typical version: a couplet series, then a shorter, lower couplet series

Common variant: single couplet series

Complex notes, medium-low pitch create distinctive rich, burry tone

SONG   *pp. 557, 558, 570*

Mostly May–July, by males; females not known to sing. Variable but stereotyped. Second series sometimes higher than first. Single-series versions are more common in western half of range; rare versions have 3 series, or 2 series followed by a single note. Tone quality generally distinctive.

Complex series and warbles, ending in typical Song

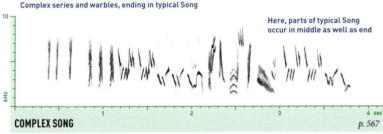

Here, parts of typical Song occur in middle as well as end

COMPLEX SONG   *p. 567*

Mostly June–July, by males during infrequent flight displays. Not well known; only available recording is shown. Males reportedly ascend with series of Chwits, then sing Complex Song during rapid descent; sometimes end with more Chwits. Individuals likely have a single version.

Polyphonic, upslurred

Clear to finely buzzy

SREET   *p. 501*

All year, including by night migrants. Lower than most other warbler Sreets.

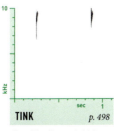

TINK   *p. 498*

Possibly all year, in high alarm.

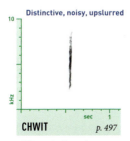

Distinctive, noisy, upslurred

CHWIT   *p. 497*

All year, in alarm. In agitation, sometimes run into rapid Chatter.

Connecticut Warbler

Oporornis agilis

Local and uncommon breeder in spruce bogs. On migration, usually secretive; stays on or near ground, walking like a thrush. Note thin, complete eye-ring.

Individuals have a single songtype, repeated with occasional truncations

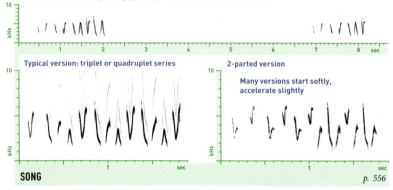

Typical version: triplet or quadruplet series

2-parted version

Many versions start softly, accelerate slightly

SONG

p. 556

Mostly May–Aug. Not known whether females sing. Song is usually loud, low, emphatic, like Northern Waterthrush; different note lengths create distinctive herky-jerky rhythm. Acceleration highly distinctive when present. Complex Song likely exists, but has not been reported.

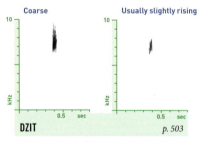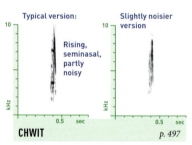

Coarse

Usually slightly rising

DZIT

p. 503

All year, often in flight, including by night migrants. Very similar to Dzits of Yellow and Blackpoll Warblers.

Typical version:

Rising, seminasal, partly noisy

Slightly noisier version

CHWIT

p. 497

All year, in mild alarm. Typical version highly distinctive, but plastic; in agitation, grades into higher, clearer Pik, or into Dzit.

Bachman's Warbler

Vermivora bachmanii

Extinct; last documented in 1962 in South Carolina. Locally common as late as 1920 in dry thickets adjoining swamps. Recordings of two birds exist, both males singing probable First-Category Songs: monotone series of buzzes, sometimes ending in a complex whistled note. Calls described as buzzy Zeep or Zip.

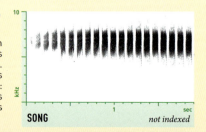

SONG

not indexed

American Redstart

Setophaga ruticilla

Common in deciduous woods. Highly active from ground level to treetops, often flycatching. Fans tail frequently, possibly to startle insect prey.

Males have 1 First-Category (repeated) songtype; 2–8 Second-Category (alternated) songtypes

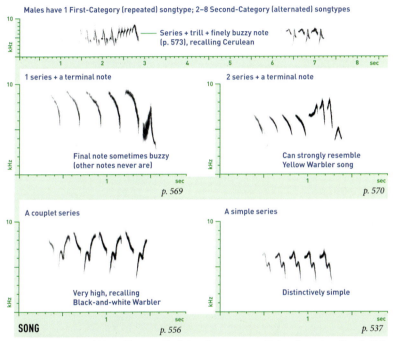

Series + trill + finely buzzy note (p. 573), recalling Cerulean

1 series + a terminal note

Final note sometimes buzzy (other notes never are)

p. 569

2 series + a terminal note

Can strongly resemble Yellow Warbler song

p. 570

A couplet series

Very high, recalling Black-and-white Warbler

SONG p. 556

A simple series

Distinctively simple

p. 537

Perhaps the most variable warbler song. Mostly Apr.–Aug.; stereotyped in breeding males, often plastic in migrants. Femalelike birds that sing are likely young males. Two song categories: First-Category Songs are always repeated; Second-Category Songs are continually switched. First-Category Songs have accented endings in about 80 percent of males; these recall Yellow Warbler, but tend to be simpler and often higher and slower. Second-Category Songs are diverse and easily confused with many other warbler species, but the constant switching of songtypes is a good identification clue.

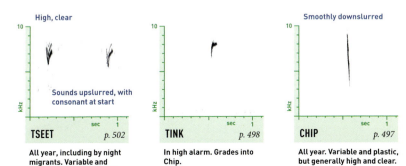

High, clear

Sounds upslurred, with consonant at start

TSEET p. 502

All year, including by night migrants. Variable and plastic, but fairly distinctive.

TINK p. 498

In high alarm. Grades into Chip.

Smoothly downslurred

CHIP p. 497

All year. Variable and plastic, but generally high and clear.

Cerulean Warbler

Setophaga cerulea

Local in mature deciduous forests in riverbottoms and on slopes and ridges. Forages high in the canopy. Population has declined in many areas.

Males with 2 Second-Category Songs usually alternate them

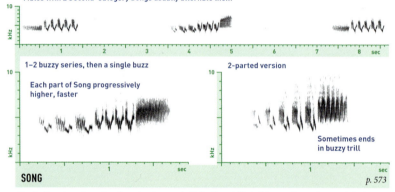

1–2 buzzy series, then a single buzz

Each part of Song progressively higher, faster

2-parted version

Sometimes ends in buzzy trill

SONG *p. 573*

Mostly May–Aug., by males. Females sing rarely. Males have 1 First-Category Song, given early in the season, and 1–2 Second-Category Songs, given mostly later in the season; identical songtypes used by different males in different categories.

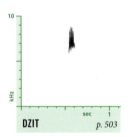

DZIT *p. 503*

All year, including by night migrants.

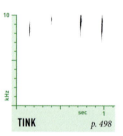

TINK *p. 498*

Possibly all year, in high alarm. Very plastic; grades into Chip.

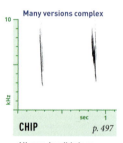

Many versions complex

CHIP *p. 497*

All year, in mild alarm. Plastic. Similar notes given in rapid series in aggression.

BLUE-WINGED × GOLDEN-WINGED WARBLER HYBRIDS

These two species hybridize frequently, creating a wide variety of intermediate plumage types, including the forms called "Brewster's Warbler" and "Lawrence's Warbler."

The Short Songs of hybrids typically match those of one of the parent species. A few sing abnormal or intermediate songs. Very rarely, a pure bird or a hybrid may sing the songs of both species, either alternating them or repeating one before switching.

Because of the regularity of hybrids and "wrong-song" singers in these two species, it may not be safe to identify a Blue-winged or Golden-winged Warbler solely by ear.

"Brewster's" Warbler

"Lawrence's" Warbler

BLACK-THROATED BLUE WARBLER

Setophaga caerulescens

Breeds in dense forest understory. Male retains distinctive plumage all year; female shows small white patch at base of primaries.

In both First- and Second-Category singing, consecutive songs tend to be similar

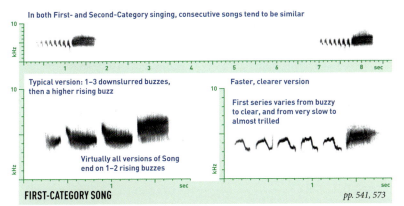

Typical version: 1–3 downslurred buzzes, then a higher rising buzz

Faster, clearer version

First series varies from buzzy to clear, and from very slow to almost trilled

Virtually all versions of Song end on 1–2 rising buzzes

FIRST-CATEGORY SONG

pp. 541, 573

Mostly May–Aug., by males. Females sing occasionally. Most males have a single First-Category songtype; a few have 2, but sing just 1 most of the time. Most versions distinctive; monotone first series eliminates Prairie Warbler, and rising buzzy final note eliminates Cerulean.

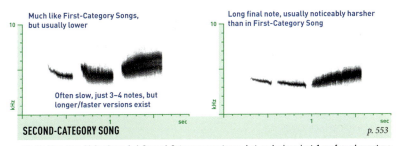

Much like First-Category Songs, but usually lower

Long final note, usually noticeably harsher than in First-Category Song

Often slow, just 3–4 notes, but longer/faster versions exist

SECOND-CATEGORY SONG

p. 553

Mostly May–Aug. Males have 1–6 Second-Category songtypes, but each sings just 1 preferred songtype most of the time. Unlike many other warblers, does not generally give strings of calls between Second-Category Songs. Songs from both categories sometimes delivered very softly, mostly in aggression.

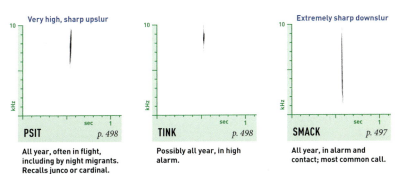

Very high, sharp upslur

Extremely sharp downslur

PSIT *p. 498*

TINK *p. 498*

SMACK *p. 497*

All year, often in flight, including by night migrants. Recalls junco or cardinal.

Possibly all year, in high alarm.

All year, in alarm and contact; most common call.

NORTHERN PARULA

Setophaga americana

Tiny and short-tailed. Common in both southern swamps and boreal forests; nests primarily in clumps of hanging Spanish moss or beard lichens.

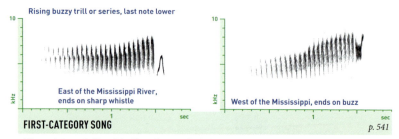

Rising buzzy trill or series, last note lower

East of the Mississippi River, ends on sharp whistle

West of the Mississippi, ends on buzz

FIRST-CATEGORY SONG p. 541

All year, but mostly Mar.–July. Given almost exclusively by males, especially unmated males seeking to attract a mate. Rising buzzy trill distinctive; only Tropical Parula Song is similar. Some versions broken into rising series of short buzzy notes, like Prairie Warbler Song, but faster.

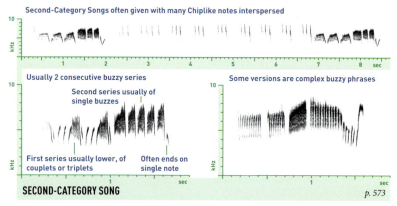

Second-Category Songs often given with many Chiplike notes interspersed

Usually 2 consecutive buzzy series

Second series usually of single buzzes

First series usually lower, of couplets or triplets

Often ends on single note

Some versions are complex buzzy phrases

SECOND-CATEGORY SONG p. 573

Mostly Mar.–July; given by males mostly at dawn or dusk. Sometimes heard later in the day, especially during territorial defense. Much more geographically variable than First-Category Song, and typically repeated at shorter intervals. Individuals have 1 songtype in each category.

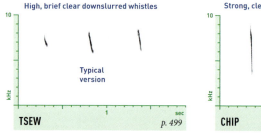

High, brief clear downslurred whistles

Typical version

TSEW p. 499

Given rather infrequently during day by perched or flying birds; also on the wing during nocturnal migration.

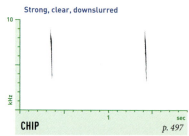

Strong, clear, downslurred

CHIP p. 497

All year, likely in alarm. Somewhat plastic, sometimes grading into Tsew; higher versions given between Second-Category Songs.

TROPICAL PARULA

Setophaga pitiayumi

Local in live oaks and deciduous woods with Spanish moss. May hybridize with Northern Parula in Texas. Breeds south to Argentina.

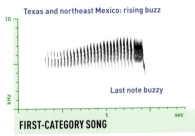

Texas and northeast Mexico: rising buzz

Last note buzzy

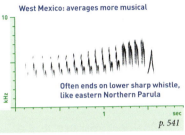

West Mexico: averages more musical

Often ends on lower sharp whistle, like eastern Northern Parula

FIRST-CATEGORY SONG *p. 541*

All year, especially Mar.–Sept., likely by males primarily to attract a mate. First-Category Songs of Texas breeders not reliably separable from those of Northern Parula. Populations farther south and west have more diverse and variable songs, mostly lacking buzzy quality.

Individual males have 1 song in each category; the 2 rarely combined (right)

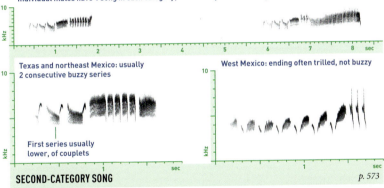

Texas and northeast Mexico: usually 2 consecutive buzzy series

First series usually lower, of couplets

West Mexico: ending often trilled, not buzzy

SECOND-CATEGORY SONG *p. 573*

Mostly Mar.–Sept. Highly variable between and within populations. Likely used by males primarily at dawn, at dusk, and in territorial disputes. Perhaps not reliably separated in the field from Second-Category Song of Northern Parula.

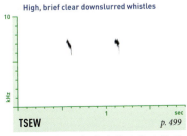

High, brief clear downslurred whistles

TSEW *p. 499*

Few recordings exist. Probably similar to Northern Parula Tsew, and given in similar situations.

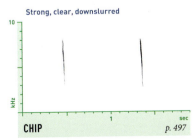

Strong, clear, downslurred

CHIP *p. 497*

Identical to Northern Parula Chip, and given in similar situations.

BLACK-THROATED GREEN WARBLER

Setophaga virens

One of our most abundant warblers, breeding primarily in coniferous and mixed forests; a few breed in Atlantic coastal cypress swamps.

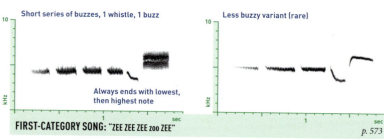

Short series of buzzes, 1 whistle, 1 buzz

Always ends with lowest, then highest note

Less buzzy variant (rare)

FIRST-CATEGORY SONG: "ZEE ZEE ZEE zoo ZEE" *p. 573*

One of our most uniform warbler songs. Mostly Mar.–Aug.; stereotyped in breeding males, often plastic in migrants. Given mostly by unpaired males. Greatest variation is in number and speed of notes in first series, ranging from 3 slow notes to 8 or more rapid ones.

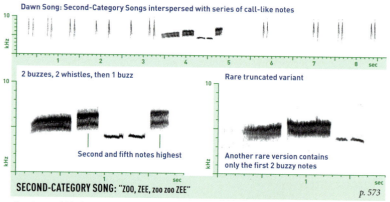

Dawn Song: Second-Category Songs interspersed with series of call-like notes

2 buzzes, 2 whistles, then 1 buzz

Second and fifth notes highest

Rare truncated variant

Another rare version contains only the first 2 buzzy notes

SECOND-CATEGORY SONG: "ZOO, ZEE, zoo zoo ZEE" *p. 573*

Even less variable than this species' First-Category Songs. Given mostly at dawn and dusk, and after males are paired; used in territorial conflicts. Often repeated at a faster rate than First-Category Songs.

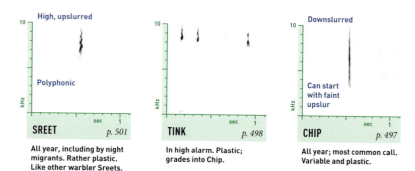

High, upslurred

Polyphonic

SREET *p. 501*

All year, including by night migrants. Rather plastic. Like other warbler Sreets.

TINK *p. 498*

In high alarm. Plastic; grades into Chip.

Downslurred

Can start with faint upslur

CHIP *p. 497*

All year; most common call. Variable and plastic.

GOLDEN-CHEEKED WARBLER

Setophaga chrysoparia

Endangered; nests only in juniper-oak woodland on the Edwards Plateau of Texas. Closely related to Black-throated Green Warbler.

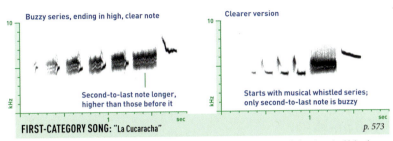

Buzzy series, ending in high, clear note

Second-to-last note longer, higher than those before it

Clearer version

Starts with musical whistled series; only second-to-last note is buzzy

FIRST-CATEGORY SONG: "La Cucaracha"   *p. 573*

Fairly uniform. Mostly Mar.–May; stereotyped in breeding males, often plastic in migrants. Males have a single First-Category Song, which they give mostly before pairing. Greatest variation is in introductory series. Many versions recall first line of the Mexican folk song "La Cucaracha."

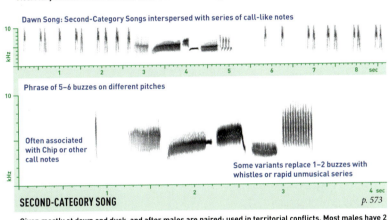

Dawn Song: Second-Category Songs interspersed with series of call-like notes

Phrase of 5–6 buzzes on different pitches

Often associated with Chip or other call notes

Some variants replace 1–2 buzzes with whistles or rapid unmusical series

SECOND-CATEGORY SONG   *p. 573*

Given mostly at dawn and dusk, and after males are paired; used in territorial conflicts. Most males have 2 Second-Category Songs, one of which is used more often than the other. Often repeated at a faster rate than First-Category Songs.

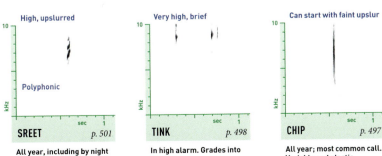

High, upslurred

Polyphonic

SREET   *p. 501*

All year, including by night migrants. Rather plastic. Like other warbler Sreets.

Very high, brief

TINK   *p. 498*

In high alarm. Grades into Chip.

Can start with faint upslur

CHIP   *p. 497*

All year; most common call. Variable and plastic.

PRAIRIE WARBLER

Setophaga discolor

Breeds in overgrown fields, dunes, and shrubby, open pine savannas; locally also in mangroves and swamps. Pumps tail up and down frequently.

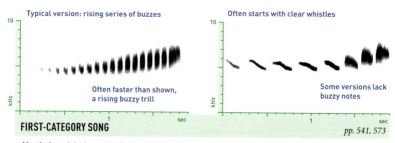

Typical version: rising series of buzzes

Often faster than shown, a rising buzzy trill

Often starts with clear whistles

Some versions lack buzzy notes

FIRST-CATEGORY SONG

pp. 541, 573

Mostly Apr.–July, by males; females sing rarely. Individuals have 1–3 versions, repeating each many times before switching. Slowest versions have only 5–6 buzzes; fastest versions are one long rising buzzy trill, still slower than most Northern Parula Songs. Clear whistled versions are likely Second-Category.

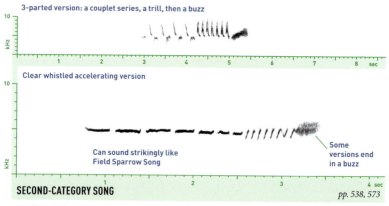

3-parted version: a couplet series, a trill, then a buzz

Clear whistled accelerating version

Can sound strikingly like Field Sparrow Song

Some versions end in a buzz

SECOND-CATEGORY SONG

pp. 538, 573

Highly variable; many resemble slow or clear First-Category Songs. Less common versions, like those shown, usually start with clear simple or couplet series, then sometimes a trill, then sometimes a buzz. Often given with call-like notes. Repertoire size unknown; songs often repeated before switching.

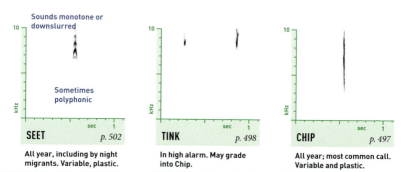

Sounds monotone or downslurred

Sometimes polyphonic

SEET *p. 502*

All year, including by night migrants. Variable, plastic.

TINK *p. 498*

In high alarm. May grade into Chip.

CHIP *p. 497*

All year; most common call. Variable and plastic.

BLACKBURNIAN WARBLER

Setophaga fusca

Breeds mostly in coniferous and mixed forests, frequenting the tops of spruces or hemlocks. At southern end of range, also found in deciduous forests.

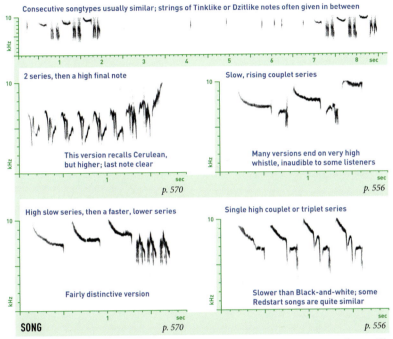

Consecutive songtypes usually similar; strings of Tinklike or Dzitlike notes often given in between

2 series, then a high final note

This version recalls Cerulean, but higher; last note clear

p. 570

Slow, rising couplet series

Many versions end on very high whistle, inaudible to some listeners

p. 556

High slow series, then a faster, lower series

Fairly distinctive version

SONG   p. 570

Single high couplet or triplet series

Slower than Black-and-white; some Redstart songs are quite similar

p. 556

Mostly May–July, by males. Females not known to sing. Males reportedly use 2 song categories, possibly with just 1 First-Category and 1 Second-Category Song per individual, but it is not clear how to distinguish the categories by ear; more study needed. Nearly any version can be given with strings of call-like notes, both at dawn and late in the day. Occasionally, males alternate 2 different songtypes.

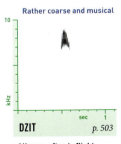

Rather coarse and musical

DZIT   p. 503

All year, often in flight, including by night migrants. Often shorter than shown.

TINK   p. 498

Possibly all year, in high alarm. Plastic, sometimes grading into other calls.

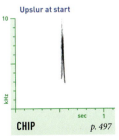

Upslur at start

CHIP   p. 497

All year, in alarm; most common call.

BLACKPOLL WARBLER

Setophaga striata

Abundant in boreal forests, and one of the most common migrant warblers in the East. Similar to Bay-breasted in vocalizations and in most plumages.

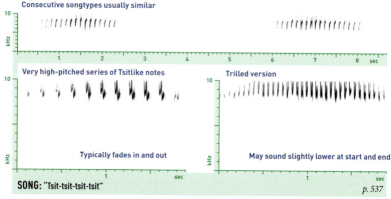

Consecutive songtypes usually similar

Very high-pitched series of Tsitlike notes

Trilled version

Typically fades in and out

May sound slightly lower at start and end

SONG: "Tsit-tsit-tsit-tsit"

p. 537

Mostly May–July. Repertoire size unknown. Typical Song is a series, but trilled versions predominate in some breeding areas, such as Nova Scotia and northern New Hampshire. Some birds have been reported switching between slow and fast songtypes, so they may have different functions. More study needed.

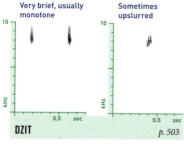

Very brief, usually monotone

Sometimes upslurred

DZIT

p. 503

All year, including by night migrants. Slightly variable and plastic. More frequent in migration than Chip.

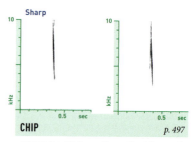

Sharp

CHIP

p. 497

All year, in mild alarm. Given rather infrequently.

DAWN SONG CALLS OF WARBLERS

At dawn, especially in the second half of the breeding season, several warbler species give Second-Category Songs with strings of call-like notes in between. The calls between the songs are variable and slightly plastic; they may resemble Chips, Tinks, Seets, or the begging calls of juveniles, but often they are subtly distinct from any of these. Two different versions from Yellow Warbler Dawn Song are shown. Many species of sparrows also give call-like notes between songs at dawn.

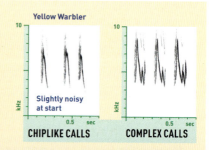

Yellow Warbler

Slightly noisy at start

CHIPLIKE CALLS

COMPLEX CALLS

Bay-breasted Warbler

Setophaga castanea

Breeds in spruce-fir forests; population closely tied to spruce budworm outbreaks, and has declined in recent decades. Sounds have been little studied.

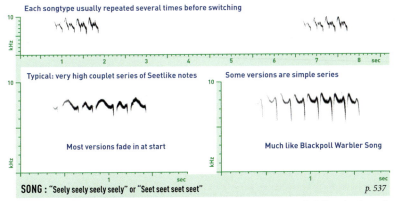

Each songtype usually repeated several times before switching

Typical: very high couplet series of Seetlike notes

Most versions fade in at start

Some versions are simple series

Much like Blackpoll Warbler Song

SONG : "Seely seely seely seely" or "Seet seet seet seet" *p. 537*

Mostly May–July, by males, though females reported to sing quietly from nest. Variable and rather soft. Couplet versions recall Black-and-white Warbler Song, but usually faster, higher, and shorter (1–1.5 seconds total). Simple versions resemble Blackpoll Warbler Song, but few Blackpoll songtypes are as slow. Repertoire size unknown, but individual males have more than 1 songtype and may deploy the songtypes in different ways. More study needed.

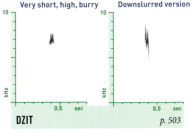

Very short, high, burry

Downslurred version

DZIT *p. 503*

Likely all year, including by night migrants. Rather variable and plastic.

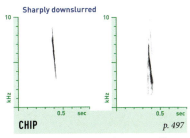

Sharply downslurred

CHIP *p. 497*

All year. Variable and plastic; not generally separable from Blackpoll Chip.

"Mangrove" Yellow Warbler

Recently, a few "Mangrove" Yellow Warblers have been found in coastal mangrove swamps of south Texas. This tropical group is considered a separate species by many authorities. Males have striking chestnut heads, and their songs are much lower, slower, and more complex than those of northern Yellow Warblers, often including short phrases. Dzit calls may average lower and coarser.

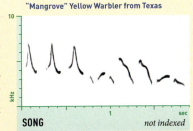

"Mangrove" Yellow Warbler from Texas

SONG *not indexed*

Cape May Warbler

Setophaga tigrina

Breeds in relatively mature forest with spruces or firs. Generally uncommon, but can become locally abundant during spruce budworm outbreaks.

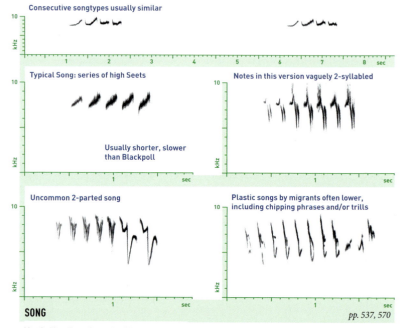

Consecutive songtypes usually similar

Typical Song: series of high Seets

Usually shorter, slower than Blackpoll

Notes in this version vaguely 2-syllabled

Uncommon 2-parted song

Plastic songs by migrants often lower, including chipping phrases and/or trills

SONG

pp. 537, 570

Mostly May–June, by males. Females not known to sing. Repertoire size poorly known, but 1 male documented to sing 3 different songtypes. Not known whether birds use 2 categories of song. Some versions are couplet series, very similar to Bay-breasted Warbler Song. Spring migrants sing a variety of songtypes, many of which are easily confused with other species.

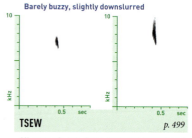

Barely buzzy, slightly downslurred

TSEW

p. 499

All year, often in flight, including by night migrants.

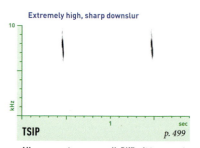

Extremely high, sharp downslur

TSIP

p. 499

All year; most common call. Difficult to separate from Tink calls of other species. Apparently does not give a lower-pitched Chip.

Yellow-throated Warbler

Setophaga dominica

Breeds in pines, wooded swamps, and mixed forests. Often forages on trunks and along large limbs, much like Black-and-white Warbler.

More complex songs given with Chips may represent Second-Category singing; more study needed

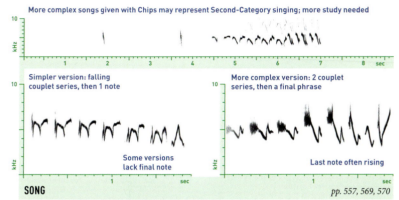

Simpler version: falling couplet series, then 1 note

Some versions lack final note

More complex version: 2 couplet series, then a final phrase

Last note often rising

SONG   *pp. 557, 569, 570*

Mostly Mar.–June, by males. Females not known to sing. Males reported to have a single songtype, but 2 song categories may exist; more study needed. Variable; first series sometimes simple. Overall pitch of song always falls, but last note often rising. End less complex than Louisiana Waterthrush Song.

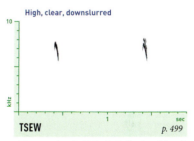

High, clear, downslurred

TSEW   *p. 499*

All year, often in flight, including by night migrants. Fairly distinctive; like Pine Warbler Tsew, but more audibly downslurred.

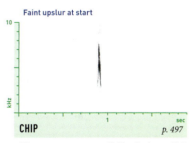

Faint upslur at start

CHIP   *p. 497*

All year; most common call. May also have a Tink call given in high alarm, but no recordings available.

"Golden" Yellow Warbler

The resident population of Yellow Warblers in the mangrove swamps of the Florida Keys belongs to the Caribbean group known as "Golden" Yellow Warbler, considered a separate species by some authorities, or part of "Mangrove" Warbler. Songs of this form are consistently lower, slower, and simpler than those of northern Yellow Warblers. Dzit calls may average lower and coarser.

"Golden" Yellow Warbler from Florida

SONG   *p. 570*

Yellow Warbler

Setophaga petechia

Our most common and widespread warbler, found in a variety of habitats from woodland edges, parks, and second growth to suburban backyards.

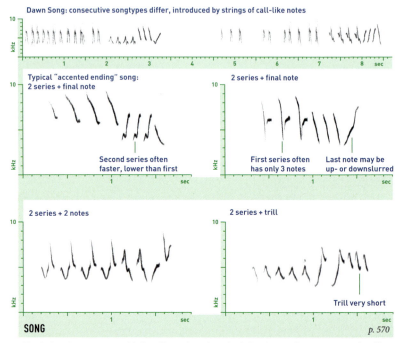

Dawn Song: consecutive songtypes differ, introduced by strings of call-like notes

Typical "accented ending" song: 2 series + final note

Second series often faster, lower than first

2 series + final note

First series often has only 3 notes

Last note may be up- or downslurred

2 series + 2 notes

2 series + trill

Trill very short

SONG

p. 570

Mostly Apr.–Aug.; stereotyped in breeding males, often plastic in migrants. Two song categories: First-Category Songs are always repeated; Second-Category Songs are usually alternated. Males have 1, rarely 2 First-Category Songs, 5–17 different Second-Category Songs. First-Category Songs usually have 2 series plus 1 terminal note, but identical songtypes sometimes used by different males in different categories.

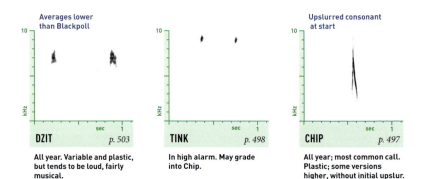

Averages lower than Blackpoll

DZIT *p. 503*

All year. Variable and plastic, but tends to be loud, fairly musical.

TINK *p. 498*

In high alarm. May grade into Chip.

Upslurred consonant at start

CHIP *p. 497*

All year; most common call. Plastic; some versions higher, without initial upslur.

Chestnut-sided Warbler

Setophaga pensylvanica

Common in deciduous brush and woodland edges. Fall birds are lime green above, pearl gray below, with yellowish wingbars.

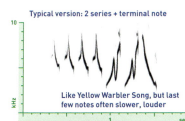

Typical version: 2 series + terminal note

Like Yellow Warbler Song, but last few notes often slower, louder

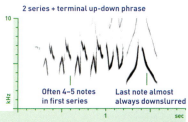

2 series + terminal up-down phrase

Often 4–5 notes in first series

Last note almost always downslurred

FIRST-CATEGORY SONG

p. 570

Mostly Apr.–Aug. Males have up to 4 First-Category songtypes, but each sings just 1 preferred songtype most of the time; consecutive songs are usually the same. First-Category Songs are fairly uniform across range; lower than Yellow Warbler Song on average, with slower, more emphatic ending.

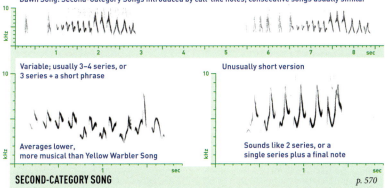

Dawn Song: Second-Category Songs introduced by call-like notes; consecutive songs usually similar

Variable; usually 3–4 series, or 3 series + a short phrase

Averages lower, more musical than Yellow Warbler Song

Unusually short version

Sounds like 2 series, or a single series plus a final note

SECOND-CATEGORY SONG

p. 570

Mostly Apr.–Aug. Males have 2–10 (usually about 5) Second-Category songtypes, but each sings just 1 preferred songtype most of the time; consecutive songs are usually the same. Second-Category Songs are fairly variable across the range; many can be difficult to distinguish from Yellow Warbler songs.

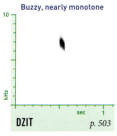

Buzzy, nearly monotone

DZIT

p. 503

All year, often in flight, including by night migrants. Slightly plastic, variable.

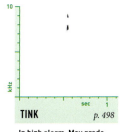

TINK

p. 498

In high alarm. May grade into Chip.

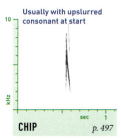

Usually with upslurred consonant at start

CHIP

p. 497

All year; most common call. Plastic; some versions higher, without initial upslur.

Yellow-rumped Warbler (Myrtle)

Setophaga coronata (*coronata* group)

Breeds mostly in mature coniferous forest. Common to abundant in migration and winter throughout much of the East.

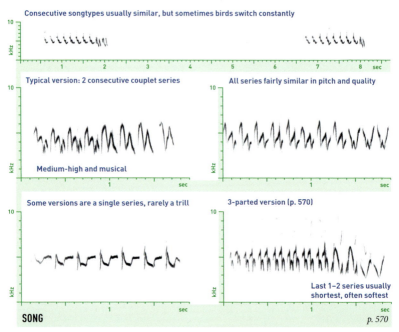

Consecutive songtypes usually similar, but sometimes birds switch constantly

Typical version: 2 consecutive couplet series

Medium-high and musical

All series fairly similar in pitch and quality

Some versions are a single series, rarely a trill

3-parted version (p. 570)

Last 1–2 series usually shortest, often softest

SONG *p. 570*

Mostly Apr.–Aug., by males. Females not known to sing. Not known whether males have 2 song categories. Repertoire size unknown, but males apparently have multiple songtypes. Much overlap with Audubon's song, but usually faster, sometimes trilled at end; series more often simple.

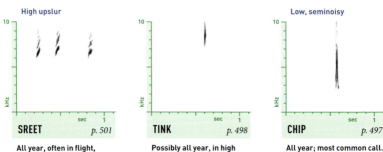

High upslur

Low, seminoisy

SREET *p. 501*

All year, often in flight, including by night migrants. Like Audubon's Sreet.

TINK *p. 498*

Possibly all year, in high alarm. Plastic; grades into Sreet.

CHIP *p. 497*

All year; most common call. Fairly distinctive.

Yellow-rumped Warbler (Audubon's)

Setophaga coronata (*auduboni* group)

Breeds in western mountains; rare in most of the East. Formerly considered a separate species; hybridizes with Myrtle in a narrow contact zone in Canada.

Consecutive songtypes usually similar, but sometimes birds switch constantly

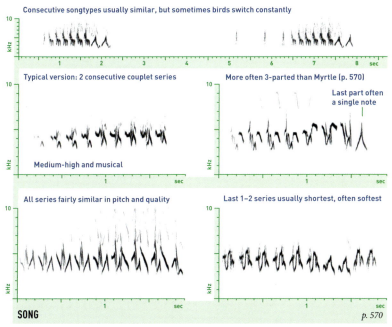

Typical version: 2 consecutive couplet series

Medium-high and musical

More often 3-parted than Myrtle (p. 570)

Last part often a single note

All series fairly similar in pitch and quality

Last 1–2 series usually shortest, often softest

SONG
p. 570

Mostly Apr.–Aug., by males. Females not known to sing. Not known whether males have 2 song categories. Repertoire size unknown, but males apparently have multiple songtypes. Averages slower than Myrtle's Song, rarely including trills, most series usually of couplets; almost never a single series.

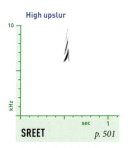

High upslur

SREET *p. 501*

All year, often in flight, including by night migrants. Like Myrtle Sreet.

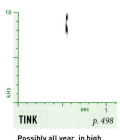

TINK *p. 498*

Possibly all year, in high alarm. Plastic; grades into Sreet.

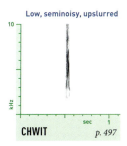

Low, seminoisy, upslurred

CHWIT *p. 497*

All year; most common call. Subtly but distinctively different from Myrtle's Chip.

HOODED WARBLER

Setophaga citrina

Breeds in dense understory of forests and swamps, especially around gaps and edges. Frequently flicks tail open, showing extensive white tail spots.

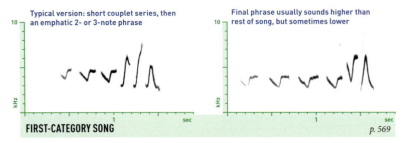

Typical version: short couplet series, then an emphatic 2- or 3-note phrase

Final phrase usually sounds higher than rest of song, but sometimes lower

FIRST-CATEGORY SONG

p. 569

Mostly Apr.–Aug., by males. Males have a single First-Category songtype, repeated mostly prior to pairing. Magnolia Warbler and American Redstart have similar pattern, but note overall lower pitch of Hooded, and tendency for final phrase to be slightly faster than first series.

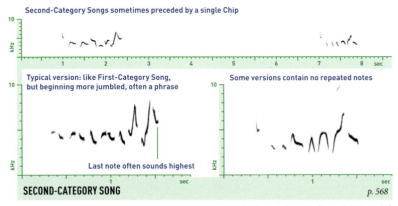

Second-Category Songs sometimes preceded by a single Chip

Typical version: like First-Category Song, but beginning more jumbled, often a phrase

Some versions contain no repeated notes

Last note often sounds highest

SECOND-CATEGORY SONG

p. 568

Mostly Apr.–Aug., by males; 1 female has been documented to sing Second-Category Song. Males have 2–4 (rarely up to 8) versions; consecutive songs often different, but can be same for long periods. Songs like those shown here are used as First-Category Songs by up to 5 percent of males.

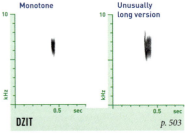

Monotone

Unusually long version

DZIT

p. 503

All year, often in flight, including by night migrants. Averages shorter than Chestnut-sided's Dzit; nearly identical to Kirtland's.

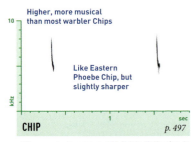

Higher, more musical than most warbler Chips

Like Eastern Phoebe Chip, but slightly sharper

CHIP

p. 497

All year; most common call. Fairly distinctive. In high aggression, reportedly runs into rapid Twitter followed by low, harsh buzz.

Magnolia Warbler

Setophaga magnolia

Breeds mostly in dense stands of young conifers. In all plumages, note yellow rump and underparts, white square spots near base of tail.

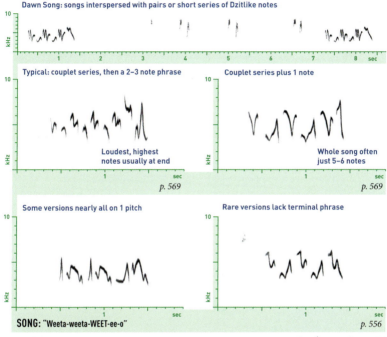

Dawn Song: songs interspersed with pairs or short series of Dzitlike notes

Typical: couplet series, then a 2–3 note phrase

Loudest, highest notes usually at end

p. 569

Couplet series plus 1 note

Whole song often just 5–6 notes

p. 569

Some versions nearly all on 1 pitch

Rare versions lack terminal phrase

SONG: "Weeta-weeta-WEET-ee-o"

p. 556

Mostly May–Aug., by males. Females not known to sing. Males reportedly have 1 First-Category Song and 1 Second-Category Song, but identical songtypes likely used by different males in different categories; more study needed. Song distinctively short, usually 1–1.5 seconds; usually slightly lower than American Redstart's and higher than Hooded Warbler's, but can be difficult to separate from both.

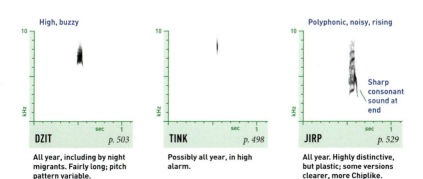

High, buzzy

DZIT *p. 503*

All year, including by night migrants. Fairly long; pitch pattern variable.

TINK *p. 498*

Possibly all year, in high alarm.

Polyphonic, noisy, rising

Sharp consonant sound at end

JIRP *p. 529*

All year. Highly distinctive, but plastic; some versions clearer, more Chiplike.

Palm Warbler

Setophaga palmarum

Breeds in boreal bogs; forages mostly near ground. Pumps tail frequently. Breeders from Quebec east are extensively yellow below; western birds paler.

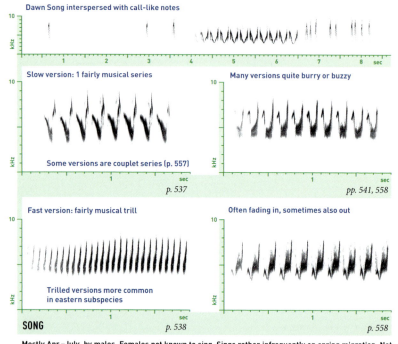

Dawn Song interspersed with call-like notes

Slow version: 1 fairly musical series

Some versions are couplet series (p. 557)

p. 537

Many versions quite burry or buzzy

pp. 541, 558

Fast version: fairly musical trill

Trilled versions more common in eastern subspecies

Often fading in, sometimes also out

SONG

p. 538

p. 558

Mostly Apr.–July, by males. Females not known to sing. Sings rather infrequently on spring migration. Not known whether males sing with 2 song categories. Repertoire size also unknown, but consecutive songs usually similar. Slow versions of song are fairly distinctive, but can be confused with single-series songs of "Myrtle" Yellow-rumped Warbler. Fast versions of song can be quite difficult to distinguish from musical versions of Chipping Sparrow and Dark-eyed Junco Songs.

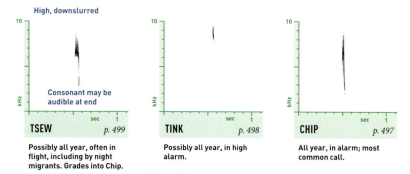

High, downslurred

Consonant may be audible at end

TSEW

p. 499

Possibly all year, often in flight, including by night migrants. Grades into Chip.

TINK

p. 498

Possibly all year, in high alarm.

CHIP

p. 497

All year, in alarm; most common call.

PINE WARBLER

Setophaga pinus

True to its name, found primarily in pines. Unlike other warblers, winters mostly in the southern U.S., and sometimes visits bird feeders.

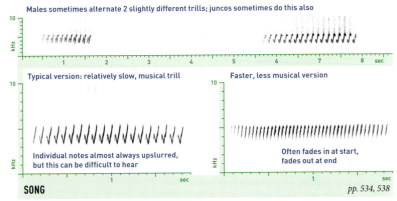

Males sometimes alternate 2 slightly different trills; juncos sometimes do this also

Typical version: relatively slow, musical trill

Faster, less musical version

Individual notes almost always upslurred, but this can be difficult to hear

Often fades in at start, fades out at end

SONG   *pp. 534, 538*

All year, but especially Mar.–June, by males. Females not known to sing. Unknown whether males have 2 song categories; most apparently have more than 1 trill type, but consecutive songs usually similar. Dawn Song often interspersed with Chips. Fast versions like Worm-eating, but usually more musical.

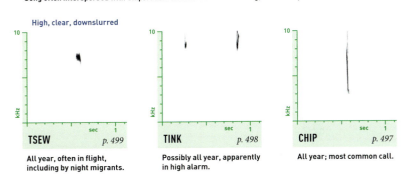

High, clear, downslurred

TSEW   *p. 499*

All year, often in flight, including by night migrants.

TINK   *p. 498*

Possibly all year, apparently in high alarm.

CHIP   *p. 497*

All year; most common call.

AGGRESSIVE TWITTERS AND RATTLES OF WARBLERS

Several species of warblers give plastic Twitters or cowbirdlike Rattles during aggressive interactions or in extreme agitation. Such sounds are heard particularly often from Black-and-white Warbler, Black-throated Blue Warbler, Magnolia Warbler, and Pine Warbler, but they have been recorded or reported in a number of other species, and it is likely that most or all warblers occasionally give similar sounds.

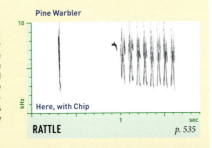

Pine Warbler

Here, with Chip

RATTLE   *p. 535*

KIRTLAND'S WARBLER

Setophaga kirtlandii

Rare and local breeder in young jack pine forests. Population depends on intensive habitat management and cowbird control. Pumps tail frequently.

3–4 rapid short series

Start sometimes accelerating

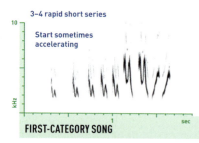

Almost always ends on upslurred notes

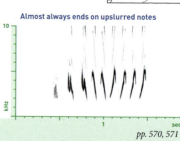

FIRST-CATEGORY SONG

pp. 570, 571

Mostly May–July, by males. Females not known to sing. Somewhat variable but generally stereotyped; males apparently have a single First-Category Song. Loud and emphatic, like Northern Waterthrush Song in pitch and pattern, but shorter, just 1–1.5 seconds in length, and usually lowest at start, not end.

Second-Category Song often interspersed with call-like notes

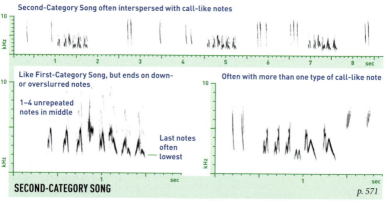

Like First-Category Song, but ends on down- or overslurred notes

1–4 unrepeated notes in middle

Last notes often lowest

Often with more than one type of call-like note

SECOND-CATEGORY SONG

p. 571

Not well studied, but apparently given in the same contexts as other Second-Category Songs of warblers. Repertoire size unknown, but consecutive songs usually similar. Notes generally complex, low in pitch. Often compared to House Wren Song, but far slower, lacking trills; individual notes always distinct.

Short, buzzy

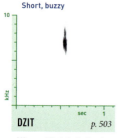

DZIT
p. 503

All year, likely including by night migrants. Rather high and fine.

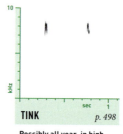

TINK
p. 498

Possibly all year, in high alarm.

Clear, downslurred

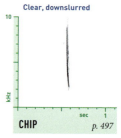

CHIP
p. 497

All year, in mild alarm; most common call.

CANADA WARBLER

Cardellina canadensis

Breeds in dense forest understory. White eye-ring and undertail coverts are distinctive. Tail often cocked or flicked while foraging.

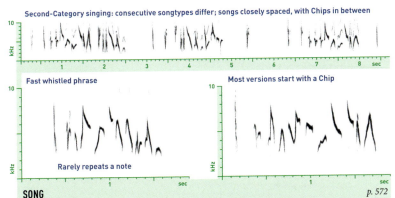

Second-Category singing: consecutive songtypes differ; songs closely spaced, with Chips in between

10 kHz

1 2 3 4 5 6 7 8 sec

Fast whistled phrase

Most versions start with a Chip

10

Rarely repeats a note

1 sec

SONG

10

1 sec

p. 572

All year, especially May–July. One report of Song from female. Males recombine 7–16 stereotyped phrases into many similar songtypes. Two song categories; First-Category singing lacks Chips between songs, and switches songtypes less frequently. Identical songs used by different males in different categories.

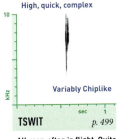
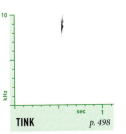

High, quick, complex

10

Variably Chiplike

kHz

sec 1

TSWIT   *p. 499*

All year, often in flight. Quite variable in pitch and length.

10

kHz

sec 1

TINK   *p. 498*

Apparently infrequent, in high alarm, mixed with Chips.

10

kHz

sec 1

CHIP   *p. 497*

All year, in alarm; most common call. Relatively low.

PLASTIC SONGS OF SPRING MIGRANT WARBLERS

In spring migration, warblers less than a year old (and perhaps some older birds) often sing highly plastic songs. This is typical of many passerines, but it can be particularly confusing in spring warblers. Young birds' attempts at trills may sound more like warbles, and their attempts at series may sound more like phrases. Hints of the tone quality and pattern of adult birds are usually present, but some songs at an early stage of development may be unidentifiable.

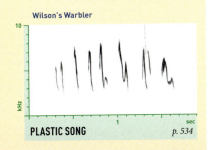

Wilson's Warbler

10

kHz

1 sec

PLASTIC SONG   *p. 534*

WILSON'S WARBLER

Cardellina pusilla

Breeds in swampy thickets in bogs and wet meadows. Female like female Yellow Warbler, but longer-tailed, with more contrast between crown and face.

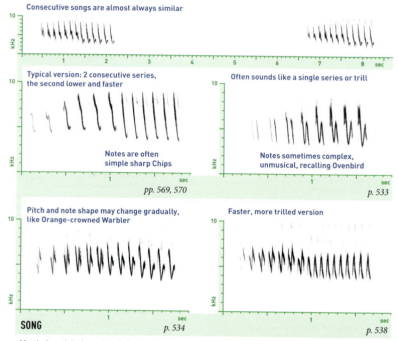

Consecutive songs are almost always similar

Typical version: 2 consecutive series, the second lower and faster

Notes are often simple sharp Chips

pp. 569, 570

Often sounds like a single series or trill

Notes sometimes complex, unmusical, recalling Ovenbird

p. 533

Pitch and note shape may change gradually, like Orange-crowned Warbler

Faster, more trilled version

SONG

p. 534

p. 538

Mostly Apr.–July, by males. At least 1 female has been recorded singing a slower, rising series of Seetlike notes. Quite variable. Males reported to have a single songtype, but 1 male has been recorded alternating 2 songtypes with call-like notes in between. Rarely adds a third series or single note at start or end. Usually sounds less musical than Nashville Warbler; never includes couplet series. Orange-crowned Warbler is almost always faster (trilled from the start).

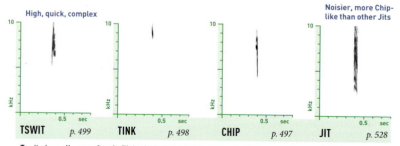

High, quick, complex

Noisier, more Chip-like than other Jits

TSWIT *p. 499*

TINK *p. 498*

CHIP *p. 497*

JIT *p. 528*

Tswit given all year, often in flight, including by night migrants. Tink apparently rare, in high alarm. High Chip given at least in late summer by juveniles, likely also sometimes in fall. Jit is most common and distinctive call; recalls Ruby-crowned Kinglet Jit but higher, briefer, sharper.

YELLOW-BREASTED CHAT

Icteria virens

A unique bird, apparently unrelated to warblers. Loud but skulking, found in second growth and clearcuts in the East and riparian thickets in the West.

Song: widely spaced single notes, series, and trills, each with a very different tone quality

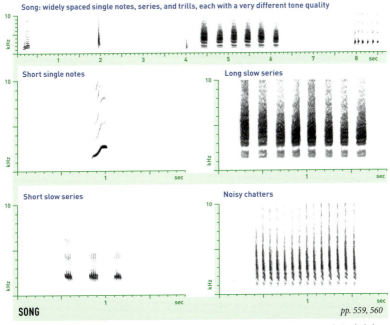

Short single notes

Long slow series

Short slow series

Noisy chatters

SONG

pp. 559, 560

Mostly Apr.–July, reportedly only by males. Each male has about 60 songtypes on average, but only 4–6 are typically used in any given five minutes of singing. Each songtype comprises either 1 note or a series of similar notes. Consecutive songs differ so greatly, and are separated by such long pauses, that they are easily mistaken for the sounds of different species; some are in fact imitations of other species. Tone quality may be whistled, nasal, noisy, or polyphonic. Song in slow-flapping display flight like regular song, but with slightly shorter intervals filled with the audible beating of wings.

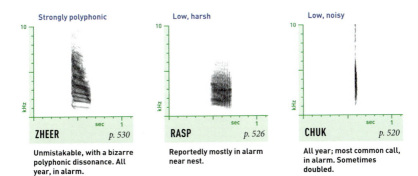

Strongly polyphonic

ZHEER *p. 530*

Unmistakable, with a bizarre polyphonic dissonance. All year, in alarm.

Low, harsh

RASP *p. 526*

Reportedly mostly in alarm near nest.

Low, noisy

CHUK *p. 520*

All year; most common call, in alarm. Sometimes doubled.

SEEDEATERS (Family Thraupidae, Subfamily Sporophilinae)

The seedeaters are tiny, finchlike birds formerly considered members of the sparrow family (Emberizidae). However, genetic evidence shows that they are more closely related to certain tropical tanagers. Only one species, White-collared Seedeater, is found as far north as south Texas.

SPARROWS, JUNCOS, AND TOWHEES (Family Emberizidae)

This family includes mostly small to medium-sized birds with conical bills and brownish or grayish plumage. Representatives of the family can be found in almost any habitat, but most spend their time on or near the ground, often in dense cover. The towhees in the genus *Pipilo* make up a distinctive group; they are larger and more boldly patterned than other sparrows, with long tails and distinctive calls.

Though they look and act quite different, the sparrows are closely related to the warblers (Parulidae), and this close affinity is evident in their vocalizations. Sparrow songs, which are apparently learned, can in many species be classified into typical Songs and Complex Songs, just as in some warbler species. In fact, the Complex Songs of warblers and sparrows often sound quite similar, and they occur in similar contexts, especially during infrequent flight displays, sometimes introduced by series of high-pitched call notes. Call repertoires are also similar in many warblers and sparrows, including a Seet or Dzeet call in contact or in flight (including during nocturnal migration), a high brief Tink in high alarm, and a Chip in mild alarm.

In addition, many sparrows also give various snarls, twitters, or whines in different kinds of interactions. The call repertoires of several species are poorly known, and it is likely that a number of sparrows give calls not described in this book. More study needed.

Pair Reunion Duets

In some sparrow species, mated pairs greet one another with complex, plastic "pair reunion duets." These duets, made up of call-like notes, tend to start with a series of high, Seetlike sounds and end with rapid, lower chips or twitters. Pair reunion duets are best known in certain groups of species from the southwestern U.S. and Central America, of which the only representative in the East is the Rufous-crowned Sparrow. However, similar behaviors may occur at times in the Olive Sparrow, the *Pipilo* towhees, and possibly a few other species.

Winter Singing

Many species of sparrow that winter in the U.S., including Song Sparrow, Fox Sparrow, Lark Sparrow, and the species in the genus *Zonotrichia*, can be heard singing nearly year-round. In fall and early winter, the majority of these are subsongs, presumably of first-year males. In late winter and early spring, the majority are plastic songs with a closer resemblance to the typical breeding songs of the species. Stereotyped breeding songs are rare before late spring.

White-collared Seedeater

Sporophila torqueola

Highly local in summer in tall grass with a few scattered bushes or trees, usually close to water. Formerly more widespread and common in south Texas.

Long Song (p. 564) is extended version of Short Song, continuing with various series, buzzes, and chatters

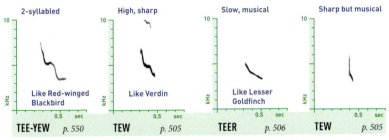

Short Song: 1–3 consecutive musical series

In Texas, first series usually higher, second lower, with upslurred notes

Version with 2 couplet series

SONG *p. 570*

Mostly Mar.–Oct., by males. Most males have a single Song. Full version (Long Song) given only at height of breeding season or when excited. Short Songs are first 2–4 series of Long Song; given repeatedly early and late in breeding season. Some versions like Yellow Warbler Song, but lower, slower.

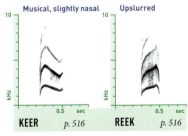

2-syllabled

Like Red-winged Blackbird

TEE-YEW *p. 550*

High, sharp

Like Verdin

TEW *p. 505*

Slow, musical

Like Lesser Goldfinch

TEER *p. 506*

Sharp but musical

TEW *p. 505*

A highly variable, somewhat plastic group of whistled calls given all year, by both sexes; most common call. Individuals may have more than one version. Usually given with a few Jirps mixed in, in contact and mild alarm, in a pattern which can recall the rather similar-looking Lesser Goldfinch.

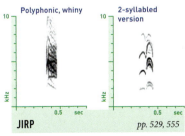

Musical, slightly nasal

KEER *p. 516*

Upslurred

REEK *p. 516*

Polyphonic, whiny

2-syllabled version

JIRP *pp. 529, 555*

A little-known set of seminasal notes; sole available examples given by male in reaction to presence of human near nest. Highly plastic.

All year, with whistled calls or by roosting flocks. Variable and slightly plastic. May recall House Finch or House Sparrow.

OLIVE SPARROW

Arremonops rufivirgatus

A primarily Central American species, common in dense brush, but more often heard than seen. All calls are plastic and poorly understood.

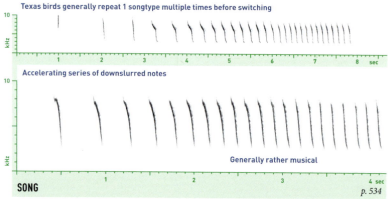

Texas birds generally repeat 1 songtype multiple times before switching

Accelerating series of downslurred notes

Generally rather musical

SONG
p. 534

Mostly Mar.–Sept., often from a hidden perch. Unknown whether female sings. Repertoire size unknown; at least some birds have multiple songtypes, all quite similar. May also have Complex Song of various single call-like notes, like towhees.

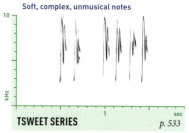

Soft, complex, unmusical notes

TSWEET SERIES
p. 533

Function little known; interspersed with various other calls. May be used in pair reunion duets. Plastic.

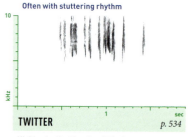

Often with stuttering rhythm

TWITTER
p. 534

Highly plastic. Function little known; may occur during pair interactions. Often mixed with other calls.

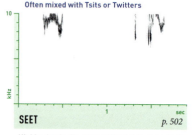

Often mixed with Tsits or Twitters

SEET
p. 502

Highly plastic. Function little known; reported in duets by foraging pairs, but also by lone birds.

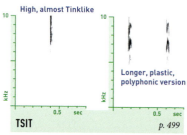

High, almost Tinklike

Longer, plastic, polyphonic version

TSIT
p. 499

All year; most common call. In alarm, typical Tsits may be run into rapid twitter. Function of polyphonic version unknown.

Green-tailed Towhee

Pipilo chlorurus

Fairly common in semiarid habitats with diverse mature shrubs. Often skulks in deep cover, but can be quite conspicuous during breeding season.

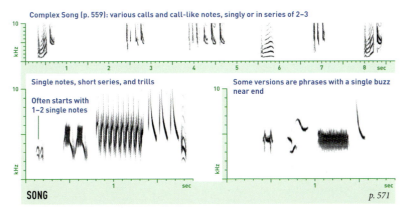

Complex Song (p. 559): various calls and call-like notes, singly or in series of 2–3

Single notes, short series, and trills

Often starts with 1–2 single notes

Some versions are phrases with a single buzz near end

SONG *p. 571*

Mostly Apr.–July, by males. Males have 5–12 songtypes; consecutive songs almost always different. In California, compare song of "Thick-billed" Fox Sparrow, which averages fewer trills and more slurred whistles, but much overlap. Soft, disjointed Complex Song often given after territorial encounters.

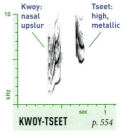

Kwoy: nasal upslur

Tseet: high, metallic

KWOY-TSEET *p. 554*

In courtship or alarm, mostly by females. Both notes also singly, Tseet in series.

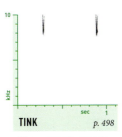

TINK *p. 498*

All year, by both sexes, in high alarm.

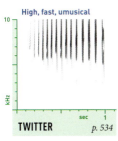

High, fast, unmusical

TWITTER *p. 534*

Mostly by courting females. More rattling, cowbirdlike than in other towhees.

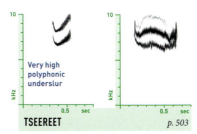

Very high polyphonic underslur

TSEEREET *p. 503*

All year, by both sexes; second most common call. Plastic and variable, but usually long, underslurred, and polyphonic; often buzzy.

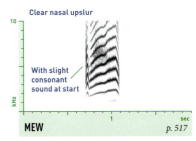

Clear nasal upslur

With slight consonant sound at start

MEW *p. 517*

All year; most common call. Somewhat plastic, but rather uniform rangewide; rarely burry or noisy. A few versions are nearly monotone.

EASTERN TOWHEE

Pipilo erythrophthalmus

Common in shrubby open woodlands. Eye is red in most of range, but white in Florida, where songs and calls differ slightly; many birds are intermediate.

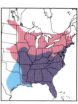

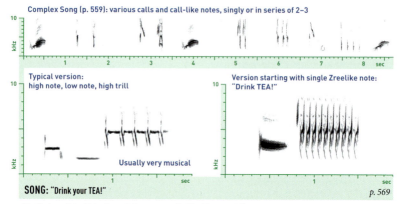

Complex Song (p. 559): various calls and call-like notes, singly or in series of 2–3

Typical version: high note, low note, high trill

Usually very musical

Version starting with single Zreelike note: "Drink TEA!"

SONG: "Drink your TEA!"

p. 569

Mostly Apr.–July. Males have 2–5 songtypes (in Florida, up to 11), each usually repeated many times before switching. Females sing very rarely. Highly variable: 1–3 (rarely up to 5) intro notes are followed by 1 trill (rarely 2). Soft, disjointed Complex Song given at dawn, or after territorial encounters.

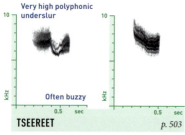

Soft, low, upslurred

ZOIT   p. 527

One of several little-known calls, given with Zrees; may serve in courtship.

Notes polyphonic, Tiplike

TINK   p. 498

All year, by both sexes, in high alarm. Sometimes run into rapid Twitter.

TWITTER   p. 535

Mostly by females in courtship, but also reported in male-male interactions.

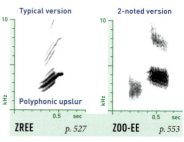

Very high polyphonic underslur

Often buzzy

TSEEREET   p. 503

All year, by both sexes; second most common call. Plastic and variable, but usually long, underslurred, and polyphonic; often buzzy.

Typical version   2-noted version

Polyphonic upslur

ZREE   p. 527   ZOO-EE   p. 553

All year, by both sexes; most common call. Highly variable, but most versions are rising, metallic, 1-noted or only vaguely 2-noted.

Spotted Towhee

Pipilo maculatus

Formerly lumped with Eastern Towhee. Plumage and sounds vary geographically; birds at eastern end of range sound more similar to Eastern Towhee.

Complex Song (p. 559): various calls and call-like notes, singly or in series of 2–3

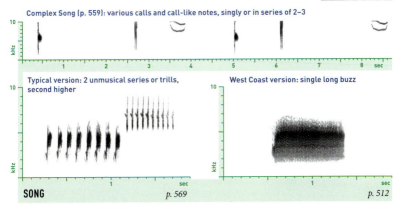

Typical version: 2 unmusical series or trills, second higher

West Coast version: single long buzz

SONG *p. 569* *p. 512*

Mostly Feb.–July. Males have 4–9 songtypes; 1 may be repeated many times before switching, or 2 may be alternated for long periods. Females sing very rarely. Highly variable. Hybrids with Eastern Towhee may sing like either parent. Soft, disjointed Complex Song given at dawn, or after territorial encounters.

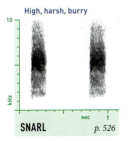

High, harsh, burry

SNARL *p. 526*

One of several little-known calls; this version given during midair tussle.

TINK *p. 498*

All year, by both sexes, in high alarm. Sometimes run into rapid Twitter.

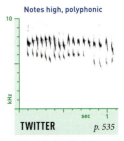

Notes high, polyphonic

TWITTER *p. 535*

Mostly by females in courtship; also in male-male interactions. Highly plastic.

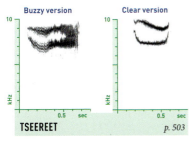

Buzzy version

Clear version

TSEEREET *p. 503*

All year, by both sexes; second most common call. Plastic and variable, but usually long, underslurred, and polyphonic; often buzzy.

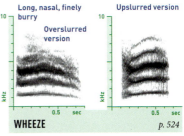

Long, nasal, finely burry

Overslurred version

Upslurred version

WHEEZE *p. 524*

All year, by both sexes; most common call. Highly variable and plastic; some individuals give both over- and upslurred versions.

RUFOUS-CROWNED SPARROW

Aimophila ruficeps

Locally common on rocky hillsides with scattered dense shrubs. Usually nests on the ground. Not closely related to other North American sparrows.

Songs often introduced by 1–2 Keer notes

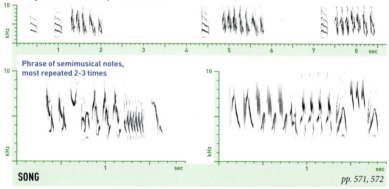

Phrase of semimusical notes, most repeated 2-3 times

SONG   *pp. 571, 572*

Mostly Mar.–Aug. May recall a fast Indigo Bunting Song, or the beginning of a House Wren Song. Males have repertoires of up to 14 songtypes; consecutive songtypes often similar, though number of repetitions in each series may vary. Typical Song is sometimes given during courtship flight display.

1 or both birds start with Tseets

1 bird ends with a chipping Twitter

PAIR REUNION DUET   *p. 566*

Given during pair reunions. May be given with and grade into Keer and Chatter.

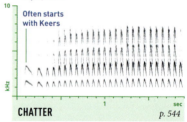

Rapid, nasal

Often starts with Keers

CHATTER   *p. 544*

All year, in alarm. Grades into Keers. Uniform but plastic; often unsteady in speed and pitch.

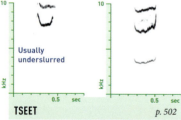

Very high, polyphonic

Usually underslurred

TSEET   *p. 502*

All year, likely in family contact; also perhaps in mild alarm. Quite plastic. Usually shorter, less burry than similar calls of towhees.

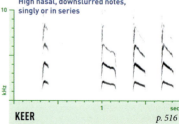

High nasal, downslurred notes, singly or in series

KEER   *p. 516*

All year, by both sexes, in mild alarm; most common call. Distinctive. Uniform but plastic; grades into Chatter.

Bachman's Sparrow

Peucaea aestivalis

A lovely singer, resident in the grassy understory of mature pine forests, as well as in freshly clearcut fields. Rather secretive when not singing.

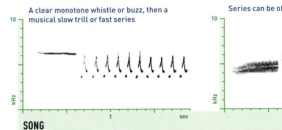

A clear monotone whistle or buzz, then a musical slow trill or fast series

Series can be of couplets

Compare Eastern Towhee Song

SONG

p. 569

Mostly Feb.–Aug., apparently by males. Individuals have 20–40 songtypes; consecutive songs usually differ, unlike in Eastern Towhee, but sometimes a songtype may be repeated 4–5 times before switching. Variable; rare versions include those without introductory notes, or with 2 trills at end.

Complex Song often immediately preceded or followed by louder typical Songs

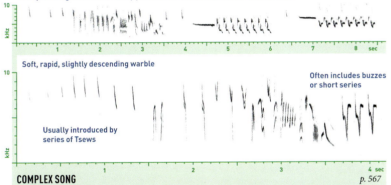

Soft, rapid, slightly descending warble

Often includes buzzes or short series

Usually introduced by series of Tsews

COMPLEX SONG

p. 567

Mostly Feb.–Aug., apparently by males, in agitation, including during aggressive encounters and in response to playback. Sometimes given in flight. Individuals likely have a single Complex Song type, but more study needed. Partial versions are common; some may grade into Chitter.

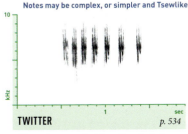

Notes may be complex, or simpler and Tsewlike

TWITTER

p. 534

All year, in aggressive interactions, including response to playback, and possibly during pair reunions. Rather plastic.

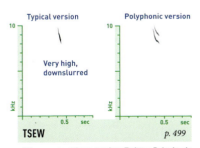

Typical version

Polyphonic version

Very high, downslurred

TSEW

p. 499

All year, sometimes run into Twitter. Polyphonic version little known; sole example is from bird also giving typical version.

Cassin's Sparrow

Peucaea cassinii

Frequents arid grasslands dotted with mesquite, rabbitbrush, or other shrubs. Abundance can vary greatly from year to year at any one location.

1–3 whistles and a trill on nearly the same pitch, then 4 whistles in a high-low, high-low pattern

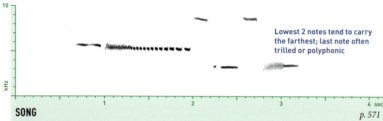

Lowest 2 notes tend to carry the farthest; last note often trilled or polyphonic

SONG *p. 571*

Highly distinctive and musical. Mostly Apr.–July in Texas, July–Sept. in Arizona, often in display flight as male glides downward, usually to a different perch. Individuals have 1–3 songtypes, all quite similar, each usually repeated 2–4 times before switching.

Complex Song often embedded in long twittering series of Psitlike notes

Various buzzy and complex notes strung together, most in short series

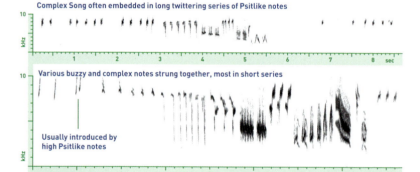

Usually introduced by high Psitlike notes

COMPLEX SONG *p. 567*

Mostly during breeding season by excited males, apparently in both close courtship and territorial encounters, sometimes in flight. Individuals may have a single Complex songtype, but repertoire size needs more study.

High Twitter: often polyphonic

Low Twitter: notes Chiplike

Very high, upslurred

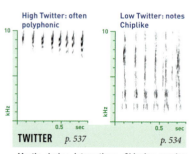
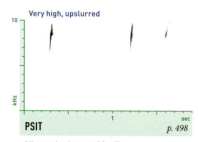

TWITTER *p. 537*

p. 534

PSIT *p. 498*

Mostly during interactions. Chipping version less frequent, usually introduced by and developing out of high version.

All year, in alarm and family contact; most common call. Highly plastic. May not be separable from Botteri's Psit.

BOTTERI'S SPARROW

Peucaea botterii

Locally common in healthy semiarid grasslands, open oak woods, and open thorn scrub. Best distinguished from Cassin's Sparrow by voice.

A varied series of 1- or 2-syllabled notes, then a series of Chips accelerating into a trill

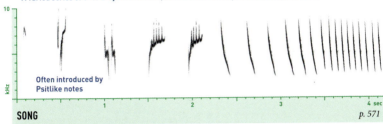

Often introduced by Psitlike notes

SONG *p. 571*

Mostly prior to egg-laying, Mar.–June in Texas, May–July in Arizona, reportedly by males only. Sometimes given in flight, usually during territorial disputes. Excited birds may double the accelerating trill at end, or omit it. Repertoire size unknown; consecutive songs tend to be similar.

Complex Song: Highly varied, but certain short combinations of notes may be repeated

Various short high notes, singly or in series of 2–3

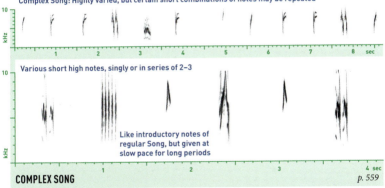

Like introductory notes of regular Song, but given at slow pace for long periods

COMPLEX SONG *p. 559*

Mostly after egg-laying, May–July in Texas, July–Sept. in Arizona, reportedly by males only; if nest fails, male reverts to regular Song. Psits, or long Twitters of Psitlike notes, often interspersed. Number of phrase types per male unknown, but apparently large.

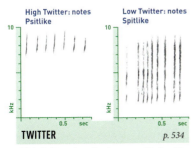

High Twitter: notes Psitlike

Low Twitter: notes Spitlike

TWITTER *p. 534*

High Twitter given in alarm or with Complex Song. Low Twitter given in aggression, in high alarm, and during pair reunions.

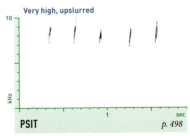

Very high, upslurred

PSIT *p. 498*

All year, in alarm and family contact; most common call. Highly plastic. May not be separable from Cassin's Psit.

AMERICAN TREE SPARROW

Spizelloides arborea

A common winter visitor in our area, found in flocks in weedy fields and open woods. Readily visits feeders. Note bi-colored bill and central breast spot.

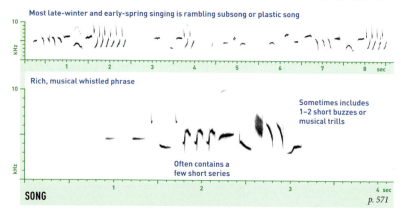

Most late-winter and early-spring singing is rambling subsong or plastic song

Rich, musical whistled phrase

Sometimes includes 1–2 short buzzes or musical trills

Often contains a few short series

SONG   *p. 571*

Mostly on breeding grounds, but wintering birds often give subsong, plastic song, and occasionally typical Song before migrating north. Breeding males have a single songtype; easily confused with Song of Fox Sparrow, but higher and less likely to contain long, slurred whistles.

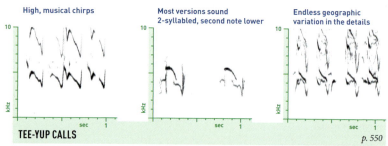

High, musical chirps

Most versions sound 2-syllabled, second note lower

Endless geographic variation in the details

TEE-YUP CALLS   *p. 550*

Often given by flocks in migration or winter; apparently rather uniform within a given flock, but often different between flocks. Plastic; perhaps flocks or individuals change calls over time, or have multiple versions. More study needed.

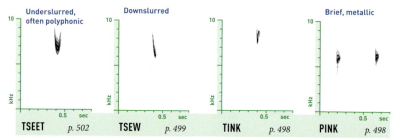

Underslurred, often polyphonic

Downslurred

Brief, metallic

TSEET   *p. 502*   **TSEW**   *p. 499*   **TINK**   *p. 498*   **PINK**   *p. 498*

Single-note calls are confusing; not all variants shown. All are plastic. Tseet is most common call during day and in night migration. Grades into Tsew, a common variant. Tink recorded from alarmed birds near nest in summer. Pink given infrequently all year, apparently in alarm. More study of calls needed.

CHIPPING SPARROW

Spizella passerina

A common and familiar breeder in open woods, parks, and towns. Prefers to nest in conifers. Outside the breeding season, often forms large flocks.

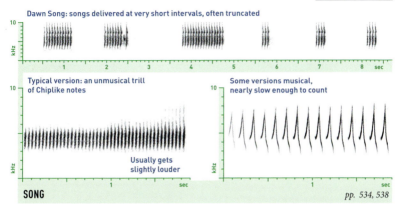

Dawn Song: songs delivered at very short intervals, often truncated

Typical version: an unmusical trill of Chiplike notes

Usually gets slightly louder

Some versions musical, nearly slow enough to count

SONG *pp. 534, 538*

Mostly May–July. Highly variable. Females not known to sing. Most males have only 1 songtype; those with 2 never alternate. Rare variants include trills that change quality, pitch, or speed partway through. Dawn Song often given by multiple birds on ground at close range in complex interaction.

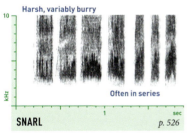

Harsh, variably burry

Often in series

SNARL *p. 526*

All year, in close-range aggressive interactions. Highly variable and plastic. Other *Spizella* sparrows have similar calls.

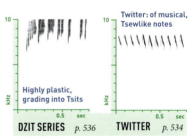

Highly plastic, grading into Tsits

Twitter: of musical, Tsewlike notes

DZIT SERIES *p. 536*   **TWITTER** *p. 534*

Mostly May–June. Dzit Series in courtship, in agitation, or by begging juvenile. Twitter before and during copulation, reportedly by female.

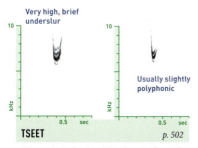

Very high, brief underslur

Usually slightly polyphonic

TSEET *p. 502*

All year, including by night migrants. Somewhat variable and plastic, but fairly distinctive once learned.

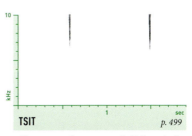

TSIT *p. 499*

All year; most common call. Rather plastic, sometimes grading into Tseet. Often run into rapid Dzit Series by agitated birds.

CLAY-COLORED SPARROW

Spizella pallida

A common breeder in prairie shrubs, particularly snowberry, as well as second growth and thickets near water. Winters in dry, grassy or weedy areas.

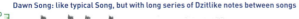

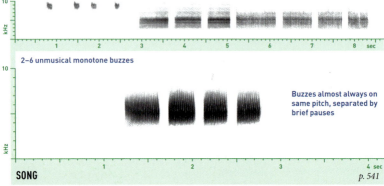

Dawn Song: like typical Song, but with long series of Dzitlike notes between songs

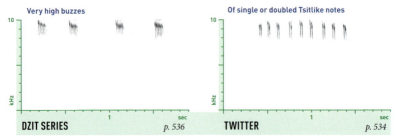

2–6 unmusical monotone buzzes

Buzzes almost always on same pitch, separated by brief pauses

SONG *p. 541*

Mostly May–July, including on spring migration. Most males have a single songtype, but some males have up to 3; each songtype is repeated many times before switching. Most Songs contain only 1 type of buzz, but rarely the type of buzz will change partway through the Song.

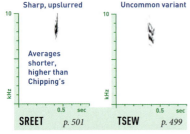

Very high buzzes

Of single or doubled Tsitlike notes

DZIT SERIES *p. 536*

TWITTER *p. 534*

Gives a variety of twittering sounds, mostly during breeding season. Dzit Series likely associated with alarm or agitation; similar sounds given in Dawn Song. Function of Twitter not clear. Likely also has musical version used in courtship as well as long snarling notes given in aggression.

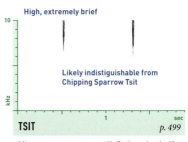

Sharp, upslurred

Averages shorter, higher than Chipping's

Uncommon variant

High, extremely brief

Likely indistiguishable from Chipping Sparrow Tsit

SREET *p. 501*

TSEW *p. 499*

TSIT *p. 499*

All year, including by night migrants. Variable, plastic; most are Sreetlike.

All year; most common call. Rather plastic. Very often given in rapid series; also run into rapid twitters by agitated birds.

BREWER'S SPARROW

Spizella breweri

Rare at the western edge of our area, mostly as a migrant. Breeds in western sagebrush steppe; often migrates in flocks with Chipping Sparrows.

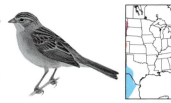

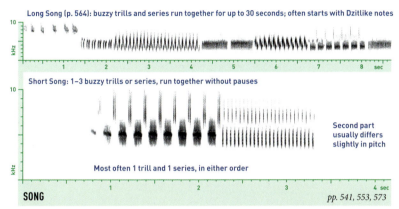

Long Song (p. 564): buzzy trills and series run together for up to 30 seconds; often starts with Dzitlike notes

Short Song: 1–3 buzzy trills or series, run together without pauses

Second part usually differs slightly in pitch

Most often 1 trill and 1 series, in either order

SONG *pp. 541, 553, 573*

Mostly May–June, but subsong or plastic song may be heard all year. Most males have only 1 Short Song type, but some have 2, each repeated many times before switching. Short Song given at regular intervals during day by breeding males; Long Song at dawn, in courtship, and in territorial encounters.

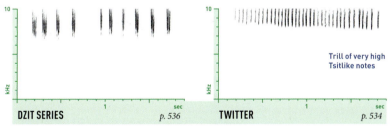

DZIT SERIES *p. 536*

TWITTER *p. 534*

Trill of very high Tsitlike notes

Gives a variety of twittering sounds, mostly during breeding season. Dzit Series likely associated with alarm or agitation; function of Twitter not clear. Likely also has musical version used in courtship as well as long snarling notes given in aggression, like Chipping Sparrow.

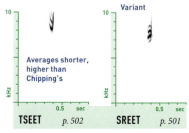

Variant

Averages shorter, higher than Chipping's

TSEET *p. 502*   **SREET** *p. 501*

All year, including by night migrants. Quite variable and somewhat plastic.

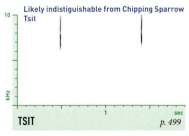

Likely indistiguishable from Chipping Sparrow Tsit

TSIT *p. 499*

All year; most common call. Rather plastic. Very often given in rapid series; also run into rapid twitters by agitated birds.

FIELD SPARROW

Spizella pusilla

Fairly common in brushy fields, prairie edges, and early second growth. Nests at or near ground level. Note pinkish bill and legs.

Complex Song (p. 564): 1–2 series and 2–3 trills on nearly the same pitch; interspersed with call-like notes

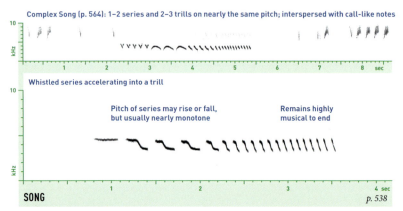

Whistled series accelerating into a trill

Pitch of series may rise or fall, but usually nearly monotone

Remains highly musical to end

SONG *p. 538*

Mostly May–July. Simple, accelerating pattern distinctive in most of range. Males apparently have 1 regular and 1 Complex songtype, often shared by neighboring birds. Complex Song given mostly at dawn, especially later in season, and in territorial encounters. Females not known to sing.

Notes very high, downslurred, Tsewlike

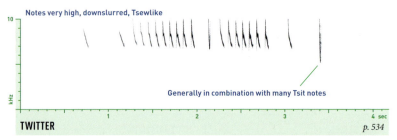

Generally in combination with many Tsit notes

TWITTER *p. 534*

At least May–Aug. The version shown apparently used in male-male aggression; has been recorded in stereotyped form at regular intervals in late August, suggesting a type of "autumn song." Similar calls, likely representing several vocalizations, reported in courtship-related contexts.

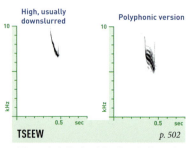

High, usually downslurred

Polyphonic version

TSEEW *p. 502*

All year, including by night migrants. Somewhat variable and plastic, but fairly distinctive.

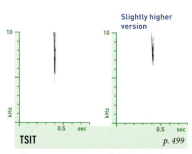

Slightly higher version

TSIT *p. 499*

All year; most common call. Rather plastic.

VESPER SPARROW

Pooecetes gramineus

A bird of open grassy fields with a few low bushes; sometimes also cultivated fields. White outer tail feathers are an excellent field mark.

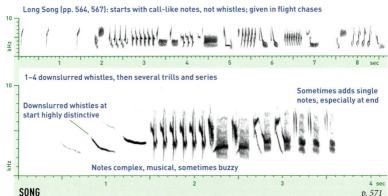

Long Song (pp. 564, 567): starts with call-like notes, not whistles; given in flight chases

1–4 downslurred whistles, then several trills and series

Downslurred whistles at start highly distinctive

Sometimes adds single notes, especially at end

Notes complex, musical, sometimes buzzy

SONG   *p. 571*

Mostly Apr.–Aug. Introductory whistles same in all Songs of 1 bird, and similar among neighbors, but regionally variable. Rest of Song highly variable; individuals combine 40 or more trills/series into 200 or more songtypes. Consecutive songtypes may have identical or different endings.

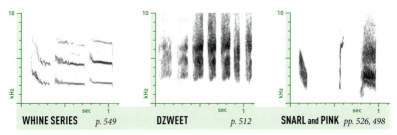

WHINE SERIES   *p. 549*   **DZWEET**   *p. 512*   **SNARL and PINK**   *pp. 526, 498*

A wide variety of calls given in high-intensity interactions during the breeding season. Function of each little known; more study needed. Dzweets and Snarls likely intergrade; some very high-pitched snarllike calls have also been recorded.

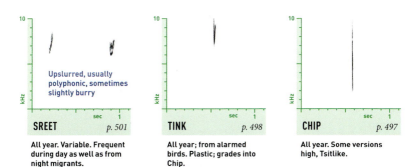

Upslurred, usually polyphonic, sometimes slightly burry

SREET   *p. 501*   **TINK**   *p. 498*   **CHIP**   *p. 497*

All year. Variable. Frequent during day as well as from night migrants.

All year; from alarmed birds. Plastic; grades into Chip.

All year. Some versions high, Tsitlike.

LARK SPARROW

Chondestes grammacus

Found in open shrubby habitats, with or without scattered trees. Common in the West; quite local in the East. Note long, rounded tail with white corners.

Long Song (p. 564): like typical Song, but continues without pauses for long periods

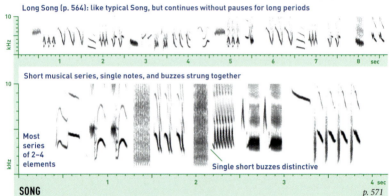

Short musical series, single notes, and buzzes strung together

Most series of 2–4 elements

Single short buzzes distinctive

SONG

p. 571

All year; especially Apr.–July, by males. Females not known to sing. Winter Songs often plastic. Repertoire size unknown; consecutive songs usually different, often with many of the same components rearranged. Courting males sing before females with wings drooped, tail raised and spread.

Twitter: extremely high, polyphonic

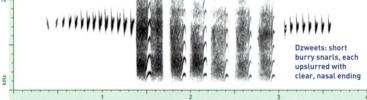

Dzweets: short burry snarls, each upslurred with clear, nasal ending

TWITTER and DZWEET

pp. 534, 512

Distinctive. Given in aggressive encounters during breeding season. These two vocalizations very often heard together, with Twitter immediately before or after Dzweets, but also given separately. Dzweets usually in short series, but sometimes given singly.

Extremely high, brief

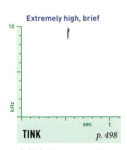

TINK

p. 498

In high alarm, near nest or young. Grades into Tsip.

Very high, downslurred

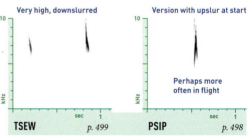

Version with upslur at start

Perhaps more often in flight

TSEW

p. 499

PSIP

p. 498

All year; most common calls, in contact and alarm. Plastic, grading into Tink; 2 most common variants shown. No Seet call known; night migrants are not known to call.

Black-throated Sparrow

Amphispiza bilineata

Common in southwestern deserts. Breeds in a variety of arid shrubby habitats, particularly in creosote bush in the south and sagebrush in the north.

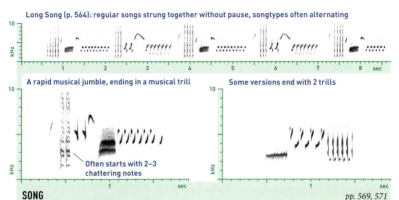

Long Song (p. 564): regular songs strung together without pause, songtypes often alternating

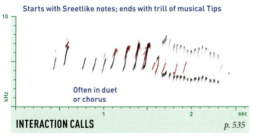

A rapid musical jumble, ending in a musical trill

Often starts with 2–3 chattering notes

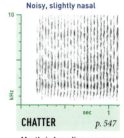

Some versions end with 2 trills

SONG *pp. 569, 571*

Mostly Mar.–Aug. High, tinkling, never very loud. Males have 5–9 songtypes; consecutive songtypes sometimes similar, but more often different; 2 songtypes often alternated for a while before switching to another 2. Truncated songs sometimes given for long periods.

Starts with Sreetlike notes; ends with trill of musical Tips

Often in duet or chorus

INTERACTION CALLS *p. 535*

Mostly in breeding season, during chases and other high-intensity aggressive interactions between two or more birds. Highly plastic.

Noisy, slightly nasal

CHATTER *p. 547*

Mostly in breeding season, during aggressive encounters.

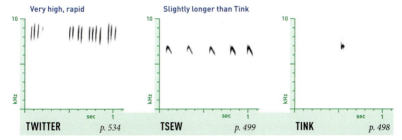

Very high, rapid

TWITTER *p. 534*

Slightly longer than Tink

TSEW *p. 499*

TINK *p. 498*

These 3 calls are usually distinct, but intergrade. Twitter given infrequently; function unknown. Tsew usually given in short series, while Tink is usually given singly, but the two are often interspersed. Tsew and Tink are most common calls, given all year in a variety of situations.

LARK BUNTING

Calamospiza melanocorys

A characteristic species of short-grass prairies. Nonbreeding males resemble females, but with more black in plumage. Forms flocks in winter.

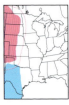

Males have 1 to a few songtypes, each comprising several series and trills in a particular order

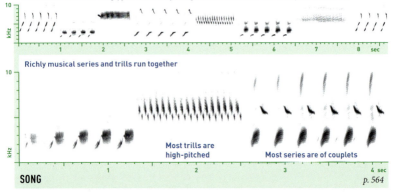

Richly musical series and trills run together

Most trills are high-pitched

Most series are of couplets

SONG *p. 564*

Mostly Apr.–June, by males in slow-flapping flight display. Consecutive songs usually similar, but many are truncated. Song often includes series of cardinal-like slurred whistles. In aggressive flight chases involving 2 or more males, song reportedly contains fewer repeated elements.

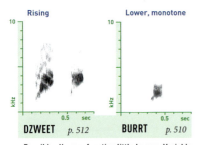

Rising

Lower, monotone

Soft, noisy notes

DZWEET *p. 512*   **BURRT** *p. 510*

Possibly all year; function little known. Variable and/or plastic, apparently grading into Snarl.

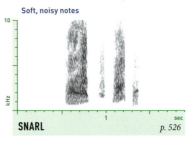

SNARL *p. 526*

Possibly all year; at least during nonbreeding season, by birds in flocks, perhaps in aggression. Highly plastic.

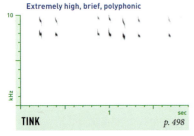

Extremely high, brief, polyphonic

TINK *p. 498*

Possibly all year; given in conjunction with Wert by birds in flocks, at least from spring to late summer. Slightly plastic.

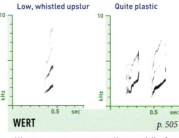

Low, whistled upslur

Quite plastic

WERT *p. 505*

All year; most common call, especially from flocks. Usually not very loud.

WHITE-THROATED SPARROW

Zonotrichia albicollis

Breeds in boreal forests; winters in flocks in brushy habitats. Head stripes may be white or tan in either sex; birds of one color morph prefer mates of the other morph.

Most males have 1 songtype; about 10% have 2, one of which is given more often than the other

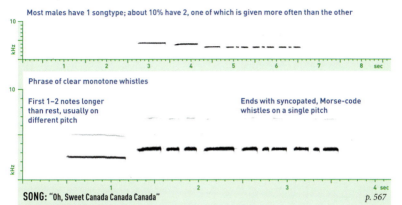

Phrase of clear monotone whistles

First 1–2 notes longer than rest, usually on different pitch

Ends with syncopated, Morse-code whistles on a single pitch

SONG: "Oh, Sweet Canada Canada Canada"   *p. 567*

A striking and easily recognized song, though quite variable. Mostly Mar.–July, by both sexes, though tan-striped females sing rarely. Female song tends to be shorter, with notes less steady in pitch. Subsong and plastic song given Aug.–Apr.

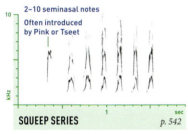

2–10 seminasal notes

Often introduced by Pink or Tseet

SQUEEP SERIES   *p. 542*

All year, in aggression, often with head extended forward and wings fluttering; also by mated pairs at the nest. Variable and plastic.

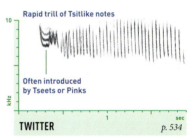

Rapid trill of Tsitlike notes

Often introduced by Tseets or Pinks

TWITTER   *p. 534*

Mostly Apr.–June, by females prior to copulation and by males in aggressive display, with head forward and wings fluttering. Quite plastic.

High, distinctly underslurred

Usually polyphonic; sometimes slightly burry

TSEET   *p. 503*

All year, in contact and in flight, including by night migrants. Plastic.

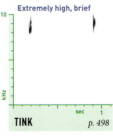

Extremely high, brief

TINK   *p. 498*

Mostly May–Aug., in alarm near nest, often with Pinks. Plastic.

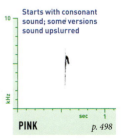

Starts with consonant sound; some versions sound upslurred

PINK   *p. 498*

All year, in alarm and other situations. Plastic.

HARRIS'S SPARROW

Zonotrichia querula

Breeds at the interface between boreal forest and tundra; winters in brushy or weedy areas of the Great Plains, often flocking with White-crowned Sparrows.

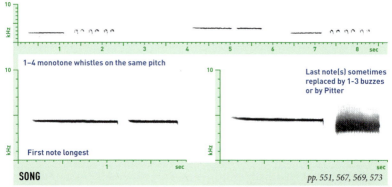

Males have 1–3 songtypes; consecutive songtypes usually differ

1–4 monotone whistles on the same pitch

Last note(s) sometimes replaced by 1-3 buzzes or by Pitter

First note longest

SONG *pp. 551, 567, 569, 573*

Mostly Mar.–July, by males. Song by spring migrants is frequent and usually plastic. The relationship of the Pitter call to the Song needs more study; it apparently forms part of some songtypes, but can also be deployed by itself between Songs.

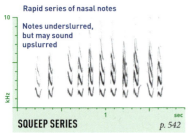

Rapid series of nasal notes

Notes underslurred, but may sound upslurred

SQUEEP SERIES *p. 542*

Possibly all year; given frequently by birds in winter flocks. Quite plastic. Much like Squeep Series of other *Zonotrichia* species.

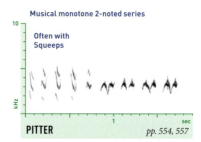

Musical monotone 2-noted series

Often with Squeeps

PITTER *pp. 554, 557*

In summer, given in between or appended to end of Songs; in winter, given by flocks with Cheep Series, Pinks, and series of Pweews.

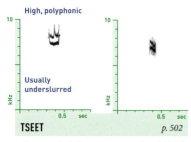

High, polyphonic

Usually underslurred

TSEET *p. 502*

All year, including presumably by night migrants. Highly plastic. This species may also have a Tink, but no recordings are available.

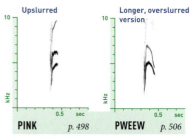

Upslurred

Longer, overslurred version

PINK *p. 498*   **PWEEW** *p. 506*

All year. Variable, but apparently rather stereotyped within individuals. Longer, more polyphonic than other *Zonotrichia* Pinks.

WHITE-CROWNED SPARROW

Zonotrichia leucophrys

Breeds in shrubby subalpine meadows and stunted conifers near treeline. In migration and winter, found in brush and weedy fields, often in flocks.

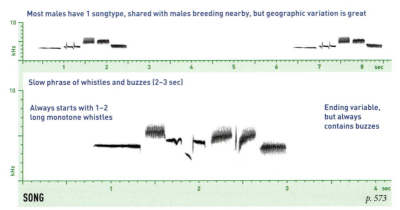

Most males have 1 songtype, shared with males breeding nearby, but geographic variation is great

Slow phrase of whistles and buzzes (2–3 sec)

Always starts with 1–2 long monotone whistles

Ending variable, but always contains buzzes

SONG *p. 573*

Mostly Mar.–July, by males; females occasionally sing, more softly and variably than males. Subsong and plastic song given Aug.–Apr. Most males have a single songtype, but those breeding at dialect boundaries may sing both dialects. In spring migration, can be mistaken for a warbler Song.

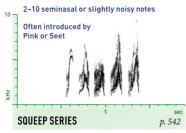

2–10 seminasal or slightly noisy notes

Often introduced by Pink or Seet

SQUEEP SERIES *p. 542*

All year, by both sexes, in mild agitation, often from flocks under feeding stations. Variable and highly plastic.

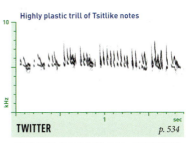

Highly plastic trill of Tsitlike notes

TWITTER *p. 534*

Mostly May–July, by both sexes, often with fluttering wings. By females in courtship, males in territorial defense.

High, rising

Usually polyphonic; rarely slightly burry

SREET *p. 501*

All year, in contact and in flight, including by night migrants. Plastic.

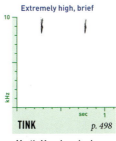

Extremely high, brief

TINK *p. 498*

Mostly May–Aug., in alarm near nest, often with Pinks. Plastic.

Starts with consonant

PINK *p. 498*

All year, in alarm and other situations; plastic. Some versions sound upslurred.

Song Sparrow

Melospiza melodia

Widespread and familiar, frequenting a wide variety of brushy habitats and wetlands. Plumage and size vary geographically.

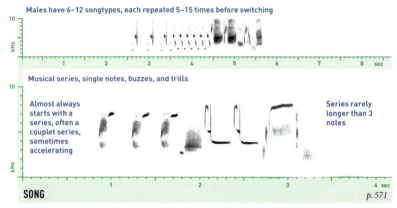

Males have 6–12 songtypes, each repeated 5–15 times before switching

Musical series, single notes, buzzes, and trills

Almost always starts with a series, often a couplet series, sometimes accelerating

Series rarely longer than 3 notes

SONG *p. 571*

All year, but mostly Mar.–Aug., by males; females rarely sing. Extremely variable, but recognizable rangewide once learned. Alternating high and low notes are typical, as are alternating buzzy and musical tones; occasionally some parts are polyphonic. Song length 2–4 seconds.

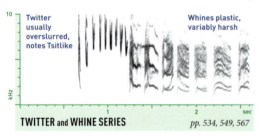

Twitter usually overslurred, notes Tsitlike

Whines plastic, variably harsh

TWITTER and WHINE SERIES *pp. 534, 549, 567*

All year, in high agitation. Twitters and Whine Series often given separately, mostly by females during breeding season, by males on winter territories. Single Tsitlike or Tsewlike notes may be given.

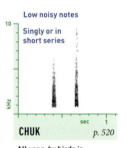

Low noisy notes

Singly or in short series

CHUK *p. 520*

All year; by birds in aggressive interactions. Also a harsh Dzweet (p. 512).

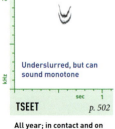

Underslurred, but can sound monotone

TSEET *p. 502*

All year; in contact and on nocturnal migration. Very like Fox Sparrow Tseet.

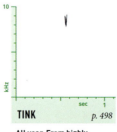

TINK *p. 498*

All year. From highly alarmed birds, often with Twitters and Whine Series.

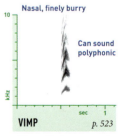

Nasal, finely burry

Can sound polyphonic

VIMP *p. 523*

All year. Most common and distinctive call; variable but easily recognizable.

Lincoln's Sparrow

Melospiza lincolnii

Breeds in swampy willow thickets and boggy areas. Rather secretive in migration and winter, skulking in a variety of brushy habitats, usually near water.

1–6 songtypes, repeated many times before switching, though endings frequently vary

Musical series and trills in rising and falling pattern

Starts low and soft, rises in pitch and volume

Highest and loudest on 1–2 trills in middle

SONG *p. 572*

Mostly May–July. Most likely to be mistaken for House Wren Song, but richer and more rigidly structured, especially at start. Sometimes given in flight display after introductory series of Dzeets, but no separate Complex Song as in Swamp Sparrow.

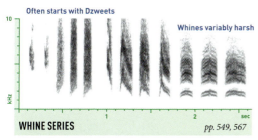

Often starts with Dzweets

Whines variably harsh

WHINE SERIES *pp. 549, 567*

Mostly May–July; highly plastic. Similar to Swamp Sparrow Whine Series and likely used in similar circumstances. Sometimes starts with slow twitter of Tsitlike notes, much like Song Sparrow.

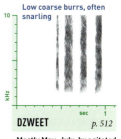

Low coarse burrs, often snarling

DZWEET *p. 512*

Mostly May–July, by agitated birds on territory. Plastic.

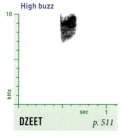

High buzz

DZEET *p. 511*

All year, including from night migrants. Rather variable and plastic.

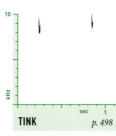

TINK *p. 498*

All year. Plastic; grades smoothly into Chip.

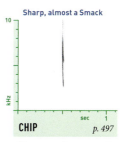

Sharp, almost a Smack

CHIP *p. 497*

All year; most common call. Highly plastic in pitch and quality.

SWAMP SPARROW

Melospiza georgiana

Breeds in cattail marshes as well as salt marshes and cedar swamps. Can be rather shy in migration and winter, frequenting brushy habitats near water.

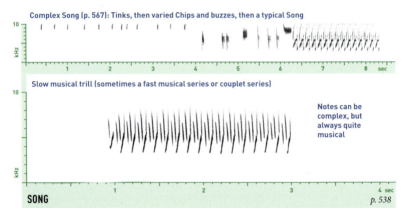

Complex Song (p. 567): Tinks, then varied Chips and buzzes, then a typical Song

Slow musical trill (sometimes a fast musical series or couplet series)

Notes can be complex, but always quite musical

SONG *p. 538*

Mostly Apr.–Aug. Males sing 1–4 songtypes, each repeated several times before switching. More musical and slower than most other single-trill songs. Complex Song rarely heard, mostly in early spring flight displays when males are establishing territories.

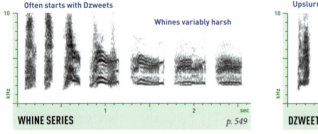

Often starts with Dzweets

Whines variably harsh

WHINE SERIES *p. 549*

Mostly Apr.–July, by males and possibly females, in territorial disputes. No equivalent of Song Sparrow Twitter known in Swamp, though females leaving the nest give quick series of Chips or Tinks.

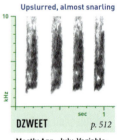

Upslurred, almost snarling

DZWEET *p. 512*

Mostly Apr.–July. Variable; like Lincoln's Sparrow Dzweets.

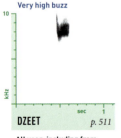

Very high buzz

DZEET *p. 511*

All year, including from night migrants. Monotone or barely underslurred.

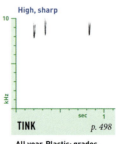

High, sharp

TINK *p. 498*

All year. Plastic; grades smoothly into Chip.

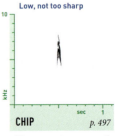

Low, not too sharp

CHIP *p. 497*

All year; most common call. Variable and plastic but fairly distinctive.

Fox Sparrow

Passerella iliaca

One of our largest and most colorful sparrows. Breeds in boreal swamps; winters and migrates in flocks in parks and woodlands.

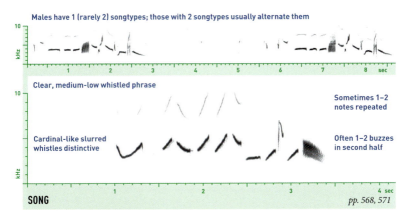

Males have 1 (rarely 2) songtypes; those with 2 songtypes usually alternate them

Clear, medium-low whistled phrase

Cardinal-like slurred whistles distinctive

Sometimes 1–2 notes repeated

Often 1–2 buzzes in second half

SONG *pp. 568, 571*

Mostly Apr.–July, by males; female Song less frequent. Subsong and plastic song common Feb.–Mar. Song may rarely incorporate 1–2 imitations of calls of other species. On breeding grounds, strings of call-like notes sometimes given between Songs; Complex Song consists mostly of such notes.

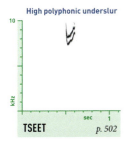 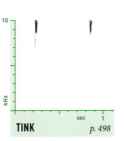 

High polyphonic underslur

TSEET *p. 502*

All year, including by night migrants. Very like Song Sparrow Tseet.

TINK *p. 498*

By highly agitated birds near nest; sometimes in response to playback.

Very sharp, loud

SMACK *p. 497*

All year, in alarm; most common call. Uniform but slightly plastic.

GEOGRAPHIC VARIATION IN FOX SPARROWS AND DARK-EYED JUNCOS

Across their range, Fox Sparrows and Dark-eyed Juncos show extensive variation in plumage as well as some variation in sounds. Some forms have been treated as separate species. This volume treats only the forms commonly encountered in the East: "Red" Fox Sparrow and "Slate-colored" Junco.

In general, western Fox Sparrows sing with more buzzes and imitations than eastern birds. Some calls differ.

Vocalizations of most forms of Dark-eyed Junco are very similar, but "Red-backed" birds in Arizona sing shorter and more complex Songs.

Dark-eyed Junco

Junco hyemalis

A familiar "snowbird" that readily visits feeders in winter. Breeds in various wooded and brushy habitats. Note white outer tail feathers.

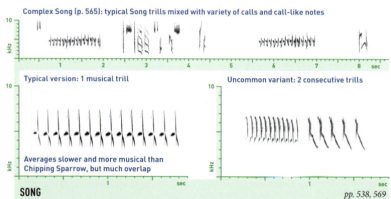

Complex Song (p. 565): typical Song trills mixed with variety of calls and call-like notes

Typical version: 1 musical trill

Averages slower and more musical than Chipping Sparrow, but much overlap

Uncommon variant: 2 consecutive trills

SONG   pp. 538, 569

Mostly Feb.–July, by males. Stereotyped but highly variable. Individuals have 1–7 songtypes, repeated many times before switching. Complex Song softer, by both sexes in close courtship during early breeding season. Subsong frequent Jan.–Mar.; like Complex Song but less stereotyped.

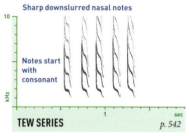

Sharp downslurred nasal notes

Notes start with consonant

TEW SERIES   p. 542

All year, in aggressive encounters. Often given with other calls, especially Twitter.

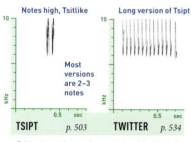

Notes high, Tsitlike

Long version of Tsipt

Most versions are 2–3 notes

TSIPT   p. 503   **TWITTER**   p. 534

Tsipt given all year, in contact and in flight, including by night migrants. Twitter given in aggression, and by female prior to copulation.

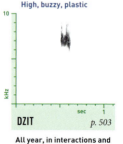

High, buzzy, plastic

DZIT   p. 503

All year, in interactions and short flights; not known from night migrants.

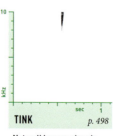

TINK   p. 498

Not well known; given in response to playback and possibly in high alarm.

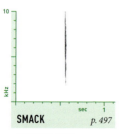

SMACK   p. 497

All year, in alarm. Distinctively sharp, almost ticking.

GRASSHOPPER SPARROW

Ammodramus savannarum

Locally common but inconspicuous in tall grass with patches of bare ground. All vocalizations can be inaudible to those with high-pitched hearing loss.

Complex Song (p. 565): a very high long rapid warble, sometimes with a few short trills

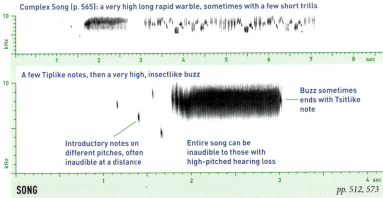

A few Tiplike notes, then a very high, insectlike buzz

Buzz sometimes ends with Tsitlike note

Introductory notes on different pitches, often inaudible at a distance

Entire song can be inaudible to those with high-pitched hearing loss

SONG
pp. 512, 573

Mostly Apr.–July, by males. Quite uniform. Each male has 2 songtypes, 1 simple and 1 Complex; simple song mostly territorial. Complex Song may strengthen the pair bond; given regularly, especially late in season, often right after typical Song or alternating with it, but rare from unpaired males.

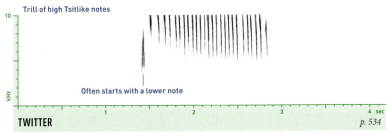

Trill of high Tsitlike notes

Often starts with a lower note

TWITTER
p. 534

Mostly May–July, by both sexes, to maintain pair bond and possibly to signal approach to nest. Sometimes considered a type of Song. Twitter of female often associated with Complex Song of male. Male and female versions may differ slightly; more study needed.

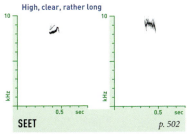

High, clear, rather long

SEET
p. 502

All year, including by night migrants. Plastic. Often sounds slighty upslurred, sometimes monotone.

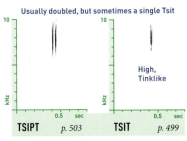

Usually doubled, but sometimes a single Tsit

High, Tinklike

TSIPT
p. 503

TSIT
p. 499

All year, in mild alarm or family contact. Usually delivered in fairly distinctive pattern of rapid doublets or triplets.

Savannah Sparrow

Passerculus sandwichensis

Breeds in grassy fields, wet meadows, and other well-watered open habitats. Some subspecies look distinctive, but vocalizations apparently similar.

Males have 1 songtype (very rarely 2), repeated with little variation

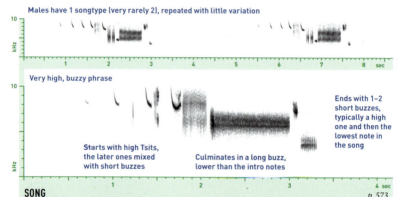

Very high, buzzy phrase

Starts with high Tsits, the later ones mixed with short buzzes

Culminates in a long buzz, lower than the intro notes

Ends with 1–2 short buzzes, typically a high one and then the lowest note in the song

SONG

p. 573

Mostly Apr.–July. Fairly variable rangewide, but usually adhering closely to the pattern shown. Long buzz near end is the loudest part of Song and carries farthest; at a distance, beware confusion with Le Conte's Sparrow Song.

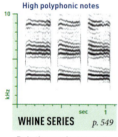

High polyphonic notes

WHINE SERIES *p. 549*

By both sexes in same-sex aggression, or in response to predators.

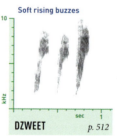

Soft rising buzzes

DZWEET *p. 512*

Reportedly by males after losing a fight, or by females rejecting a male.

High, unmusical, chipping

TWITTER *p. 534*

Function unknown; more study needed.

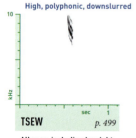

High, polyphonic, downslurred

TSEW *p. 499*

All year, including by night migrants. Most versions clear, but some are buzzy.

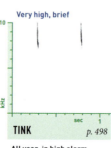

Very high, brief

TINK *p. 498*

All year, in high alarm, reportedly more often by male. Grades into Chip.

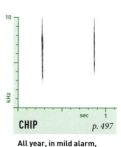

CHIP *p. 497*

All year, in mild alarm, reportedly more often by female. Grades into Tink.

Baird's Sparrow

Ammodramus bairdii

Uncommon and local in native prairies with dense grass and few or no shrubs. Secretive and difficult to detect when not singing.

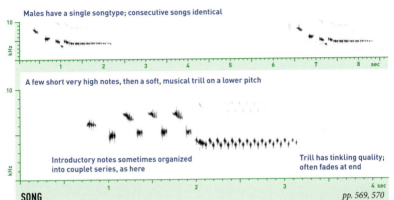

Males have a single songtype; consecutive songs identical

A few short very high notes, then a soft, musical trill on a lower pitch

Introductory notes sometimes organized into couplet series, as here

Trill has tinkling quality; often fades at end

SONG　　　　　　　　　　　　　　　　　　*pp. 569, 570*

Mostly May–Aug., by males. Females not known to sing. Slightly variable but stereotyped; much more musical than any other sparrow Song in its range. Some versions end in single lower note after trill. Rare flight song reported: a few Tsiplike notes as bird ascends, then a typical Song as it descends.

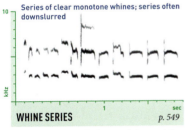

Series of clear monotone whines; series often downslurred

WHINE SERIES　　　　*p. 549*

May–Aug., likely by both sexes, in intense interactions and territorial disputes. Rather plastic; grades into Tsweet Series.

Rapid series of upslurred notes

Notes plastic, sometimes partly buzzy

TSWEET SERIES　　　*p. 533*

May–Aug., possibly by both sexes, in intense interactions and territorial disputes. Agitated birds occasionally give single Tsweets.

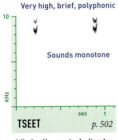

Very high, brief, polyphonic

Sounds monotone

TSEET　　*p. 502*

Likely all year, including by night migrants. Plastic; rarely longer than shown.

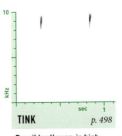

TINK　　*p. 498*

Possibly all year, in high alarm. Grades into Chip.

CHIP　　*p. 497*

Possibly all year, in mild alarm. Grades into Tink.

SEASIDE SPARROW

Ammodramus maritimus

As its name indicates, found exclusively in coastal saltmarshes. Populations are local and scattered; some are extinct or endangered (see p. 453).

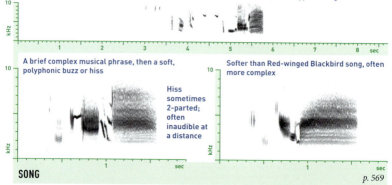

Complex Song (p. 567): varied Tink- or Chuplike notes and short buzzes, then a typical Song

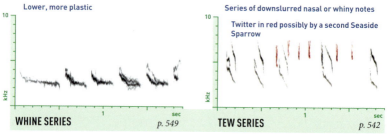

A brief complex musical phrase, then a soft, polyphonic buzz or hiss

Hiss sometimes 2-parted; often inaudible at a distance

Softer than Red-winged Blackbird song, often more complex

SONG

p. 569

All year, but mostly Apr.–July. Males have 2–4 songtypes, each usually repeated many times before switching. Females reportedly give Songlike sounds in close courtship. Complex Song given infrequently, mostly in flight display; may serve a courtship function.

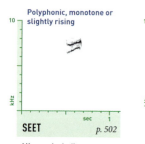

Lower, more plastic

WHINE SERIES   p. 549

Series of downslurred nasal or whiny notes

Twitter in red possibly by a second Seaside Sparrow

TEW SERIES   p. 542

Several different series calls and twitter calls have been recorded and/or described, apparently serving different functions. Most common sounds shown: sharp Tew Series given during interactions; lower, longer Whine Series possibly in alarm. More study needed.

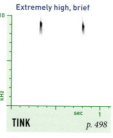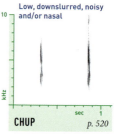

Polyphonic, monotone or slightly rising

SEET   p. 502

All year, including presumably by night migrants. Plastic.

Extremely high, brief

TINK   p. 498

All year, in alarm and aggression. Most common call. Plastic.

Low, downslurred, noisy and/or nasal

CHUP   p. 520

All year, in alarm and aggression, often with Tinks. Plastic. Rather distinctive.

LE CONTE'S SPARROW

Ammodramus leconteii

Rather secretive and subtly plumaged, but strikingly beautiful when seen up close. Frequents wet meadows, sedge fields, and marsh edges.

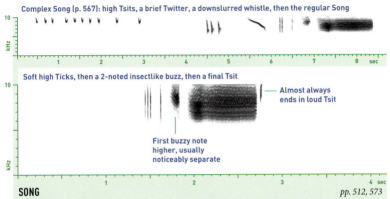

Complex Song (p. 567): high Tsits, a brief Twitter, a downslurred whistle, then the regular Song

Soft high Ticks, then a 2-noted insectlike buzz, then a final Tsit

Almost always ends in loud Tsit

First buzzy note higher, usually noticeably separate

SONG   *pp. 512, 573*

Mostly May–July, reportedly only by males, sometimes at night. Fairly soft and easily overlooked, but highly uniform. Each male apparently has a single simple Song and a single Complex Song, given infrequently by excited birds, often during flight display.

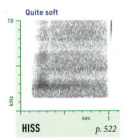

Quite soft

HISS   *p. 522*

Poorly known; sole example is from agitated male responding to playback.

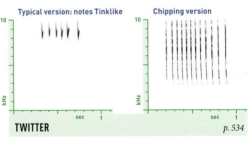

Typical version: notes Tinklike

Chipping version

TWITTER   *p. 534*

Perhaps all year, in high alarm; plastic. Grades into single Tinks. Sole example of chipping version (right) is from agitated male responding to playback. Also gives a Whine Series (p. 549).

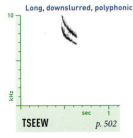

Long, downslurred, polyphonic

TSEEW   *p. 502*

All year, including by night migrants. Distinctive, but beware variant Henslow's.

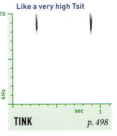

Like a very high Tsit

TINK   *p. 498*

All year, in alarm; most common call, at least on breeding grounds.

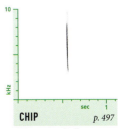

CHIP   *p. 497*

All year; may indicate greater alarm than Tsit.

Henslow's Sparrow

Ammodramus henslowii

Uncommon, highly local, and secretive. Requires tall grasslands with standing dead vegetation. Breeding distribution tends to change from year to year.

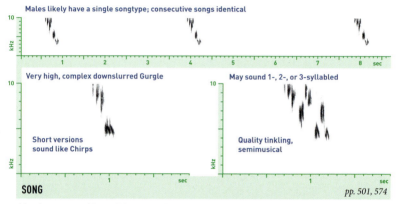

Males likely have a single songtype; consecutive songs identical

Very high, complex downslurred Gurgle

Short versions sound like Chirps

May sound 1-, 2-, or 3-syllabled

Quality tinkling, semimusical

SONG *pp. 501, 574*

May–Aug., presumably mostly by males; often given at night. Slightly variable but highly stereotyped. Easily overlooked; usually given from a low perch, often one obscured by grass. Some authors have reported a Complex Song given in flight, but its existence remains unconfirmed.

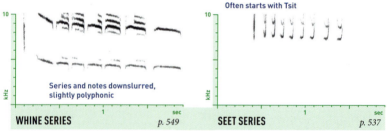

Often starts with Tsit

Series and notes downslurred, slightly polyphonic

WHINE SERIES *p. 549*

SEET SERIES *p. 537*

Both types of series probably given mostly Apr.–Aug., by both sexes, during pair interactions and territorial disputes. Rather plastic. Often introduced by Tsit or a soft rapid Chatter. Differences in function not known.

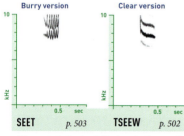
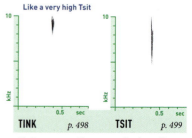

Burry version

Clear version

Like a very high Tsit

SEET *p. 503*

TSEEW *p. 502*

TINK *p. 498*

TSIT *p. 499*

All year, including by night migrants. Apparently quite plastic; at least some individuals give both burry and clear versions, as well as intergrades.

Most common calls, apparently given all year, usually in alarm. Somewhat plastic; many versions are intermediate between Tink and Tsit.

Saltmarsh Sparrow

Ammodramus caudacutus

Very similar to Nelson's Sparrow, and formerly lumped with it under the name Sharp-tailed Sparrow. Breeds and winters in Atlantic coastal salt marshes.

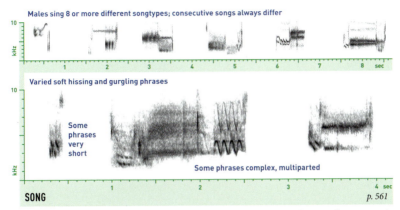

Males sing 8 or more different songtypes; consecutive songs always differ

Varied soft hissing and gurgling phrases

Some phrases very short

Some phrases complex, multiparted

SONG

p. 561

Mostly May–July, by males. Carries only a short distance. By far the most complex song of any sparrow in the genus *Ammodramus*; individual songtypes can resemble songs of Nelson's or Seaside Sparrows, but the constant switching of songtypes eliminates those species.

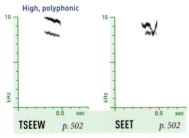

High, polyphonic

TSEEW *p. 502*   **SEET** *p. 502*

All year, including by night migrants. Plastic, but often sounds slightly downslurred

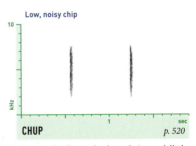

Low, noisy chip

CHUP *p. 520*

Apparently all year, in alarm, but especially by agitated birds near a nest. May also give a Tink in high alarm.

DISTINCTIVE SUBSPECIES OF SEASIDE SPARROW

The Seaside Sparrow breeds along the Atlantic and Gulf Coasts in local, often isolated populations, many quite distinct in plumage. Sounds of all populations are similar, but local dialects are distinctive. Two subspecies from Florida deserve special mention:

Dusky Seaside Sparrow (EXTINCT)
Ammodramus maritimus nigrescens

The darkest subspecies, formerly found on the central Atlantic coast of Florida; driven to extinction in the 1980s by habitat destruction.

Cape Sable Seaside Sparrow (ENDANGERED)
Ammodramus maritimus mirabilis

The palest subspecies, found only in southwest Florida, on the prairies of the Everglades and the Big Cypress Swamp.

NELSON'S SPARROW

Ammodramus nelsoni

Rather shy and local. Breeds in grassy freshwater marshes of the northern Great Plains, in sedges ringing Hudson Bay, and in North Atlantic salt marshes.

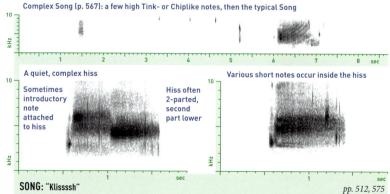

Complex Song (p. 567): a few high Tink- or Chiplike notes, then the typical Song

A quiet, complex hiss

Sometimes introductory note attached to hiss

Hiss often 2-parted, second part lower

Various short notes occur inside the hiss

SONG: "Klissssh"

pp. 512, 575

Mostly May–July, by males, including at night. Females rarely sing. Distinctive, but does not carry far; only slightly variable between populations. Males likely sing a single regular and a single Complex Song. Complex Song usually given in display flight.

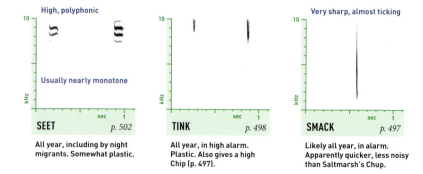

High, polyphonic

Usually nearly monotone

SEET
p. 502

All year, including by night migrants. Somewhat plastic.

TINK
p. 498

All year, in high alarm. Plastic. Also gives a high Chip (p. 497).

Very sharp, almost ticking

SMACK
p. 497

Likely all year, in alarm. Apparently quicker, less noisy than Saltmarsh's Chup.

CARDINALS, GROSBEAKS, BUNTINGS, AND THEIR ALLIES (Family Cardinalidae) *following pages*

This diverse family is closely related to the sparrows (family Emberizidae) and the finches (family Fringillidae), and taxonomists have shuffled some species among these families more than once. Recently, the North American tanagers were added to the Cardinalidae. Most species in the family share the traits of brightly colored plumage, at least in breeding males; thick, often conical bills, used in many species for cracking hard seeds; and complex, musical Songs, which are apparently learned. The buntings, Blue Grosbeak, and Dickcissel have very similar call repertoires, which resemble those of many warblers and sparrows.

DICKCISSEL

Spiza americana

A seminomadic bunting of weedy prairies and agricultural areas. In any area, may be common one year and absent the next. Forms large flocks in winter.

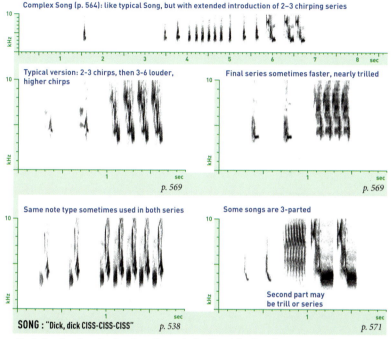

Complex Song (p. 564): like typical Song, but with extended introduction of 2–3 chirping series

Typical version: 2-3 chirps, then 3-6 louder, higher chirps
p. 569

Final series sometimes faster, nearly trilled
p. 569

Same note type sometimes used in both series

Some songs are 3-parted
Second part may be trill or series

SONG : "Dick, dick CISS-CISS-CISS" *p. 538*
p. 571

Mostly Apr.–Aug., by males. Females not known to sing. Repertoire size unknown; most males repeat a single songtype for long periods, but at least some males can alternate 2 songtypes. Number of notes, and number of types of notes, may increase with excitement level, grading smoothly into Complex Song, sometimes given in flight. Not very musical.

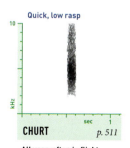

Quick, low rasp

CHURT *p. 511*

All year, often in flight, including by night migrants. Distinctive.

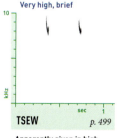

Very high, brief

TSEW *p. 499*

Apparently given in high alarm, possibly near nest.

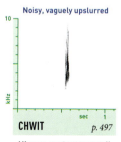

Noisy, vaguely upslurred

CHWIT *p. 497*

All year; most common call. May sound vaguely 2-syllabled.

SCARLET TANAGER

Piranga olivacea

Breeds in mature deciduous and mixed forests. Male is scarlet only in spring and summer; in fall, turns greenish like female, but retains black wings.

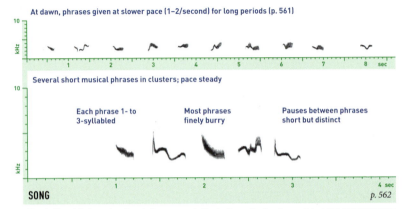

At dawn, phrases given at slower pace (1–2/second) for long periods (p. 561)

Several short musical phrases in clusters; pace steady

Each phrase 1- to 3-syllabled

Most phrases finely burry

Pauses between phrases short but distinct

SONG *p. 562*

Mostly May–July, by males. Female Songs reportedly shorter and softer than male's; pairs sometimes sing together during nest construction. Males have 10 individual phrases on average, which can appear in any order within a given series. Very like American Robin Song, but note burry quality.

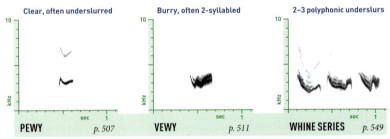

Clear, often underslurred

Burry, often 2-syllabled

2–3 polyphonic underslurs

PEWY *p. 507*   **VEWY** *p. 511*   **WHINE SERIES** *p. 549*

A confusing set of intergrading calls; most common variants shown. Pewy and Vewy highly plastic, both given softly in family contact. Pewy, upslurred Pwee, and likely other variants also given by night migrants. Vewy also given by begging juveniles. Whine series given by females in close courtship.

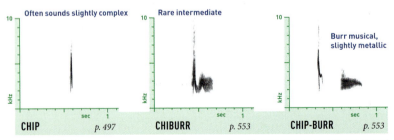

Often sounds slightly complex

Rare intermediate

Burr musical, slightly metallic

CHIP *p. 497*   **CHIBURR** *p. 553*   **CHIP-BURR** *p. 553*

Chip and Chip-Burr are most common calls; given all year by both sexes in alarm. Slightly variable and plastic. Chip without Burr may signal higher agitation. Chiburr poorly known; given at least by courting females and possibly in other situations.

SUMMER TANAGER

Piranga rubra

Breeds in southeastern pine-oak woods and deciduous forests. Adult male is red all year. First-year males and some females are mottled red and green.

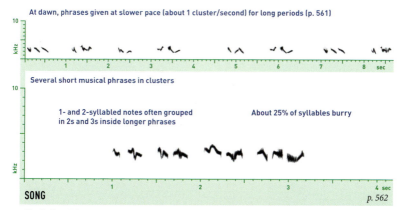

At dawn, phrases given at slower pace (about 1 cluster/second) for long periods (p. 561)

Several short musical phrases in clusters

1- and 2-syllabled notes often grouped in 2s and 3s inside longer phrases

About 25% of syllables burry

SONG

p. 562

Mostly May–July, by males. Female sings infrequently; Song reportedly shorter and softer than male's, with less pause between notes. Easy to confuse with American Robin Song, but averages slightly burrier, with more herky-jerky rhythm created by groupings of notes inside phrases.

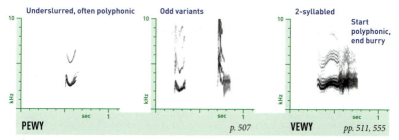

Underslurred, often polyphonic

Odd variants

2-syllabled

Start polyphonic, end burry

PEWY

VEWY

p. 507

pp. 511, 555

A confusing set of plastic, intergrading calls. Pewy and Vewy are most common variants, given softly in family contact. Pewy, clear upslurred Pwee, and likely other variants given by night migrants. Vewy also given by begging juveniles. Courting female likely gives Whine Series like Scarlet Tanager.

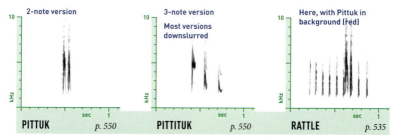

2-note version

3-note version

Most versions downslurred

Here, with Pittuk in background (red)

PITTUK

p. 550

PITTITUK

p. 550

RATTLE

p. 535

All year; most common call. Usually 3–4 notes, but highly plastic; number of notes tends to increase with agitation level, grading into Rattle. First note usually highest, but tendency decreases in high alarm near nest or young. Single-note Pik calls very rare.

NORTHERN CARDINAL

Cardinalis cardinalis

A familiar, beloved resident of shrubby woods, deserts, and backyards. Often visits feeders. Range has expanded northward over the past century.

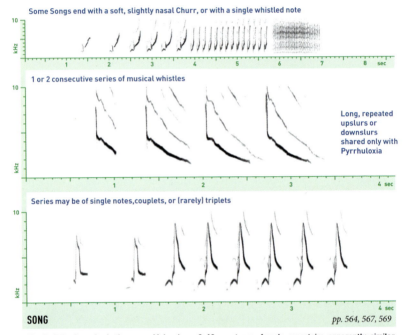

Some Songs end with a soft, slightly nasal Churr, or with a single whistled note

1 or 2 consecutive series of musical whistles

Long, repeated upslurs or downslurs shared only with Pyrrhuloxia

Series may be of single notes, couplets, or (rarely) triplets

SONG *pp. 564, 567, 569*

Mostly Feb.–Aug., by both sexes. Males have 8–10 songtypes; female repertoire apparently similar. Females sometimes sing from the nest. Each Song usually repeated several times before switching, but Songs are often truncated. Highly variable; some Songs have 3 or more consecutive series, especially when birds are excited. Each series usually faster than the last, with later series often trills. First series sometimes accelerating.

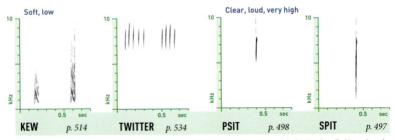

Soft, low

Clear, loud, very high

KEW *p. 514*   **TWITTER** *p. 534*   **PSIT** *p. 498*   **SPIT** *p. 497*

Kew given softly in family contact, and between mates during encounters with rival males. Twitter given in response to hawks or other potential predators; also sometimes in response to playback. Psit is most common call; grades into lower Spit, which is rarer, and possibly more aggressive.

PYRRHULOXIA

Cardinalis sinuatus

Found in scrubby southwestern deserts and arid backyards, often side by side with the related Northern Cardinal; the two species rarely interact.

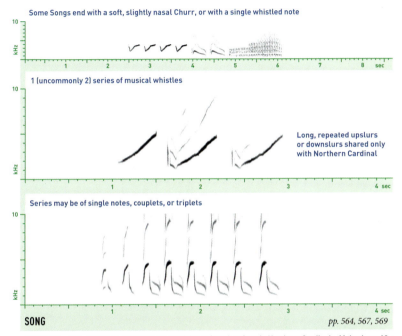

Some Songs end with a soft, slightly nasal Churr, or with a single whistled note

1 (uncommonly 2) series of musical whistles

Long, repeated upslurs or downslurs shared only with Northern Cardinal

Series may be of single notes, couplets, or triplets

SONG   *pp. 564, 567, 569*

Mostly Feb.–Aug., by males; females sing much less often than female Northern Cardinals. Males have 10–14 songtypes. Each Song usually repeated several times before switching, but Songs are often truncated. Songs generally simpler and slower than Northern Cardinal: usually one series, rarely faster than 5 notes/second, often just 2–3 notes total; but Northern Cardinal often difficult or impossible to rule out by Song alone.

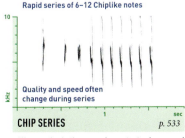

Rapid series of 6–12 Chiplike notes

Quality and speed often change during series

CHIP SERIES   *p. 533*

All year, by both sexes, in contact, alarm, and aggression. Quite plastic, but distinctive; Northern Cardinal rarely sounds similar.

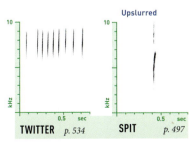

TWITTER *p. 534*   **SPIT** *p. 497*

Upslurred

Twitter like Northern Cardinal's. Spit plastic but generally lower, less sharp, more obviously whistled than cardinal's Psit.

Rose-breasted Grosbeak

Pheucticus ludovicianus

Breeds in deciduous and mixed forests, woodland edges, and residential areas with mature trees. Regularly visits feeders. Female has pale pinkish bill.

Usually 10–20 phrases per Song, significantly more than American Robin or tanagers

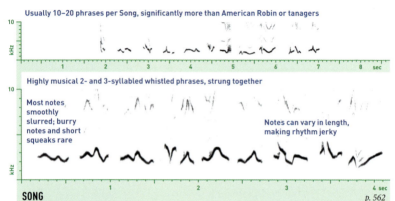

Highly musical 2- and 3-syllabled whistled phrases, strung together

Most notes smoothly slurred; burry notes and short squeaks rare

Notes can vary in length, making rhythm jerky

SONG p. 562

Mostly Apr.–July, by both sexes, sometimes from the nest. Individual birds use about 15–25 different phrases in their Songs. In excitement, Song becomes longer, sometimes including softer, varied portions with cardinal-like whistles. Reported to give Complex Song in flight display; no recordings available.

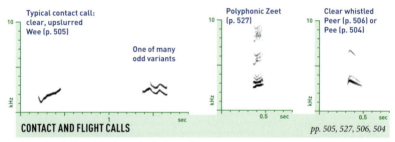

Typical contact call: clear, upslurred Wee (p. 505)

One of many odd variants

Polyphonic Zeet (p. 527)

Clear whistled Peer (p. 506) or Pee (p. 504)

CONTACT AND FLIGHT CALLS pp. 505, 527, 506, 504

A hugely variable and plastic set of calls given softly during the day, apparently by both sexes, in close pair contact. One common night flight call is a monotone whistled Pee like the version shown at right, like Hermit Thrush Teer but shorter; some nocturnal versions apparently a burry Veet like that of Wood Thrush.

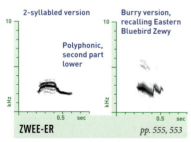

2-syllabled version

Burry version, recalling Eastern Bluebird Zewy

Polyphonic, second part lower

ZWEE-ER pp. 555, 553

Mostly June–Sept., by begging juveniles. Likely also sometimes by adult females in courtship. Quite variable.

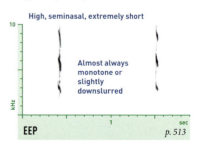

High, seminasal, extremely short

Almost always monotone or slightly downslurred

EEP p. 513

All year; most common call. Fairly uniform but slightly plastic.

Black-headed Grosbeak

Pheucticus melanocephalus

Breeds in riparian deciduous woods or drier coniferous forests. Sometimes hybridizes with Rose-breasted Grosbeak. Female has grayish, bicolored bill.

Complex Song: phrases strung together almost without pause; some trills or repeated phrases

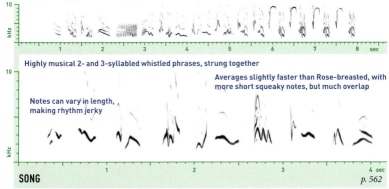

Highly musical 2- and 3-syllabled whistled phrases, strung together

Averages slightly faster than Rose-breasted, with more short squeaky notes, but much overlap

Notes can vary in length, making rhythm jerky

SONG
p. 562

Mostly Apr.–July, by both sexes, sometimes from the nest. Males have about 25 different phrases, females about 13. In excitement, Song becomes longer, sometimes including softer, varied portions with cardinal-like whistles. Complex Song infrequent, in flight display with exaggerated wingbeats.

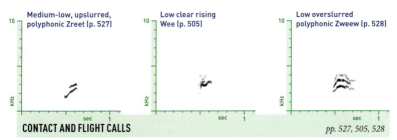

Medium-low, upslurred, polyphonic Zreet (p. 527)

Low clear rising Wee (p. 505)

Low overslurred polyphonic Zweew (p. 528)

CONTACT AND FLIGHT CALLS
pp. 527, 505, 528

A poorly known, hugely variable and plastic category of calls. Versions like those shown given softly during the day, apparently by both sexes in close pair contact. Nocturnal flight calls little known, but likely similar to Rose-breasted Grosbeak's; perhaps more reliably upslurred.

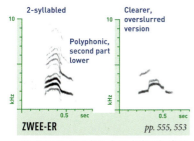

2-syllabled

Clearer, overslurred version

Polyphonic, second part lower

ZWEE-ER
pp. 555, 553

Similar calls given June–Sept., by begging juveniles and sometimes by adult females, likely in courtship. Quite variable.

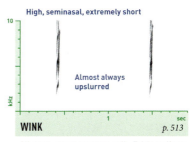

High, seminasal, extremely short

Almost always upslurred

WINK
p. 513

All year; most common call. Fairly uniform, slightly plastic. Difference from Rose-breasted Eep is subtle but consistent.

Indigo Bunting

Passerina cyanea

Common in overgrown fields and wood-land edges. Nonbreeding males be-come mostly brown. Female like female Lazuli, but slightly streaked below.

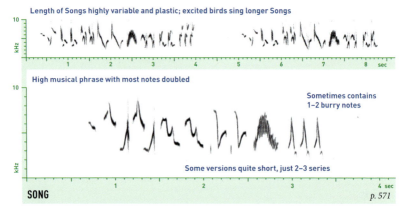

Length of Songs highly variable and plastic; excited birds sing longer Songs

High musical phrase with most notes doubled

Sometimes contains 1–2 burry notes

Some versions quite short, just 2–3 series

SONG *p. 571*

Apr.–Aug., by males only. Most adult males have a single songtype, which they repeat with minor varia-tions, usually truncations; a few have 2 songtypes. Doubling of most phrases is fairly distinctive. Quality more musical than most warbler Songs; like American Goldfinch, but slower with shorter series.

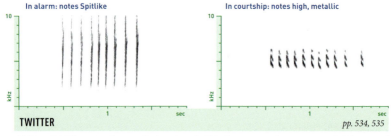

In alarm: notes Spitlike

In courtship: notes high, metallic

TWITTER *pp. 534, 535*

Twitter of Spitlike notes given mostly June–Aug., in high alarm near nest or young, usually interspersed among many single Spit calls. Twitter of polyphonic Tsewlike notes given Apr.–Aug., by females in close courtship, and by highly agitated males in territorial encounters, usually with fluttering wings.

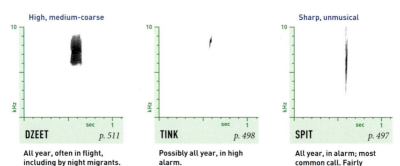

High, medium-coarse

Sharp, unmusical

DZEET *p. 511*

TINK *p. 498*

SPIT *p. 497*

All year, often in flight, including by night migrants. Rarely a Dzweet (p. 512).

Possibly all year, in high alarm.

All year, in alarm; most common call. Fairly uniform.

Lazuli Bunting

Passerina amoena

Breeds in shrubby areas, riparian woods, and chapparal. Nonbreeding males much like breeding males, with some patchy brown above.

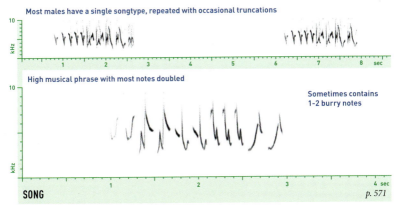

Most males have a single songtype, repeated with occasional truncations

High musical phrase with most notes doubled

Sometimes contains 1-2 burry notes

SONG *p. 571*

Apr.–Aug., by males only. Very similar to Indigo Bunting's song, but averages higher and faster, and thus more likely to be mistaken for a warbler; somewhat more likely than Indigo Bunting to include longer series of 3–5 notes. Some versions quite short, just 2–3 series.

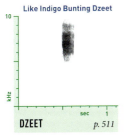

Like Indigo Bunting Dzeet

DZEET *p. 511*

All year, often in flight. May average barely higher than Indigo's.

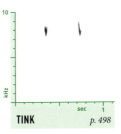

TINK *p. 498*

Possibly all year, in high alarm.

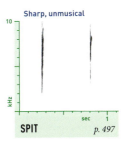

Sharp, unmusical

SPIT *p. 497*

All year, in alarm; most common call. Much like Indigo Bunting's Spit.

SONGS OF INDIGO BUNTING, LAZULI BUNTING, AND HYBRIDS

The Songs of Indigo and Lazuli Bunting are very similar, but in areas where only one species occurs, the Songs contain only species-specific note types, and each species usually ignores playback of the other's Song. However, in areas where both species occur, they respond aggressively to each other's Songs, and some individuals sing with note types from both species. Hybrids, which occur frequently in the zone of overlap, may also sing with both species' note types; more study needed. For these reasons, it is frequently difficult or impossible to identify Indigo and Lazuli Buntings by sound alone, at least in areas where both occur.

VARIED BUNTING

Passerina versicolor

Breeds in dense arid scrub, in washes or adjacent to riparian areas. Females resemble other female buntings, but lack wingbars; bill slightly decurved.

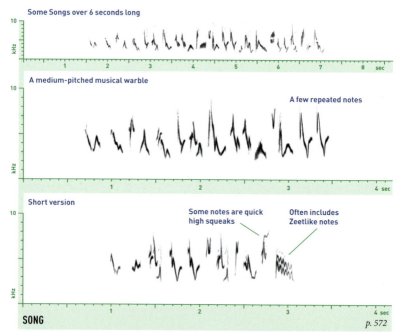

Some Songs over 6 seconds long

A medium-pitched musical warble

A few repeated notes

Short version

Some notes are quick high squeaks

Often includes Zeetlike notes

SONG *p. 572*

Mostly Apr.–Aug., by males only. Males appear to have a single basic songtype, which they deliver with minor variations, usually truncations, at any time of day. Much like Blue Grosbeak Song, but slightly higher-pitched; combination of high squeaks and lower burry notes may recall western Warbling Vireo Song. Long, soft versions of Song reportedly given during courtship.

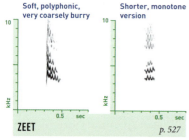

Soft, polyphonic, very coarsely burry

Shorter, monotone version

ZEET *p. 527*

All year, by both sexes, in contact and while foraging, often mixed with Pinks or Song. Highly plastic, and not very loud.

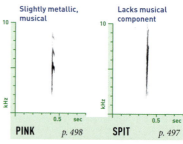

Slightly metallic, musical

Lacks musical component

PINK *p. 498*　　**SPIT** *p. 497*

All year, by both sexes, in contact and alarm. Varies from Pink to Spit; slightly plastic. Higher Tinklike note reported in high alarm.

Painted Bunting

Passerina ciris

Unmistakable. Found in semi-open brushy fields, coastal scrub, and sometimes residential areas. First-year males resemble females.

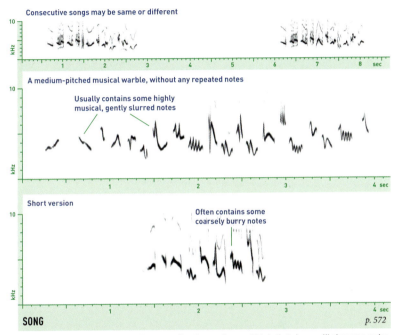

Consecutive songs may be same or different

A medium-pitched musical warble, without any repeated notes

Usually contains some highly musical, gently slurred notes

Short version

Often contains some coarsely burry notes

SONG   *p. 572*

Mar.–Aug., by males; females not known to sing. Green-plumaged birds that sing are likely young males. Individuals estimated to have at least 3–5 songtypes, each of which may be delivered with minor variations, especially in the middle of the song. Length of Song quite variable; young birds may sing shorter songs on average. Similar to Varied Bunting Song, but most notes more musical, more obviously whistled; some versions recall House Finch Song.

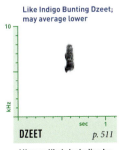

Like Indigo Bunting Dzeet; may average lower

DZEET   *p. 511*

All year, likely including by night migrants, but apparently infrequent.

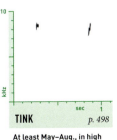

TINK   *p. 498*

At least May–Aug., in high agitation. Plastic.

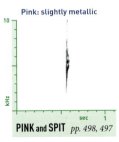

Pink: slightly metallic

PINK and SPIT   *pp. 498, 497*

All year, in contact and alarm. Varies from Pink to Spit, like in Varied Bunting.

Blue Grosbeak

Passerina caerulea

Found in overgrown fields and scrub with scattered trees. Larger than Indigo Bunting, with chestnut wingbars and a much thicker, bicolored bill.

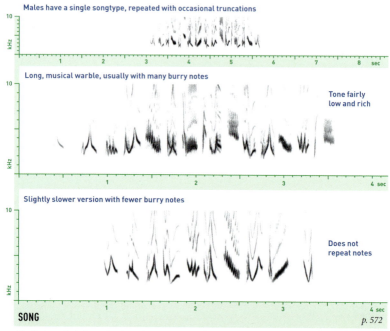

Males have a single songtype, repeated with occasional truncations

Long, musical warble, usually with many burry notes

Tone fairly low and rich

Slightly slower version with fewer burry notes

Does not repeat notes

SONG

p. 572

Mostly Apr.–Sept., by males. Females not known to sing; brown-plumaged singers may be first-spring males. Older males average longer songs than younger males. Much like Songs of Varied Bunting and western Warbling Vireo, but lower-pitched; lacks squeaky notes. Lower, slower, and burrier than Song of Purple Finch; notes often slow enough to count.

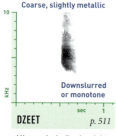

Coarse, slightly metallic

Downslurred or monotone

DZEET *p. 511*

All year, including by night migrants. Distinctly lower than Indigo Bunting Dzeet.

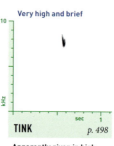

Very high and brief

TINK *p. 498*

Apparently given in high alarm and in response to playback.

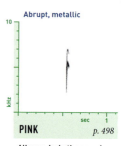

Abrupt, metallic

PINK *p. 498*

All year, by both sexes, in alarm; most common call.

BLACKBIRDS, MEADOWLARKS, GRACKLES, COWBIRDS, AND ORIOLES (Family Icteridae)

These birds, often called the "icterids," comprise a highly diverse New World family of several distinct lineages. The blackbirds and grackles are dark-plumaged, highly social birds of wetlands and woodland edges. The orioles are colorful arboreal birds that hang their intricately woven nests from tree branches. The meadowlarks and the Bobolink are ground-nesting birds of fields and prairies. And the cowbirds are brood parasites, laying their eggs in the nests of other species.

The icterids have astonishingly varied vocal repertoires. Songs are learned and highly complex, usually whistled or polyphonic. A few species of oriole incorporate imitations of other birds into their songs. In some icterids, different categories of song serve different functions: for example, the Complex Songs of meadowlarks and the Whistled Songs of cowbirds are given mostly in excitement or in flight. In a few icterids, including Rusty and Yellow-headed Blackbirds, certain categories of song can form part of the alarm response to aerial predators.

Songlike Calls

In this family, the boundary between songs and calls is particularly blurry. Many calls are learned, and individuals often have repertoires of multiple call types. For example, the orioles have repertoires of short whistled or burry calls that they give for long periods, often switching call types, much as singing birds might switch songtypes. The function of these calls is not entirely clear.

Other icterid vocalizations that share characteristics of songs and calls include the whistled and burry calls of Bobolink, the alert calls of Red-winged Blackbird, the Pink calls of cowbirds, and the series calls of Great-tailed and Boat-tailed Grackles.

Female Songs

In most oriole species, both sexes sing. In nonmigratory southern species, the sexes tend to look and sound fairly similar, and females sing frequently, in some cases more frequently than males. In migratory northern species, the sexes tend to look different, and female songs tend to be slightly less complex and less frequent.

In the meadowlarks, the cowbirds, and Red-winged Blackbird, male-female pairs often perform synchronized duets, the female's Rattle starting halfway through the male's Song. These duets can be a good way to identify mated pairs. By some criteria, the Rattles that females use in these situations can be considered a type of song.

Red-winged Blackbird

Agelaius phoeniceus

One of our most abundant native birds, especially in wetland habitats, where it forms dense breeding colonies. Vocal communication unique and complex.

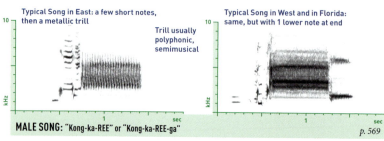

Typical Song in East: a few short notes, then a metallic trill

Trill usually polyphonic, semimusical

Typical Song in West and in Florida: same, but with 1 lower note at end

MALE SONG: "Kong-ka-REE" or "Kong-ka-REE-ga"

p. 569

All year, especially Apr.–July. Territorial males spread wings slightly during each Song to display red shoulders. Males have 3–7 songtypes, each usually repeated many times before switching, but males in close courtship may cycle through their Song repertoire, interspersing Twitter calls.

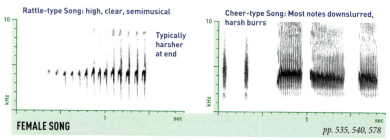

Rattle-type Song: high, clear, semimusical

Typically harsher at end

Cheer-type Song: Most notes downslurred, harsh burrs

FEMALE SONG

pp. 535, 540, 578

Mostly Apr.–July. Rattle-type Songs, often started halfway through mate's Song, tend to reinforce pair bond. Cheer-type Songs more aggressive, territorial. Many intermediate Songs begin with Rattles and end with Cheers. Chuklike notes often incorporated.

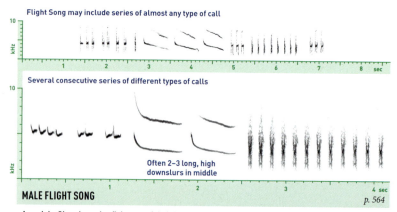

Flight Song may include series of almost any type of call

Several consecutive series of different types of calls

Often 2–3 long, high downslurs in middle

MALE FLIGHT SONG

p. 564

Apr.–July. Given by males flying out of their breeding territories, and occasionally upon return, apparently to notify mates of departure and arrival. Extremely geographically variable; level of variation within individual males unknown.

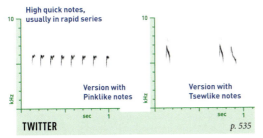

High quick notes, usually in rapid series

Version with Pinklike notes

Version with Tsewlike notes

TWITTER *p. 535*

Variable; by both sexes before copulation, by males between Songs at dawn and when other blackbirds fly over. Female version usually lower and clearer.

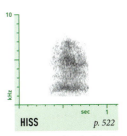

HISS *p. 522*

By males confronting intruders or showing potential nest sites to females.

ALERT CALLS OF RED-WINGED BLACKBIRDS

Individual male Red-winged Blackbirds have repertoires of 12–20 different Alert Calls, including some from each category below. Birds in a breeding colony tend to share call types, but regional variation is tremendous. Neighboring males call almost continuously, all giving the same call type. An individual switches call types upon noticing potential danger, and the others change calls to match. Thus, it is the change in call rather than the type of call that propagates the danger signal. Highly agitated males may switch frequently between 2 or 3 different call types. Females give only 2–3 such calls.

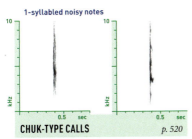

1-syllabled noisy notes

CHUK-TYPE CALLS *p. 520*

Brief and usually low, though pitch may vary. Different versions serve as pair contact calls, in mobbing predators, etc.

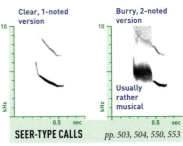

Clear, 1-noted version

Burry, 2-noted version

Usually rather musical

SEER-TYPE CALLS *pp. 503, 504, 550, 553*

Males have 3-7 of these calls, usually given in high alarm. May be 1- or 2-syllabled, clear or burry. Usually downslurred.

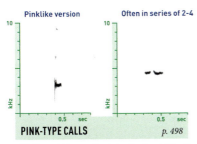

Pinklike version

Often in series of 2-4

PINK-TYPE CALLS *p. 498*

Highly variable, grading into all other types of Alert Call. Some versions burry.

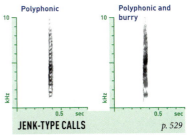

Polyphonic

Polyphonic and burry

JENK-TYPE CALLS *p. 529*

Calls in this category are abrupt and polyphonic. May be clear or burry; grade into Chuck-type calls.

EASTERN MEADOWLARK

Sturnella magna

Common in pastures and other open, grassy habitats. In flight, shows distinctive "football shape" and white outer tail feathers. Forms flocks in winter.

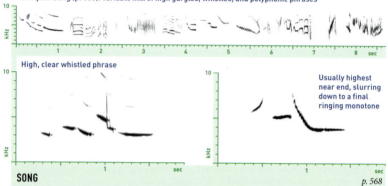

Complex Song (p. 563): variable mix of high gurgles, whistles, and polyphonic phrases

High, clear whistled phrase

Usually highest near end, slurring down to a final ringing monotone

SONG

p. 568

Mostly Mar.–Aug., by males; female not known to sing. Individuals have 50–100 songtypes, each usually repeated before switching. Higher and simpler than Western Meadowlark Song. Florida birds sometimes include burry notes. Complex Song (top) given infrequently by excited males, usually in flight.

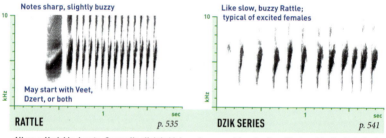

Notes sharp, slightly buzzy

May start with Veet, Dzert, or both

RATTLE

p. 535

Like slow, buzzy Rattle; typical of excited females

DZIK SERIES

p. 541

All year. Variable, innate. Generally slightly higher, slower, and buzzier than Western Meadowlark Rattle. Often given by female halfway through her mate's Song, but also from both sexes in response to threats. Sometimes shortened to just Veet or Dzert plus 2–3 rattling notes.

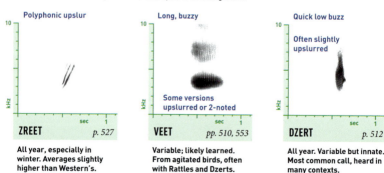

Polyphonic upslur

ZREET

p. 527

All year, especially in winter. Averages slightly higher than Western's.

Long, buzzy

Some versions upslurred or 2-noted

VEET

pp. 510, 553

Variable; likely learned. From agitated birds, often with Rattles and Dzerts.

Quick low buzz

Often slightly upslurred

DZERT

p. 512

All year. Variable but innate. Most common call, heard in many contexts.

WESTERN MEADOWLARK

Sturnella neglecta

The voice of the western prairie. Looks nearly identical to Eastern, but prefers shorter grass and drier habitats, and vocalizations differ markedly.

Complex Song (p. 563): variable, rich polyphonic warbles like Bobolink Song, often following Teer calls

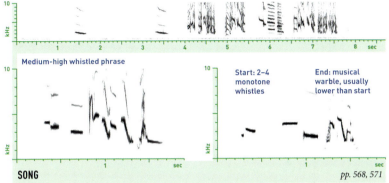

Medium-high whistled phrase

Start: 2–4 monotone whistles

End: musical warble, usually lower than start

SONG   *pp. 568, 571*

Mostly Mar.–Aug., by males; females not known to sing. Very loud; can carry for over a mile. Males average 7 songtypes, each usually repeated. Rarely a Western may learn some Eastern songtypes. Complex Song (top) given infrequently by excited males, usually in flight; faster, lower than Eastern's.

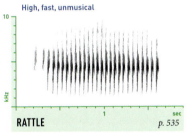

High, fast, unmusical

RATTLE   *p. 535*

All year. Variable; innate. Often given by female halfway through her mate's Song. Rare from adult males. Often introduced by Chup.

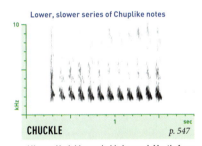

Lower, slower series of Chuplike notes

CHUCKLE   *p. 547*

All year. Variable; probably learned. Mostly from adult males. May sometimes intergrade with Rattle. Also a little-known harsh Churr (p. 524).

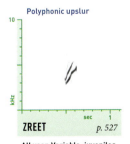

Polyphonic upslur

ZREET   *p. 527*

All year. Variable; juveniles in late summer give a high short monotone version.

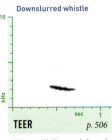

Downslurred whistle

TEER   *p. 506*

All year. Mellow and clear; in response to predators or prior to Complex Song.

Low, clucking

Often slightly buzzy

CHUP   *p. 520*

All year. Variable; innate. Longest, buzziest versions can approach Eastern Dzert.

GREAT-TAILED GRACKLE

Quiscalus mexicanus

A Mexican border bird as late as 1960, but now a widespread fixture of suburbs, parks, and wetlands. Gathers in deafening flocks, often in city parks.

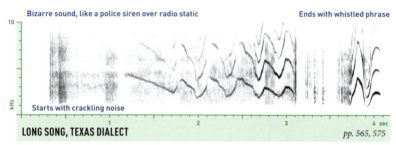

Bizarre sound, like a police siren over radio static

Ends with whistled phrase

Starts with crackling noise

LONG SONG, TEXAS DIALECT

pp. 565, 575

By males in Texas, northeast Mexico, and likely the eastern Great Plains. Sometimes alternated with Short Songs and Series. Diagnostic central phrase a sirenlike rising and falling whistle produced simultaneously with a soft hiss. Short, whistled terminal phrases variable; also given separately.

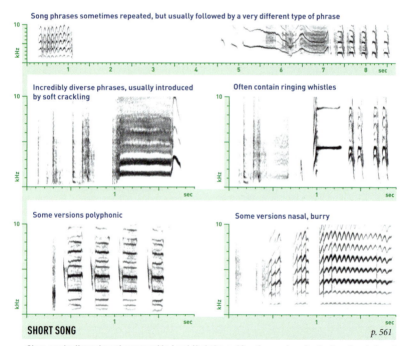

Song phrases sometimes repeated, but usually followed by a very different type of phrase

Incredibly diverse phrases, usually introduced by soft crackling

Often contain ringing whistles

Some versions polyphonic

Some versions nasal, burry

SHORT SONG

p. 561

Given nearly all year by males rangewide. Loud. Variation and function poorly understood; each male has at least 5 Short Songs that can be alternated or combined with each other or with Series. Most Short Songs begin with whispered "stick-breaking" sounds or whistling audible only at close range.

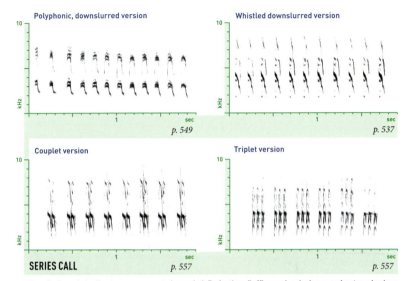

Polyphonic, downslurred version p. 549

Whistled downslurred version p. 537

Couplet version p. 557

Triplet version p. 557

SERIES CALL

A confusing catch-all category; more study needed. By feather-fluffing males during song bouts or in close courtship of females; also upon leaving a song perch and in other situations. Individuals have many series types, which vary geographically. One series type may switch to another without pause.

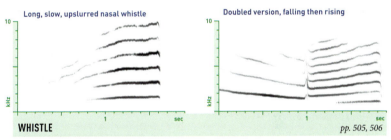

Long, slow, upslurred nasal whistle

Doubled version, falling then rising

WHISTLE pp. 505, 506

Distinctive. Given rangewide by males during song bouts, often accompanied by feather fluffing; sometimes appended to the start of a Short Song. Most versions are upslurred, but some are downslurred or combinations. No similar sound reported from Boat-tailed Grackle.

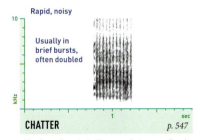

Rapid, noisy

Usually in brief bursts, often doubled

CHATTER p. 547

Versions given by both sexes, but more often by females, at the approach of a courting male and in disputes with other females.

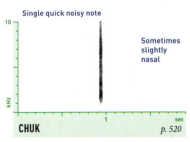

Single quick noisy note

Sometimes slightly nasal

CHUK p. 520

By both sexes, in alarm and in flight. Somewhat variable; versions of males and females reportedly differ.

Boat-tailed Grackle

Quiscalus major

Breeds in coastal marshes and Florida uplands. Nearly identical to the closely related Great-tailed Grackle, but voice is quite different.

Long Song may last a minute, with a Crackle every 10 seconds or so

2-syllabled Jeerlike notes in series, regularly punctuated by brief Crackles

LONG SONG   *p. 565*

Only minor variation throughout range. Given by breeding males, often in small groups that take turns; one bird sings, fluttering wings above back during each Crackle, while others listen with bills pointed skyward, sometimes joining in briefly after the Crackles. Crackling apparently a vocal sound.

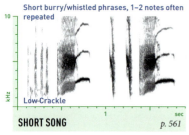

Short burry/whistled phrases, 1–2 notes often repeated

Low-Crackle

SHORT SONG   *p. 561*

Like Great-tailed Grackle Short Song, but much less frequent and less varied, with consecutive songtypes more often the same.

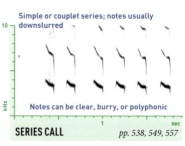

Simple or couplet series; notes usually downslurred

Notes can be clear, burry, or polyphonic

SERIES CALL   *pp. 538, 549, 557*

Less varied than Great-tailed Series; given in similar situations. Single Peerlike notes also given by males in alarm and in flight.

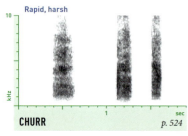

Rapid, harsh

CHURR   *p. 524*

By females, at least in alarm; males may give similar sounds. Some versions reportedly match Great-tailed Grackle Chatter.

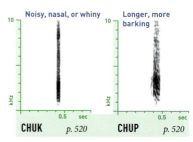

Noisy, nasal, or whiny

CHUK   *p. 520*

Longer, more barking

CHUP   *p. 520*

Variable. Can match Great-tailed's Chuk, and given in similar situations, but more likely than Great-tailed to sound nasal or whiny.

COMMON GRACKLE

Quiscalus quiscula

Abundant in open woods, fields, towns, parks, and lawns; has expanded range westward in recent years. Forms large, loud flocks, often with other blackbirds.

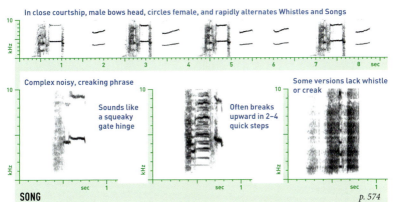

In close courtship, male bows head, circles female, and rapidly alternates Whistles and Songs

Complex noisy, creaking phrase

Sounds like a squeaky gate hinge

Often breaks upward in 2–4 quick steps

Some versions lack whistle or creak

SONG *p. 574*

All year, especially Mar.–June, by both sexes, but primarily males. Female sings mostly in response to mate, both birds fluffing feathers and spreading wings. Each individual reportedly has 1 unique songtype, given repeatedly or mixed with Jeers and other calls.

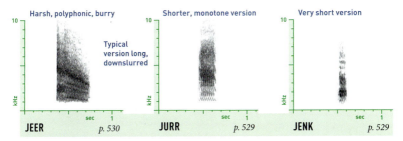

Harsh, polyphonic, burry

Typical version long, downslurred

Shorter, monotone version

Very short version

JEER *p. 530*   **JURR** *p. 529*   **JENK** *p. 529*

Given frequently by females, rarely by males; highly plastic. Versions shown are common, but the full range of intermediates may be given by any bird. Given in a variety of situations, including upon arrival and departure from a colony and during disputes.

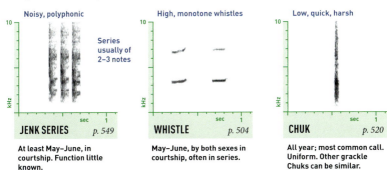

Noisy, polyphonic

Series usually of 2–3 notes

High, monotone whistles

Low, quick, harsh

JENK SERIES *p. 549*   **WHISTLE** *p. 504*   **CHUK** *p. 520*

At least May–June, in courtship. Function little known.

May–June, by both sexes in courtship, often in series.

All year; most common call. Uniform. Other grackle Chuks can be similar.

Rusty Blackbird

Euphagus carolinus

Breeds in bogs; winters in swamps, often in large flocks. In migration, found in wetlands and riparian woods. Uncommon and apparently declining.

Gurgle-creaks and Gurgles sometimes alternated, but either may be repeated many times

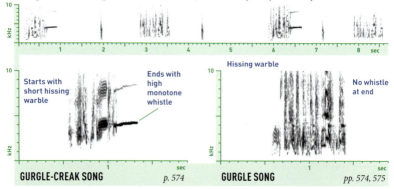

Starts with short hissing warble

Ends with high monotone whistle

GURGLE-CREAK SONG *p. 574*

Hissing warble

No whistle at end

GURGLE SONG *pp. 574, 575*

Individuals of both sexes apparently have one Gurgle-creak and one Gurgle. Both types of Song given all winter, plastic in fall, becoming stereotyped by May. Gurgle apparently not given after first 2 weeks on breeding grounds; may serve courtship function. Gurgle-creak continues throughout summer.

Starts with brief cluck or bark

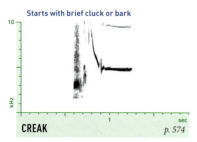

CREAK *p. 574*

Apparently only on breeding grounds, in territorial disputes, in alarm, and when mobbing predators, often with Gurgle-Creaks.

Notes unmusical, Chiplike

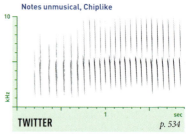

TWITTER *p. 534*

Function unknown; given on breeding grounds in response to playback of Song, and sometimes in spring migration. More study needed.

Harsh, snarling

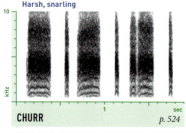

CHURR *p. 524*

Given by both sexes in high alarm near nest. Variable, especially in length. Usually interspersed with Chuks.

Noisy

Lower, softer

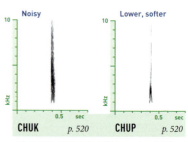

CHUK *p. 520* **CHUP** *p. 520*

Chuk given all year, by both sexes, in mild alarm or contact. Some versions clearer, more nasal than shown. Chup often given prior to flight.

Brewer's Blackbird

Euphagus cyanocephalus

Closely related to Rusty Blackbird, but favors more open country at all seasons. Originally a western species; has greatly expanded range eastward.

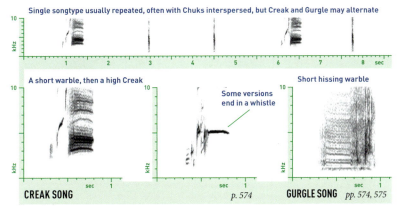

Single songtype usually repeated, often with Chuks interspersed, but Creak and Gurgle may alternate

A short warble, then a high Creak

Some versions end in a whistle

Short hissing warble

CREAK SONG *p. 574*

GURGLE SONG *pp. 574, 575*

All year, by both sexes. Males reported to have one Creak and one Gurgle each; both given with wing-spreading display in summer, mostly without display in winter. Songs have both territorial and courtship functions; differences between Creak and Gurgle need more study.

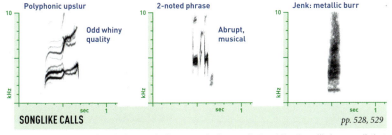

Polyphonic upslur

Odd whiny quality

2-noted phrase

Abrupt, musical

Jenk: metallic burr

SONGLIKE CALLS *pp. 528, 529*

A catch-all category of confusing sounds. Some may be Song variants, while others likely serve distinct functions; more study needed. All appear stereotyped within individuals, but variable between birds. Not illustrated: a downslurred 2-syllabled whistle reportedly given in response to predators.

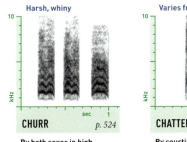

Harsh, whiny

CHURR *p. 524*

By both sexes in high agitation; often in series. Plastic.

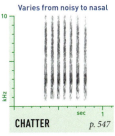

Varies from noisy to nasal

CHATTER *p. 547*

By courting males or fighting females; speed and quality vary situationally.

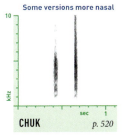

Some versions more nasal

CHUK *p. 520*

All year; most common call. Variable, but rather like other blackbird Chuks.

YELLOW-HEADED BLACKBIRD

Xanthocephalus xanthocephalus

Highly distinctive both visually and vocally. Breeds in tall marsh vegetation, nesting over deeper water than Redwinged Blackbird. Often forages in agricultural areas.

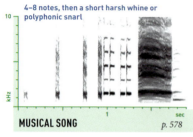

4–8 notes, then a short harsh whine or polyphonic snarl

MUSICAL SONG *p. 578*

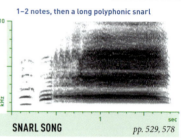

1–2 notes, then a long polyphonic snarl

SNARL SONG *pp. 529, 578*

All year, but mostly May–July. Males have 2 categories of song. During Musical Song, male usually lifts head and both wings. During Snarl Song, used especially during territorial disputes, lifts wings slightly and turns head to left. Snarl sometimes omitted from both types of song, or given by itself, sounding like Jurr.

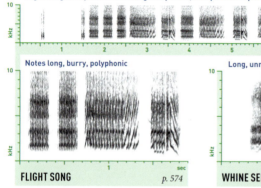

Flight Song: Complex series of long buzzy notes, with phrases repeating several times

Notes long, burry, polyphonic

FLIGHT SONG *p. 574*

May–Aug., usually in flight, by males driving off other males, or by groups of males mobbing hawks or herons. Also during copulation.

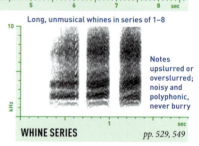

Long, unmusical whines in series of 1–8

Notes upslurred or overslurred; noisy and polyphonic, never burry

WHINE SERIES *pp. 529, 549*

May–Aug., by females leaving nest or confronting other females. Head and wings sometimes lifted. Rarely begun during mate's Song.

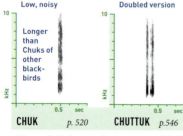

Low, noisy

Longer than Chuks of other blackbirds

Doubled version

CHUK *p. 520*

CHUTTUK *p.546*

All year, by both sexes. Most common call. Variable; breeding males often give doubled version Chuttuk (right).

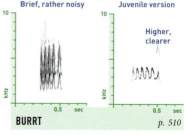

Brief, rather noisy

Juvenile version

Higher, clearer

BURRT *p. 510*

Mostly May–Aug., by females or by begging juveniles. Function little known; more study needed.

BROWN-HEADED COWBIRD

Molothrus ater

Once limited to following bison herds on the high plains; now common in many habitats. A brood parasite, laying its eggs in other species' nests.

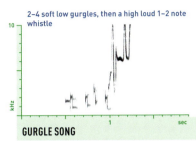

2–4 soft low gurgles, then a high loud 1–2 note whistle

GURGLE SONG

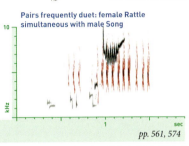

Pairs frequently duet: female Rattle simultaneous with male Song

pp. 561, 574

Nearly all year; gives plastic songs in winter, becoming stereotyped by spring; sings little in fall. Individual males average 4 Gurgle songtypes in the East, more in the West; consecutive songs may be same or different. When directed at other cowbirds, given in display with bowed head, arched wings.

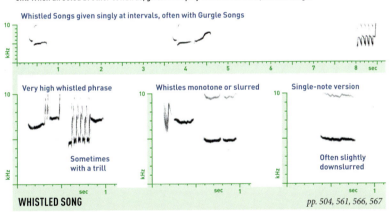

Whistled Songs given singly at intervals, often with Gurgle Songs

Very high whistled phrase

Sometimes with a trill

Whistles monotone or slurred

Single-note version

Often slightly downslurred

WHISTLED SONG

pp. 504, 561, 566, 567

Nearly all year, mostly or exclusively by males; given prior to and during flight, in songlike contexts, and in alarm. A learned sound, like Gurgle Song, with local dialects. Most males have 2 versions, 1 multinoted and 1 single-noted; the multinoted version is often truncated.

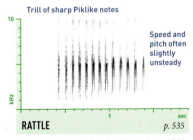

Trill of sharp Piklike notes

Speed and pitch often slightly unsteady

RATTLE *p. 535*

All year, mostly by females. Somewhat variable and quite plastic, especially in female; male's version is quite monotone.

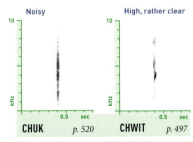

Noisy

High, rather clear

CHUK *p. 520*

CHWIT *p. 497*

Possibly all year; rarely heard. Chuk by both sexes; Chwit little known. Juveniles beg with high, coarse Cheet (p. 511).

BRONZED COWBIRD

Molothrus aeneus

Found in semi-open country. Note red eye and thick ruff of feathers on back of male's neck. A brood parasite, laying its eggs in the nests of other bird species.

Soft, slow series of rising musical whines, each ending in a gurgle

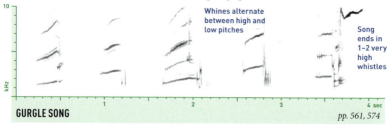

Whines alternate between high and low pitches

Song ends in 1–2 very high whistles

GURGLE SONG

pp. 561, 574

All year, especially May–July, apparently only by males. Given from perch, often in display with bowed head, arched wings, and fluffed neck feathers; sometimes preceded by hovering display and followed by Whistled Song in circling flight. Repertoire size unknown.

Gurgle Song sometimes culminates in much louder Whistled Song

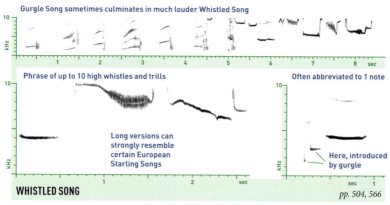

Phrase of up to 10 high whistles and trills

Often abbreviated to 1 note

Long versions can strongly resemble certain European Starling Songs

Here, introduced by gurgle

WHISTLED SONG

pp. 504, 566

All year, mostly or exclusively by males; given in courtship during both hover display and circling flight display, as well as from perch. Likely also given in other contexts. A learned sound, with regional dialects; individuals have a single version, frequently truncated.

Trill of sharp Piklike notes

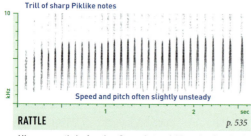

Speed and pitch often slightly unsteady

RATTLE

p. 535

All year, mostly by females. Somewhat variable and quite plastic. Very similar to Brown-headed Cowbird Rattle.

Quick, rather noisy

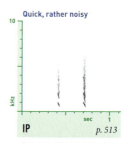

IP

p. 513

Possibly all year; rare. Juveniles beg with high polyphonic Zeer (p. 528).

SHINY COWBIRD

Molothrus bonariensis

A South American and Caribbean bird, now local in small numbers in southern Florida. A brood parasite, laying its eggs in the nests of other bird species.

Very soft gurgling clucks, then a high loud whistled phrase

Introductory notes slower, more clucking than in Brown-headed

GURGLE SONG

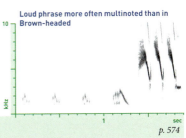

Loud phrase more often multinoted than in Brown-headed

p. 574

Apparently only by males, to females in courtship and to other males in aggression. Repertoire size unknown, but consecutive songs tend to be similar. When other cowbirds are near, given in display with bowed head, arched wings.

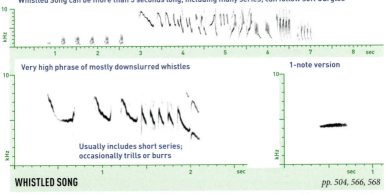

Whistled Song can be more than 5 seconds long, including many series; can follow soft Gurgles

Very high phrase of mostly downslurred whistles

1-note version

Usually includes short series; occasionally trills or burrs

WHISTLED SONG

pp. 504, 566, 568

Nearly all year, mostly or exclusively by males; appears to be given more often and in more contexts than Gurgle Song. Repertoire size unknown, but consecutive songs tend to be similar, albeit frequently truncated. Single-note version given prior to or during flight; may function as a separate call.

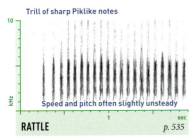

Trill of sharp Piklike notes

Speed and pitch often slightly unsteady

RATTLE *p. 535*

All year, mostly by females. Somewhat variable and quite plastic. Very similar to Brown-headed Cowbird Rattle.

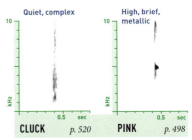

Quiet, complex

High, brief, metallic

CLUCK *p. 520*   **PINK** *p. 498*

Possibly all year; infrequent. Cluck reportedly by both sexes. Pink highly variable. Juveniles beg with high polyphonic notes.

ORCHARD ORIOLE

Icterus spurius

Our smallest oriole. Nests in shelter-belts and woodland edges, especially near water. Young males resemble females, but with black bib.

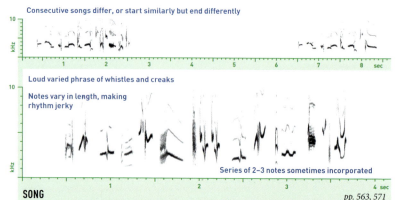

Consecutive songs differ, or start similarly but end differently

Loud varied phrase of whistles and creaks

Notes vary in length, making rhythm jerky

Series of 2–3 notes sometimes incorporated

SONG   *pp. 563, 571*

Mostly Mar.–June. Highly variable; consecutive songs can vary tremendously. Chuklike noisy notes occasionally included. Apparently incorporates imitations only rarely, and possibly only in quiet subsong or whisper song. First-year males sing often.

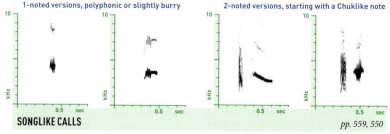

1-noted versions, polyphonic or slightly burry

2-noted versions, starting with a Chuklike note

SONGLIKE CALLS   *pp. 559, 550*

All year. Variable but stereotyped; individuals appear to have 3–4 versions, usually given singly, in random order, often interspersed with Chuks. Components of 2-noted versions may also be given separately, but quite distinctive when heard together.

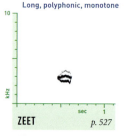

Long, polyphonic, monotone

May be long or short

Recalls Red-winged Blackbird

ZEET   *p. 527*

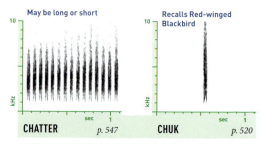

CHATTER   *p. 547*

CHUK   *p. 520*

All year, possibly in pair contact. Somewhat variable and plastic.

All year. Chatters infrequent. Gives single Chuks often, but with a complete range of intergrades in between.

Hooded Oriole

Icterus cucullatus

Breeds in arid regions with scattered trees, including suburbs; particularly fond of ornamental palms. Young males resemble females, but with black bib.

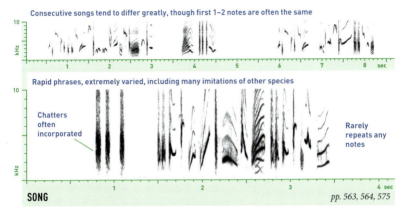

Consecutive songs tend to differ greatly, though first 1–2 notes are often the same

Rapid phrases, extremely varied, including many imitations of other species

Chatters often incorporated

Rarely repeats any notes

SONG　　　　　　　　　　　　　　　　　　　*pp. 563, 564, 575*

Apparently mostly Mar.–May. Hooded Orioles have a reputation for singing less than other orioles, but song is perhaps overlooked because it is rather soft and jumbled. Imitations of other bird species often make up the majority of song. Short Chatters often given between or within songs.

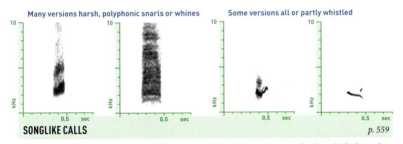

Many versions harsh, polyphonic snarls or whines　　　Some versions all or partly whistled

SONGLIKE CALLS　　　　　　　　　　　　　　　　*p. 559*

All year. Variable but stereotyped; individuals appear to have 2–4 versions, usually given singly, in random order, interspersed with many Chatters and sometimes Zreets. Burry or noisy polyphonic versions typical and fairly distinctive; clear whistled versions generally less common.

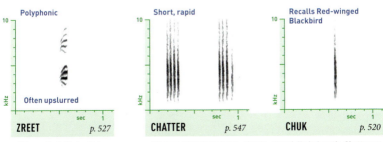

Polyphonic　　　　　Short, rapid　　　　　Recalls Red-winged Blackbird

Often upslurred

ZREET　　*p. 527*　　**CHATTER**　　*p. 547*　　**CHUK**　　*p. 520*

All year. Variable in pitch, and rather plastic. Some versions monotone (p. 527).

All year. Slightly variable; plastic, especially in length. Chatter averages shorter than in other orioles; given frequently. Single Chuks infrequent, usually mixed with Chatters.

Spot-breasted Oriole

Icterus pectoralis

Native to Central America; introduced to Florida in the 1940s, where a small urban population is established, though numbers have apparently declined.

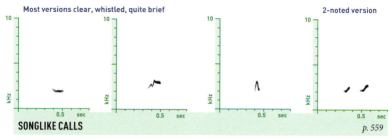

Song often includes slow whistled phrases and series, with sections repeated

A slow phrase or series of musical whistles, notes often nearly monotone

Here, a slow whistled triplet series

SONG *p. 568*

Mostly Mar.–May in Florida. Females reported to sing roughly as often as males, apparently with similar song. Song may include simple series, complex series, and/or phrases. Some versions can recall Song of Northern Cardinal, but generally slower, longer, and more varied.

Most versions clear, whistled, quite brief

2-noted version

SONGLIKE CALLS *p. 559*

All year. Variable but fairly stereotyped; individuals appear to have 3–4 versions, usually given singly, in random order, interspersed with Zreeklike notes and occasionally Chatters. Brief, 1-noted versions are typical, but 2-noted versions (or doubling of 1-note versions) are not uncommon.

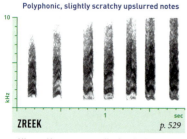

Polyphonic, slightly scratchy upslurred notes

ZREEK *p. 529*

All year. Most common call; often in slow series.

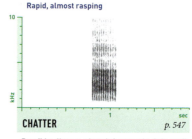

Rapid, almost rasping

CHATTER *p. 547*

Possibly all year; given infrequently. Sole available example is shown.

Altamira Oriole

Icterus gularis

Our largest oriole, found in scrubby woodlands. Resembles Hooded Oriole, but plumage differs slightly. Weaves a hanging nest, like other orange orioles.

 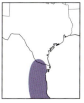

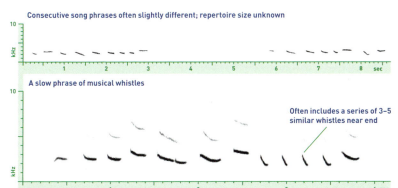

Consecutive song phrases often slightly different; repertoire size unknown

A slow phrase of musical whistles

Often includes a series of 3–5 similar whistles near end

SONG *p. 568*

Apparently all year. Not known whether females sing, but seems likely. Song often takes form of discrete phrases, 3–4 seconds in length; but some recordings are of shorter series repeated without pause, or longer run-on series that include occasional nasal or noisy notes. More study needed.

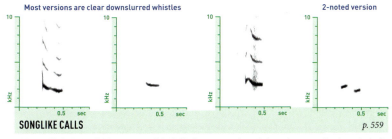

Most versions are clear downslurred whistles

2-noted version

SONGLIKE CALLS *p. 559*

All year. Variable but fairly stereotyped; individuals appear to have 2–4 versions, usually given singly, in random order, sometimes interspersed with Zreeklike notes. 2-noted versions are not uncommon.

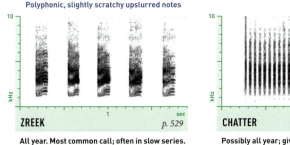

Polyphonic, slightly scratchy upslurred notes

ZREEK *p. 529*

All year. Most common call; often in slow series.

CHATTER *p. 547*

Possibly all year; given rather infrequently.

Audubon's Oriole

Icterus graduacauda

In Texas, found in dense riparian thickets as well as thorn scrub and live-oak woods. Builds a hidden, cup-shaped nest, like Scott's and Orchard Orioles.

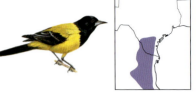

Repertoire size unknown; consecutive songtypes may be same or different

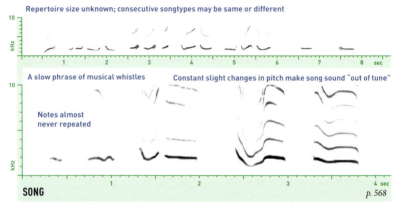

A slow phrase of musical whistles

Constant slight changes in pitch make song sound "out of tune"

Notes almost never repeated

SONG

p. 568

All year, by both sexes. Frequently compared to the sound of a person learning to whistle, an impression often strengthened by irregular pauses, which make rhythm seem halting. Little pitch change within or between whistles. Some songs of Altamira Oriole are quite similar.

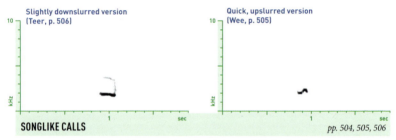

Slightly downslurred version (Teer, p. 506)

Quick, upslurred version (Wee, p. 505)

SONGLIKE CALLS

pp. 504, 505, 506

All year, reportedly in contact between members of a pair. Less variable than in other orioles; fairly stereo-typed; individuals may have 1–2 versions, usually given singly. Consecutive calls often the same. Some-times interspersed with Zreeklike notes or bits of Song. Only available recordings are of 1-noted versions.

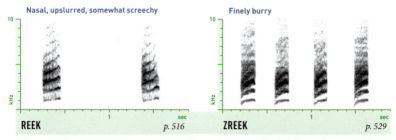

Nasal, upslurred, somewhat screechy

Finely burry

REEK

p. 516

ZREEK

p. 529

All year. Reek and Zreek are common variants of the most common call; often given in slow series. At least in west Mexico, also has a rapid short Chatter like Spot-breasted's, given rarely and softly in altercations.

SCOTT'S ORIOLE

Icterus parisorum

Found in arid habitats, especially where desert transitions into pinyon-juniper or oak woodlands. Often nests in palms, tall yuccas, or junipers.

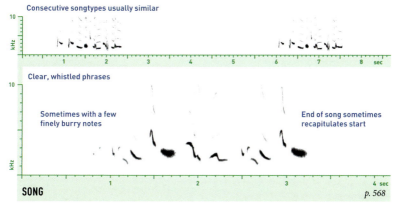

Consecutive songtypes usually similar

10 kHz ... 1 2 3 4 5 6 7 8 sec

Clear, whistled phrases

Sometimes with a few finely burry notes

End of song sometimes recapitulates start

10 kHz ... 1 2 3 4 sec

SONG *p. 568*

All year. Males likely have multiple songtypes, but repertoire size unknown. Females also sing, sometimes from the nest; not known whether female song differs consistently from male song. Sometimes contains 1–2 repeated notes.

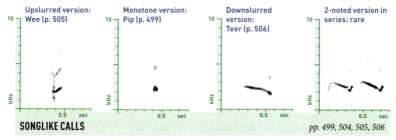

Upslurred version: Wee (p. 505)

Monotone version: Pip (p. 499)

Downslurred version: Teer (p. 506)

2-noted version in series: rare

10 kHz ... 0.5 sec | 10 ... 0.5 sec | 10 ... 0.5 sec | 10 ... 0.5 sec

SONGLIKE CALLS *pp. 499, 504, 505, 506*

Possibly all year. Individuals have perhaps 1–2 versions; consecutive versions tend to be the same. Very short, 1-noted, upslurred or monotone versions are typical. The 2-noted version in series is unusual; may be part of an aberrant songtype, or an undescribed vocalization.

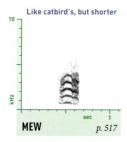

Like catbird's, but shorter

10 kHz ... sec 1

MEW *p. 517*

Not well known. Variable and plastic; some versions fairly harsh.

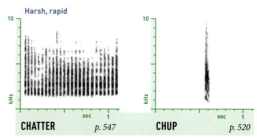

Harsh, rapid

10 kHz ... sec 1

CHATTER *p. 547*

10 kHz ... sec 1

CHUP *p. 520*

All year. Chatter apparently rare; sole available example shown. Chup much more common; quite variable, ranging from quite harsh to a polyphonic Jit (p. 528). May grade into Mew.

BALTIMORE ORIOLE

Icterus galbula

Familiar and highly popular for its bright plumage and cheerful song. Nests in woodland edges and mature shade trees in urban parks.

Long Song (p. 564): soft, continuous, and highly varied, with occasional chatters; likely in close courtship

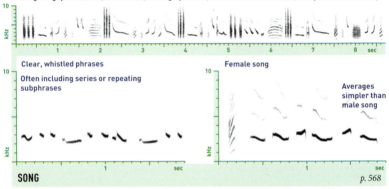

Clear, whistled phrases

Often including series or repeating subphrases

Female song

Averages simpler than male song

SONG　　　　　　　　　　　　　　　　　　*p. 568*

Mostly Apr.–Aug. Males have 2–12 similar songtypes, rarely repeated; some males sing mostly short versions of 4–6 notes, others longer versions of 12–14 notes. Female repertoire size unknown. Females sing mostly near mate; most Songs are short series of downslurs, but some are as complex as males'.

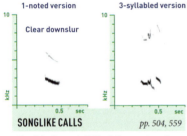

1-noted version

Clear downslur

3-syllabled version

SONGLIKE CALLS　　　　　　*pp. 504, 559*

Possibly all year. Individuals have perhaps 3 versions, given singly in random order; most are clear and whistled, 1- to 3-syllabled.

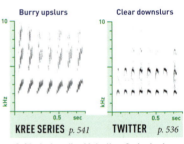

Burry upslurs

Clear downslurs

KREE SERIES *p. 541*　　**TWITTER** *p. 536*

Softly, by breeding birds; Kree Series in chases and pair interactions, Twitter possibly in pair contact.

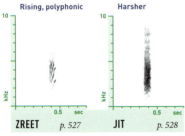

Rising, polyphonic

Harsher

ZREET *p. 527*　　**JIT** *p. 528*

Possibly all year. Zreet infrequent; apparently used rarely by night migrants. Jit mixed with Songlike Calls or Chatters.

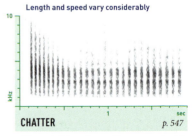

Length and speed vary considerably

CHATTER *p. 547*

All year, in interactions and in response to various types of threat. Plastic; perhaps rarely shortened to a single Chuk.

Bullock's Oriole

Icterus bullockii

Breeds in riparian woods. Interbreeds with Baltimore where ranges overlap; hybrids may have any combination of parental plumage and vocalizations.

Long Song (p. 564): soft, continuous, and highly varied, with occasional chatters; likely in close courtship

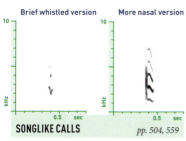

Short phrases, notes often varied in quality

Usually starts with short, syncopated chatter

Often includes a series

Female Song

Averages simpler than male Song

SONG

pp. 571, 575

Mostly Apr.–July. Notes may have almost any tone quality, but whistles or slightly nasal notes usually predominate. Females sing regularly, mostly near mate; repertoire size unknown in both sexes, but at least some males have multiple songs, usually repeated many times before switching.

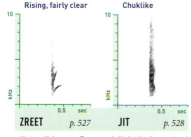

Brief whistled version | More nasal version

Burry upslurs | Clear downslurs

SONGLIKE CALLS *pp. 504, 559*

KREE SERIES *p. 541* | **TWITTER** *p. 536*

Possibly all year. Individuals have perhaps 3 versions, given singly in random order; some are nasal or noisy, 1- or 2-noted.

Softly, by breeding birds; Kree Series in chases and pair interactions, Twitter possibly in pair contact.

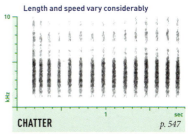

Rising, fairly clear | Chuklike

Length and speed vary considerably

ZREET *p. 527* | **JIT** *p. 528*

CHATTER *p. 547*

Not well known. Zreet and Jit both given mixed with Songlike Calls, and possibly best classified in that category.

All year, in interactions and in response to various types of threat. Plastic; rarely shortened to a single Chuk.

BOBOLINK

Dolichonyx oryzivorus

Nests in mid- to tall-grass prairie and hayfields. Leaves as early as July for marshlands where birds molt before continuing south to South America.

Males typically have 2 similar songtypes, delivered in any order

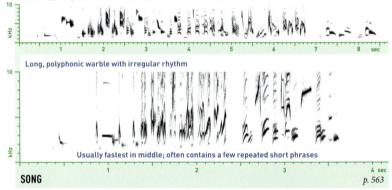

Long, polyphonic warble with irregular rhythm

Usually fastest in middle; often contains a few repeated short phrases

SONG   *p. 563*

Mostly Apr.–June, often from highest perch in territory, or during song flights of up to 1 minute in length. Full Songs are generally 4–6 seconds long, but birds very frequently repeat just the first few notes, sometimes for long periods. Neighboring birds tend to share songtypes.

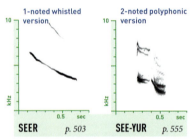

1-noted whistled version

2-noted polyphonic version

SEER   *p. 503*   **SEE-YUR**   *p. 555*

Mostly Apr.–July, by singing males, singly or appended to start of Song. Highly variable; may simply represent extremely abbreviated Song.

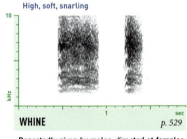

High, soft, snarling

WHINE   *p. 529*

Reportedly given by males, directed at females, usually after a song flight. Plastic; usually not very loud.

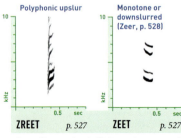

Polyphonic upslur

Monotone or downslurred (Zeer, p. 528)

ZREET   *p. 527*   **ZEET**   *p. 527*

Zreet all year, often in flight; most common call during migration, including at night. Zeet reportedly by breeding female.

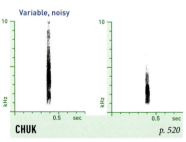

Variable, noisy

CHUK   *p. 520*

All year. Quite variable; different versions reported in different contexts, but much variation is likely individual or geographic.

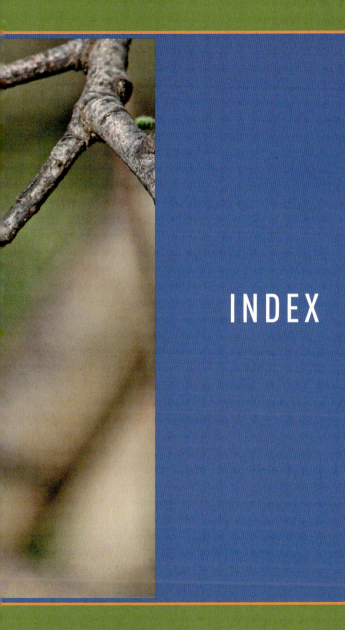

INDEX

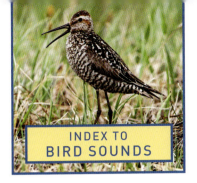

INDEX TO
BIRD SOUNDS

HOW TO USE THE VISUAL INDEX

This visual index lists every sound in the species accounts in a group with similar sounds. Each group is indicated by a name, a description, and a symbol that indicates roughly how the sounds in the group appear on the spectrogram. Page numbers in the index refer to the species accounts; page numbers in the species accounts refer to the index.

How to look up a sound

- **By characteristics of the sound:** Use the Quick Index on the inside back cover of the book. Find the description that best matches the sound, and then turn to the corresponding pages in the visual index.
- **By similarity with a known sound of a familiar species.** Turn to the species account of the familiar bird. Find the similar sound in the species account, and then use the page number listed with it to turn to the corresponding pages in the visual index.

Organization of the index

The index is split into seven major parts by sound pattern:
1. Single-note sounds (p. 496)
2. Sounds made of repeated similar notes (p. 531)
3. 2- and 3-syllabled phrases (p. 550)
4. Complex series (p. 556)
5. Songs of different separate notes, series, or phrases (p. 559)
6. Very long complex songs (p. 563)
7. Medium to long phrases and combinations (p. 566)

Within these categories, sounds are generally listed according to the following principles:
1. Short to long (or simple to complex)
2. High to low
3. Slow to fast

The basic tone qualities (p. TK) generally appear in the following order:

1. Ticking or mechanical
2. Whistled
3. Hooting, cooing, or growling
4. Burry or buzzy
5. Nasal
6. Harsh
7. Combinations of burry/buzzy, nasal, and harsh
8. Polyphonic

The basic pitch patterns (p. TK) generally appear in the following order:

1. Monotone
2. Upslurred
3. Downslurred
4. Overslurred
5. Underslurred

Bird sounds are complex and variable; many exceptions to these guidelines occur.

INDEX PART 1: A SINGLE-NOTE SOUND

A single note given by itself, or repeated after a pause

A CLICK, TAP, OR SNAP
A nearly instantaneous, wholly unmusical note that shows on the spectrogram as a completely vertical line.

TICK
Extremely brief, noisy clicklike note without any change in pitch.

Soft
Mountain Plover, p. 128
Buff-breasted Sandpiper,
   p. 137

Louder, often in irregular series
Forster's Tern, p. 176
Buff-bellied Hummingbird,
   p. 105
Tree Swallow, p. 304

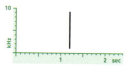

Common Myna, p. 352

TOCK
Like Tick, but more resonant and sometimes more musical. Often quite loud.

American Coot, p. 117
Common Gallinule, p. 116

Sora, p. 113
Limpkin, p. 120

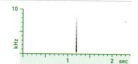

A CHIPLIKE NOTE
An extremely brief, extremely sharp note, slightly more musical than a Click.

WHIT
Extremely brief, sharply upslurred single note. Pitch medium to high.

Solitary Sandpiper, p. 145
Least Flycatcher, p. 255
Willow Flycatcher, p. 259

Lower
Wild Turkey (juvenile), p. 67

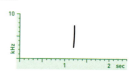

SPIT
Like Whit, but sharper, nearly a snapping sound.

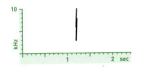

Bushtit, p. 312
Rock Wren, p. 320
Northern Cardinal, p. 458
Pyrrhuloxia, p. 459

Indigo Bunting, p. 462
Lazuli Bunting, p. 463
Varied Bunting, p. 464
Painted Bunting, p. 465

CHWIT
Like Whit, but slightly noisier and longer.

Gray Catbird (juvenile), p. 345
Mourning Warbler, p. 393
Yellow-rumped Warbler
  (Audubon's), p. 411
Dickcissel, p. 455

Averages squeakier
Connecticut Warbler, p. 394
Northern Red Bishop, p. 358

Lower, mostly noisy
Eastern Bluebird, p. 341

Slightly more ringing
Brown-headed Cowbird,
  p. 479

CHIP
An extremely brief, sharply downslurred note. Pitch medium to high.

Fairly sharp, rather high
Sedge Wren, p. 326
Ovenbird, p. 379
Worm-eating Warbler, p. 382
Swainson's Warbler, p. 384
Tennessee Warbler, p. 388
Nashville Warbler, p. 390
Black-and-white Warbler,
  p. 385
Blue-winged Warbler, p. 386
Golden-winged Warbler,
  p. 387
Cerulean Warbler, p. 396
Northern Parula, p. 398
Tropical Parula, p. 399
Black-throated Green Warbler,
  p. 400
Golden-cheeked Warbler,
  p. 401
Prairie Warbler, p. 402
Blackburnian Warbler, p. 403
American Redstart, p. 395

Blackpoll Warbler, p. 404
Bay-breasted Warbler, p. 405
Yellow-throated Warbler,
  p. 407
Yellow Warbler, p. 408
Chestnut-sided Warbler,
  p. 409
Palm Warbler, p. 414
Pine Warbler, p. 415

High, slightly more musical
Gray Catbird (juvenile), p. 345
Wilson's Warbler (juvenile),
  p. 418

High, ringing/metallic
Hooded Warbler, p. 412
(*see also* Tsit, p. 499)

Lower, slightly musical
Sharp-shinned Hawk, p. 201
Swamp Sparrow, p. 444
Eastern Phoebe, p. 261

Lower, fairly sharp
Marsh Wren (juvenile), p. 324
Kentucky Warbler, p. 392
Kirtland's Warbler, p. 416
Canada Warbler, p. 417

Fairly low, slightly complex
Scaled Quail, p. 69
Scarlet Tanager, p. 456
Yellow-rumped Warbler
  (Myrtle), p. 410

Very sharp, almost smacking
Bohemian Waxwing, p. 354
Vesper Sparrow, p. 435
Lincoln's Sparrow, p. 443
Savannah Sparrow, p. 448
Baird's Sparrow, p. 449
Le Conte's Sparrow, p. 451
Nelson's Sparrow, p. 454

SMACK
Like Chip, but sharper, almost clicking, like the smacking of lips.

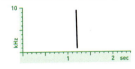

Rather high, thin
Buff-bellied Hummingbird,
  p. 105
Winter Wren, p. 318
Black-throated Blue Warbler,
  p. 397

Ovenbird, p. 379
Fox Sparrow, p. 445
Dark-eyed Junco, p. 446

Lower, slightly harsher, Chuklike
Brown Thrasher, p. 346
Long-billed Thrasher, p. 347

Slightly lower
Nelson's Sparrow, p. 454

TIP
Extremely brief whistled note with no change in pitch. Pitch medium.

McCown's Longspur, p. 376
Purple Finch, p. 360

TINK

Like Tip (p. 497), but extremely high pitched. More musical than Chip (p. 497) or Tsit. Most versions are quite plastic and variable.

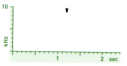

Ovenbird, p. 379
Worm-eating Warbler, p. 382
Louisiana Waterthrush, p. 381
Common Yellowthroat, p. 391
Kentucky Warbler, p. 392
Mourning Warbler, p. 393
Black-and-white Warbler, p. 385
Blue-winged Warbler (possibly), p. 386
Golden-winged Warbler, p. 387
Cerulean Warbler, p. 396
Black-throated Blue Warbler, p. 397
Black-throated Green Warbler, p. 400
Golden-cheeked Warbler, p. 401
Prairie Warbler, p. 402
Blackburnian Warbler, p. 403
American Redstart, p. 395

Yellow Warbler, p. 408
Chestnut-sided Warbler, p. 409
Kirtland's Warbler, p. 416
Yellow-rumped Warbler (Myrtle), p. 410
Yellow-rumped Warbler (Audubon's), p. 411
Magnolia Warbler, p. 413
Palm Warbler, p. 414
Pine Warbler, p. 415
Canada Warbler, p. 417
Wilson's Warbler, p. 418
Eastern Towhee, p. 424
Spotted Towhee, p. 425
Green-tailed Towhee, p. 423
American Tree Sparrow, p. 430
Vesper Sparrow, p. 435
Lark Sparrow, p. 436
Black-throated Sparrow, p. 437

Lark Bunting, p. 438
White-throated Sparrow, p. 439
White-crowned Sparrow, p. 441
Song Sparrow, p. 442
Lincoln's Sparrow, p. 443
Swamp Sparrow, p. 444
Fox Sparrow, p. 445
Dark-eyed Junco, p. 446
Savannah Sparrow, p. 448
Baird's Sparrow, p. 449
Seaside Sparrow, p. 450
Le Conte's Sparrow, p. 451
Henslow's Sparrow, p. 452
Nelson's Sparrow, p. 454
Indigo Bunting, p. 462
Lazuli Bunting, p. 463
Painted Bunting, p. 465
Blue Grosbeak, p. 466

PSIT

Very high, very sharp upslurred whistle. Like Spit (p. 497), but higher.

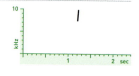

Buff-bellied Hummingbird, p. 105
Cassin's Sparrow, p. 428
Botteri's Sparrow, p. 429

Even sharper
Sharp-shinned Hawk, p. 201
Black-throated Blue Warbler, p. 397
Northern Cardinal, p. 458

PINK

Extremely brief single note, sharply upslurred at start, but main part of note monotone, slightly ringing, often metallic. Pitch medium to high.

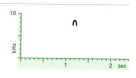

Solitary Sandpiper, p. 145
Vesper Sparrow, p. 435
White-throated Sparrow, p. 439
White-crowned Sparrow, p. 441
Varied Bunting, p. 464
Painted Bunting, p. 465

Blue Grosbeak, p. 466
Shiny Cowbird, p. 481

Averages longer
Harris's Sparrow, p. 440

Often higher, sharper, Spitlike
Northern Waterthrush, p. 380
Nashville Warbler, p. 390

High, metallic, Tinklike
American Tree Sparrow, p. 430

Lower, more musical
Red-winged Blackbird, pp. 468–469

PSIP

Like Psit, but more musical. Shorter than Pseep (p. 500).

Singly or in series
Boreal Chickadee, p. 307
Black-capped Chickadee, p. 308
Carolina Chickadee, p. 309

Slightly sharper
Lark Sparrow, p. 436

Very high
Golden-crowned Kinglet, p. 331

TSIT

Higher than Chip (p. 497), sharp, like an extremely high Smack (p. 497). Can be confused with Tink or with Psit.

Orange-crowned Warbler,
p. 389
Nashville Warbler, p. 390
Olive Sparrow, p. 422
Chipping Sparrow, p. 431
Clay-colored Sparrow, p. 432

Brewer's Sparrow, p. 433
Field Sparrow, p. 434
Grasshopper Sparrow, p. 447
Henslow's Sparrow, p. 452

Slightly more ringing
Prothonotary Warbler, p. 383

Slightly lower
Orange-crowned Warbler,
p. 389

TSWIT

Like Sreet (p. 501), but with stronger consonant sound at start; generally lower, sharper, and more Chiplike. See Tseet (p. 502), which is slightly longer.

Canada Warbler, p. 417

Wilson's Warbler, p. 418

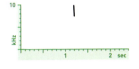

TSIP

Like Chip (p. 497), but higher. Sharper than Tsew. Some versions of Tink are similar.

Tufted Titmouse, p. 310
Black-crested Titmouse,
p. 311

Cape May Warbler, p. 406

TSEW

Like Tsip, but less sharp, more musical. Shorter than Tseew (p. 502).

Northern Parula, p. 398
Tropical Parula, p. 399
Yellow-throated Warbler,
p. 407
Bachman's Sparrow, p. 427
American Tree Sparrow,
p. 430
Lark Sparrow, p. 436
Palm Warbler, p. 414
Dickcissel, p. 455

Slightly longer, more Seetlike
Brown Creeper, p. 317
Bachman's Sparrow, p. 427
Pine Warbler, p. 415
Cape May Warbler, p. 406

*Slightly longer, polyphonic
(though this is difficult to hear)*
Savannah Sparrow, p. 448
Clay-colored Sparrow, p. 432

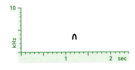

Often in series
Black-throated Sparrow,
p. 437

A PEEP- OR CHIRPLIKE NOTE

Very brief notes that start and end abruptly lower, creating the impression that they begin and/or end with the consonants "p" or "k."

PIP (WHISTLED)

Very brief, sharply overslurred whistle or seminasal note.

Sharper, often doubled
Wilson's Plover, p. 124

Less sharp at start, almost a Wip
Red-cockaded Woodpecker,
p. 237
Willow Flycatcher, p. 259

Medium pitch
Black-bellied Plover, p. 129
Ash-throated Flycatcher,
p. 264

Scott's Oriole, p. 487
Pine Grosbeak, p. 372

Low, rarely single
Eared Grebe, p. 80
Red Knot, p. 134
Red-necked Phalarope,
p. 150

Medium pitch, seminasal
Scissor-tailed Flycatcher,
p. 272

Western Kingbird, p. 270
Couch's Kingbird, p. 269

*Briefer, slightly more nasal;
compare Vimp, p. 523*
Carolina Wren, p. 322

A PEEP- OR CHIRPLIKE NOTE, CONTINUED

WIP
Like Pip (p. 499), but with a weaker initial consonant sound.

Medium pitch
Great Crested Flycatcher, p. 265

Brown-crested Flycatcher, p. 266

PEEP
Higher than Pip (p. 499), and often longer. Clearer than Cheep.

Very high, short
Red Phalarope, p. 150

High, short
Buff-breasted Sandpiper, p. 137
Long-billed Dowitcher, p. 141

Alder Flycatcher, p. 258

Medium-high to high, slightly longer
Black-legged Kittiwake, p.156
American Oystercatcher, p. 122

Semipalmated Plover, p. 125
Piping Plover, p. 126
Purple Sandpiper, p. 136
Wild Turkey (juvenile), p. 67

PWIK
Like Pip (p. 499), but much sharper and less musical.

Eastern Wood-Pewee, p. 253

PSEEP
Like Peep, but higher, less sharp, with weaker consonant sounds. Longer than Psip (p. 498).

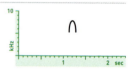

Blue-winged Teal (male), p. 50
Ring-necked Duck (male), p. 56

Yellow-bellied Flycatcher, p. 256
Acadian Flycatcher, p. 257
Vermilion Flycatcher, p. 263

Usually in duet with other calls
Couch's Kingbird, p. 269

PIK
Like Pip (p. 499), but sharper, with stronger consonant sounds at start and end. Lower than Peek.

Downy Woodpecker, p. 234
Ladder-backed Woodpecker, p. 236

Lower
American Three-toed Woodpecker, p. 238

PEEK
Like Pik, but higher.

Hairy Woodpecker, p. 235

KIP
Like Pik, but sharper and much briefer. Like Eep (p. 513), but with a much stronger consonant sound at start.

High
Common Tern, p. 174
Roseate Tern, p. 177
Merlin, p. 244

Lower
Black-backed Woodpecker, p. 239
Boreal Owl, p. 220

CHIRP (TYPICAL)

A brief semimusical note, like Peep or Cheep but more complex. Pitch medium to high. Usually downslurred. Lacks polyphonic quality of Jirp (p. 529).

High, stereotyped
Yellow-bellied Flycatcher, p. 256
Henslow's Sparrow, p. 452

Medium pitch, stereotyped
Hutton's Vireo, p. 286

Medium pitch, stereotyped or plastic
Red-whiskered Bulbul, p. 330
House Finch, p. 361
House Sparrow, p. 356
Eurasian Tree Sparrow, p. 357

Medium pitch, stereotyped or plastic, sometimes finely buzzy
Tree Swallow, p. 304

Medium-low, finely buzzy
Evening Grosbeak, p. 371

CHEEP

High, plastic, often slightly screechy complex notes. Higher than Chirp; usually harsher and more complex than Peep. Higher than screechy Bark (p. 515).

High, plastic
White-rumped Sandpiper, p. 138
Horned Lark, p. 299

Medium-high, plastic
Snowy Plover, p. 123
Semipalmated Sandpiper, p. 140
Western Sandpiper, p. 140
Red Phalarope, p. 150

Swainson's Warbler, p. 384
Scaly-breasted Munia, p. 358

Medium-high, abrupt, often stereotyped
Townsend's Solitaire, p. 334
American Robin, pp. 342–343

Medium pitch, plastic
Purple Sandpiper, p. 136
Ruddy Turnstone, p. 134

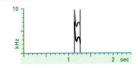

Medium to low pitch, almost a Squeak or screechy Bark
White-winged Scoter, p. 58
Roseate Spoonbill (juvenile), p. 195
Stilt Sandpiper, p. 135
Buff-breasted Sandpiper, p. 137
Red-necked Phalarope, p. 150

SQUEEP

A high Cheeplike note, but clearer and more stereotyped. Often sounds upslurred.

Sprague's Pipit, p. 354
American Robin, pp. 342–343

Some versions vaguely 2-syllabled
Barn Swallow, p. 302

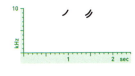

A HIGH SEETLIKE NOTE
Very high musical whistled notes. Sometimes polyphonic or buzzy, though this can be difficult to hear.

SREET

Extremely high, short, slightly upslurred whistled or polyphonic note.

Tennessee Warbler, p. 388
Orange-crowned Warbler, p. 389
Nashville Warbler, p. 390
Black-throated Green Warbler, p. 400
Golden-cheeked Warbler, p. 401
Clay-colored Sparrow, p. 432
Brewer's Sparrow, p. 433

Slightly longer
Prothonotary Warbler, p. 383

Swainson's Warbler, p. 384
Vesper Sparrow, p. 435
White-crowned Sparrow, p. 441

Sometimes slightly buzzy (though this can be difficult to hear); see Dzit, p. 503
Ovenbird, p. 379
Blue-winged Warbler, p. 386
Golden-winged Warbler, p. 387

Slightly lower
Mourning Warbler (sometimes slightly buzzy), p. 393
Yellow-rumped Warbler (Myrtle), p. 410
Yellow-rumped Warbler (Audubon's), p. 411

SEET (NOT BURRY)
Extremely high, nearly monotone whistled or polyphonic note.

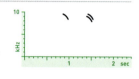

Short, polyphonic (though this can be difficult to hear)
Prairie Warbler, p. 402
Olive Sparrow, p. 422

Fairly long, slightly lower, more overslurred
Brown-headed Nuthatch, p. 316

Fairly long
Grasshopper Sparrow, p. 447

Like Grasshopper Sparrow, but often polyphonic (though this can be difficult to hear)
Henslow's Sparrow, p. 452
Seaside Sparrow, p. 450
Saltmarsh Sparrow, p. 453
Nelson's Sparrow, p. 454

Similar but even longer, lower, polyphony easier to hear
Northern Mockingbird (juvenile), p. 349

Bewick's Wren, p. 323

Very long, very high, not polyphonic
Cedar Waxwing, p. 353

TSEEW
Extremely high, short, slightly downslurred whistled or polyphonic note. Like Tsew (p. 499), but slightly longer.

Brown Creeper, p. 317
Field Sparrow, p. 434

More downslurred, often doubled
Pine Siskin, p. 370

Longer, more monotone, usually polyphonic (though this can be difficult to hear)
Henslow's Sparrow, p. 452
Le Conte's Sparrow, p. 451

Saltmarsh Sparrow, p. 453

TSEET
Extremely high, short, slightly underslurred whistled or polyphonic note. Like Seet, but with stronger consonant sound at start. Shorter than Tseereet.

Shortest
Brewer's Sparrow, p. 433
Baird's Sparrow, p. 449

Slightly longer, with particularly strong consonant sounds
American Tree Sparrow, p. 430
Chipping Sparrow, p. 431

Slightly longer still, with weaker consonant sounds
American Redstart, p. 395
Northern Red Bishop, p. 358

Even longer, more monotone, with weaker consonant sounds
White-throated Sparrow, p. 439

Harris's Sparrow, p. 440
Song Sparrow, p. 442
Fox Sparrow, p. 445

Very long, clear
Rufous-crowned Sparrow, p. 426

TSWEET
High, rising complex note; starts with consonant, sounds squeaky.

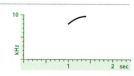

Hutton's Vireo, p. 286
Ovenbird, p. 379

Northern Waterthrush, p. 380

SREE
Extremely high, long, upslurred whistle.

Long
Veery, p. 335

Short
Black Guillemot, p. 152

SEER
Extremely high, long, downslurred whistle.

Gray-cheeked Thrush, p. 336
Bicknell's Thrush, p. 337
Swainson's Thrush, p. 338
Hermit Thrush, p. 339
Wood Thrush, p. 340
American Robin, pp. 342–343
Bohemian Waxwing, p. 354

Sometimes in series
Buff-bellied Hummingbird, p. 105
Tufted Titmouse, p. 310
Black-crested Titmouse, p. 311

Lower, often longer
Bobolink, p. 490

Red-winged Blackbird, pp. 468–469

Finely burry
Black Guillemot, p. 152

TSEEREET
Like Tseet, but longer, and often more noticeably underslurred.

Clear to finely buzzy
Eastern Towhee, p. 424

Spotted Towhee, p. 425
Green-tailed Towhee, p. 423

TSIPT
Very brief coarse note made of 2–3 high sharp Tsit- or Tsiplike notes. Shorter, coarser, and less musical than Dzit.

Dark-eyed Junco, p. 446

Grasshopper Sparrow, p. 447

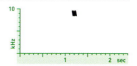

DZIT
Very short, very high buzzy note. Shorter and usually finer than Burry Seet.

Worm-eating Warbler, p. 382
Connecticut Warbler, p. 394
Blue-winged Warbler, p. 386
Golden-winged Warbler, p. 387
Cerulean Warbler, p. 396
Blackburnian Warbler, p. 403
Blackpoll Warbler, p. 404
Bay-breasted Warbler, p. 405
Yellow Warbler, p. 408
Magnolia Warbler, p. 413
Dark-eyed Junco, p. 446

Often slightly rising; see Sreet, p. 501
Northern Waterthrush, p. 380
Louisiana Waterthrush, p. 381
Kentucky Warbler, p. 392
Black-and-white Warbler, p. 385

Slightly lower
Bushtit (usually in rapid series), p. 312

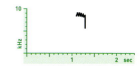

Chestnut-sided Warbler, p. 409
Kirtland's Warbler, p. 416
Hooded Warbler, p. 412

Lower still
Common Yellowthroat, p. 391

TSEERP
Like Tseet, but longer, burry, and downslurred.

American Robin, pp. 342–343

SEET (BURRY)
Very coarse, very high, musical burry whistles. Longer than Dzit. Compare Very High Trills (p. 537).

Ruby-crowned Kinglet (juvenile), p. 332
Brown Creeper, p. 317
American Robin, pp. 342–343

Gray Catbird (juvenile), p. 345
Henslow's Sparrow, p. 452
White-throated Sparrow Tseet, p. 439

Rhythm sometimes stuttering
Golden-crowned Kinglet, p. 331

A WHISTLE
Medium to high pitched, rather musical notes.

VARIABLE SINGLE WHISTLES
These species give a huge variety of single whistled notes, usually 1-syllabled, which vary individually and geographically. In this group, consecutive whistles are usually the same. See also the group "Mostly Single 1-Syllabled Notes," p. 575, in which consecutive sounds are usually different.

Medium-low, brief, usually monotone or downslurred
    Audubon's Oriole Songlike Calls, p. 486
    Scott's Oriole Songlike Calls, p. 487

    Baltimore Oriole Songlike Calls, p. 488
    Bullock's Oriole Songlike Calls, p. 489

Medium-low, usually monotone or upslurred, slightly polyphonic
    Pine Grosbeak, p. 372

Medium to high, usually monotone or downslurred
    Red-winged Blackbird Alert Calls, p. 469

HEE
A brief clear, ringing monotone whistle, without any strong consonant sounds.

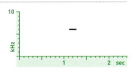

High, breathy, often preceded or followed by rustling or quacking
    Gadwall, p. 46
    Mallard, p. 48
    American Black Duck, p. 49
    Mottled Duck (likely), p. 49
    Northern Pintail, p. 52
    Ruddy Duck (female), p. 63

High, extremely plastic
    Sabine's Gull, p. 157

High, clear, ringing
    Townsend's Solitaire, p. 334

High, slightly burry
    Green-winged Teal, p. 53

Lower, slightly burry
    Northern Pintail, p. 52

Even lower, fairly mellow
    Lesser Prairie-Chicken, p. 77
    Pied-billed Grebe, p. 83

PEE
Like Hee, but with stronger consonant sound at start. See also Pip or Peep Series (p. 436).

    Short-billed Dowitcher, p. 141

    Rose-breasted Grosbeak, p. 460

A LONGER MONOTONE WHISTLE
Long clear monotone whistles, like Hee but longer.

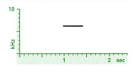

High, thin
    Common Grackle, p. 475
    Brown-headed Cowbird, p. 479
    Bronzed Cowbird, p. 480
    Shiny Cowbird, p. 481

Medium-high
    Western Wood-Pewee, p. 252
    Say's Phoebe, p. 262
    (occasionally truncated songs from other species)

Low, mellow, highly plastic
    Western Screech-Owl, p. 222

WEET
Like Whit (p. 496), but higher and more musical. Higher and much sharper than Wee.

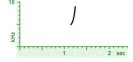

    Wilson's Plover, p. 124
    American Pipit, p. 355

    Smith's Longspur, p. 374

WEE
A short, sligtly upslurred whistle, sometimes nearly monotone. Medium pitch.

Medium-high
Swainson's Thrush, p. 338
Rose-breasted Grosbeak, p. 460

Black-headed Grosbeak, p. 461

Medium-low
Audubon's Oriole, p. 486

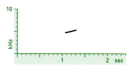

Scott's Oriole, p. 487

WERT
A short upslurred whistle, lower and sharper than Wee. Longer and more musical than Pwut (p. 519).

Lark Bunting, p. 438

Lower
Mountain Plover, p. 128
Snowy Plover, p. 123

PWEE OR DEET
Like Wee, but with stronger consonant sound at start; often slightly underslurred.

Short
Alder Flycatcher, p. 258

Averages slightly longer, higher
Yellow-bellied Flycatcher, p. 256

Usually even longer
Eastern Wood-Pewee, p. 253

Longer, higher, shriller
Killdeer Deet, p. 127

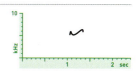

Low, variable
American Golden-Plover, p. 129

WEEP
Like Wee, but with stronger consonant sound at end. Longer and higher than Wip (p. 500).

Medium-high
Great Crested Flycatcher, p. 265
Olive-sided Flycatcher, p. 254

High, often in series (Weep-weep)
Hudsonian Godwit, p. 133
Baird's Sandpiper, p. 138
Stilt Sanpiper, p. 135

Very high, shrill
Sharp-shinned Hawk, p. 201

A LONGER UPSLURRED WHISTLE
A rising whistle, longer than Weet or Weep.

Piping Plover Weet, p. 126
Whimbrel Curree, p. 131

More nasal
European Starling, pp. 350–351

Great-tailed Grackle, pp. 472–473

Sometimes broken. See also Rising Broken Nasal Note (p. 519).
Snowy Plover Ter-weet, p. 123

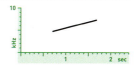

TEW (HIGH)
A high, sharply downslurred whistle; less sharp and more musical than Chip (p. 497).

Very high, sharp
Northern Goshawk, p. 203

Longer, with subtle break in middle
Verdin, p. 313

High, not very sharp
Black Phoebe, p. 260

Lower, varying in pitch and sharpness
White-tailed Kite, p. 199
Sharp-shinned Hawk, p. 201
Osprey, p. 211

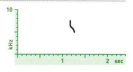

Golden Eagle, p. 210
White-collared Seedeater, p. 421

TEW (LOW)
Like sounds in the preceding group, but lower and more musical.

Purple Martin, p. 305
Common Myna, p. 352

Sharper, audible at close range
Northern Bobwhite, p. 68

Less sharp
Northern Mockingbird,
p. 349
Brown Thrasher, p. 346
Long-billed Thrasher, p. 347

TEER
Musical downslurred whistle, longer and less sharp than Tew.

Quite high, long, often vaguely multisyllabic
Lesser Goldfinch, p. 369

High, thin, mostly rather sharp
Black Phoebe, p. 260
Horned Lark, p. 299
Chestnut-collared Longspur,
p. 375
Smith's Longspur, p. 374

Lapland Longspur, p. 373
Snow Bunting, p. 377
White-collared Seedeater,
p. 421

Medium pitch, nearly monotone
Hermit Thrush, p. 339

Low, brief, nearly monotone
Audubon's Oriole, p. 486

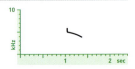

Low, loud, often nearly monotone
Western Meadowlark, p. 471
Scott's Oriole, p. 487
(Songlike Calls of some orioles)

PEW
Like Tew, but with different consonant sound at start.

Black-bellied Whistling-
Duck, p. 37

Lesser Yellowlegs, p. 147
Stilt Sandpiper, p. 135
Hill Myna, p. 351

PEER
Lower and longer than Tew, much less sharp, with different consonant sound at start.

Yellow-bellied Flycatcher,
p. 256
Rose-breasted Grosbeak,
p. 460

Much longer, slightly higher
Black-bellied Plover, p. 129
Say's Phoebe, p. 262

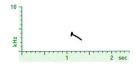

Long, quite high
Northern Beardless-
Tyrannulet, p. 251

A LONGER DOWNSLURRED WHISTLE
A long downslurred whistle, longer than Peer or Teer, usually without strong consonant sounds.

High, clear, variable
European Starling,
pp. 350–351
Great-tailed Grackle,
pp. 472–473

Low, mellow, not very loud
Montezuma Quail, p. 70

PWEEW
Longer, distinctly overslurred version of Pip (p. 499); lower than Pseep (p. 500).

Acadian Flycatcher, p. 257
Alder Flycatcher, p. 258

Willow Flycatcher, p. 259
Harris's Sparrow, p. 440

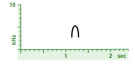

PWEER
Like Pweew, but more drawn out at end.

Alder Flycatcher, p. 258

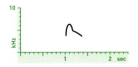

WEEW (WHISTLED)
Medium-high overslurred whistle, more drawn out than Pweew. Clearer than Wheew (p. 508). Lacks polyphonic quality of Zweew (p. 528). Sometimes vaguely multisyllabic.

Purple Finch, p. 360

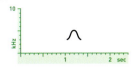

A LONGER OVERSLURRED WHISTLE
A long overslurred whistle, longer than Weew, usually without strong consonant sounds. See also Mellow Wail Series (p. 539).

High, often broken
Black-bellied Whistling-Duck
 Sweeoo, p. 37

Variable, often mellow
Gray Jay, p. 294

PEWY
Brief underslurred whistle, usually with consonant at start. Medium pitch.

American Golden-Plover,
 p. 129
Yellow-bellied Flycatcher,
 p. 256

Alder Flycatcher, p. 258
Summer Tanager, p. 457
Scarlet Tanager, p. 456

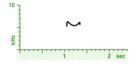

PYOOWEE OR TEWEE (WHISTLED)
Like Pewy, but longer. See also Seminasal Tewee, p. 516.

Medium-low, long (Pyoowee)
Black-bellied Plover, p. 129
Long-billed Curlew, p. 132
Eastern Wood-Pewee, p. 253

High, thin (Tewee)
Barn Swallow, p. 302
Tree Swallow, p. 304

VARIABLE SHORT BREATHY WHISTLES
A catch-all category of variable Wheewlike notes, including short low Whews and Whips, upslurred Wheeps, downslurred Wheers, and burry Whirrs. Each species listed here gives several different sounds in this category.

Fulvous Whistling-Duck,
 p. 36

Black-bellied Whistling-
 Duck, p. 37
Wood Duck, p. 45

Greater Scaup, p. 55
Lesser Scaup, p. 55
Ring-necked Duck, p. 56

WHEEP
High, breathy nasal upslur. Longer and more nasal than Weep (p. 505).

Wood Duck, p. 45

A WHISTLE, CONTINUED

WHEEW
High, breathy nasal or seminasal overslur; like Wow, but much higher. Shriller or more nasal than seminasal Weew (p. 507).

High, variable, clear to shrill
  Northern Harrier, p. 197

High, sometimes burry at start or end
  Wood Duck, p. 45

  Eurasian Wigeon, p. 47

Medium pitch
  Northern Pintail, p. 52

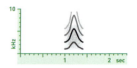

Quite variable
  Black-bellied Whistling-Duck, p. 37

A LOW-PITCHED NOTE
A low whistle, such as a Whoop, Hoot, or Coo, or a low burry or harsh note, such as a Growl or Groan.

OOIT
A low, mellow, upslurred whistle, longer than Pwut (p. 519), very slightly nasal.

  Sora, p. 113

Softer, shorter, slightly lower
  Boreal Owl, p. 220

Even shorter
  Long-eared Owl, p. 218
  Greater Prairie-Chicken, p. 76

WHOOP
A low, mellow whistle or slightly nasal note, medium-long, slightly overslurred.

Usually loud
  Tundra Swan, p. 43
  Mute Swan, p. 42

Rather soft
  Common Loon, p. 180

WOW
Longer than Wheew; lower, more nasal, and often longer than Whoop.

  Redhead, p. 56

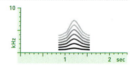

WHUP
Low, short whistled note, often slightly complex, like a low-pitched Chirp (p. 501). Usually plastic.

Whistled to breathy
  Muscovy Duck, p. 44

Slightly screechy
  Canvasback, p. 54

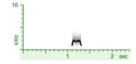

HOOT/COO
Low, clear, medium to long notes.

Nearly monotone
  Sharp-tailed Grouse, pp. 74–75
  Long-eared Owl, p. 218

Upslurred
  Common Eider, p. 56
  Common Ground-Dove, p. 88
  Great Gray Owl, p. 217

Overslurred
  Common Eider, p. 56
  Snowy Owl, p. 214

Overslurred, broken, sometimes hoarse
  Mourning Dove Growl Song, p. 90

Downslurred
  Barred Owl Hoo-wah, p. 216

GRIP
Like Pwut (p. 519), but burry. Mostly audible at close range.

Common Poorwill, p. 100   Chuck-will's-widow, p. 102

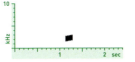

GROWL
A very low, nearly monotone burry note. Usually audible only at close range.

Greater Roadrunner, p. 95   Eastern Whip-poor-will,
Common Pauraque, p. 103   p. 101
Chuck-will's-widow, p. 102

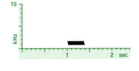

GROAN
A very low, nearly monotone, partly noisy note. Usually audible only at close range. See also *Croaking Groan (p. 527).*

Long, very low
    Red-billed Pigeon, p. 86

Long, higher, nasal
    Red-billed Pigeon, p. 64
    Brown Pelican, p. 184

American White Pelican,
    p. 184
Little Blue Heron, p. 190

Short, very low
    Montezuma Quail, p. 70

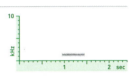

Common Nighthawk, p. 98

A BURRY OR BUZZY NOTE
A note that rises and falls very rapidly in pitch, creating a trilled effect. More musical versions are called burry; less musical versions are called buzzy. Both burry and buzzy notes can be either coarse (meaning the up-and-down changes in pitch are slower) or fine (meaning the changes are faster).

TREMOLO
A musical, mellow whistled note that repeatedly rises and falls in pitch, slowly enough that each change in pitch is clearly audible.

Medium-high, seminasal
    Long-billed Curlew, p. 132
    Least Tern, p. 178

Low, monotone or upslurred
    Common Loon, p. 180

Even lower, long, downslurred
    Eastern Screech-Owl, p. 223

Similar but monotone or upslurred, tremolo subtle
    "McCall's" Eastern
    Screech-Owl, p. 223

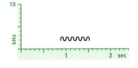

BREET
A medium-high, coarsely burry, nearly monotone musical note. Can also be considered a short trill.

Often slightly rising
    Pectoral Sandpiper, p. 137
    Baird's Sandpiper, p. 138
    Least Sandpiper, p. 139

Higher, usually downslurred
    Western Sandpiper, p. 140

High, coarse, musical
    Sandhill Crane (juvenile),
    p. 118

Whooping Crane (juvenile),
    p. 119

Slighly lower, less musical
    Great Crested Flycatcher,
    p. 265
    Evening Grosbeak, p. 371
    Bronze Mannikin, p. 357

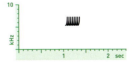

Slightly lower, often slightly downslurred
    Ash-throated Flycatcher,
    p. 264

BURRT
Like Breet (p. 509), but lower.

Montezuma Quail, p. 70
Least Flycatcher, p. 255
Great Crested Flycatcher, p. 265
Brown-crested Flycatcher, p. 266
Loggerhead Shrike, p. 274
Northern Shrike, p. 275

Common Myna, p. 352
Lower, less musical, Churtlike
Snowy Plover, p. 123
Wilson's Plover, p. 124
Pectoral Sandpiper, p. 137
Curve-billed Thrasher, p. 348
Yellow-headed Blackbird, p. 478

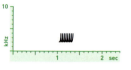

Even lower, quite musical
Purple Martin, p. 305
Lark Bunting, p. 438

BEERT
Like Burrt, but downslurred. See also Downslurred Trill, p. 539.

Carolina Wren, p. 322

House Wren, p. 319

BREER
Like Burrt, but longer and distinctly overslurred. Coarser and lower than Vreer.

Brown-crested Flycatcher, p. 266
Alder Flycatcher, p. 258
Couch's Kingbird, p. 269

Short, coarse, plastic
Olive-sided Flycatcher, p. 254

VEET
Finely burry musical monotone note.

Medium-high
Canyon Wren, p. 321

Slightly lower
Wood Thrush, p. 340
Veery, p. 335
Eastern Meadowlark, p. 470

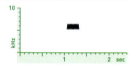

VEW
Like Veet, but sharply downslurred.

Medium pitch
Lapland Longspur, p. 373
Chestnut-collared Longspur, p. 375

High, sharply downslurred
Horned Lark, p. 299

VEER (TYPICAL)
Like Vew, but longer.

Hutton's Vireo, p. 286
Purple Martin, p. 305
Bicknell's Thrush, p. 337
Gray-cheeked Thrush, p. 336

Veery, p. 335
Wood Thrush, p. 340
Red-whiskered Bulbul, p. 330

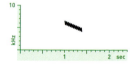

VEER (LONG)
Like Veer, but lower and much longer. Medium-low pitch. Occasionally broken into a few similar but shorter whistles in slightly falling series.

Montezuma Quail Male Song, p. 70

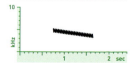

VREER
Like Veet, but overslurred. Finer and higher than Breer.

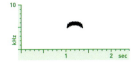

Long, high
Bicknell's Thrush, p. 337
Gray-cheeked Thrush, p. 336

Shorter, slightly lower
Smith's Longspur, p. 374

VEWY
Like Pewy, but finely burry.

Veery, p. 335
Scarlet Tanager, p. 456

Summer Tanager, p. 457

CHEET
Like Breet (p. 509), but less musical. Coarser than Dzeet.

Medium-high, slightly musical
Western Sandpiper, p. 140
Loggerhead Shrike, p. 274
Northern Shrike, p. 275
Northern Rough-winged
Swallow, p. 303
Tree Swallow, p. 304

Brown-headed Cowbird
(juvenile), p. 479

Medium pitch, coarse, less musical
Yellow-bellied Flycatcher,
p. 256

CHURT
Like Burrt, but less musical. Coarser than Dzert (p. 512). Lower than Cheet. Higher than Rasp (p. 526).

Medium-low
Cliff Swallow, p. 301
Bank Swallow, p. 302
Northern Rough-winged
Swallow, p. 303

Nearly toneless
Dickcissel, p. 455
Snow Bunting, p. 377

Low, coarse, rough
Sedge Wren, p. 326

Common Yellowthroat, p. 391

CHEWT
Like Churt, but distinctly downslurred.

Low, fine, rough
Northern Mockingbird,
p. 349

Higher, longer
Green Kingfisher, p. 227
Verdin, p. 313

SHEET
A semimusical, coarsely burry Chirplike note with a distinctive lisping or sibilant quality; more complex and musical than Cheet.

Tree Swallow, p. 304
Northern Rough-winged
Swallow, p. 303

House Sparrow, p. 356

DZEET
Like Dzit (p. 503), but longer. Lower and more abrupt than Burry Seet (p. 503). Usually nearly monotone. Shorter than Buzz (p. 512).

Medium-high, fine
Swamp Sparrow, p. 444
Lincoln's Sparrow, p. 443

Medium pitch, fairly coarse
Indigo Bunting, p. 462
Lazuli Bunting, p. 463
Painted Bunting, p. 465

Medium-low, fairly coarse
Blue Grosbeak, p. 466

DZERT
Like Dzeet (p. 511), but lower.

Bewick's Wren, p. 323
Eastern Meadowlark, p. 470

Smith's Longspur, p. 374

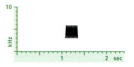

BUZZ
One long monotone buzzy note. Longer than Dzeet (p. 511) or Dzert.

Loggerhead Shrike, p. 274
Northern Shrike, p. 275
Common Redpoll, p. 366
Hoary Redpoll, p. 367

Longer, sometimes with low consonant at start
Spotted Towhee, p. 425

High, insectlike, with introductory notes that can be inaudible at a distance
Grasshopper Sparrow, p. 447
Le Conte's Sparrow, p. 451
Nelson's Sparrow, p. 454

DZIK
Very brief buzzy note, medium in pitch, often slightly nasal. Higher and less musical than Grip (p. 509).

Very brief
Western Grebe, p. 79
Clark's Grebe, p. 78
Common Nighthawk, p. 98

Antillean Nighthawk, p. 100

Longer, variable
Bewick's Wren, p. 323

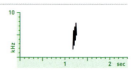

DZWEET
Like Dzert, but upslurred. Sometimes in series; see Dzik Series (p. 541).

Medium coarse, soft, high
Savannah Sparrow, p. 448
Indigo Bunting, p. 462

Finer, higher
Black-and-white Warbler, p. 385

Fine, soft, but harsher, more snarling
Vesper Sparrow, p. 435
Lark Sparrow, p. 436
Song Sparrow, p. 442
Swamp Sparrow, p. 444

Lincoln's Sparrow, p. 443
Lark Bunting, p. 438

DZREE
Like Dzweet, but longer and less obviously upslurred.

Coarse; upslur slight
Western Wood-Pewee, p. 252

DZEER
Like Dzert, but downslurred.

Usually given singly; very high
Eastern Kingbird, p. 271

Usually given singly; longer, lower, coarser
Western Wood-Pewee, p. 252

Usually in series or mixed with other calls
Alder Flycatcher, p. 258
Couch's Kingbird, p. 269
Scissor-tailed Flycatcher, p. 272

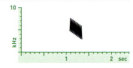

PEENT
Medium-pitched, extremely fine buzzy note with some nasal quality.

Short, monotone, from ground
American Woodcock, p. 143

Downslurred, often 2-syllabled
Common Nighthawk, p. 98

Lower, longer
Sandwich Tern, p. 173
Royal Tern, p. 172
Forster's Tern, p. 176

A SHORT NASAL NOTE
A brief seminasal or nasal note, broken or unbroken.

EEP
Extremely brief seminasal to nasal notes with almost no change in pitch. Pitch medium-high. Consonant sounds weak or absent. The nasal version of Tip (p. 497).

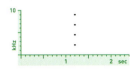

Soft, often repeated
Ruby-throated Hummingbird, p. 106
Black-chinned Hummingbird, p. 107

Red-breasted Nuthatch, p. 314
White-breasted Nuthatch, p. 315

Loud, quite high
Rose-breasted Grosbeak, p. 460

IP
Extremely brief seminasal to nasal notes. Sharply downslurred, though this can be difficult to hear. The nasal version of Chip (p. 497).

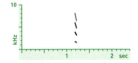

Medium-high
Ruby-throated Hummingbird, p. 106
Black-chinned Hummingbird, p. 107

Medium-low
Bronzed Cowbird, p. 480

WINK
Extremely brief seminasal to nasal notes. Sharply upslurred, though this can be difficult to hear. The nasal version of Whit (p. 496).

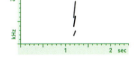

Sanderling, p. 135
Brown-headed Nuthatch, p. 316

Black-headed Grosbeak, p. 461

WIK (HIGH)
Like Wip (p. 500), but seminasal to fairly nasal. Briefer and sharper than High Reek (p. 516).

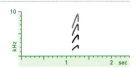

High, clear
Least Grebe, p. 84
Long-eared Owl (juvenile), p. 218

Medium-high, more nasal
Eared Grebe, p. 80

Lower, fairly nasal
Black-necked Stilt, p. 121

KEW
Like Pew (p. 506), but seminasal to fairly nasal. Briefer and sharper than high Keer.

Medium pitch, sharp
Boreal Owl, p. 220

Higher, less sharp
Semipalmated Plover, p. 125
Arctic Tern, p. 175
Bonaparte's Gull, p. 158
Black-headed Gull, p. 158

Slightly lower, even less sharp
Forster's Tern, p. 176

Even lower and less sharp
Northern Saw-whet Owl, p. 221
Boreal Owl, p. 220
Elf Owl, p. 224

Fairly low, fairly sharp
Limpkin, p. 120
Little Gull, p. 157
Black Tern, p. 179

Quite low, mellow
Black Rail, p. 109
Eastern Screech-Owl, p. 223
Western Screech-Owl, p. 222
Northern Cardinal, p. 458

SQUEAK
A brief broken seminasal note. Higher than Yelp (see Yelp Series, p. 543); higher and clearer than Screechy Barks (p. 520).

Least Bittern, p. 188

Virginia Rail, p. 112

CHIRP (SQUEALING)
A high, rapid, complex seminasal note, usually broken.

Budgerigar, p. 246
Monk Parakeet, p. 246
Green Parakeet, p. 246
Mitred Parakeet, p. 247
Red-masked Parakeet, p. 247
White-eyed Parakeet, p. 247

Blue-crowned Parakeet, p. 247
White-winged Parakeet, p. 248
Yellow-chevroned Parakeet, p. 248
Lilac-crowned Parrot, p. 249

White-fronted Parrot, p. 249
Orange-winged Parrot, p. 249

KLEE
A brief high whistled or seminasal note, starting at one pitch, then almost immediately breaking upward to a monotone pitch.

American Avocet, p. 122

Aplomado Falcon, p. 244

KLIP
A brief upslurred nasal note that breaks upward. Lower than Klee, with more abrupt end. Medium pitch.

Burrowing Owl, p. 225

Aplomado Falcon, p. 244

TSOOK
A low, swallowed Bark that begins as a brief high Squeak, the two tone qualities nearly simultaneous.

Northern Shoveler, p. 51
Blue-winged Teal, p. 50
Sharp-tailed Grouse, pp. 74–75

Golden Eagle, p. 210
Peregrine Falcon, p. 242
Groove-billed Ani, p. 96

HONK (UNBROKEN)
Brief monotone nasal note, low to medium pitched. Listed from highest to lowest pitch within groups.

High, some approaching Keek in pitch
> Snow Goose, p. 39
> Least Grebe, p. 84
> Black-legged Kittiwake, p.156
> Ross's Goose, p. 39

Slightly higher
> Cackling Goose, p. 40

Pitch variable; clear to burry
> Common Raven, p. 290

Medium-low pitch
> Trumpeter Swan, p. 43

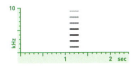

Low, long; like a truck horn
> Greater White-fronted Goose, p. 41
> Brant, p. 41
> White Ibis, p. 195

HONK (BROKEN)
Brief nasal note, starting low, then almost immediately breaking upward to a nearly monotone pitch.

> Brant, p. 41
> Cackling Goose, p. 40
> Canada Goose, p. 40
> Greater White-fronted Goose, p. 41

Longer, more ringing
> Whooping Crane, p. 119

Distinctly 2 syllabled, second note wailing
> Indian Peafowl, p. 65

KEEK
High, brief monotone nasal note.

Extremely brief
> Purple Gallinule, p. 115
> Common Gallinule, p. 116

Slightly longer
> Sora, p. 113

Variably noisy, often in series
> Golden-fronted Woodpecker, p. 230

YAP
Higher than Nasal Bark; much like Keek, but longer and more noticeably overslurred, with weaker consonant sounds.

> Harlequin Duck, p. 57
> Bridled Tern, p. 170
> Royal Tern, p. 172

> Sandwich Tern, p. 173
> Black Tern, p. 179

YIP
Like Yap but slightly briefer and noisier.

> European Starling, pp. 350–351

BARK (NASAL)
Brief nasal note of medium pitch.

Very brief, sounding like "Kek"
> Cooper's Hawk, p. 202
> Common Gallinule, p. 116
> Black-necked Stilt, p. 121
> Black Skimmer, p. 168

Longer, usually clear, highly nasal
> Long-tailed Duck, p. 57
> Harlequin Duck, p. 57
> Cattle Egret, p. 192
> American Coot, p. 117
> Black Skimmer, p. 168

> Great Horned Owl, p. 213
> Fish Crow Aw, p. 289

A SHORT NASAL NOTE, CONTINUED

WIK (LOW)
A brief upslurred nasal note, lower than High Wik (p. 513); shorter and sharper than Low Reek.

Black-billed Magpie, p. 298

A LONGER NASAL NOTE
A fairly long nasal or seminasal note, broken or unbroken, but lacking a harsh quality.

REEK (HIGH) OR REEP
High, upslurred seminasal note. Longer than High Wik (p. 513); higher than Low Reek. Sometimes in series (Reek-reek).

Wood Duck, p. 45
Red Knot, p. 134
Marbled Godwit, p. 133
Great Kiskadee, p. 267

Gray Jay, p. 294
White-collared Seedeater, p. 421
Audubon's Oriole, p. 486

KEER (HIGH)
High downslurred seminasal note. Longer than Kew (p. 514); higher than Low Reek. See also Jeer (p. 517) and Single Notes of Gulls (p. 518).

Ruddy Turnstone, p. 134
Bonaparte's Gull, p. 158
Black-headed Gull, p. 158
Roseate Tern, p. 177
Elf Owl, p. 224
White-breasted Nuthatch, p. 315

Blue Jay, pp. 292–293
Gray Jay, p. 294
White-collared Seedeater, p. 421
Rufous-crowned Sparrow, p. 426

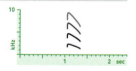

Higher, longer
Long-eared Owl (juvenile), p. 218

WEEW (SEMINASAL)
High, overslurred seminasal note without strong consonant sounds; longer and higher than Yap (p. 515). Higher than Reer.

American Oystercatcher, p. 122

Hudsonian Godwit, p. 133

TEWEE (SEMINASAL)
A medium-low, seminasal underslurred note. See also Whistled Pyoowee or Tewee (p. 507).

Variable and plastic
Brown Thrasher, p. 346
Long-billed Thrasher, p. 347

Curve-billed Thrasher, p. 348

REER
Medium-high, medium-long, slightly nasal overslurred note. Slightly higher and less nasal than Mew. Lower than Weew.

Common Raven, p. 290

MEW

Medium-high, medium-long, highly nasal notes, variable in pitch pattern. Longer than Yap (p. 515); shorter than Wail (p. 518); higher than Yank. See also Single Notes of Gulls (p. 518).

High, usually upslurred
Green-tailed Towhee, p. 423

Slightly lower, whinier, usually upslurred
Loggerhead Shrike, p. 274
Northern Shrike, p. 275

Slightly lower, more nasal, usually downslurred
Yellow-bellied Sapsucker, p. 233
Cooper's Hawk, p. 202

Low, nasal, usually overslurred, sometimes harsh
Gray Catbird, p. 345

Low, nasal, brief, variably harsh
Scott's Oriole, p. 487

REEK (LOW)

A brief upslurred nasal note, lower and more nasal than High Reek. Longer than Low Wik.

Wilson's Phalarope, p. 151
Blue-headed Vireo, p. 158

Black-billed Magpie, p. 298

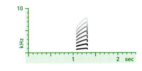

KEER (LOW)

High downslurred seminasal note. Longer than Kew (p. 514); higher than Low Reek (p. 517).

Willet (Eastern), p. 149
Willet (Western), p. 148

Brown Jay, p. 295

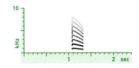

JEER (NASAL)

A medium-long, slightly downslurred nasal note, without any burry quality. Lower and more nasal than High Keer.

Shrill, sometimes metallic
Blue Jay, pp. 292–293

YANK

Medium-low, medium-long nasal notes, usually nearly monotone, but not as monotone as Honk (p. 515). Lower than Mew.

Plastic, nearly monotone, medium in pitch and length
Least Grebe, p. 84
Wild Turkey, p. 67
American Avocet, p. 122

Upland Sandpiper, p. 130
Red-masked Parakeet, p. 247
Mitred Parakeet, p. 247

Red-breasted Nuthatch, p. 314

SCREAM (NASAL)

A long, medium- to high-pitched seminasal or nasal note. Higher than Wail (p. 518). Some versions broken.

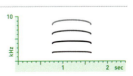

Monotone, very high, whistled
Sharp-shinned Hawk, p. 201

Downslurred
Gray Hawk, p. 205
Short-tailed Hawk, p. 207
Ferruginous Hawk, p. 209

Rough-legged Hawk, p. 209

Overslurred
Zone-tailed Hawk, p. 208

WAIL

A long, medium-pitched nasal note. Pitch, pattern, and length are plastic in all species, but in general, Wails of owls tend to be overslurred; Wails of hawks tend to be downslurred; and Wails of gulls are often monotone, broken, and/or in series. In many bird species, tremolo versions may be given in agitation.

Pitch quite variable
Great Black-backed Gull, p. 167
Purple Gallinule, p. 115
Ring-billed Gull, p. 161
Herring Gull, p. 163
Glaucous Gull, p. 165
Boreal Owl, p. 220

Medium-high pitch
Ferruginous Hawk, p. 209
Franklin's Gull, p. 160
Rough-legged Hawk, p. 209

Northern Saw-whet Owl, p. 221
Northern Goshawk, p. 203

Medium pitch
Laughing Gull, p. 159
Sora, p. 113
American Coot, p. 117
Red-throated Loon, p. 181
Eastern Screech-Owl, p. 223
Barred Owl (usually upslurred), p. 216

Low, sometimes nearly moaning and/or slightly hoarse
Ruffed Grouse, p. 72
Black-legged Kittiwake, p.156
Lesser Black-backed Gull, p. 166
Great Horned Owl, p. 213
Long-eared Owl, p. 218
Crested Caracara, p. 245

MOAN

A very low-pitched nasal sound, usually long and nearly monotone.

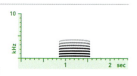

Trumpeter Swan, p. 43
Wood Stork (juvenile), p. 183
Northern Gannet, p. 183
Snowy Egret, p. 190
Atlantic Puffin, p. 153

Medium-high
Harlequin Duck, p. 57

Very short, audible at close range
Mountain Plover, p. 128

SINGLE NOTES OF GULLS

Extremely plastic, individually variable Yelps, Squeals, Keers, and broken Wails, usually 1-syllabled.

Very high, squealing
juveniles of many gull species

Medium-high to high, variable
Ring-billed Gull, p. 161
Herring Gull, p. 163
Glaucous Gull, p. 165

Medium-high, clear and nasal
Franklin's Gull, p. 160

Medium-low, clear and nasal
Laughing Gull, p. 159

Medium-low, yelping, sometimes hoarse
California Gull, p. 162
Lesser Black-backed Gull, p. 166

Low, hoarse, squawking
Great Black-backed Gull, p. 167

KLEER

Like Keer (p. 517), but breaking downward to a lower pitch near start.

Northern Flicker, p. 240

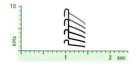

SQUEAL (TYPICAL)

A high, long, broken seminasal note. Higher than Broken Wail. See also Single Notes of Gulls.

Rough-legged Hawk, p. 209
Cooper's Hawk (juvenile), p. 202
Northern Goshawk, p. 203
Mississippi Kite, p. 200

Often multisyllabled, with several voice breaks per note
Greater White-fronted Goose, p. 41

SQUEAL (LONG)
Like Squeal, but longer. Higher than Broken Wail.

Short
    many gull species (begging
      calls; see p. 155)

Averaging longer
    Red-tailed Hawk, p. 204
    Short-tailed Hawk,
      p. 207

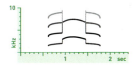

A RISING BROKEN NASAL NOTE
*An upslurred nasal or seminasal note that breaks to a higher pitch at
end. See also 2-Syllabled Nasal Note (p. 554) and Longer Upslurred
Whistles (p. 505).*

Wood Duck Reek, p. 45
Stilt Sandpiper Weep, p. 135
Smooth-billed Ani Weelip,
    p. 97

Mellow, whooping
Mute Swan Whoop, p. 42

WAIL (BROKEN)
*A Wail that breaks at least once to a different pitch. Note that most
vocalizations in the Wail category can occasionally break. See also
Single Notes of Gulls.*

Medium pitch
    Purple Gallinule, p. 115

Low pitch
    California Gull, p. 162

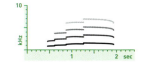

WAIL (LONG, BROKEN)
*A long, mellow, slightly nasal whistle that breaks upward 1–3 times,
sometimes downward at end. See also 2-Syllabled Whistles (p. 551).*

Common Loon, p. 180

A SHORT HARSH NOTE
*Very brief noisy notes, nasal notes, or low sharp whistles, with little musical quality. Longer than Clicks or
Taps, and often lower in pitch.*

PWUT
*Very low, brief upslurred whistle, sharp and not very musical; may
recall the sound of a drip of water.*

Surf Scoter, p. 58
Plain Chachalaca, p. 66
Sharp-tailed Grouse, pp. 74–75
Black-crowned Night-Heron,
    p. 193

Eastern Whip-poor-will,
    p. 101
Common Pauraque, p. 103
Swainson's Thrush, p. 338

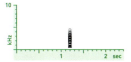

KUK
*A quick low nasal note, lower and shorter than Bark (p. 515). More
nasal than Cluck (p. 520).*

Quite nasal
    Tundra Swan, p. 43
    Scaled Quail, p. 69
    Lesser Nighthawk, p. 99
    Spruce Grouse, p. 73

Less nasal, less noisy
    American Robin, pp. 342–343

CLUCK
Like Kuk (p. 519), but harsher, sometimes more complex. Often in series.

Blue-winged Teal, p. 50
Lesser Scaup, p. 55
Greater Scaup, p. 55
Harlequin Duck, p. 57
Red-billed Pigeon, p. 64

Wild Turkey, p. 67
Shiny Cowbird, p. 481

Higher, upslurred
Ruffed Grouse, p. 72

(Red Squirrel)

CHIT
A brief high noisy note, sometimes partly nasal. Higher than Chuk.

Gray Partridge, p. 71
Downy Woodpecker, p. 234
Gray Vireo, p. 283

House Wren, p. 319
Marsh Wren (Eastern), p. 325
Marsh Wren (Western), p. 324

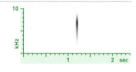

CHUK
A brief low noisy note, sometimes partly nasal. Lower than Chit.

Often fairly nasal
Golden-fronted Woodpecker, p. 230
Ladder-backed Woodpecker (rare), p. 236

Usually lower and harsher
Least Bittern, p. 188
Limpkin, p. 120
Cactus Wren, p. 327
Brown Thrasher, p. 346
Long-billed Thrasher, p. 347

Song Sparrow, p. 442
Bobolink, p. 490
Red-winged Blackbird, pp. 468–469
Great-tailed Grackle, pp. 472–473
Boat-tailed Grackle, p. 474
Common Grackle, p. 475
Rusty Blackbird, p. 476
Brewer's Blackbird, p. 477
Brown-headed Cowbird, p. 479

Orchard Oriole, p. 482

Often slightly longer, grating
Yellow-headed Blackbird, p. 478

Shorter, more Keklike
Yellow-breasted Chat, p. 419
Hooded Oriole, p. 483

CHWUT
A brief, low upslurred noisy note.

Gray Catbird, p. 345

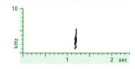

CHUP
Like Chip (p. 497) but lower, slightly nasal or noisy, downslurred.

Medium pitch
Common Yellowthroat, p. 391
Seaside Sparrow, p. 450
Saltmarsh Sparrow, p. 453

Medium-low
Hermit Thrush, p. 339

Curve-billed Thrasher, p. 348
Western Meadowlark, p. 471
Rusty Blackbird, p. 476

Rather barking quality
Gyrfalcon, p. 243
Scott's Oriole, p. 487

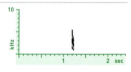

Boat-tailed Grackle, p. 474

A LONGER HARSH NOTE

BARK (SCREECHY)
A very brief, high, harsh nasal note. Lower than Cheep (p. 501).

White-winged Scoter, p. 58
Red-necked Grebe, p. 81
Red-throated Loon, p. 181

Virginia Rail, p. 112
Least Tern, p. 178

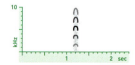

BARK (ROUGH)
A brief note that is both nasal and noisy. Pitch medium.

Averaging noisier, less nasal
Bufflehead, p. 59
Greater Prairie-Chicken,
  p. 76
Lesser Prairie-Chicken,
  p. 77
Purple Swamphen, p. 114
Short-eared Owl, p. 219
Long-eared Owl, p. 218
Snowy Owl, p. 214

High, noisy, sharply downslurred
Red-bellied Woodpecker,
  p. 231

Quite harsh, downslurred, often in flight
Lilac-crowned Parrot, p. 249
Red-crowned Parrot, p. 249
White-fronted Parrot, p. 249

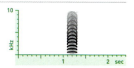

High, clear, almost a Yap
Great Gray Owl, p. 217

Often short, partly noisy or burry
Ruffed Grouse, p. 72
Spruce Grouse, p. 73
Common Murre, p. 154

SQUAWK
A brief, medium-pitched nasal and noisy note. Lower and shorter than Screech; harsher than Bark; like a harsh version of Honk (p. 515).

American Bittern, p. 187
Black-crowned Night-Heron,
  p. 193
Yellow-crowned Night-Heron,
  p. 193

Great Blue Heron, p. 189
Snow Goose, p. 39

Harsh, very nasal
Smooth-billed Ani, p. 97

GRUNT (TYPICAL)
A low, brief noisy note, usually with a slight nasal quality. Longer and lower than Chuk. Compare burry versions of Grunt (p. 25).

Slightly nasal
Canada Goose, p. 40
Ross's Goose, p. 39
Muscovy Duck, p. 44
Common Merganser, p. 62
Northern Fulmar, p. 182
Northern Gannet, p. 183
Double-crested Cormorant,
  p. 185

Neotropic Cormorant, p. 186
American White Pelican,
  p. 184
Brown Pelican, p. 184
American Bittern, p. 187
Roseate Spoonbill, p. 195

Slightly nasal and 2-syllabled
Reddish Egret, p. 191

Slightly harsher
Caspian Tern, p. 171
Florida Scrub-Jay, p. 297

Voiceless, huffing
Black Vulture, p. 196

SCREECH
A high, medium-long nasal and noisy note. Usually plastic, but generally not broken or burry. Higher than Squawk; shorter than Shriek (p. 522).

Green-winged Teal, p. 53
Scaled Quail Song, p. 69
White Ibis, p. 195
White-tailed Hawk, p. 208
Virginia Rail, p. 112
Clapper Rail, p. 110
King Rail, p. 111
Purple Gallinule, p. 115
Purple Swamphen, p. 114
American Coot, p. 117

Wilson's Snipe, p. 142
Common Murre, p. 154

Long, plastic, variably nasal
Great Horned Owl, p. 213
Great Gray Owl, p. 217
Long-eared Owl, p. 218

Singly, by juveniles
Common Raven, p. 290
Chihuahuan Raven, p. 291

Often in long series, by juveniles
Common Tern, p. 174
Arctic Tern, p. 175
Forster's Tern, p. 176
Roseate Tern, p. 177
Least Tern, p. 178
Brown Noddy, p. 170
Black Tern, p. 179

REECH
Like Screech, but upslurred.

Plain Chachalaca, p. 66

SKEWCH
Like Screech (p. 521), but downslurred.

Green Heron, p. 192

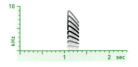

SHRIEK
Long, high, partly nasal harsh note. Longer than Screech (p. 521).

Usually monotone
Barn Owl, p. 212
Chestnut-fronted Macaw,
  p. 248

Usually upslurred
Great Horned Owl, p. 213
Snowy Owl, p. 214

Great Gray Owl, p. 217
Barn Owl (juvenile), p. 212
Northern Hawk Owl, p. 215
Barred Owl, p. 216
Short-eared Owl, p. 219

Usually with short Kuk notes
Green Heron, p. 192

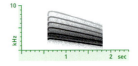

Usually overslurred, broken
Wood Duck, p. 45

SCREAM (HARSH)
Long harsh note, at least slightly nasal or seminasal. Harsher than Wail (p. 518); more nasal than Shriek (p. 522).

Very harsh, snarling, monotone
Harris's Hawk, p. 205

Downslurred
Red-tailed Hawk, p. 204

Swainson's Hawk, p. 207
Cooper's Hawk, p. 202

Overslurred
Common Raven, p. 290

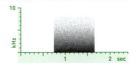

HISS
A rushing of air, a toneless burst of noise without any musical or nasal quality. Usually softer than Snarl (p. 526); not grating.

Canada Goose, p. 40
Mute Swan, p. 42
Wood Stork, p. 183
Turkey Vulture, p. 196
Barn Owl, p. 212

Le Conte's Sparrow, p. 451
Red-winged Blackbird,
  pp. 468–469

Lower, roaring
Black Vulture, p. 196

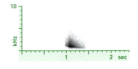

Brief, usually in steady series
Muscovy Duck, p. 44

KHOW
A brief, very low, downslurred whispered hiss.

Green Heron, p. 192

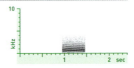

FOOM
A low, downslurred rushing of air, often slightly nasal, with abrupt start.

Common Nighthawk, p. 98   Antillean Nighthawk, p. 100

WHOOSH
A swishing sound made by opening and closing the spread tail feathers. Not very loud.

Spruce Grouse, p. 73

CRACKLE
A burst of noise containing ticking and knocking notes, recalling the sound of a breaking stick. See also Long Songs Punctuated by Crackles (p. 565).

Soft, short
Long-tailed Duck, p. 57

Longer, fairly loud
Anhinga, p. 186
Crested Caracara, p. 245

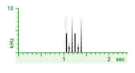

A HARSH BURRY NOTE
A sound with some combination of nasal, noisy, and burry qualities.

VIMP
Brief, finely burry nasal note, sometimes slightly harsh. Can be difficult to distinguish from Jit (p. 528).

Medium-high, clear
Winter Wren, p. 318
Song Sparrow, p. 442

VRIT
Like Vimp, but lower, harsher, and often coarser.

Blue Jay, pp. 292–293

Medium-low, slightly harsh
Gray Jay, p. 294
Woodhouse's Scrub-Jay,
p. 296

Red-vented Bulbul, p. 330

Sharply downslurred
Great Kiskadee, p. 267

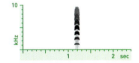

KRIK
Like Grate (p. 525) but very brief. Coarser and harsher than Vimp.

Medium-low, coarse, grating
Mountain Plover, p. 128
Bonaparte's Gull, p. 158
Black-headed Gull, p. 158
Royal Tern, p. 172

Sandwich Tern, p. 173

Fairly high, coarse, grating
Red-cockaded Woodpecker,
p. 237

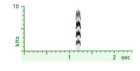

YANK (BURRY)
A burry nasal note. Higher and clearer than Caw (p. 524).

White-breasted Nuthatch,
p. 315

CAW
Brief overslurred nasal note, always at least slightly burry and at least slightly harsh. Pitch low to medium. Finer than Churr.

American Crow, p. 288

More nasal
Fish Crow, p. 289

Deeper, coarser
Common Raven, p. 290
Chihuahuan Raven, p. 291

CHURR
Like Grate, but lower and harsher. More nasal than Snarl (p. 526), and less harsh. Finer than Churring Rattle (p. 531). Longer than burry versions of Grunt. See Chatter Duets (p. 544).

Coarser, more nasal
Plain Chachalaca, p. 66
Black-necked Stilt, p. 121
Sooty Tern, p. 169

Finer, usually harsher
Western Meadowlark, p. 471
Rusty Blackbird, p. 476
Brewer's Blackbird, p. 477

Boat-tailed Grackle, p. 474

KWIRR
A brief nasal burr, very coarse, almost a tremolo. Higher, coarser, and clearer than Churr. Lower and more musical than Grate.

High, brief, coarse, monotone
Red-bellied Woodpecker, p. 231
Golden-fronted Woodpecker, p. 230

Softer, even coarser, less nasal
Northern Flicker, p. 240

Low, fine, down- or overslurred
Horned Grebe, p. 82

KWEEAH
Like Kwirr, but higher and finer, often harsher. Lower and finer than Grate.

Red-headed Woodpecker, p. 232

Horned Grebe, p. 82

WHEEZE
A long, finely burry or hoarse nasal note. Easily confused with Whine (p. 529), but not polyphonic (the difference can be hard to hear).

Medium-high, up- or overslurred, slightly harsh
Spotted Towhee, p. 425

Lower, overslurred, finely burry
Eurasian Collared-Dove, p. 87

More monotone, hoarser
White-winged Dove, p. 91
American Woodcock (distraction), p. 143

PURR
A very low burry note, usually somewhat nasal. Finer than Bleat (p. 544), higher and coarser than Growl (p. 509).

Sharp-tailed Grouse, pp. 74–75
Ruffed Grouse, p. 72
Spruce Grouse, p. 73

Lower, coarser, slightly longer
Anhinga, p. 186
Least Bittern Song, p. 188

Very low, almost a Growl
Plain Chachalaca, p. 66

High, nasal, almost a Churr
Wild Turkey, p. 67

GRATE

A high-pitched coarsely burry nasal note, often harsh. Can be considered the burry version of a Screech (p. 521). Higher than Churr; longer than Krik (p. 523).

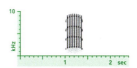

High, coarse
Clark's Grebe Song, p. 78
Western Grebe Long Grate, p. 79
Semipalmated Plover, p. 125
Mountain Plover, p. 128
Willet (Eastern), p. 149
Willet (Western), p. 148
Nanday Parakeet, p. 248

High, fine, almost screeching
Black Tern, p. 179
Wilson's Plover, p. 124
Red Phalarope, p. 150
Dunlin, p. 136

Medium pitch
Black-necked Stilt, p. 121
Sabine's Gull, p. 157
Bonaparte's Gull, p. 158

Black-headed Gull, p. 158
Nanday Parakeet, p. 248
Monk Parakeet, p. 246
White-eyed Parakeet, p. 247
Blue-crowned Parakeet, p. 247
Orange-winged Parrot, p. 249

KEER (GRATING)

A distinctly downslurred Grate.

Mountain Plover, p. 128
Royal Tern, p. 172
Common Tern, p. 174
Forster's Tern, p. 176

Black Tern, p. 179
Green Parakeet, p. 246
Bank Swallow, p. 302

WAIL (GRATING)

A variable, often plastic mix of a Wail (p. 518) and a Grate or a Cranelike Rattle (p. 531).

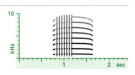

Grating quality rather fine
American Coot, p. 117
Common Gallinule, p. 116
Purple Gallinule, p. 115

Sooty Tern, p. 169

Grating quality rather rattling
Mute Swan, p. 42
Limpkin Kreew, p. 120

A burry wail or screech
Horned Grebe Song, p. 82

QUACK

A low, nasal note with a finely grating quality.

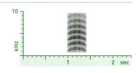

Rather grating and nasal, variably harsh
Mallard (female), p. 48
American Black Duck (female), p. 49
Mottled Duck (female), p. 49
Blue-winged Teal, p. 50
Cinnamon Teal, p. 51
Northern Shoveler, p. 51
Northern Pintail, p. 52
Lesser Scaup, p. 55

Tinnier, more grating, less nasal
Mallard (male), p. 48
American Black Duck (male), p. 49
Mottled Duck (male), p. 49

Higher, squeaky or screechy
Green-winged Teal, p. 53
Snowy Owl, p. 214

Low, nasal, clear
White-faced Ibis, p. 194
Glossy Ibis, p. 194

Short, nasal, monotone
Gadwall, p. 46

Nasal to croaking
Red-throated Loon, p. 181

Short, slightly squawking
Yellow-crowned Night-Heron, p. 193

GRUNT (BURRY)

Like Quack, but harsher and coarser. Generally lower and harsher than Churr. See typical Grunt (p. 521).

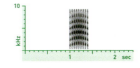

Rather fine and quacking
Gadwall, p. 46

Harsh, coarse, often churring
American Wigeon, p. 47
Eurasian Wigeon, p. 47
Redhead, p. 56
Canvasback, p. 54
Ring-necked Duck, p. 56

Greater Scaup, p. 55
Lesser Scaup, p. 55
Common Goldeneye, p. 60
Barrow's Goldeneye, p. 61
Bufflehead, p. 59
Red-breasted Merganser, p. 62
Hooded Merganser, p. 63

Red-necked Grebe, p. 81
Common Eider, p. 56
Smooth-billed Ani, p. 97

Rather high, nasal, coarse
Northern Harrier, p. 197

SNARL

A very harsh noisy note, like a loud Hiss (p. 522), but harsher, generally with a burry or grating quality; like Whine (p. 529) or Jurr (p. 529), but not polyphonic.

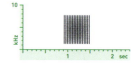

Medium-high
Budgerigar, p. 246
Tree Swallow, p. 304
Marsh Wren (Eastern), p. 325
Chipping Sparrow, p. 431
Vesper Sparrow, p. 435
Lark Bunting, p. 438
Snow Bunting, p. 377

Medium-high, often slightly grating
Roseate Tern, p. 177
Common Tern, p. 174
Arctic Tern, p. 175
Gull-billed Tern, p. 171

Medium-low
White-tailed Kite, p. 199

Red-breasted Nuthatch, p. 314
Black-tailed Gnatcatcher, p. 328
Purple Martin, p. 305
Cactus Wren, p. 327
European Starling, pp. 350–351
Common Myna, p. 352
Northern Mockingbird, p. 349
Spotted Towhee, p. 425

Sometimes more nasal
Bewick's Wren, p. 323

Slightly more nasal and churring
Pectoral Sandpiper, p. 137
Brown Thrasher, p. 346

Long-billed Thrasher, p. 347
Black-billed Magpie, p. 298

Long, like rattlesnake rattle
Burrowing Owl, p. 225

Medium-low, short, rather soft
Ladder-backed Woodpecker, p. 236
Woodhouse's Scrub-Jay, p. 296

Long, low, loud, sometimes slightly croaking
Great Blue Heron, p. 189
Black-crowned Night-Heron, p. 193

RASP

Like Snarl, but coarser and more grating. See also *Whine (p. 529).*

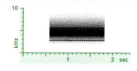

Medium pitch
Sedge Wren, p. 326

Low and harsh
Black-capped Vireo, p. 282
Marsh Wren (Eastern), p. 325

Yellow-breasted Chat, p. 419

Even lower and harsher
American Crow, p. 288
Cactus Wren, p. 327

SHRIEK (GRATING)

Like Shriek (p. 522), but grating.

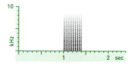

Glossy Ibis (juvenile), p. 194
White-faced Ibis (juvenile), p. 194

White Ibis (juvenile), p. 195
Roseate Spoonbill (juvenile), p. 195

CROAK (MONOTONE)

A note made of very rapid repeated Tocklike or knocking notes, ranging from toneless to nasal.

Medium to high, resonant, slightly musical
American Coot, p. 117
Purple Swamphen, p. 114

Medium-high, nasal, fine, almost quacking
Tamaulipas Crow, p. 287

Medium-low, harsh, coarse
American White Pelican, p. 184
Anhinga, p. 186
Brown Noddy, p. 170

Medium-low, harsh, rather fine
Great Blue Heron, p. 189
Great Egret, p. 189
Snowy Egret, p. 190
Little Blue Heron, p. 190
Reddish Egret, p. 191
Tricolored Heron, p. 191

Low, harsh, coarse, rattling
Limpkin, p. 120

Very low, resonant, rather musical
Common Raven, p. 290

Very low, like the snorts of a hog
Common Goldeneye, p. 60
Barrow's Goldeneye, p. 61
Double-crested Cormorant, p. 185
Neotropic Cormorant, p. 186

Similar but usually longer
Razorbill, p. 153

CROAK (SPEECHLIKE)

A sound made of rapid clicklike notes whose loudest (darkest) frequencies change over the course of the call, creating an effect vaguely reminiscent of human speech.

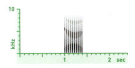

Tricolored Heron, p. 191
Razorbill, p. 153
Snail Kite, p. 200
Tamaulipas Crow, p. 287

Fish Crow, p. 289

Coarser, more rattling, usually in series
Crested Caracara, p. 245

CROAK (TICKING)

A short Croak made of soft Ticklike notes rather than knocking or resonant notes.

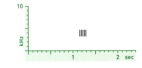

Common Goldeneye, p. 60

Barrow's Goldeneye, p. 61

GROAN (CROAKING)

Like low Groan (p. 509), but higher, more nasal, more croaking, and usually longer.

Northern Fulmar, p. 182
Great Blue Heron, p. 189
Great Egret, p. 189
Snowy Egret, p. 190
Little Blue Heron, p. 190

Reddish Egret, p. 191
Tricolored Heron, p. 191
Cattle Egret, p. 192

Usually higher, more Cawlike
American Crow, p. 288

Fish Crow, p. 289

A POLYPHONIC NOTE

A note created by the mixing of two simultaneous sounds, one generated by each side of a bird's syrinx, resulting in a characteristic dissonant, metallic, or whiny quality.

ZEET (TYPICAL)

A brief, clear, nearly monotone polyphonic note.

Medium-high
Rose-breasted Grosbeak, p. 460

Bobolink, p. 490
Orchard Oriole, p. 482
Hooded Oriole, p. 483

ZEET (BURRY)

Like Zeet, but coarsely burry.

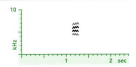

Varied Bunting, p. 464

ZREET

A clear, upslurred polyphonic note.

High to medium-high
Bobolink, p. 490
Eastern Meadowlark, p. 470
Western Meadowlark, p. 471
Black-headed Grosbeak, p. 461
Hooded Oriole, p. 483

White-winged Crossbill, p. 365

Medium-low
Cave Swallow, p. 300
Baltimore Oriole, p. 488
Bullock's Oriole, p. 489

Slightly lower, longer (Zree)
Eastern Towhee, p. 424

Still lower, shorter (Zoit)
Eastern Towhee, p. 424

A POLYPHONIC NOTE, CONTINUED

ZREE
A medium-long upslurred polyphonic note.

High, clear, long
American Goldfinch, p. 368

Medium-high, fairly long
Brewer's Blackbird, p. 477
Pine Siskin, p. 370
Common Redpoll, p. 366

Hoary Redpoll, p. 367

Medium-low, long
Lesser Goldfinch, p. 369

Medium-low, short, almost a Zreet
House Finch, p. 361

ZEER
A high, clear downslurred polyphonic note.

Cliff Swallow, p. 301
Cave Swallow, p. 300
Bank Swallow, p. 302

Bronzed Cowbird (juvenile),
p. 480

Higher, clearer
Bobolink, p. 490

PZEER
Like Zeer, but with distinct consonant sound at start.

Evening Grosbeak, p. 371

ZWEEW
Like seminasal Weew (p. 516), but polyphonic.

Rather high, clear
Pine Siskin, p. 370

Lower, slightly harsher
Black-headed Grosbeak,
p. 461

House Finch, p. 361
House Sparrow, p. 356
Eurasian Tree Sparrow,
p. 357

ZEWY
A musical, slightly polyphonic underslurred whistle.

Eastern Bluebird, p. 341

JIT
A short polyphonic note, often slightly noisy. Like Vimp (p. 523), but polyphonic instead of burry and nasal (the difference can be hard to hear). See also Finch Flight Call Group (p. 533).

Medium-high, unmusical, downslurred, Chiplike
Wilson's Warbler, p. 418

Lower, monotone, almost a Chit
Ruby-crowned Kinglet,
p. 332
Black-capped Vireo, p. 282

Medium-low, monotone
Bell's Vireo, p. 281
Scott's Oriole, p. 487

Low, seminoisy, upslurred
Barn Swallow, p. 302
Baltimore Oriole, p. 488
Bullock's Oriole, p. 489

Similar but clearer
House Finch, p. 361

Similar but more sharply upslurred
Hutton's Vireo, p. 286
Philadelphia Vireo, p. 278

Red-eyed Vireo, p. 276
Black-whiskered Vireo,
p. 277
White-eyed Vireo, p. 280
Warbling Vireo, p. 279

Averaging clearer
Yellow-throated Vireo, p. 284

JIRP

A polyphonic version of Chirp (p. 501). More complex and usually slightly harsher than Jit. See also Finch Flight Call Group (p. 533).

White-collared Seedeater, p. 421
House Finch, p. 361
House Sparrow, p. 356

Eurasian Tree Sparrow, p. 357

Slightly sharper
Magnolia Warbler, p. 413

WHINE

A long, harsh, polyphonic note, often finely burry. Medium pitch. Lke a polyphonic Snarl (p. 526). Compare Wheeze (p. 524).

High, long
Greater Roadrunner, p. 95

Usually not burry to finely burry
Loggerhead Shrike, p. 274
Northern Shrike, p. 275
Hermit Thrush, p. 339
Townsend's Solitaire, p. 334
Gray Catbird, p. 345

House Wren, p. 319
Cactus Wren, p. 327
Woodhouse's Scrub-Jay, p. 296
Florida Scrub-Jay, p. 297
Gray Jay, p. 294
Bobolink, p. 490
Yellow-headed Blackbird, p. 478

Often coarsely burry. See also Rasp (p. 526).
Hutton's Vireo, p. 286
White-eyed Vireo, p. 280
Gray Vireo, p. 283
Warbling Vireo, p. 279

SHEER

Like Whine, but higher and distinctly downslurred.

High, whiny
Blue-gray Gnatcatcher, p. 329

Lower, harsher
Black-tailed Gnatcatcher, p. 328

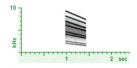

JENK

A brief, monotone burry polyphonic note, rather harsh, with a metallic ring.

Louder
Red-winged Blackbird, pp. 468–469

Common Grackle, p. 475
Brewer's Blackbird, p. 477

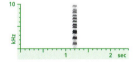

JURR

A harsh, metallic, monotone polyphonic snarl. Longer than Jenk; more monotone and more metallic than a burry Whine.

Florida Scrub-Jay, p. 297
Common Grackle, p. 475

Yellow-headed Blackbird, p. 478

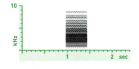

ZREEK

An upslurred, slightly harsh polyphonic note. Always at least slightly burry. Often in series.

Medium pitch, short, fine
Spot-breasted Oriole, p. 484
Audubon's Oriole, p. 486
Altamira Oriole, p. 485

Medium pitch, long, coarse
Woodhouse's Scrub-Jay, p. 296
Florida Scrub-Jay, p. 297

ZHREE
A long, upslurred polyphonic snarl.

    Pine Siskin, p. 370

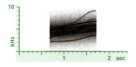

ZHEER
A medium-long downslurred burry polyphonic note. Variably harsh.

Medium-high, fine
    Red-eyed Vireo, p. 276
    Black-whiskered Vireo,
       p. 277

Yellow-green Vireo, p. 277
Philadelphia Vireo, p. 278
Yellow-breasted Chat, p. 419

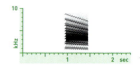

JEER (BURRY)
Like Zheer but lower, coarser, and more metallic.

    Common Grackle, p. 475

European Starling,
    pp. 350–351

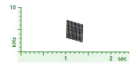

CHURR (POLYPHONIC)
A low, polyphonic coarse burr or short trill.

**Medium-low, sometimes slightly
noisy**
    Gray-cheeked Thrush, p. 336
    Bicknell's Thrush, p. 337
    Veery, p. 335

**Lower, clearer, audible only at
close range**
    Wood Duck, p. 45

Loud, harsh
    Common Grackle, p. 475

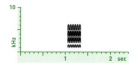

INDEX PART II: A SOUND MADE OF REPEATED SIMILAR NOTES

The same note repeated, without a significant pause. Includes series that change in speed or pitch, and irregular Twitters.

A TICKING OR TAPPING SERIES

A series or trill of Ticklike vocalizations, or of tapping or clapping sounds, including woodpecker drums and other mechanical sounds made by the bill or wings.

RATTLE (TICKING)

A trill or rapid series of Ticklike or clicking notes, coarser than Croaks (p. 527).

Voiceless, coarse
  Mountain Plover, p. 128
  American Crow, p. 288
  Fish Crow, p. 289
  Tamaulipas Crow, p. 287

Often slower, almost a series
  Florida Scrub-Jay, p. 297

Woodhouse's Scrub-Jay, p. 296

Fast, high
  Blue Jay, pp. 292–293
  Green Jay, p. 294
  Common Goldeneye, p. 60

Plastic, irregular in rhythm
  Western Grebe, p. 79

RATTLE (CROAKING)

Like a Croak (p. 527), but coarser; like a Ticking Rattle, but lower, with longer, noisier notes. Coarser, harsher, and less resonant than Cranelike Rattle and Knocking Rattle.

Plastic, sometimes crackling
  Crested Caracara, p. 245

Short, monotone
  Mute Swan, p. 42
  Cinnamon Teal, p. 51

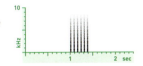

RATTLE (CRANELIKE)

A grating note with a distinctive resonant knocking quality. Lower and more knocking than Grate (p. 525).

Sandhill Crane, p. 118

Limpkin Kreew, p. 120

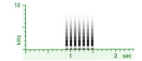

RATTLE (KNOCKING)

A short series of very low knocking notes. Lower and less grating than Cranelike Rattle. See also Croaking Rattle and Knocking or Purring Phrase (p. 552).

Common Raven, p. 290
Chihuahuan Raven, p. 291
Florida Scrub-Jay, p. 297
Woodhouse's Scrub-Jay, p. 296

Slower, longer, notes more Keklike
  Yellow-billed Cuckoo, p. 92
  Mangrove Cuckoo, p. 94

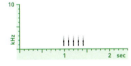

RATTLE (CHURRING)

More nasal than Croaking or Cranelike Rattles; coarser than Churr (p. 524).

Helmeted Guineafowl, p. 65
Wilson's Snipe, p. 142

Ringed Kingfisher, p. 228

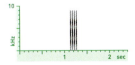

A SERIES OF SNAPS OR CLAPS
Made by the wings or bill.

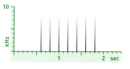

Wing claps in flight
most dove species (see
p. 85)
Barn Owl, p. 212

Eastern Whip-poor-will,
p. 101
Chuck-will's-widow, p. 102

A STEADY MECHANICAL DRUM OR RATTLE
A trill of taps, snaps, or claps made by wings or bill.

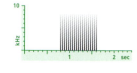

Drum of bill against hard surface
most woodpecker species,
(see p. 229)

Short quick rattle of bill snaps
Greater Roadrunner, p. 95

Short quick rattle of wing claps
Short-eared Owl, p. 219

A STEADY WING FLUTTER OR WHIRR
Rapid flutter of wings, generally audible only at close range.

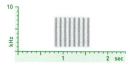

Without any musical quality
Spruce Grouse, p. 73
Sharp-tailed Grouse,
pp. 74–75
Inca Dove, p. 88
Common Ground-Dove,
p. 88

(other dove species
occasionally)

*With a whirring hum, during
flight display*
Scissor-tailed Flycatcher,
p. 272

AN ACCELERATING MECHANICAL DRUM OR RATTLE
Made by the wings or bill.

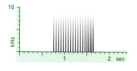

*Drum of bill against hard
surface; acceleration subtle*
American Three-toed
Woodpecker, p. 238
Black-backed Woodpecker,
p. 239
Pileated Woodpecker, p. 241

Bill snaps
Vermilion Flycatcher, p. 263

*Clicklike notes, then a quick low
buzz*
Ruddy Duck Bubble Display,
p. 63

Extremely low, beating thumps
Ruffed Grouse, p. 72

A DECELERATING MECHANICAL DRUM OR RATTLE
Made by the wings or bill.

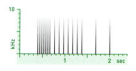

Drum of bill against hard surface; deceleration obvious
Yellow-bellied Sapsucker, p. 233

A PLASTIC, IRREGULAR MECHANICAL DRUM OR RATTLE
Made by the wings or bill.

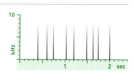

*Bill clatter, given at nesting
colonies*
Wood Stork, p. 183
(several heron species)

*Wing claps, single or in irregular
series*
Long-eared Owl, p. 218

OTHER WINGBEAT SOUNDS

Snort snapping rattle
Trumpeter Swan, p. 43
Tundra Swan, p. 43
Limpkin, p. 120

Nasal or polyphonic hum
Mute Swan, p. 42
Black Vulture, p. 196

Rock Pigeon, p. 85

A CHIP SERIES OR UNMUSICAL TRILL

CHIP SERIES

A series of Chips or similar notes. See also Finch Flight Call Group.

Verdin, p. 313
White-tailed Kite, p. 199

Wilson's Warbler, p. 418
Pyrrhuloxia, p. 459

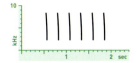

TSWEET SERIES

A series of rising, complex Chiplike notes.

Olive Sparrow, p. 422

Baird's Sparrow, p. 449

FINCH FLIGHT CALL GROUP

"Flight calls" of finches are variable but stereotyped short Chiplike, Piplike, or polyphonic notes, usually given in series of 3–4 (not necessarily in flight). Red Crossbill call types are indexed here for reference; field identification can be difficult (see p. 362).

Chip Chip Chip
Red Crossbill Type 3 flight call, p. 363

Jeet Jeet Jeet (polyphonic)
Red Crossbill Type 3 excitement call, p. 363
Lesser Goldfinch, p. 369

Jirp Jirp Jirp (polyphonic)
Common Redpoll, p. 366
Hoary Redpoll, p. 367
Pine Siskin, p. 370

Jit Jit Jit (polyphonic)
White-winged Crossbill, p. 365

Kip Kip Kip
Red Crossbill Type 1 flight call, p. 362
Red Crossbill Type 2 flight call, p. 363

Kwit Kwit Kwit
Red Crossbill Type 4 flight call, p. 364

Peep Peep Peep
Red Crossbill Type 1 excitement call, p. 362

Peer Peer Peer
Red Crossbill Type 10 excitement call, p. 364

Pip Pip Pip
Olive-sided Flycatcher, p. 254
Red Crossbill Type 2 excitement call, p. 363

Pitter Pitter Pitter (see also Pitter, p. 554)
Olive-sided Flycatcher, p. 254

Red Crossbill Type 10 excitement call, p. 364

Pwip Pwip Pwip
Red Crossbill Type 4 excitement call, p. 364

Tew Tew Tew
American Goldfinch, p. 368

Trit Trit Trit
Red Crossbill Type 8 excitement call, p. 365

Tyip Tyip Tyip
Red Crossbill Type 8 flight call, p. 365

Whit Whit Whit
Red Crossbill Type 10 flight call, p. 364

RATCHET

A short series of doubled Smacklike notes, each pair of notes so close together that they sound like a single note. Very abrupt and harsh.

Gray Catbird, p. 345
Brown Thrasher, p. 346

Long-billed Thrasher, p. 347
Curve-billed Thrasher, p. 348

AN ACCELERATING SERIES OF CHIPS
A series of Chiplike notes that accelerates, sometimes into a trill.

Olive Sparrow, p. 422

TRILL (CHIPPING)
A single unmusical trill of Chiplike or Tsitlike notes.

Chipping Sparrow, p. 431
Worm-eating Warbler, p. 382
Pine Warbler, p. 415

RATTLE (SMACKING)
A rapid trill of Smacklike notes; faster and more ticking than classic Chatter (p. 547).

Long
Carolina Wren, p. 322
Common Yellowthroat, p. 391

Short, mixed with single notes
Winter Wren, p. 318

AN UNSTEADY TRILL OR SERIES
A slow trill or rapid series of semimusical Chiplike notes that changes gradually in pitch and/or quality; even stereotyped versions may sound plastic.

Orange-crowned Warbler,
p. 389

Wilson's Warbler, p. 418

TWITTER (CHIPPING)
A highly plastic series of Chiplike or Spitlike notes, often accelerating briefly into a trill.

Almost always from high overhead
Chimney Swift, p. 104

In agitation, usually from perch
Song Sparrow, p. 442
Le Conte's Sparrow, p. 451

Savannah Sparrow, p. 448
Rusty Blackbird, p. 476
Verdin, p. 313

Notes Spitlike or slightly complex
Indigo Bunting, p. 462

Botteri's Sparrow, p. 429
Cassin's Sparrow, p. 428

TWITTER (OF TSITLIKE NOTES)
A highly plastic series of medium- to high-pitched Tsitlike or Spitlike notes, often accelerating briefly into a trill. Can be tough to distinguish from Chipping Twitters. Less musical than Twitters of Tiplike notes.

Notes musical, almost Seetlike
Chipping Sparrow, p. 431
Field Sparrow, p. 434
Lesser Goldfinch, p. 369
Botteri's Sparrow, p. 429

Notes musical to complex
Bachman's Sparrow, p. 427

Notes very sharp, Spitlike
Northern Cardinal, p. 458

Pyrrhuloxia, p. 459

Notes very sharp, Tsitlike
Buff-bellied Hummingbird,
p. 105
Green-tailed Towhee, p. 423
Olive Sparrow, p. 422
Lark Sparrow, p. 436
Clay-colored Sparrow, p. 432
Brewer's Sparrow, p. 433

Black-throated Sparrow,
p. 437
Dark-eyed Junco, p. 446
White-throated Sparrow,
p. 439
White-crowned Sparrow,
p. 441
Grasshopper Sparrow, p. 447
Le Conte's Sparrow, p. 451

TWITTER (OF TIPLIKE NOTES)

A highly plastic series of semimusical, medium- to high-pitched Tiplike notes, often accelerating briefly into a trill. See also Kree Series (p. 541).

Medium-pitch, rather slow
Acadian Flycatcher, p. 257
Vermilion Flycatcher, p. 263
Brown-crested Flycatcher, p. 266

Medium-pitch, rapid, trilled
Eared Grebe, p. 80
Killdeer, p. 127
Alder Flycatcher, p. 258
Willow Flycatcher, p. 259
Least Flycatcher, p. 255

Yellow-bellied Flycatcher, p. 256
Say's Phoebe, p. 262
Marsh Wren (Eastern), p. 325

High, musical, sometimes polyphonic. Compare Twitter of Psip or Seetlike notes (p. 537).
Black-throated Sparrow Interaction Calls, p. 437
Eastern Towhee, p. 424
Spotted Towhee, p. 425
Indigo Bunting, p. 462

High, musical, notes often Pinklike
Red-winged Blackbird, pp. 468–469

High, less musical
White-rumped Sandpiper, p. 138
Gray Kingbird, p. 273
Eastern Kingbird, p. 271
Tropical Kingbird, p. 268

CHITTER (SCREECHY OR COMPLEX)

Variable, usually plastic twitters of high-pitched, usually unmusical notes, often with a screechy quality.

Eastern Screech-Owl, p. 223
Western Screech-Owl, p. 222
Elf Owl, p. 224
Snowy Owl, p. 214
Great Horned Owl, p. 213
Great Gray Owl, p. 217
Northern Hawk Owl, p. 215

Northern Saw-whet Owl, p. 221
Boreal Owl, p. 220
Barn Owl, p. 212
Ferruginous Pygmy-Owl, p. 226
Short-eared Owl, p. 219

Peregrine Falcon, p. 242
(likely other falcon species)

A PEEPING SERIES OR RATTLE

RATTLE (TYPICAL)

A coarse trill of sharp, unmusical Pik- or Pwiklike notes.

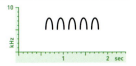

Higher, longer, more sputtering
Brown-headed Cowbird, p. 479
Bronzed Cowbird, p. 480
Shiny Cowbird, p. 481
several warbler species (see p. 378)

Lower, shorter
Brown-headed Nuthatch, p. 316
Lapland Longspur, p. 373
Smith's Longspur, p. 374
Chestnut-collared Longspur, p. 375
McCown's Longspur, p. 376
Snow Bunting, p. 377

Lower, longer, less musical, sometimes slightly buzzy
Eastern Meadowlark, p. 470
Western Meadowlark, p. 471

Low, clucking, downslurred
Summer Tanager, p. 457

Short, peeping
Long-billed Dowitcher, p. 141

High, slow, slightly more musical or metallic at start
Red-winged Blackbird, pp. 468–469
Black Phoebe, p. 260

High, fast, slightly screechy
Belted Kingfisher High Rattle, p. 228

American Kestrel, p. 245

Medium-low, harsh, slightly screechy
Belted Kingfisher Low Rattle, p. 228
Hairy Woodpecker, p. 235
American Three-toed Woodpecker, p. 238
Black-backed Woodpecker, p. 239

Individual notes Kriklike
Red-cockaded Woodpecker, p. 237

PSEEP SERIES

Long strings of Pseeplike notes, usually from begging juvenile birds.

most grebe species, (see pp. 78–84)
Swallow-tailed Kite, p. 199

American Coot, p. 117
Scaly-breasted Munia, p. 358

A PEEPING SERIES OR RATTLE, CONTINUED

PIP OR PEEP SERIES
A series of Piplike or Peeplike notes; slower and more stereotyped than Peeping Twitters. See also Finch Flight Call Group (p. 533).

Series of 2–4 Pee notes
Spotted Sandpiper, p. 144

Long series of Pips
Spotted Sandpiper, p. 144

Short series of 5–7 Pips
Whimbrel, p. 131
Ash-throated Flycatcher, p. 264

White-eyed Vireo, p. 280

Long stereotyped series of Peeps
Groove-billed Ani, p. 96
Smooth-billed Ani, p. 97

Long plastic series of Peeps
Royal Tern, p. 172
Sandwich Tern, p. 173

Arctic Tern, p. 175
Roseate Tern, p. 177
Least Tern, p. 178

A SHORT TRILL OF PIPLIKE NOTES
Like a Pipping Trill, but usually shorter, often just 2–3 notes.

Usually 2 identical sharp Pips
Wilson's Plover Piddip, p. 124
Red-necked Phalarope Piddip, p. 150

Usually 3 rather mellow Pips
Upland Sandpiper Quiddyquit, p. 130

TRILL (PIPPING)
A trill of medium-low Piplike notes. Often plastic. See Peeping Twitter.

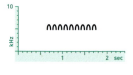

Least Sandpiper, p. 139
Western Sandpiper, p. 140
Solitary Sandpiper, p. 145
Ferruginous Pygmy-Owl, p. 226

Gray Vireo, p. 283
Yellow-throated Vireo, p. 284
Blue-headed Vireo, p. 158
Black-capped Vireo, p. 282
Pine Grosbeak, p. 372

TWITTER (PEEPING)
A highly plastic series of medium- to high-pitched Pip- or Peeplike notes, often accelerating briefly into a trill.

Often slightly chattering, with some noisy notes
Red Phalarope, p. 150
Red-necked Phalarope, p. 150

Long, with large changes in pitch
American Oystercatcher, p. 122

Most notes near same pitch
Fulvous Whistling-Duck, p. 36
Black-bellied Whistling-Duck, p. 37

Long-billed Dowitcher, p. 141
Short-billed Dowitcher, p. 141
Least Sandpiper, p. 139
Semipalmated Sandpiper, p. 140
Western Sandpiper, p. 140
Baird's Sandpiper, p. 138
Sanderling, p. 135
Dunlin, p. 136
Red-cockaded Woodpecker, p. 237
Brown-headed Nuthatch, p. 316

Rather soft, medium-low; notes musical Pips
Baltimore Oriole, p. 488
Bullock's Oriole, p. 489

Short, low, slightly nasal; first note usually highest
Chestnut-collared Longspur Kiddle, p. 375

A VERY HIGH SERIES OR TRILL

DZIT OR DZEET SERIES
A series of very high burry or buzzy notes.

Buff-bellied Hummingbird, p. 105
Bushtit, p. 312

Chipping Sparrow, p. 431
Clay-colored Sparrow, p. 432
Brewer's Sparrow, p. 433

SEET SERIES
A high to very high monotone series of Seetlike notes.

Stereotyped, regularly repeated
Bay-breasted Warbler, p. 405
Blackpoll Warbler, p. 404
Cape May Warbler, p. 406
Golden-crowned Kinglet,
p. 331

Plastic, given irregularly
Black-capped Chickadee,
p. 308

Carolina Chickadee, p. 309
Boreal Chickadee, p. 307
Tufted Titmouse, p. 310
Black-crested Titmouse,
p. 311
Henslow's Sparrow, p. 452

Plastic, series often quite long
House Wren, p. 319

Notes downslurred, well spaced
Buff-bellied Hummingbird
Song, p. 105

4-6 notes with consonant at start
Brown Creeper Psee Series,
p. 317

A VERY HIGH TRILL
*A fairly musical trill of Psip- or Seetlike notes. Compare Burry
Seet (p. 503).*

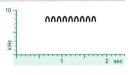

Fairly musical
Cedar Waxwing, p. 353

*Less musical; trills often
downslurred*
Bohemian Waxwing, p. 354

*Musical, extremely coarse; notes
sometimes slow enough to count*
Bushtit Seet Trill, p. 312
(some ground squirrel
species)

Notes sharp, trill cricketlike
Ferruginous Pygmy-Owl
(juvenile), p. 226
Blackpoll Warbler, p. 404

TWITTER (OF PSIP- OR SEETLIKE NOTES)
*A highly plastic series of very high-pitched notes, often accelerating
briefly into a trill.*

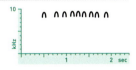

Golden-crowned Kinglet,
p. 331

Cassin's Sparrow, p. 428

A MUSICAL SERIES OR TRILL

A SERIES OF VARIABLE WHISTLES
*These species give a huge variety of whistled notes in series. The notes in any given series are identical, but
the series vary individually and geographically.*

*Short series, often repeated;
notes medium-high*
Tufted Titmouse Song, p. 310
Black-crested Titmouse
Song, p. 311

*Short series, fast or slow, often
slightly plastic*
Gray Jay, p. 294

Short series; notes high, musical
American Redstart, p. 395
Palm Warbler, p. 414

*Short series; notes musical to
metallic*
Great-tailed Grackle Series
Calls, p. 472

*Long series; notes high, musical
to chirping*
American Pipit Song, p. 355

*Long series; notes shrill, usually
nearly monotone*
Killdeer, p. 127

*Long series; notes mellow,
usually up- or overslurred*
Smooth-billed Ani Pip Series,
p. 97
Groove-billed Ani Weep
Series, p. 96

A SERIES OF HEELIKE NOTES
A series of high monotone whistles.

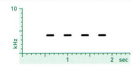

*5–15 medium-high monotone
whistles*
Montezuma Quail Female
Song, p. 70

*Highly plastic, unsteady in pitch
and rhythm*
Western Wood-Pewee, p. 252
Eastern Wood-Pewee, p. 253

A SERIES OF UPSLURRED WHISTLES
Slower than Sputter (p. 539). See also Northern Cardinal Song Group (p. 567).

Notes medium-low, variable, short
Spotted Sandpiper
Pee-pwee-pwee, p. 144
Solitary Sandpiper
Pwee-pwee-pwee, p. 145

Notes higher, Weeplike, long
Baird's Sandpiper Weep Series, p. 138
See also Weep, p. 505

Notes high, Weeplike, short
Olive-sided Flycatcher Weep Series, p. 254

Higher, thinner
Prothonotary Warbler Song, p. 383

A SERIES OF DOWNSLURRED WHISTLES
See also Northern Cardinal Song Group (p. 567) and Finch Flight Call Group (p. 533).

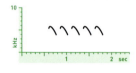

3–5 medium-low Pewlike notes, often in flight
Lesser Yellowlegs Pew Series, p. 147
Greater Yellowlegs Pew Series, p. 146
Short-billed Dowitcher Pew Series, p. 141

3–5 medium-high Peerlike notes
Tufted Titmouse Song, p. 310
Black-crested Titmouse Song, p. 311

5–7 high Peerlike notes, series downslurred
Northern Beardless-Tyrannulet Day Song, p. 251

Longer series of Peerlike notes
Lesser Yellowlegs Peer Series, p. 147
Boat-tailed Grackle Series Calls, p. 474

A DOWNSLURRED SERIES OF WHISTLES
A downslurred series of whistled notes, sometimes changing in speed and quality.

Notes nearly monotone
Montezuma Quail Female Song, p. 70
Northern Beardless-Tyrannulet Day Song, p. 251

Notes variable; series slowing
Canyon Wren Male Song, p. 321

AN ACCELERATING SERIES OF WHISTLES OR CHIRPS
A series of similar whistled or chirping notes that increases in speed.

All notes musical whistles
Field Sparrow, p. 434
Prairie Warbler, p. 402

All notes complex, semimusical
Dickcissel, p. 455

TRILL (MUSICAL, WHISTLED)
A single musical to semimusical trill or rapid series.

High, whistled; notes mostly sharp or complex
Pine Warbler, p. 415
Palm Warbler, p. 414
Chipping Sparrow, p. 431
Dark-eyed Junco, p. 446

Swamp Sparrow, p. 444

Usually slightly unsteady in pitch
Orange-crowned Warbler, p. 389
Wilson's Warbler, p. 418

Medium-low, highly musical
Black-crested Titmouse Song, p. 311

SPUTTER OR WEET SERIES
High, semimusical slow trills or fast series, the individual notes usually upslurred Weetlike notes. Less nasal than Laugh (p. 542).

Slightly slower (Weet Series)
Wood Thrush, p. 340
Curve-billed Thrasher
(eastern), p. 348

Slightly faster (Sputters)
Northern Beardless-
Tyrannulet, p. 251
Brown-crested Flycatcher,
p. 266

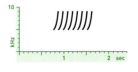

Western Kingbird, p. 270

WING WHISTLES
A trill or rapid series of breathy, musical, nearly monotone whistles. Sometimes sounds slightly polyphonic, if left and right wing are not precisely in tune. For unmusical wing sounds, see p. 532.

Short (given mostly at takeoff)
Mourning Dove, p. 90
sometimes other doves (see
p. 85)

Short, repeated (given at bottom of display dives)
Ruby-throated Hummingbird,
p. 106
Black-chinned Hummingbird,
p. 107

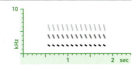

Long (given throughout flight)
Common Goldeneye, p. 60
Barrow's Goldeneye, p. 61
Surf Scoter, p. 58
White-winged Scoter, p. 58
Hooded Merganser, p. 63

TRILL (DOWNSLURRED)
A downslurred musical to semimusical trill. See also Beert (p. 510).

Very high, not very musical
Bohemian Waxwing, p. 354

Medium-low, musical to semimusical
Carolina Wren Beert, p. 322
House Wren Beert, p. 319

A LOW-PITCHED SERIES OR TRILL

A SERIES OF MELLOW MONOTONE WHISTLES
Lower than Series of Heelike Notes (p. 537).

3–5 notes, series often repeated
Black-billed Cuckoo Song,
p. 93

Long continuous series
Northern Saw-whet Owl
Song, p. 221

Long series, notes sometimes upslurred
Ferruginous Pygmy-Owl
Song, p. 226

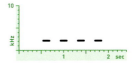

KLEW SERIES
Series of rather low, mellow downslurred notes, each of which breaks to a lower pitch.

Hook-billed Kite, p. 198

Black-billed Cuckoo, p. 93

MELLOW WAIL SERIES
Series of long overslurred mellow whistles, sometimes slightly nasal, sometimes broken. See also Longer Overslurred Whistles (p. 507).

Gray Hawk, p. 205

Groove-billed Ani, p. 96
Smooth-billed Ani, p. 97

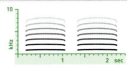

A SOUND MADE OF REPEATED SIMILAR NOTES 539

HOOTING OR COOING SERIES
A steady series of single hooting or cooing notes.

Slow; individual notes upslurred
  Common Ground-Dove Song,
  p. 88

*Slow; individual notes monotone
or overslurred*
  Rock Pigeon Song, p. 85
  Great Gray Owl Song, p. 217
  Snowy Owl Song, p. 214

Slow; series slightly downslurred
  Yellow-billed Cuckoo Song,
  p. 92
  Greater Roadrunner Song,
  p. 95

Fast; individual notes monotone
  Short-eared Owl Song, p. 219

*Series fairly high, fast, slightly
upslurred*
  Boreal Owl Song, p. 220

TRILL (HOOTING)
One low, hooting trill.

Long (10 seconds or more)
  Lesser Nighthawk Trill,
  p. 99
  Boreal Owl Song, p. 220
  Northern Hawk Owl Song,
  p. 215

(trills of some toad species)

*Shorter, rhythm often slightly
irregular*
  Eastern Screech-Owl Trill,
  p. 223

WILSON'S SNIPE GROUP
*A rapid, upslurred or barely overslurred series of low, almost hooting
whistles.*

  Wilson's Snipe Winnow, p. 142   Boreal Owl Song, p. 220

AN ACCELERATING SERIES OF HOOTS
An accelerating series of hooting notes or very low-pitched whistles.

  Western Screech-Owl Song, p. 222

GROWL SERIES
A series of very low-pitched burry notes. Not very loud.

  Wilson's Phalarope, p. 151   Greater Roadrunner, p. 95
  Common Pauraque, p. 103

A SERIES OF BURRS OR BUZZES

A SERIES OF CHEERLIKE NOTES
A series of coarsely burry, rather unmusical notes.

  Red-winged Blackbird Female
  Song, p. 468

DZIK SERIES
A series of high, brief, sharply upslurred buzzes. See Dzweet (p. 512).

Eastern Meadowlark, p. 470   Lesser Goldfinch, p. 369
American Goldfinch, p. 368

KREE SERIES
A series of soft, medium-pitched upslurred burry notes. Some Twitters of Tiplike notes (p. 535) can sound similar. See also Screechy Chitters (p. 535).

Baltimore Oriole, p. 488   Bullock's Oriole, p. 489

A MONOTONE SERIES OF BUZZES
A series of 2 or more fairly long buzzes, all similar in pitch and quality.

Slower, usually 2–6 notes   *Faster, sometimes almost a trill,*
Clay-colored Sparrow, p. 432   *sometimes a couplet series*
                                   Palm Warbler, p. 414
                                   Brewer's Sparrow, p. 433

PEENT SERIES
A rapid series of extremely fine buzzy notes with a slight nasal quality.

American Woodcock Cackle,   Antillean Nighthawk
   p. 143   Pikkity-kik, p. 100

A RISING SERIES OF BUZZES
A rising series of buzzes, all similar in quality.

Black-throated Blue Warbler,   Prairie Warbler, p. 402
   p. 397   Northern Parula, p. 398

TRILL (RISING, BUZZY)
A rising trill of buzzy notes, last note sometimes different.

Prairie Warbler, p. 402   *Last note lower, sometimes not*
                                  buzzy
                                    Northern Parula, p. 398
                                    Tropical Parula, p. 399

AN ACCELERATING BUZZY SERIES
A series of buzzy notes, getting faster, usually all on the same pitch.

Bank Swallow Song, p. 302

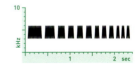

A DECELERATING BUZZY SERIES
A series of buzzy notes, slowing and usually falling in pitch at end.

Series usually overslurred   *Series usually falling*
Canyon Wren Female Song,   Mountain Plover Song, p. 128
   p. 321   Dunlin Song, p. 136

A SERIES OF SHORT NASAL NOTES

KEEK SERIES
A series of high short abrupt monotone nasal notes. Like Whinny but more monotone.

Northern Harrier, p. 197
Whimbrel, p. 131
Long-billed Curlew, p. 132
Northern Hawk Owl, p. 215
Northern Saw-whet Owl,
    p. 221
Aplomado Falcon, p. 244

Higher, notes downslurred
Sharp-shinned Hawk, p. 201
Sora, p. 113

High, series often overslurred
Swallow-tailed Kite, p. 199
Bald Eagle, p. 210

Notes slightly Piklike
American Three-toed
    Woodpecker, p. 238

Notes short; series long, slow
Willet (Eastern), p. 149
Willet (Western), p. 148

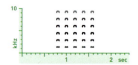

Notes longer, lower, series slow
Red-shouldered Hawk,
    p. 206

Notes nasal, Keklike; series long
Wood Duck, p. 45
Northern Flicker, p. 240
Pileated Woodpecker, p. 241

LAUGH
A rapid series or slow trill of nasal, often upslurred notes. Pitch medium. More nasal than Sputter (p. 539).

Medium-high
Semipalmated Sandpiper,
    p. 140

Fairly high, often quite slow
Willet (Eastern), p. 149

Willet (Western), p. 148

Medium-low
American Avocet, p. 122
African Collared-Dove, p. 87
Hutton's Vireo, p. 286

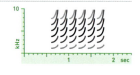

A SERIES OF SQUEEPS OR PWEEWS
A series of high complex Cheeplike notes, each usually upslurred.

Medium-high, usually 2–3 notes
American Robin Squee-
    squeep, p. 343

Medium-high, first note Whitlike
Least Flycatcher Pweew
    Series, p. 255

Medium-low, plastic, sometimes noisy
White-throated Sparrow
    Squeep Series, p. 439
Harris's Sparrow Squeep
    Series, p. 440

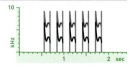

White-crowned Sparrow
    Squeep Series, p. 441

WIK SERIES (HIGH)
A series of fairly high, seminasal upslurred Wik notes.

Northern Flicker, p. 240
Hairy Woodpecker, p. 235

Pileated Woodpecker, p. 241

TEW SERIES
Series of high, sharply downslurred seminasal notes.

American Pipit, p. 355
Seaside Sparrow, p. 450

Dark-eyed Junco, p. 446

A SHORT SERIES OF YAPS
A rapid series of 2-5 Yaps, the notes sometimes running together. See also 2- or 3-Syllabled Nasal Note (p. 554).

Usually just 2 or 3 notes
Laughing Gull Pup-pup,
    p. 159
Franklin's Gull Pup-pup,
    p. 160

Royal Tern, p. 172

Often longer
Gull-billed Tern, p. 171

CHWEEK SERIES
Series of high, sharply underslurred seminasal notes. Compare Klee Series.

Slow
  Ferruginous Pygmy-Owl,
    p. 226

Fast
  American Three-toed
    Woodpecker, p. 238

Green Kingfisher, p. 227

BARK SERIES (NASAL)
A series of brief medium-low nasal notes. Less harsh than Quacking Bark Series (p. 548); more nasal than Kek Series (p. 546), and notes often longer.

Notes brief, Keklike; series fast
  Cooper's Hawk, p. 202
  Bald Eagle, p. 210

Notes brief, Keklike; series slow
  Black-legged Kittiwake,
    p.156

Series monotone, often long
  Common Goldeneye, p. 60

Barrow's Goldeneye, p. 61
Black-necked Stilt, p. 121
Wilson's Snipe, p. 142
Wilson's Phalarope, p. 151
Black Tern, p. 179
Western Screech-Owl, p. 222

Notes monotone or downslurred
  Northern Harrier, p. 197

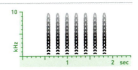

Notes longer, plastic
  Sooty Tern, p. 169

Notes low, mellow; first and last notes sometimes lower
  Greater Roadrunner, p. 95

KLEE SERIES AND SIMILAR SOUNDS
Like Whinny, but with broken notes. Compare Chweek Series.

Long, slow series, only some notes broken
  Red-tailed Hawk, p. 204
  Swainson's Hawk, p. 207
  American Avocet, p. 122

Shorter, faster series, most notes broken
  Golden Eagle, p. 210
  Osprey, p. 211

Swallow-tailed Kite, p. 199
Solitary Sandpiper Song,
    p. 145
Greater Yellowlegs, p. 146
Least Tern, p. 178
American Kestrel, p. 245

Series fast, notes slightly noisy, Peeklike
  Hairy Woodpecker, p. 235

Downy Woodpecker, p. 234
Ladder-backed Woodpecker,
    p. 236

Very loud, screeching
  Plain Chachalaca, p. 66

YELP SERIES
A series of brief, medium-pitched broken nasal notes. See also Large Gull Long Calls (p. 576).

  Wild Turkey, p. 67

  Northern Bobwhite, p. 68

WHINNY (OVERSLURRED)
A rapid series or slow trill of Keek- or Keklike notes, the first and last 1–2 notes usually slightly lower than the rest. Higher and coarser than Bleats. Compare Screechy Whinny (p. 546) and Keek Series.

High, semimusical
  American Robin, pp. 342–343

Medium-high, nasal
  Hook-billed Kite, p. 198
  Pileated Woodpecker, p. 241

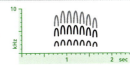

Medium-high, seminasal, slow
  Elf Owl, p. 224

WHINNY (DOWNSLURRED)
A rapid series or slow trill of Keek-, Kek-, or Piklike notes, the series generally falling and sometimes slowing.

High, downslurred; notes Piklike
  Downy Woodpecker, p. 234
  Ladder-backed Woodpecker,
    p. 236

High, decelerating; notes Keklike
  Sora, p. 113

Lower, slowing; notes Keklike
  Common Gallinule, p. 116

A SOUND MADE OF REPEATED SIMILAR NOTES   543

WHINNY-RASP
A whinny of Keeklike notes that accelerate into a harsh rasp.

Black-backed Woodpecker, p. 239

CHATTER (NASAL)
Rapid, highly plastic rattle of seminasal, semimusical Piplike notes. Higher and more plastic than Nasal Trill, See also *Chatter Duet.*

Rufous-crowned Sparrow, p. 426

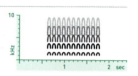

CHATTER DUET
A rapid series of nasal Piklike notes, usually by 2 birds in highly synchronized chorus, often sounding like a single bird. See also *Churr (p. 524) and Nasal Chatter.*

Slower, more laughing
Pied-billed Grebe, p. 83

Faster, churring
Least Grebe, p. 84
Black Rail, p. 109

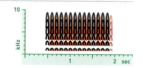

TRILL (NASAL)
A rapid trill of brief, highly nasal notes. Faster and more monotone than Whinnies (p. 543); more stereotyped than nasal Chatter (p. 544).

Red-breasted Nuthatch Fast
Song, p. 314

White-breasted Nuthatch
Fast Song, p. 315

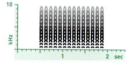

TWITTER (NASAL)
A highly plastic rapid series or trill of very brief nasal notes, often irregular in rhythm.

Soft, often trilled
Red-breasted Nuthatch,
p. 314

Usually irregular single notes
White-breasted Nuthatch,
p. 315

BLEAT (HIGH)
A fairly high burry note, trill, or tremolo, with a nasal to screechy quality.

Extremly coarse, screechy
Northern Hawk Owl, p. 215

A long high coarse nasal burr
Horned Grebe, p. 82

Similar but finer
Eared Grebe, p. 80

GARGLE
Comical nasal gibberish, like the language of cartoon aliens.

Snowy Egret, p. 190

BLEAT (LOW)
A burry wail or moan. Coarser than Churr (p. 524).

High, rather clear
  Tundra Swan, p. 43

Medium-high
  Bridled Tern, p. 170

Medium pitch, very nasal
  Lesser Nighthawk, p. 99

Medium pitch, noisier
  Brant, p. 41

Sandhill Crane, p. 118
Whooping Crane, p. 119
California Gull, p. 162
Common Murre, p. 154

Low, rather fine
  Glossy Ibis, p. 194
  White-faced Ibis (probably),
    p. 194

Even lower, coarser
  Great Cormorant, p. 185
  Least Bittern Song, p. 188

GOBBLE
A short rapid series of Kuklike notes. Coarser and higher than Purr (p. 524).

Short, monotone
  Sharp-tailed Grouse,
    pp. 74–75

Longer, downslurred
  Wild Turkey, p. 67

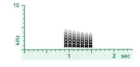

A SERIES OF LONGER NASAL NOTES

REEK OR YANK SERIES
A series of rising, medium- to low-pitched nasal notes.

Series of 5–15 notes
  Cooper's Hawk, p. 202
  White-breasted Nuthatch
    Slow Song, p. 315

*Longer series of single
well-spaced notes*
  Red-breasted Nuthatch
    Slow Song, p. 314

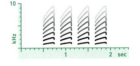

KLEER SERIES
A series of high downslurred seminasal or nasal notes that break to a lower pitch.

Quite slow, nasal
  Red-shouldered Hawk,
    p. 206

Quite fast, seminasal
  Greater Yellowlegs, p. 146

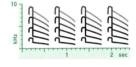

SCREAM SERIES
A series of long, medium- to high-pitched seminasal or nasal notes.

*Screams very high, monotone or
upslurred*
  Sharp-shinned Hawk, p. 201
  American Kestrel, p. 245

*Screams upslurred, high,
seminasal to nasal*
  Wood Duck, p. 45

Northern Goshawk, p. 203
Bald Eagle, p. 210
Osprey, p. 211
Merlin, p. 244
Aplomado Falcon, p. 244
Peregrine Falcon, p. 242
Prairie Falcon, p. 242

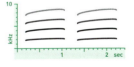

Gyrfalcon, p. 243
Crested Caracara, p. 245

SQUEAL SERIES
A series of high seminasal notes that break to a higher pitch and back down, quality clear to harsh.

3–5 harsh, hoarse squeals
  Yellow-bellied Sapsucker,
    p. 233

Longer series, notes mostly clear
  Red-tailed Hawk, p. 204

Red-shouldered Hawk,
  p. 206
Zone-tailed Hawk, p. 208
Short-tailed Hawk, p. 207
Harris's Hawk, p. 205

many raptor species in
copulation (see p. 198)

WAIL SERIES
A series of long overslurred wails.

Medium-low
    Indian Peafowl, p. 65

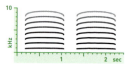

MOAN SERIES
A series of low nasal moans.

    Purple Swamphen, p. 114

A SERIES OF SHORT HARSH NOTES

WHINNY (SCREECHY)
A series of high, variably screechy Keeklike notes, the first and last 1–2 notes in the series often slightly lower. Compare Keek Series (p. 542) and Overslurred Whinny (p. 543).

    Red-necked Grebe, p. 81   Merlin, p. 244

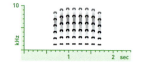

KEK SERIES
A series of very brief noisy Chuklike notes.

    Green Heron, p. 192   Purple Gallinule, p. 115
    Clapper Rail, p. 110   Common Gallinule, p. 116
    King Rail, p. 111   Marsh Wren (Western), p. 324

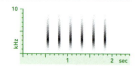

CHITTIT AND SIMILAR SOUNDS
A rapid pair of identical Chitlike or Chuklike notes.

High, slightly nasal Jiddit
    Ruby-crowned Kinglet, p. 332
    Black-capped Vireo, p. 282
Harsher
    White-winged Crossbill Chittit,
    p. 365

Snow Bunting Chip-it, p. 377
Low
    Ringed Kingfisher Chuttut,
    p. 228
    Yellow-headed Blackbird
    Chuttuk, p. 478

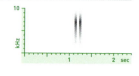

CHITTER (NOISY)
Like Chatter, but higher-pitched, made of Chitlike, not Chuklike notes. Usually plastic.

    Downy Woodpecker, p. 234
    Chimney Swift (juv.), p. 104
    European Starling,
    pp. 350–351
    Barn Swallow, p. 302
Notes rather nasal
    Wood Duck, p. 45
Notes rather chipping
    Canyon Wren, p. 321
Short, rather sputtering
    Boreal Chickadee, p. 307

Black-capped Chickadee,
p. 308
Carolina Chickadee, p. 309
Notes usually nasal, slightly buzzy
    Ruddy Turnstone, p. 134
    Bewick's Wren, p. 323
    Eastern Phoebe, p. 261
    Western Kingbird, p. 270
    Scissor-tailed Flycatcher,
    p. 272

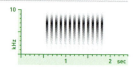

Slightly screechy
    Pectoral Sandpiper, p. 137
    Stilt Sandpiper, p. 135
    Ash-throated Flycatcher,
    p. 264
    Great Kiskadee, p. 267
Quite high
    Tropical Kingbird, p. 268

CHATTER (NOISY)
A rapid series or slow trill of Chuklike noisy notes. Faster than Snarl Series.

Brown Noddy, p. 170
Lesser Yellowlegs, p. 147
Red-headed Woodpecker, p. 232
Yellow-bellied Sapsucker, p. 233
White-winged Parakeet, p. 248
Green Kingfisher, p. 227
Green Jay, p. 294
House Wren, p. 319
Sedge Wren, p. 326
Cactus Wren, p. 327
Black-tailed Gnatcatcher, p. 328
Eastern Bluebird, p. 341
Curve-billed Thrasher, p. 348

Hooded Oriole, p. 483
Orchard Oriole, p. 482
Altamira Oriole, p. 485
Scott's Oriole, p. 487
Baltimore Oriole, p. 488
Bullock's Oriole, p. 489
Great-tailed Grackle, pp. 472–473
Brewer's Blackbird, p. 477
House Sparrow, p. 356
Eurasian Tree Sparrow, p. 357
Bronze Mannikin, p. 357

Usually in middle of long barking series
Egyptian Goose, p. 38

Loud, slow, clucking
Plain Chachalaca, p. 66

Rather slow and nasal
Black-billed Magpie, p. 298

Fast to slow
Gray Jay, p. 294

Fast, almost rasping or churring
Yellow-chevroned Parakeet, p. 248
Winter Wren, p. 318
Black-throated Sparrow, p. 437
Spot-breasted Oriole, p. 484

CHUCKLE
A rapid series of grunts, low barks, or Kuklike notes. Lower and more nasal than Chatters, and often slower. Less croaking than Kruk Series (p. 548). Faster than Grunt Series (p. 548).

Short series of rather harsh grunts
Snow Goose, p. 39
Ross's Goose, p. 39
Greater White-fronted Goose, p. 41
Mallard, p. 48
Gadwall, p. 46
Lesser Scaup, p. 55
Common Eider, p. 56
Common Merganser, p. 62
Roseate Spoonbill, p. 195
Snowy Owl, p. 214

Short series of low, swallowed Kuklike notes
Least Bittern Song, p. 188

Short series of grunting to screeching Kuklike notes
California Gull, p. 162
Ring-billed Gull, p. 161
Herring Gull, p. 163

Lesser Black-backed Gull, p. 166
Great Black-backed Gull, p. 167
Glaucous Gull, p. 165

Short series of very nasal honks
Trumpeter Swan, p. 43

Short fast series, almost churring
Northern Pintail, p. 52
Green-winged Teal, p. 53
Gull-billed Tern, p. 171
Wood Thrush, p. 340

Medium-long series of nasal Kuklike notes
Sharp-tailed Grouse, pp. 74–75
Greater Prairie-Chicken, p. 76
Lesser Prairie-Chicken, p. 77

Rather long series of harsh downslurred barks
Red-bellied Woodpecker, p. 231

Rather long series of low Chuklike notes
Cactus Wren Song, p. 327

Low, harsh; rhythm irregular
Red-necked Grebe, p. 81
Northern Fulmar, p. 182

Notes grating
Sabine's Gull, p. 157

Notes slightly screechy
Burrowing Owl, p. 225

Notes variable, semimusical
Piping Plover, p. 126
Upland Sandpiper, p. 130
Western Meadowlark, p. 471

A SERIES OF LONGER HARSH NOTES

SNARL SERIES
A series of very harsh noisy notes. Slower than Noisy Chatter.

Egyptian Goose, p. 38
Carolina Wren, p. 322

Cactus Wren, p. 327
Black-tailed Gnatcatcher, p. 328

A SERIES OF LONGER HARSH NOTES, CONTINUED

SCREECH SERIES
A series of medium-long screechy notes.

Horned Grebe, p. 82
Least Bittern, p. 188
Northern Goshawk, p. 203
juveniles of some tern
 species, pp. 169–179
Northern Hawk Owl, p. 215
Peregrine Falcon, p. 242

Prairie Falcon, p. 242
Gyrfalcon, p. 243

Highly plastic; sometimes in screechy couplets
Yellow-bellied Sapsucker, p. 233

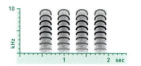

Repeated short series
Tricolored Heron, p. 191

GRUNT SERIES
A series of low, fairly brief harsh notes, like Chuckle (p. 547) but slower. See also Rail Long Calls (p. 577).

Egyptian Goose, p. 38
Northern Gannet, p. 183

Great Cormorant, p. 185

A HARSH BURRY SERIES

BARK SERIES (QUACKING)
A series of odd Quacklike barking notes.

Notes long, finely burry, rather quacking
Mangrove Cuckoo, p. 94

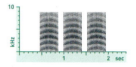

CHURR SERIES
A series of low, harsh nasal burry notes.

Black Rail, p. 109
Ringed Kingfisher, p. 228
Red-headed Woodpecker, p. 232

Red-bellied Woodpecker, p. 231
Golden-fronted Woodpecker, p. 230

KRUK SERIES
Rapid, often plastic series of very brief harsh croaking notes.

Low, grating
Great Blue Heron, p. 189
Snowy Egret, p. 190
Little Blue Heron, p. 190

Tricolored Heron, p. 191
Cattle Egret, p. 192

Medium-low, nasal
Great Egret, p. 189

RASP SERIES
A series of coarse harsh grating notes.

Series short, typically starting with a Kip note
Black-backed Woodpecker, p. 239

CROAK SERIES
A series of low Croaks.

Fast, rather high and ticking
Snail Kite, p. 200

Fast to slow, rather low, harsh
Anhinga, p. 186

Glossy Ibis, p. 194
White-faced Ibis (probably), p. 194
Brown Noddy, p. 170

AN ACCELERATING SERIES OF CROAKS
A series of low, harsh croaking notes that increases in speed.

Spruce Grouse, p. 73

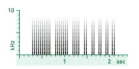

A POLYPHONIC SERIES
See Finch Flight Call Group, p. 533.

WHINE SERIES (HARSH)
A series of noisy polyphonic notes. Slower than Whiny Chatter.

Each note rising
Tufted Titmouse Dee Series, p. 310
Black-crested Titmouse Dee Series, p. 311

Loggerhead Shrike, p. 274
Black-tailed Gnatcatcher, p. 328
Yellow-headed Blackbird, p. 478

Notes rather long
Northern Shrike, p. 275

Plastic, grading into a chatter
Belted Kingfisher, p. 228

White-eyed Vireo, p. 280
Bell's Vireo, p. 281
Yellow-throated Vireo, p. 284
Gray Jay, p. 294

CHATTER (WHINY)
A trill of noisy polyphonic notes. Faster than Harsh Whine Series.

Notes short, high, series rapid
Red-eyed Vireo, p. 276
Black-whiskered Vireo, p. 277
Yellow-green Vireo, p. 277

Notes longer, series slower
Bell's Vireo, p. 281

White-eyed Vireo, p. 280
Blue-headed Vireo, p. 158

First note often longest
Yellow-throated Vireo, p. 284
Black-capped Vireo, p. 282
Gray Vireo, p. 283

Rock Wren, p. 320

Last note often longest
Warbling Vireo, p. 279

METALLIC SERIES
A series of brief Jenklike notes, usually rather harsh, with a metallic ring. See also Zreek Series.

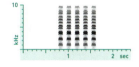

Often fairly harsh
Green Jay Jenk Series, p. 295
Common Grackle Jenk Series, p. 475

Often less harsh
Great-tailed Grackle Series Calls, p. 473

Boat-tailed Grackle Series Calls, p. 474

ZREEK SERIES
A series of upslurred, finely burry polyphonic notes, with at least some metallic quality. See also Metallic Series.

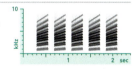

Woodhouse's Scrub-Jay, p. 296
Florida Scrub-Jay, p. 297

Notes screechier, series slower
Scaled Quail, p. 69

WHINE SERIES (HIGH)
A plastic, usually falling series of high polyphonic notes.

Vesper Sparrow, p. 435
Savannah Sparrow, p. 448
Baird's Sparrow, p. 449

Le Conte's Sparrow, p. 451
Henslow's Sparrow, p. 452

WHINE SERIES (LOW)
A plastic series of often underslurred polyphonic notes. Pitch medium.

Black-capped Vireo, p. 282
Song Sparrow, p. 442
Lincoln's Sparrow, p. 443

Swamp Sparrow, p. 444
Seaside Sparrow, p. 450
Scarlet Tanager, p. 456

INDEX PART III: A PHRASE OF 2–3 SYLLABLES

A phrase of 2 or 3 different notes, or a note of 2 or 3 syllables, by itself, or repeated after a pause.

A 2- OR 3-SYLLABLED PHRASE CONTAINING CHIPLIKE NOTES

A SHORT SHARP NOTE AND A WHISTLE
A phrase of 2 different notes, the one Chiplike or Piplike, the other whistled.

2 OR 3 SHORT SHARP NOTES (PITCHIT GROUP)
A phrase of 2 different Chiplike notes.

A 2- OR 3-SYLLABLED WHISTLED PHRASE

HIGHLY VARIABLE SHORT WHISTLED OR BURRY PHRASES
These species give a huge variety of 2- or 3-syllabled phrases that vary individually and geographically.

A 2-SYLLABLED WHISTLE
A whistle that breaks once to a different pitch. See also *Upward-breaking Wails (p. 518),* and *Pitter (p. 554).*

*Long, breaking upward;
sometimes slightly nasal*
  Eared Grebe Song, p. 80
  Whimbrel Cur-lew, p.131
  Long-billed Curlew Cur-lee,
    p. 132
  Semipalmated Plover Peweep,
    p. 125

Snowy Plover Ter-weet,
  p. 123
Smooth-billed Ani Weelip,
  p. 97

Short, breaking downward.
  Piping Plover Pee-lo, p. 126
  Red-crowned Parrot Pee-loo,
    p. 249

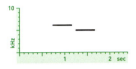

Northern Beardless-
  Tyrannulet Pee-uk, p. 251

*Highly plastic, often slightly
screechy*
  Common Murre, p. 154

2 LONG MONOTONE WHISTLES
*A pair of high monotone whistles that differ from one another in pitch
or length.*

*High; second note distinctly
lower*
  Black-capped Chickadee,
    p. 308

High; notes on same pitch
  Broad-winged Hawk Pee-tee,
    p. 206
  Harris's Sparrow, p. 440

*Low, mellow; notes on nearly
same pitch*
  Black Scoter Song, p. 59

2 SHORT SEPARATE MUSICAL WHISTLES
A pair of different whistled notes.

*Notes 1 second apart, second
upslurred*
  Northern Bobwhite
    Bob-white, p. 68

*First note upslurred, second
monotone*
  Northern Bobwhite
    Hoyp-woo, p. 68

First note longer, second shorter
  Killdeer Dee-dit, p. 127

All notes usually upslurred
  Spotted Sandpiper Pee-pwee,
    p. 144

All notes usually downslurred
  Horned Lark Tee-ter, p. 299

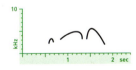

2–4 short high notes, variable
  Verdin Song, p. 313

3 SHORT SEPARATE WHISTLES
A trio of different whistled notes.

2–4 short high notes, variable
  Verdin Song, p. 313

All notes upslurred
  Spotted Sandpiper
    Pee-pwee-pwee, p. 144

*1 Pip, then 2 long slurred
whistles*
  Olive-sided Flycatcher
    Song, p. 254
  *2 Pips, then 1 long
    overslurred whistle*
  American Oystercatcher
    Pitti-weew, p. 122

*Variable, shrill, first note longest
and highest*
  Killdeer Dee-dit-dit,
    p. 127

A 2- OR 3-SYLLABLED BREATHY PHRASE
Phrase of 2–3 high, breathy, Wheewlike notes. See also *Breathy
Phrase (p. 568).*

2-noted, high
  Fulvous Whistling-Duck
    Ki-wheer, p. 36
  Black-bellied Whistling-Duck
    Ki-whew, p. 37

Eurasian Wigeon Wheew,
  p. 47

2-noted, lower, more nasal
  Red-breasted Merganser
    Whiddew, p. 62

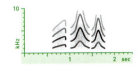

Usually 3- to 4-noted
  American Wigeon
    Whi-whee-whew, p. 47

A 2- OR 3-SYLLABLED LOW-PITCHED PHRASE

WIDDOO
Like Wow (p. 508), but 2- to 3-syllabled.

Lesser Prairie-Chicken, p. 77
Canvasback, p. 54

Lesser Scaup, p. 55
Greater Scaup, p. 55

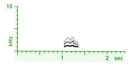

A 2- TO 3-SYLLABLED HOOTING OR COOING PHRASE
A phrase of 2 or 3 low-pitched hooting or cooing notes.

2-syllabled
Inca Dove Song, p. 88
Common Poorwill Song, p. 100
White-tipped Dove Song,
p. 89

3-syllabled
Greater Prairie-Chicken
Ooh-loo-woo, p. 76

White-tipped Dove Song,
p. 89
Eurasian Collared-Dove
Song, p. 87

*Usually longer, but sometimes
3-syllabled, first note broken*
Mourning Dove Song, p. 90

White-crowned Pigeon Song,
p. 86

A DOUBLE HOOTING TRILL OR GROWL
A 2-parted low, hooting trill, burry coo, purr, or growl.

Low, growling
Hooded Merganser Growl
Display, p. 63

Slightly higher, more musical
White-crowned Pigeon Growl
Song, p. 86

Slightly higher and slower
Western Screech-Owl
Double-trill, p. 222
Eastern Screech-Owl
(McCall's) Trill, p. 223

A GROWLING OR GROANING PHRASE
*A low 2- or 3-syllabled phrase, at least partly constructed of growling
notes.*

*Vaguely 2-syllabled growls,
often repeated*
Common Merganser G'daa,
p. 62
American Woodcock Hiccup,
p. 143
Eastern Whip-poor-will
Growl Phrase, p. 101

2 notes, both low growls
Hooded Merganser Growl
Display, p. 63

*2 notes, second long,
overslurred, and burry*
African Collared-Dove Song,
p. 87

3–4 growling or groaning notes
Eurasian Collared-Dove
Growl Song, p. 87
Inca Dove Growl Song, p. 88
Common Ground-Dove Growl
Song, p. 88

*3–4 burry notes, then 2–3 clear
notes*
White-tipped Dove Growl
Song, p. 89

A KNOCKING OR PURRING PHRASE
A brief phrase containing a rattle of very low knocking notes. **See also**
Knocking Rattle (p. 531).

Knocking
Lesser Nighthawk Chortle,
p. 99
Black-billed Cuckoo Chortle,
p. 93
Yellow-billed Cuckoo Chortle,
p. 92

Common Raven, p. 290
Chihuahuan Raven, p. 291

Purring
Rock Pigeon Growl Song,
p. 85

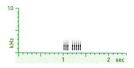

Slightly chattering
Mangrove Cuckoo Chortle,
p. 94

A 2- OR 3-SYLLABLED BURRY OR BUZZY PHRASE

VARIABLE BURRY OR BUZZY 2-NOTE PHRASES
These species give a huge variety of 2-syllabled phrases with at least one burry or buzzy note or syllable. These phrases vary individually and geographically.

Bewick's Wren Pewik, Greevit, etc., p. 323
Rose-breasted Grosbeak Zwee-er, p. 460
Black-headed Grosbeak Zwee-er, p. 461
Eastern Towhee Zoo-ee, p. 424

Eastern Meadowlark Veet, p. 470

Whistled, buzzy, burry, or any combination
Loggerhead Shrike Song, p. 274
Northern Shrike Song, p. 275

Hutton's Vireo Song, p. 286
Blue Jay Pumphandle Calls, p. 292
Tufted Titmouse Song, p. 310
Red-winged Blackbird Alert Calls, p. 469

A 2-NOTE BURRY OR BUZZY PHRASE
A 2-noted phrase with at least one part burry or buzzy.

2 different buzzes
Blue-winged Warbler, p. 386
Brewer's Sparrow, p. 433

Only first note burry
Wilson's Plover, p. 124
Eastern Phoebe Breep-it, p. 261

First note brief, not burry; second overslurred
Common Pauraque Song, p. 103
Couch's Kingbird Pip-breer, p. 269
Ash-throated Flycatcher Pip-breer, p. 264
Alder Flycatcher Pip-pweer, p. 258

First note brief, not burry; second monotone or downslurred
Snowy Plover Peer-purr, p. 123
Acadian Flycatcher Tsi-burrt, p. 257
Least Flycatcher Song, p. 255
Yellow-bellied Flycatcher Tsi-burrt, p. 256
Eastern Phoebe Fee-burr, p. 261

Like above group, but coarser
Eastern Phoebe Fee-blitty, p. 261
Ash-throated Flycatcher Ka-brik, p. 264
Gray Kingbird Pi-speer, p. 273

Notes upslurred, variably burry
Brown-crested Flycatcher Bree-burr, p. 266

First note a burry upslur, second a burry downslur
Willow Flycatcher Zree-beer, p. 259

First note a low Pwit, second a nasal monotone burr
Swainson's Thrush Pwit-burr, p. 338

First note Chiplike, second a monotone burr
Rock Wren Spit-tee, p. 320
Scarlet Tanager Chip-burr, p. 456

A 2-NOTE PEENTING PHRASE
Like Peent, but usually distinctly 2-noted

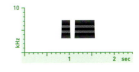

2-noted, second note longer
Common Goldeneye Pi-peent, p. 60

2-noted, second note shorter
Barrow's Goldeneye Peent-ip, p. 61

A 3-NOTE BURRY OR BUZZY PHRASE
A 3-noted phrase with at least one part burry or buzzy.

Wilson's Plover, p. 124
Alder Flycatcher Song, p. 258
Acadian Flycatcher Tsi-burrt, p. 258
Golden-winged Warbler, p. 387

Black-throated Blue Warbler, p. 397

Plastic, sometimes 2 or 4 notes
Purple Martin Veer Phrase, p. 305

Verdin Chewt Phrase, p. 313
Townsend's Solitaire Zweer-zweer-zree, p. 334

A 2- OR 3-SYLLABLED NASAL PHRASE

PITTER
Very brief medium-low whistled or seminasal note, broken into 2 distinct syllables; second may be higher or lower. Often in series. See also Finch Flight Call Group (p. 533).

Whistled
Harris's Sparrow, p. 440

Seminasal
Ruby-crowned Kinglet, p. 332

2- OR 3-SYLLABLED NASAL NOTE
A single short nasal note that breaks or changes pitch, sounding 2- or 3-syllabled. See also Rising Broken Nasal Note (p. 519) and Short Series of Yaps (p. 542).

Medium-high, usually 2-syllabled
Laughing Gull, p. 159
see also Marbled Godwit
Gur-rik Series, p. 133

High, 2-syllabled underslur
Blue-headed Vireo Zer-wee, p. 285

High, usually 3-syllabled
Gull-billed Tern Kee-erik, p. 171
Sooty Tern Kwiddy-wit, p. 169

Low, strongly nasal
Fish Crow Uh-oh, p. 289

Low, mellow, seminasal
Western Screech-Owl
Kew-du-du, p. 222

2- OR 3-NOTED NASAL PHRASE
A 2- or 3-noted phrase with at least one part nasal. See also 2- or 3-Syllabled Breathy Phrase (p. 551) and Tewee (p. 507).

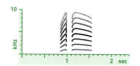

2 notes on similar pitch, second longer
Western Grebe Song, p. 79
Burrowing Owl Song, p. 225
Scaly-breasted Munia
Pi-peep, p. 358

First note short, high, squeaky
Brown-headed Nuthatch
Skew-doo, p. 316

First note upslurred, second Tseetlike
Green-tailed Towhee
Kwoy-tseet, p. 423

2- to 3-note phrase, first note highest and shortest
Great Kiskadee Grip-the-wheel, p. 267

2 or 3 unbroken honks in stereotyped pattern
Common Raven, p. 290

2 or 3 longer, keerlike notes
Willet (Eastern) Keer-wee-wee, p. 149
Willet (Western) Keer-wee-wee, p. 148

Variable, wailing or honking
Indian Peafowl, p. 65

A 2- OR 3-SYLLABLED HARSH PHRASE

2- TO 3-SYLLABLED SCREECHY PHRASE
A short phrase of screechy notes.

Of 2 notes; high, wheezy
Fulvous Whistling-Duck
Ki-wheer, p. 36

Of 2 notes; medium-low
Ring-necked Pheasant
Keek-kuk, p. 71

Gray Partridge Kee-raw, p. 71

Of 2–4 syllables; medium-high
Caspian Tern Kee-kareer, p. 171
Sooty Tern Kwiddy-wit, p. 169

Of 3–5 notes; medium-low
Domestic Chicken Crow, p. 64

A 2- OR 3-SYLLABLED HARSH BURRY PHRASE

2- OR 3-SYLLABLED GRATING OR CROAKING PHRASE
A high-pitched harsh grating or croaking sound of 2 or 3 syllables.

High, 2-syllabled, variably grating
Black Tern Chivik, p. 179
Least Tern Kideek, p. 178
Roseate Tern Chivik, p. 177

2-syllabled, harshly grating
Common Tern Kee-arr, p. 174
Arctic Tern Keeyer, p. 175
Sandwich Tern Keerik, p. 173

2-syllabled grate or 2 grating notes
Western Grebe Song, p. 79

3-syllabled grate
Black Tern Chividik, p. 179

2-syllabled ticking croak
Snail Kite Ker-wuck, p. 200

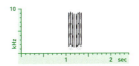

Deep 2-syllabled croak
Common Raven Kro-uk, p. 290

A WHISTLE AND A HARSH NOTE OR QUACK
A 2-noted phrase, one note a clear high musical whistle, the other a Snarl, Quack, or Grunt.

A high rising whistle, then a noisy to grating note
White-tailed Kite Wee-Snarl, p. 199

A lower rising nasal note, then a coarse, rattling snarl
Mute Swan Wee-Snarl, p. 42

A high whistled Hee, then a nasal quack
Gadwall Hee-Quack, p. 46

A high whistled Hee, then a low soft grunt
Mallard, p. 48

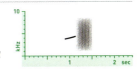

A low soft grunt, then a high whistled Hee
Northern Pintail Grunt-hee, p. 52

A wail, then a high squeaky Peep
Boreal Owl Wail-peep, p. 220

A 2- OR 3-SYLLABLED POLYPHONIC PHRASE

A 2-NOTED POLYPHONIC PHRASE
A 2-noted phrase with at least one part polyphonic.

White-collared Seedeater
Jirp (variant), p. 421
Bobolink See-yur, p. 490
American Goldfinch Bee-bee, p. 368

Red Crossbill Chit-too, p. 362
Scaly-breasted Munia
Pi-peep, p. 358

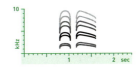

A 2- OR 3-SYLLABLED POLYPHONIC NOTE
A single polyphonic note that changes in pitch, creating the impression of 2 or 3 syllables.

Sharp-tailed Grouse
Hoo-wee-urr, p. 74
Eastern Bluebird Zewy, p. 341
Summer Tanager Vewy, p. 457

Rose-breasted Grosbeak
Zwee-er, p. 460
Black-headed Grosbeak
Zwee-er, p. 461

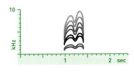

TWANGING OR METALLIC PHRASE
A phrase that includes at least one nearly monotone note with a bizarre, metallic nasal tone quality, like the sound of a harmonica.

Indian Peafowl, p. 65
Blue Jay Pumphandle Calls, p. 292

Hill Myna, p. 351

INDEX PART IV: THE SAME PHRASE REPEATED (A COMPLEX SERIES)

The same phrase repeated without a significant pause. Includes series of couplets, triplets, and more complex phrases, as well as series of series.

A COMPLEX SERIES OF CLICKING OR MECHANICAL NOTES

COMPLEX TICKING SERIES
Ticklike notes in rhythmically complex series.

Ticks in couplets
  Barrow's Goldeneye Tick-it
    Series, p. 61
  Green Kingfisher Tick-tick,
    p. 227

Ticks in alternating couplets and triplets
  Yellow Rail Song, p. 108

A COMPLEX SERIES OF CHIPLIKE NOTES

CHIPLIKE COUPLET SERIES
A repeated 2-syllabled phrase of high, sharp Chiplike notes.

  Eastern Kingbird Tick-it
    Series, p. 271

  Northern Red Bishop,
    p. 358

A VERY HIGH COMPLEX SERIES

COMPLEX SEET SERIES
Very high couplet or triplet series of Seetlike notes.

Very high, monotone couplet series
  Black-and-white Warbler,
    p. 385
  Bay-breasted Warbler, p. 405
  Blackburnian Warbler, p. 403
  American Redstart, p. 395

Similar but very long, plastic
  Black Guillemot, p. 152

Similar but notes well spaced
  Buff-bellied Hummingbird,
    p. 105

Very high rising couplet series
  Blackburnian Warbler, p. 403

Medium-high monotone couplet series
  Magnolia Warbler, p. 413

Very high monotone triplet series
  Black-and-white Warbler,
    p. 385

Very high rising triplet series
  Blackburnian Warbler, p. 403

HIGH TRILLED COUPLET SERIES
A couplet series of high, twittering trills. See also Chips and Trills, p. 534.

From overhead at night
  American Woodcock, p. 143

A COMPLEX SERIES OF WHISTLED NOTES

YELLOWTHROAT SONG GROUP
A triplet or quadruplet series of sharp but musical whistles, usually medium in pitch.

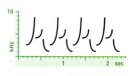

Triplet series
  Ruby-crowned Kinglet, p. 332

Triplet or quadruplet series
  Carolina Wren, p. 322

  Common Yellowthroat, p. 391

Quadruplet series
  Connecticut Warbler, p. 394

WHISTLED COUPLET SERIES

A repeated 2-syllabled phrase made of whistled notes. See also Northern Cardinal Song Group, p. 567, and Pitter, p. 554.

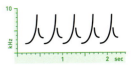

Notes sharp, rather unmusical, becoming louder
Ovenbird, p. 379

Notes sharp, but musical
Palm Warbler, p. 414
Mourning Warbler, p. 393
Common Yellowthroat, p. 391
Kentucky Warbler, p. 392
Carolina Wren, p. 322
Great-tailed Grackle Series Calls, p. 473
Boat-tailed Grackle Series Calls, p. 474

Notes sharp, but musical; series downslurred
Yellow-throated Warbler, p. 407

Notes musical, pattern variable
Tufted Titmouse Song, p. 310
Black-crested Titmouse Song, p. 311
Ruby-crowned Kinglet, p. 332
Harris's Sparrow Pitter, p. 440

Notes high, variable; series long
American Pipit Song, p. 355

Notes medium-high, series very long and rapid
Piping Plover Song, p. 126

Notes very high, slightly screechy, almost monotone
Broad-winged Hawk Pee-tee Series, p. 206

Notes shrill, almost monotone
Killdeer Song, p. 127

WILLET SONG GROUP

A triplet or quadruplet series of medium-low clear whistled notes, often continuing for long periods. Frequently given in flight.

Triplet series
Greater Yellowlegs, p. 146

Quadruplet series
Lesser Yellowlegs, p. 147

Willet (Eastern), p. 149
Willet (Western), p. 148

SERIES OF SERIES

A short series of 3-5 whistled or semimusical notes, themselves repeated in a stereotyped fashion.

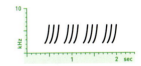

Variable, but usually not very musical
Great-tailed Grackle Series Calls, p. 473

A LOW-PITCHED COMPLEX SERIES

COMPLEX HOOTING OR COOING SERIES

A couplet, triplet, or quadruplet series of hooting or cooing notes.

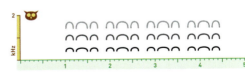

Triplet series
Eurasian Collared-Dove Song, p. 87

Quadruplet series, often introduced by long overslurred coo
White-crowned Pigeon Song, p. 86
Red-billed Pigeon Song, p. 86

Burry couplet or triplet series
African Collared-Dove Song, p. 87

GROWL-CLUCK SERIES

A couplet series of alternating Clucks and low Growls, audible at close range.

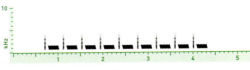

Chuck-will's-widow, p. 102

Eastern Whip-poor-will, p. 101

A LOW-PITCHED COMPLEX SERIES, CONTINUED

GULPING COMPLEX SERIES
A couplet or triplet series of very low-pitched gulping notes.

American Bittern Song, p. 187
Little Blue Heron Chuk-a Series, p. 190

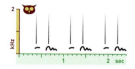

A COMPLEX SERIES OF BURRS OR BUZZES

BURRY COUPLET SERIES
A repeated 2-syllabled phrase of burry notes.

Medium-high, very finely burry
Palm Warbler, p. 414

Medium-low, coarser, musical
Kentucky Warbler, p. 392

Mourning Warbler, p. 393

Lower, coarser, and slower
Spotted Sandpiper, p. 144

A COMPLEX SERIES OF NASAL NOTES

NASAL COUPLET SERIES
A repeated 2-syllabled phrase of nasal notes.

Longer notes in each phrase lower, downslurred
Little Gull Long Call, p. 157

Longer notes in each phrase upslurred
American Coot, p. 117
Hairy Woodpecker, p. 235
Northern Flicker, p. 240
Red-bellied Woodpecker, p. 231

All notes high, rather squeaky
Red-cockaded Woodpecker, p. 237

One note squeaky, one Chiplike
Peregrine Falcon Ee-chup Series, p. 242
Prairie Falcon Ee-chup Series, p. 243

High, squeaky, fast or slow
Groove-billed Ani Cheep-weep Series, p. 96

All notes short, clipped
Willet (Eastern) Ka-lip Series, p. 149
Willet (Western) Ka-lip Series, p. 148

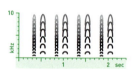

Wilson's Snipe Kik-a Series, p. 142
Purple Gallinule Keek-a Series, p. 115
Purple Swamphen, p. 114

Series of 2-syllabled notes
Marbled Godwit Gur-rik Series, p. 133
Hudsonian Godwit Poowit Series, p. 133

A COMPLEX SERIES OF HARSH NOTES

COUPLET SERIES OF KEKS
A couplet series of brief harsh noisy notes.

Virginia Rail Song, p. 112

COUPLET SERIES OF GRUNTS OR SCREECHES
A repeated 2-syllabled phrase of barking, grunting, or screechy notes.

Noisy, barking to grunting (Chuk-a Series)
Wood Stork, p. 183
Black-crowned Night-Heron, p. 193
Cattle Egret, p. 192
Reddish Egret, p. 191
Golden-fronted Woodpecker, p. 230

Yellow-bellied Sapsucker, p. 233

Low, gulping (Chuk-a Series)
Little Blue Heron, p. 190

High, slow, nasal to screechy
Scaled Quail Kuk-curr Series, p. 69

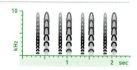

Helmeted Guineafowl Buck-wheat Series, p. 65

High, fast, screechy
Ring-necked Pheasant Kuttuk, p. 71

INDEX PART V: DIFFERENT SHORT NOTES, SERIES, OR PHRASES
A complex song made of different short notes, series, or phrases, all separated by pauses.

A SONG OF MOSTLY DIFFERENT SEPARATE SINGLE NOTES

MOSTLY SINGLE 1-SYLLABLED NOTES
Songs consisting mostly of well-spaced single 1-syllabled notes, with consecutive notes often different. See also Variable Single Whistles, p. 504, in which consecutive notes are usually the same.

Musical, many notes whistled and/or polyphonic
Osprey, p. 211
Lapland Longspur Whistled Calls, p. 373
McCown's Longspur Whistled Calls, p. 376
Orchard Oriole Songlike Calls, p. 482
Spot-breasted Oriole Songlike Calls, p. 484

Altamira Oriole Songlike Calls, p. 485
Baltimore Oriole Songlike Calls, p. 488
Bullock's Oriole Songlike Calls, p. 489

Unmusical, most notes resembling Chips or Spits, sometimes with Seets
American Woodcock, p. 143
Eastern Towhee Complex Song, p. 424

Spotted Towhee Complex Song, p. 425
Green-tailed Towhee Complex Song, p. 423
Botteri's Sparrow Complex Song, p. 429

Most notes Chiplike, often in short series, often with occasional burry phrases
Red Crossbill (all types), pp. 362–364

SINGLE NOTES WITH SOME RATTLES OR CHATTERS
Songs consisting mostly of single musical whistles mixed with short Rattles or Chatters, all sounds well spaced. See also Yellow-breasted Chat (p. 419).

Whistles and short Rattles or Chatters
Chestnut-collared Longspur Whistled Calls, p. 375

McCown's Longspur Whistled Calls, p. 376

Hooded Oriole Songlike Calls, p. 483

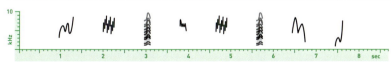

MOSTLY SINGLE CHIRPS
Songs consisting mostly of well-spaced single complex 1–2 syllabled chirping notes, with consecutive notes usually different.

Tree Swallow Dawn Song, p. 304

House Finch, p. 361
House Sparrow, p. 356

Eurasian Tree Sparrow, p. 357

A SONG OF MOSTLY DIFFERENT SEPARATE SERIES OR TRILLS

BROWN THRASHER SONG GROUP
Extremely varied short doubled phrases, separated by pauses.　　　　Brown Thrasher, p. 346

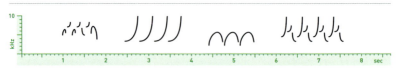

MOCKINGBIRD/CHAT SONG GROUP
Widely varied series, single notes, and/or trills, separated by pauses. Some versions include chatters.

Mostly series and trills,
semimusical; pauses short
　　Common Redpoll, p. 366
　　Hoary Redpoll, p. 367

Mostly series, often including
imitations; pauses short to long
　　Northern Mockingbird,
　　　　p. 349

Mostly 1- or 2-noted series,
some trills; all musical
　　Rock Wren, p. 320

Series, trills, chatters, and
single notes; pauses long
　　Yellow-breasted Chat, p. 419
　　(see also Songlike Calls of
　　　　orioles)

Mostly musical trills
　　Wood Thrush Trill Song,
　　　　p. 340

Series of Tinks, Chips, and
Snarls, with occasional single
snarls
　　Black-tailed Gnatcatcher,
　　　　p. 328

A SONG OF MOSTLY DIFFERENT SEPARATE SINGLE PHRASES

2 DIFFERENT SHORT PHRASES
Songs consisting of 2 different 1–3 syllabled phrases, consecutive phrases usually different.

Purely whistled, slow, drawn-out
　　Eastern Wood-Pewee Day
　　　　Song, p. 253

At least partly burry, faster
　　Eastern Phoebe Song, p. 261

Black Phoebe Song, p. 260
Western Wood-Pewee Dawn
　　Song, p. 252
Ash-throated Flycatcher
　　Dawn Song, p. 264

Great Crested Flycatcher
　　Dawn Song, p. 265
Brown-crested Flycatcher
　　Dawn Song, p. 266

2–5 OF 1 NOTE OR PHRASE, THEN A DIFFERENT PHRASE
Songs that prominently feature an AAAB pattern of well-spaced notes or phrases, sometimes with other single notes or phrases in between instances of the pattern.

All notes high, abrupt,
semimusical
　　Acadian Flycatcher Dawn
　　　　Song, p. 257

Introductory notes include
twitters of nasal Pips
　　Western Kingbird Dawn Song,
　　　　p. 270
　　Scissor-tailed Flycatcher
　　　　Dawn Song, p. 272

Another note type sometimes
between AAAB patterns
　　Northern Beardless-Tyran-
　　　　nulet Dawn Song, p. 251
　　Couch's Kingbird Dawn Song,
　　　　p. 269

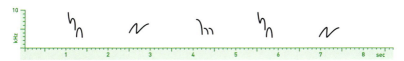

3 DIFFERENT SHORT PHRASES
Songs consisting of 3 different 1–3 syllabled phrases, consecutive phrases usually different.

Purely whistled
Eastern Wood-Pewee Dawn
Song, p. 253

Whistled to slightly burry
Say's Phoebe Dawn Song,
p. 262

All strongly burry
Willow Flycatcher Song,
p. 259

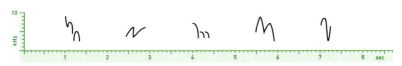

MULTIPLE SHORT DIFFERENT PHRASES
Songs consisting mostly of well-spaced single 2–4 syllabled phrases, usually without any trills or series, consecutive phrases usually different.

Phrases medium-high, usually whistled, not burry
Philadelphia Vireo, p. 278
Red-eyed Vireo, p. 276
Yellow-green Vireo, p. 227
Blue-headed Vireo, p. 285
Gray Vireo, p. 283
Townsend's Solitaire Disjunct
Song, p. 334
Red-whiskered Bulbul,
p. 330
Purple Finch Disjunct Song,
p. 360

Phrases medium-high, both whistled and burry
Acadian Flycatcher Complex
Song, p. 257
Summer Tanager Dawn Song,
p. 457

Phrases medium-high, mostly burry
Yellow-throated Vireo, p. 284
Hutton's Vireo, p. 286
Scarlet Tanager Dawn Song,
p. 456

Phrases medium-high, whistled, burry, or polyphonic
Loggerhead Shrike Song,
p. 274
Northern Shrike Song, p. 275
Red-vented Bulbul Song,
p. 330

Phrases high, tinkling, complex
Horned Lark Long Song,
p. 299

Many phrases high, musical, polyphonic
American Robin Complex
Song, p. 342
American Goldfinch Disjunct
Song, p. 368
Lesser Goldfinch Disjunct
Song, p. 369
Pine Siskin Disjunct Song,
p. 370

Phrases varying widely in pitch and tone quality
Gray Catbird, p. 345

Phrases whispered, hissing
Saltmarsh Sparrow, p. 453

Some phrases high slurred whistles, others gurgling
Brown-headed Cowbird,
p. 479
Bronzed Cowbird, p. 480

WIDELY VARIED PHRASES AND SERIES
A mix of phrases and series, often of varied lengths, sometimes with trills, single notes, or more complex patterns.

Often including polyphony, or long slurred whistles
Common Myna, p. 352
Hill Myna, p. 351
European Starling,
pp. 350–351

Great-tailed Grackle Short
Song, p. 472
Boat-tailed Grackle Short
Song, p. 474

Phrases usually contain 1–2 short semimusical chatters
Black-capped Vireo, p. 282

A SONG OF MOSTLY DIFFERENT SEPARATE PHRASES IN CLUSTERS

A ROBINLIKE SONG

Songs consisting of 1–3 syllabled phrases grouped into clusters separated by longer pauses.

Phrases high, musical, tinkling
  McCown's Longspur, p. 376

*Phrases medium-high, musical,
not burry*
  American Robin, pp. 342

*Phrases medium-high,
semimusical, not burry*
  Black-whiskered Vireo, p. 277

*Clusters long; phrases mostly
slurred whistles*
  Rose-breasted Grosbeak,
    p. 460

  Black-headed Grosbeak,
    p. 461

Phrases finely burry
  Purple Martin Dawn Song,
    p. 305
  Scarlet Tanager, p. 456
  Summer Tanager, p. 457

*Phrases of mostly rising,
polyphonic Jitlike notes*
  Barn Swallow Dawn Song,
    p. 302

Phrases of soft, hoarse whines
  Northern Rough-winged
    Swallow Song, p. 303

*Phrases include Chirps, often
with whistles and gurgles*
  Tree Swallow Dawn Song,
    p. 304

INDEX PART VI: A LONG COMPLEX SONG

A very long continuous song that lasts at least 4 seconds without a significant pause, including a variety of different note types or patterns. For long songs that consist of the same note repeated, see Index Part II (p. 531). For long songs that consist of the same phrase repeated, see Index part IV (p. 556).

A LONG SONG WITH FEW REPEATED NOTES

A VERY LONG PHRASE OR WARBLE

A long (usually 4 or more seconds), mostly musical warble or phrase, without significant pauses or repetition.

High and tinkling, all notes musical
McCown's Longspur, p. 376

Loud or soft, medium-high, with many polyphonic notes
Bobolink, p. 490
Eastern Meadowlark Complex Song, p. 470
Western Meadowlark Complex Song, p. 471

Fairly soft, medium-low, with many polyphonic notes
Eastern Bluebird Long Song, p. 341

Medium-high, mostly whistles and warbles
Townsend's Solitaire Continuous Song, p. 344

Similar but faster, with an occasional high loud Seet
House Finch Long Song, p. 361

Quite varied, with harsh notes and imitations
Gray Catbird, p. 345
Hooded Oriole, p. 483

Fairly soft, quite varied, with polyphonic notes and imitations
Blue-gray Gnatcatcher, p. 329
Bell's Vireo Complex Song, p. 281
Blue-headed Vireo Complex Song, p. 285
Gray Vireo Complex Song, p. 283

Yellow-throated Vireo Complex Song, p. 284
White-eyed Vireo Complex Song, p. 280
Woodhouse's Scrub-Jay Quiet Song, p. 296
Florida Scrub-Jay Quiet Song, p. 297
Blue Jay Quiet Song, p. 293
Gray Jay Quiet Song, p. 294
Black-billed Magpie Quiet Song, p. 298
Green Jay Quiet Song, p. 295

A LONG SONG WITH SOME REPEATED NOTES

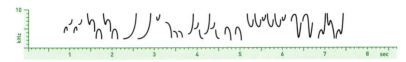

CURVE-BILLED THRASHER SONG GROUP

A long (usually 4 or more seconds) mostly musical phrase without significant pauses, but with at least some notes repeated once.

Fairly high, some notes polyphonic
Pine Siskin, p. 370

Medium-low, most notes polyphonic
Bobolink, p. 490

Medium-low, few repetitions, some polyphony, some imitations
Orchard Oriole, p. 482

Notes extremely varied, often including imitations
European Starling, pp. 350

Curve-billed Thrasher, p. 348
Long-billed Thrasher, p. 347
Pine Grosbeak Long Song, p. 372
Purple Finch Long Song, p. 360

A LONG SONG WITH SOME REPEATED NOTES, CONTINUED

A MIX OF SINGLE NOTES, CHATTERS, AND TRILLS
A long mix of single musical notes, brief chatters, and trills or short series, without significant pauses.

Black-capped Vireo Complex
Song, p. 282
Bullock's Oriole Complex
Song, p. 489

Baltimore Oriole Complex
Song, p. 488
Hooded Oriole, p. 483

Black-throated Sparrow,
p. 437
Lark Sparrow, p. 436

A LONG SONG OF MOSTLY SERIES OR TRILLS

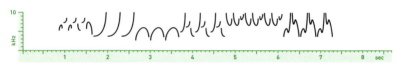

MULTIPLE CONSECUTIVE SERIES
At least 3 series of mostly musical notes, sometimes with trills and/or single notes; no significant pauses.

*Medium-pitch; all notes highly
musical, whistled*
Field Sparrow Complex Song,
p. 434

*Mostly high, musical, some
notes polyphonic*
American Goldfinch, p. 368
Prothonotary Warbler, p. 383
Worm-eating Warbler, p. 382

High, musical, with imitations
Lesser Goldfinch, p. 369

Notes extremely varied
Common Myna, p. 352
Northern Mockingbird, p. 349

Notes unmusical, chirping
Dickcissel Complex Song,
p. 455

With long, high slurred whistles
Red-winged Blackbird Flight
Song, p. 468

*With long, medium-low slurred
whistles, often trills*
Northern Cardinal, p. 458
Pyrrhuloxia, p. 459
Lark Bunting, p. 438

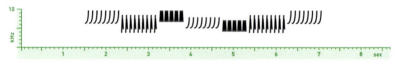

MULTIPLE CONSECUTIVE TRILLS, CHATTERS, OR BUZZY SERIES
3 or more trills or fast series, some often chattery or buzzy, without pauses in between.

*Musical series, chattery trills,
and buzzes*
White-collared Seedeater
Long Song, p. 421

*Mostly trills, semimusical to
buzzy*
White-winged Crossbill,
p. 365
Common Redpoll, p. 366
Hoary Redpoll, p. 367

Series and trills, mostly buzzy
Brewer's Sparrow, p. 433

*Mostly musical, sometimes with
harsh buzzes and single notes*
Vesper Sparrow, p. 435
Lark Sparrow, p. 436

SERIES OF CHIPLIKE NOTES AND BURRY PHRASES
*1 to several series of Chiplike notes, then a simple or complex musical series that usually includes burry
notes. Some versions more complex, culminating in multiple consecutive burry series.*

Red Crossbill (all types), pp. 362–364

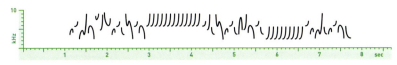

WINTER WREN SONG GROUP
A long high musical tinkling warble, with trills interspersed.

Entire song very high, musical
  Winter Wren, p. 318
  Grasshopper Sparrow, p. 447

Note types varied; song often plastic, with irregular pauses
  Dark-eyed Junco Complex Song, p. 446

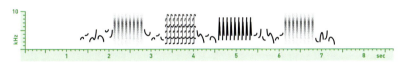

MARSH WREN LONG SONGS
Chatters or complex semimusical trills, often with some gurgling or tinkling warbles interspersed.

Trills lower, harsher, more rattling
  Marsh Wren (Eastern), p. 325

Marsh Wren (Western), p. 324

PUNCTUATED BY TICKING TRILLS
A complex, often very long song of warbles, phrases, and sometimes series, with occasional ticking trills interspersed.

Barn Swallow Song, p. 302
Cliff Swallow Song, p. 301

Cave Swallow Song, p. 300
Purple Martin Day Song, p. 305

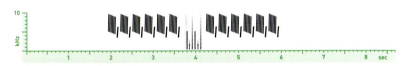

PUNCTUATED BY CRACKLES
A complex, often very long song of whistles, burrs, or other notes, with occasional crackles or "stick-breaking sounds" interspersed.

Song mostly a buzzy series of 1- or 2-syllabled notes
  Boat-tailed Grackle, p. 474

Song more varied, sometimes with sirenlike whistles
  Great-tailed Grackle, p. 472

INDEX PART VII: A COMPLEX SONG

A complex song of more than 3 notes or syllables, less than 4 seconds long. Includes all songs not indexed in the prior categories.

A VERY HIGH-PITCHED COMPLEX SONG

A VERY HIGH COMPLEX SONG
A complex, often plastic phrase of very high long whistles and trills. Often preceded by lower, complex notes or phrases.

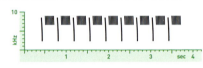

Often quite long; often with both whistles and trills
   European Starling, p. 350
   Bronzed Cowbird Whistled
      Song, p. 480
   Shiny Cowbird Whistled Song,
      p. 481

Often quite long; mostly simple and couplet series, no trills
   Black Guillemot, p. 152

Usually 4–6 notes, none very long
   Brown Creeper, p. 317

Usually 4 notes or fewer, usually including trills
   Brown-headed Cowbird
      Whistled Song, p. 479

CHIPS AND TRILLS
A mix of Chiplike notes and repeating trills, given during flight displays.

Chips nasal, trills rattling
   Ruby-throated Hummingbird
      Shuttle Display, p. 106
   Black-chinned Hummingbird
      Shuttle Display, p. 107

Chips plastic, trills run together
   American Woodcock Chip
      Series, p. 143

A COMPLEX SONG WITH SEETS OR TINKS AT START, LOWER NOTES AT END

TITMOUSE TSEET-SONGS
Highly variable short (2–4 note) songs; typically starts with a high Tseetlike note, followed by much lower, musical whistles.

   Tufted Titmouse, p. 310   Black-crested Titmouse, p. 311

SPARROW PAIR REUNION CALLS
A series of high Seets transitioning gradually or suddenly into a series of Chiplike notes. Highly plastic, often given by 2 or more birds at once.

   Rufous-crowned Sparrow, p. 426

DZEET-CHIPPITY
1–3 quick high buzzes, then a short chipping trill. Highly plastic.

   Ruby-throated Hummingbird,   Black-chinned Hummingbird,
      p. 106   p. 107

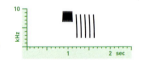

A 3-PART MUSICAL SONG STARTING WITH SEET SERIES
A unique 3-part song: an accelerating series of Seetlike notes, then a lower seminasal couplet series, then a variable complex series of musical whistles.

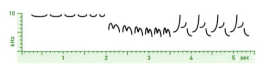

   Ruby-crowned Kinglet,
   p. 332

CHICK-A-DEE CALL GROUP

Starts with high Pseeplike notes, transitioning gradually or suddenly into a series of nasal notes or whines. Plastic; "dee" notes at end may be given separately.

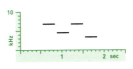

"Dee" notes rather slow, nasal
  Black-capped Chickadee,
  p. 308

"Dee" notes rather slow, hoarse
or whiny
  Boreal Chickadee, p. 307

"Dee" notes rather slow,
upslurred, often just 1
  Tufted Titmouse, p. 310

"Dee" notes rather rapid
  Carolina Chickadee, p. 309
  Black-crested Titmouse,
  p. 311

*Ends in a semimusical Sputter
instead of nasal notes*
  Golden-crowned Kinglet
  Song, p. 331

*Starts with sharp Tsits or Chips,
ends with musical whines*
  Song Sparrow Whine Series,
  p. 442
  Lincoln's Sparrow Whine
  Series, p. 443

"Dee" notes very high and nasal,
with only 1 note prior
  Brown-headed Nuthatch
  Skew-doo, p. 316

COMPLEX SONGS OF WARBLERS AND SPARROWS

A catch-all category of generally quiet, infrequent songs that often start with high-pitched call-like notes, followed by a complex, varied phrase that may include buzzes, trills, and single notes. Sometimes given in flight.

Very few repeated notes
  Louisiana Waterthrush,
  p. 381
  Ovenbird, p. 379
  Swainson's Warbler, p. 384
  Common Yellowthroat, p. 391

  Kentucky Warbler, p. 392

Some repeated notes
  Northern Waterthrush,
  p. 380
  Bachman's Sparrow, p. 427
  Cassin's Sparrow, p. 428
  Mourning Warbler, p. 393

Consisting mostly of series and trills
  Prothonotary Warbler, p. 383

  Worm-eating Warbler, p. 382
  Vesper Sparrow, p. 435

Rather short: a few Tinks, a short phrase, then a trill or buzz
  Nelson's Sparrow, p. 454
  Seaside Sparrow, p. 450
  Le Conte's Sparrow, p. 451
  Swamp Sparrow, p. 444

A COMPLEX SONG OF MOSTLY LONG MUSICAL WHISTLES

CHICKADEE AND *ZONOTRICHIA* SONG GROUP

Songs consisting almost entirely of high clear, monotone whistles.

Usually 2 notes, the second one lower
  Black-capped Chickadee,
  p. 308

*2–3 notes on the same pitch
(next song on different pitch)*
  Harris's Sparrow, p. 440

Usually 3–4 notes on 2–3 different pitches
  Carolina Chickadee, p. 309
  Brown-headed Cowbird
  Whistled Song, p. 479

Usually 5 or more notes, ending in a monotone triplet series
  White-throated Sparrow,
  p. 439

NORTHERN CARDINAL SONG GROUP

Usually 1–3 consecutive whistled series; pitch medium-low and most notes highly musical. Series of very long slurred whistles distinctive.

  Northern Cardinal, p. 458   Pyrrhuloxia, p. 459

A COMPLEX SONG OF MOSTLY LONG MUSICAL WHISTLES, CONTINUED

ALTAMIRA ORIOLE SONG GROUP
A long song of slow, medium-low, slightly slurred musical whistles on slightly different pitches. May recall a child learning to whistle.

Altamira Oriole, p. 485

Audubon's Oriole, p. 486

Spot-breasted Oriole, p. 484

A WHISTLED PHRASE
A phrase of 4 or more musical whistled notes that are not monotone.

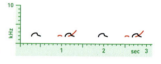

Very high and loud
European Starling, p. 350

2–4 short high notes
Verdin Song, p. 313

Variable, shrill, first note longest and highest
Killdeer Dee-dit-dit, p. 127

High, with few or no sharp notes
Eastern Meadowlark, p. 470

High, some notes sharp
Brown Creeper, p. 317

Shiny Cowbird Whistled Song, p. 481

Medium-high, some notes sharp
Louisiana Waterthrush, p. 381
Hooded Warbler, p. 412
Smith's Longspur, p. 374

Medium-low, musical
Red-vented Bulbul, p. 330
Red-whiskered Bulbul, p. 330

Medium-low, musical, usually ending in gurgle
Western Meadowlark, p. 471

Medium-low, musical, with long slurred whistles
Fox Sparrow, p. 445

Medium-low, musical; second half often repeats first half
Scott's Oriole, p. 487
Baltimore Oriole, p. 488

Lower, mellower, plastic
American Golden-Plover, p. 129

A MELLOW WHISTLED PHRASE OR CHORUS
A phrase, duet, or chorus of short, mellow whistled phrases, some broken. Most notes well spaced.

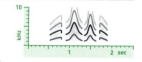

Repeated over and over at night
Eastern Whip-poor-will Song, p. 101
Chuck-will's-widow Song, p. 102

Duet or chorus; notes well spaced, some broken
Northern Bobwhite Duet, p. 68

A BREATHY PHRASE
A phrase of 4 or more high, breathy, Wheewlike notes. See also 2- to 3-Syllabled Breathy Phrase (p. 551).

Usually 3- to 4-noted
American Wigeon
Whi-whee-whew, p. 47

Usually 4 or more notes
Black-bellied Whistling-Duck
Kip-whee-witter, p. 37

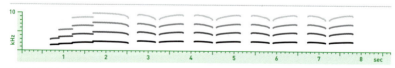

LOON SONG GROUP
A long phrase of musical whistles, some broken. Sometimes in chorus.

Common Loon, p. 180

CURLEW SONG GROUP
A long phrase of nearly monotone musical whistles and trills, usually culminating in a long overslurred whistle that breaks downward and then upward.

Black-bellied Plover, p. 129
Long-billed Curlew, p. 132

Whimbrel, p. 131

Upland Sandpiper, p. 130

A COMPLEX SONG OF MOSTLY MUSICAL SERIES OR TRILLS

Including most nonbuzzy warbler and sparrow songs.

ONE NOTE, SERIES, OR SHORT PHRASE, THEN A SERIES

1–3 single notes or a 2–3 note phrase, then a series.

*1 sharply downslurred whistle,
then a lower series of same*
  American Goldfinch
    Pee-tee-tee-tee, p. 368
  Mississippi Kite Keek-tew-
    tew-tew, p. 200

*First note often a Cheep, the rest
lower Kuks*
  American Robin Cheep/Kuk,
    p. 343

First note a Whit
  Least Flycatcher Pweew
    Series, p. 255

*Series highly musical, of single
notes or couplets*
  Bachman's Sparrow, p. 427
  Harris's Sparrow, p. 440
  Eastern Towhee, p. 424
  Northern Cardinal, p. 458
  Pyrrhuloxia, p. 459

All notes semi- to unmusical
  Sedge Wren, p. 326
  Dickcissel, p. 455
  Eastern Towhee, p. 424
  Spotted Towhee, p. 425

All notes high, seminasal
  Brown-headed Nuthatch
    Skew-doo, p. 316

ONE NOTE, SERIES, OR SHORT PHRASE, THEN A TRILL

*1–3 single notes or a 2–3 note phrase, then a trill. Some songs may end
with an additional single note after the trill. Compare the above
category.*

*1–2 whistled, polyphonic, or buzzy
notes, then a musical trill*
  Bewick's Wren, p. 323
  Eastern Towhee, p. 424
  Bachman's Sparrow, p. 427
  Black-throated Sparrow, p. 437

Baird's Sparrow, p. 449

*Trill slightly metallic, sometimes
with 1 note at end*
  Seaside Sparrow, p. 450
  Red-winged Blackbird,
    p. 468

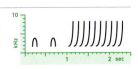

*A few notes, then a semimusical
or unmusical trill*
  Marsh Wren (Eastern), p. 325
  Marsh Wren (Western), p. 324
  Spotted Towhee, p. 425

A SERIES, THEN 1 NOTE

A short series followed by a single note; all notes high sharp whistles.

High
  American Redstart, p. 395

*Medium-high, series often of
couplets*
  Magnolia Warbler, p. 413

*Medium-high, series often of
couplets and downslurred*
  Yellow-throated Warbler,
    p. 407

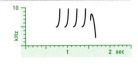

HOODED WARBLER SONG GROUP

*A short series followed by a short phrase; all notes medium-high,
fairly sharp whistles.*

*Medium-high, often starting with
couplet series*
  Magnolia Warbler, p. 413

*Medium-low, often starting with
couplet series*
  Hooded Warbler, p. 412

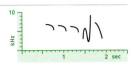

*Medium-low, often starting with
1-syllabled downslurred whistles*
  Louisiana Waterthrush, p. 381
  Swainson's Warbler, p. 384

2 TRILLS, OR A SERIES AND A TRILL

*2 consecutive musical to unmusical whistled simple trills, or 1 fast
series and 1 trill.*

*First part usually higher; all
notes fairly Chiplike*
  Wilson's Warbler, p. 418
  Orange-crowned Warbler,
    p. 389

*Second part faster and higher;
all notes rather unmusical*
  Spotted Towhee, p. 425

*2 trills, or a series and a trill in
either order, both fairly musical*
  Dark-eyed Junco, p. 446

2 SIMPLE SERIES

2 consecutive series of unmusical to musical 1-syllabled whistles.

All notes Chiplike, rather unmusical
　Wilson's Warbler, p. 418

All notes medium-high, very musical
　Yellow Warbler (Golden), p. 408
　White-collared Seedeater, p. 421

All notes high, musical, simple
　Yellow Warbler, p. 408
　Chestnut-sided Warbler p. 409

All notes high, musical, usually complex
　Yellow-rumped Warbler (Myrtle), p. 410
　Yellow-rumped Warbler (Audubon's), p. 411

All notes very high
　Blackburnian Warbler, p. 403
　Cape May Warbler, p. 406

A COUPLET SERIES, THEN A TRILL

A whistled couplet series, then a musical or unmusical trill.

Trill rather unmusical, chipping
　Nashville Warbler, p. 390
　Tennessee Warbler, p. 388

Trill very musical
　Baird's Sparrow, p. 449

2 COUPLET SERIES

2 consecutive couplet series of unmusical to musical whistles.

Medium low, usually burry, musical
　Mourning Warbler, p. 393

Medium low, musical
　Kentucky Warbler (rare), p. 392

Medium high, musical
　White-collared Seedeater, p. 421

High, musical, notes often only vaguely 2-syllabled
　Yellow-rumped Warbler (Myrtle), p. 410
　Yellow-rumped Warbler (Audubon's), p. 411

Medium low, rather unmusical, becoming louder
　Ovenbird, p. 379

Very high
　Black-and-white Warbler, p. 385

Medium high, musical, first series downslurred
　Yellow-throated Warbler, p. 407

2 SIMPLE SERIES, THEN 1 NOTE OR PHRASE

2 quick series, then a final note or 2-note phrase, all notes high clear sharp whistles.

High; all notes clear, whistled
　Yellow Warbler, p. 408
　Chestnut-sided Warbler, p. 409
　American Redstart, p. 395

Usually lower, most notes complex
　Blackburnian Warbler, p. 403
　Yellow-rumped Warbler (Audubon's), p. 411

Yellow-rumped Warbler (Myrtle), p. 410

3 OR 4 CONSECUTIVE SERIES

Usually 3, rarely 4 consecutive series of medium-low clear sharp whistles. Some series may be of couplets.

Medium-low, last part usually a series of Tews
　Northern Waterthrush, p. 380

Medium-high, last part usually a chipping trill
　Tennessee Warbler, p. 388

High, all notes clear and musical
　Yellow Warbler, p. 408

Chestnut-sided Warbler, p. 409

Middle section often just a single note, or a 2–3 note phrase
　Kirtland's Warbler, p. 416

High, notes complex and musical, first series longest
　Yellow-rumped Warbler (Audubon's), p. 411

Yellow-rumped Warbler (Myrtle), p. 410

3–4 very high series, some of couplets
　Black-and-white Warbler, p. 385

MULTIPLE CONSECUTIVE SHORT SERIES
4–7 consecutive series of musical notes, most series only 2 notes.

Indigo Bunting, p. 462
Lazuli Bunting, p. 463

Rufous-crowned
Sparrow, p. 426
Kirtland's Warbler,
p. 416

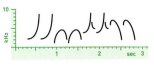

COMPLEX ACCELERATING SONGS
High, generally musical songs that prominently feature an accelerating series, with 1 or more notes before and/or after it.

High, upslurred, accelerating series ending on 1 lower note
Vermilion Flycatcher, p. 263

Chiplike notes before the series
Botteri's Sparrow, p. 429

An accelerating musical trill, then 2–4 monotone whistles
Cassin's Sparrow, p. 428

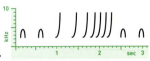

An accelerating musical trill, at or near the start of a complex musical song
Song Sparrow, p. 442

A PHRASE WITH SOME TRILLS OR BUZZES
A musical song of mostly single notes, with at least some buzzes, trills, or whistled series; some include polyphonic notes.

An accelerating musical trill, then 2–4 monotone whistles
Cassin's Sparrow, p. 428

Medium to medium-high, with long slurred musical whistles
American Tree Sparrow,
p. 430
Fox Sparrow, p. 445

Medium-high, most notes sharp; sometimes buzzes or series
Chestnut-collared Longspur,
p. 375
Snow Bunting, p. 377

Medium-high; notes can vary widely in quality
Orchard Oriole, p. 482

Bullock's Oriole, p. 489

Medium-high, many notes polyphonic
Lapland Longspur, p. 373

SONG SPARROW SONG GROUP
A musical mix of buzzes, trills, and whistled series, sometimes with single notes.

Usually starts with a single note or short phrase
Black-throated Sparrow,
p. 437
Bewick's Wren, p. 323
Lark Sparrow, p. 436
Green-tailed Towhee, p. 423

Usually starts with a musical series
Song Sparrow, p. 442

2–3 downslurred whistles, then only series or trills
Vesper Sparrow, p. 435

Short, not very musical
Dickcissel, p. 455

A COMPLEX SONG CONTAINING MUSICAL WARBLES

AN ACCELERATING PHRASE
A musical whistled phrase that speeds up, the last few notes running together into a warble.

Medium-low, first 1–2 notes nearly monotone whistles
Western Meadowlark, p. 471

High, tinkling, rising
Horned Lark, p. 299

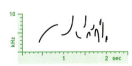

A 1- TO 4-SECOND WARBLE

Short (1–4 seconds) musical warbles of whistled, burry, and/or polyphonic notes. See also Gurgle (p. 574).

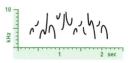

High, mostly sharp whistles
Canada Warbler, p. 417
Smith's Longspur, p. 374
Rufous-crowned Sparrow, p. 426

Medium-low, whistled, with a few short burry notes
Warbling Vireo, p. 279
Varied Bunting, p. 464
Painted Bunting, p. 465

Blue Grosbeak, p. 466

Second half of song often repeats first half
Snow Bunting, p. 377

Medium-low, whistled, usually with 1–2 long slurred buzzes
House Finch, p. 361

Lower, richly musical, usually lacking buzzes
Pine Grosbeak, p. 372

Purple Finch, p. 360

Includes polyphonic and burry notes
Bell's Vireo, p. 281
Eastern Bluebird, p. 341

Odd nasal, yelping warble
Chestnut-fronted Macaw, p. 248

WARBLE PLUS SERIES OR TRILLS

A highly musical song that starts with rapid notes, most in short series; usually culminates in high trill, then a lower trill and sometimes other notes.

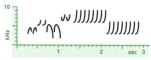

Starts fast, usually culminates in a high trill near the end
House Wren, p. 319

Similar but slower and more musical at start
Lincoln's Sparrow, p. 443

Variable, often lacking trills
Purple Finch, p. 360

Higher, with sharper notes, no trills
Rufous-crowned Sparrow, p. 426

A HOOTING, COOING, OR GROWLING COMPLEX SONG

PIED-BILLED GREBE SONG

2–3 consecutive series of low, almost hooting whistles, the later series slower, of couplets or triplets

Pied-billed Grebe, p. 83

A HOOTING OR COOING PHRASE

A phrase of 4 or more very low-pitched hooting or cooing notes.

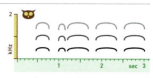

4-note phrase "WHO COOKS for YOU?"
Barred Owl Song, p. 216
White-winged Dove Short Song, p. 91

4-note phrase "UP cuppa COO"
Red-billed Pigeon Song, p. 86

Usually 3–4 notes, the first one broken
White-crowned Pigeon Song, p. 86
Mourning Dove Song, p. 90

Usually 4–6 notes, none broken
Great Horned Owl Song, p. 213

6 or more notes
Barred Owl Series Song, p. 216
White-winged Dove Long Song, p. 91

A HOOTING AND CACKLING CHORUS

A distinctive "caterwauling" duet or chorus of hooting phrases and barking nasal cackles.

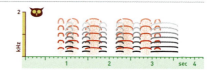

Barred Owl, p. 216

GROWLING OR GROANING PHRASES

See p. 552.

A BUZZY COMPLEX SONG

ALL-BUZZY PHRASE
A phrase of buzzy notes, buzzy series, and/or buzzy trills, at least some of which differ in pitch or quality.

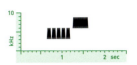

2 different buzzes
Blue-winged Warbler, p. 386
Brewer's Sparrow, p. 433

3 different buzzes
Golden-winged Warbler, p. 387
Black-throated Blue Warbler, p. 397

1 buzz, before or after 1 buzzy series or trill
Blue-winged Warbler, p. 386
Golden-winged Warbler, p. 387
Cerulean Warbler, p. 396

Black-throated Blue Warbler, p. 397
Brewer's Sparrow, p. 433

3-parted, each part higher, last one a buzz
Cerulean Warbler, p. 396

3-parted, each part higher, last one a whistle
Golden-cheeked Warbler, p. 401

"Zee-zee-zee zoo zee" or "Zoo, zee, zoo-zoo-zee"
Black-throated Green Warbler, p. 400

3- to 5-parted, quite variable
Blue-winged Warbler, p. 386
Golden-winged Warbler, p. 387
Golden-cheeked Warbler, p. 401
Northern Parula, p. 398
Tropical Parula, p. 399
Brewer's Sparrow, p. 433

A PHRASE OF WHISTLES AND BUZZES
A phrase of 3 or more fairly long notes, mostly buzzes and musical whistles, occasionally with musical trills or short series.

A series of notes starting whistled, becoming buzzy
Black-throated Blue Warbler, p. 397
Prairie Warbler, p. 402
Harris's Sparrow, p. 440

1–2 musical series, then a buzz
Prairie Warbler, p. 402
American Redstart, p. 395

First note a monotone whistle; song complex, often including musical trills or short series
White-crowned Sparrow, p. 441

5 or more whistles and buzzes mixed in various patterns
Black-throated Green Warbler, p. 400
Golden-cheeked Warbler, p. 401

SAVANNAH SPARROW SONG GROUP
Phrases of a few high Tinklike notes, then a high, insectlike buzz, sometimes with a few different notes right before or after the buzz.

Only 3–4 notes before the buzz, 0–1 notes after
Grasshopper Sparrow, p. 447
Le Conte's Sparrow, p. 451

More complex
Savannah Sparrow, p. 448

A BURRY COMPLEX SONG

A HIGH TWITTERING OR BURRY PHRASE
A high short burry or twittering phrase, introduced by an accelerating twitter.

Given almost exclusively prior to sunrise
Gray Kingbird Dawn Song, 273

Tropical Kingbird Dawn Song, p. 268

Eastern Kingbird Dawn Song, p. 271

A BURRY COMPLEX SONG, CONTINUED

BURRY COMPLEX SHOREBIRD/ FLYCATCHER SONG GROUP

Usually long, a mix of whistled notes and burry phrases, many notes and phrases repeated. Often rather plastic.

Long and fairly stereotyped
Short-billed Dowitcher Song, p. 141
Long-billed Dowitcher Song, p. 141
Wilson's Plover Song, p. 124
Semipalmated Plover Song, p. 125
Snowy Plover Ter-weet/ Peer-purr, p. 123
Eastern Phoebe Flight Song, p. 261

Say's Phoebe Flight Song, p. 262

Averaging higher, less musical, often with Whits or Pips
Least Flycatcher Flight Song, p. 255
Willow Flycatcher Flight Song, p. 259

Highly plastic interaction calls
Northern Beardless-Tyrannulet Interaction Calls, p. 251

Western Wood-Pewee Interaction Calls, p. 252
Eastern Wood-Pewee Interaction Calls, p. 253
Say's Phoebe Interaction Calls, p. 262

Mostly coarse, harsh Churrs, with few musical notes
Yellow-headed Blackbird Flight Song, p. 478

A COMPLEX SONG FEATURING GURGLES, CREAKS, CHATTERS, AND/OR IMITATIONS

A GURGLE

A very rapid brief jumble of fairly musical notes, usually 1–3 syllables total. Generally shorter and faster than Warbles (p. 572).

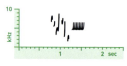

Ending in short trill, or else second half repeating first half
Black-capped Chickadee Gurgle Song, p. 308
Carolina Chickadee Gurgle Song, p. 309
Boreal Chickadee Gurgle Song, p. 307

All notes different, no trills
Blue Jay Pumphandle Calls, p. 292
Brewer's Blackbird Gurgle Song, p. 477
Rusty Blackbird Gurgle Song, p. 476

Usually given with Chirps and whistles
Tree Swallow, p. 304

High, very brief, stereotyped
Henslow's Sparrow Song, p. 452

SQUEAKY GATE HINGE GROUP

1–2 noisy or complex gurgling notes, then a high monotone whistle or creak.

Usually starts with noisy and/or polyphonic notes
Common Grackle Song, p. 475

Usually starts with 1 medium-low warble or gurgle
Brewer's Blackbird Creak Song, p. 477
Rusty Blackbird Gurgle-Creak Song, p. 476

Usually starts with 1 or more low gurgles; final whistles variable
Brown-headed Cowbird Gurgle Song, p. 479
Bronzed Cowbird Gurgle Song, p. 480
Shiny Cowbird Gurgle Song, p. 481

Usually starts with 1 low note
Loggerhead Shrike Song, p. 274

Northern Shrike Song, p. 275
Rusty Blackbird Creak Song, p. 476

Highly variable
Blue Jay Pumphandle Calls, p. 292

A PHRASE OF MUSICAL NOTES AND RATTLES

A soft phrase containing 1-3 musical notes with a ticking Rattle or Snarl.

Musical notes low, mellow
American Crow Gurgle Phrase, p. 288

Musical notes seminasal; plastic, given on wing
Cliff Swallow Songlike Calls, p. 301

A COMPLEX HISSING GURGLE
A complex phrase of rapid notes superimposed on a soft hiss.

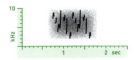

Nelson's Sparrow Song,
p. 454
Brewer's Blackbird Gurgle
Song, p. 477

Rusty Blackbird Gurgle Song,
p. 476

See also
Great-tailed Grackle, p. 472

A CHATTER, THEN A PHRASE
A short musical phrase introduced by 1–3 quick low noisy notes, usually in a syncopated pattern.

Phrase of polyphonic, burry notes
Eastern Bluebird Agitated
Song, p. 341

Phrase short, highly varied in tone quality
White-eyed Vireo Song,
p. 280

Hooded Oriole Song, p. 483
Bullock's Oriole Song, p. 489

A NOTE OR 2, THEN A CHATTER
A single note or 2–3 note series or phrase, followed by a Chatter or unmusical series.

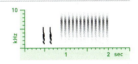

Sedge Wren, p. 326
Marsh Wren (Western), p. 324

Marsh Wren (Eastern), p. 325

A NASAL, HARSH, HARSH BURRY, OR KNOCKING COMPLEX SONG (INCLUDING "LONG CALLS")

A NASAL PHRASE
A phrase of 4 or more different nasal notes.

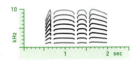

Stereotyped rising "How, how-are-you?"
Long-tailed Duck Song, p. 57

Soft, plastic sounds like a radio being tuned
Northern Bobwhite, p. 68

High, 4-syllabled, squeaky phrase
Least Tern Kideer-kiddik,
p. 178

A NASAL AND CHITTERING PHRASE
A phrase with both clear nasal notes and harsh Chitters, Churrs, or Rattles.

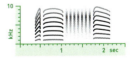

Rather stereotyped, given at dawn
Great Kiskadee Dawn Song,
p. 267

Plastic, in interactions, often in chorus
Western Kingbird Deer
Series, p. 270

Scissor-tailed Flycatcher
Peer-churr, p. 272

GODWIT SONG GROUP
1–2 consecutive, usually complex series of high seminasal notes, often with various notes before and/or after the series.

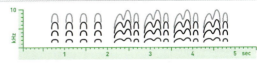

High, with Klip or Klee notes
Solitary Sandpiper, p. 145

Just 1 note before a high, seminasal 2-note series
White-tailed Hawk Wail
Phrase, p. 208

Built around a high, seminasal 3-note series
Hudsonian Godwit, p. 133

Built around a medium-low, nasal 3-note series
Marbled Godwit, p. 133

A NASAL, HARSH, HARSH BURRY, OR KNOCKING COMPLEX SONG (INCLUDING "LONG CALLS"), CONTINUED

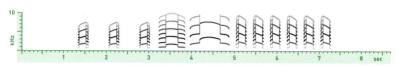

LARGE GULL LONG CALLS

1–3 high, long broken Wails or Squeals, followed by a series of shorter, lower broken notes, the whole performance often preceded by a number of introductory notes. All versions variable and plastic.

High, clear, yelping to squealing
Herring Gull, p. 163
Glaucous Gull, p. 165
Thayer's Gull, p. 164
Iceland Gull, p. 164

High, hoarse; final series rather slow
Ring-billed Gull, p. 161

Slightly lower, nasal and hoarse
California Gull, p. 162
Lesser Black-backed Gull, p. 166

Low, rough, moaning or groaning
Great Black-backed Gull, p. 167

Clear and nasal, fastest notes at start
Laughing Gull, p. 159

Clear and nasal, fast notes absent or at end
Franklin's Gull, p. 160

Lower, clear and nasal, mostly a 3-noted series
Black-legged Kittiwake, p.156

COOT AND GALLINULE LONG CALLS

Nasal to partly grating notes in 1–3 consecutive series. Many patterns possible; usually fastest at start. Toward end, notes often longer and series complex.

High, downslurred, decelerating series of high Keeklike notes
Sora, p. 113

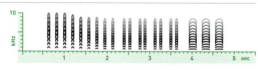

Medium-low downslurred, decelerating series becoming complex or grating
Common Gallinule, p. 116

Medium-low clear nasal couplet series, second note of each couplet longer and upslurred
American Coot, p. 117

Medium-low; all notes short, near the same pitch
Purple Gallinule, p. 115

Rather low; most notes long, fairly clear
Purple Swamphen, p. 114

GROUSE AND ALCID LONG CALLS

Low nasal notes in 1–3 consecutive series. Many patterns possible; usually fastest in middle or at end. Some versions can include high Squeals.

Often 2- or 3-parted, fastest in first or second part
Greater Prairie-Chicken, p. 76
Spruce Grouse, p. 73

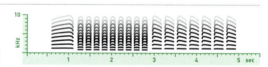

Often 1- or 2-parted, fastest near end
Sharp-tailed Grouse, pp. 74–75
Lesser Prairie-Chicken, p. 77

Lower, harsher, bleating and groaning
Common Murre, p. 154

Very low, slow, moaning
Atlantic Puffin, p. 153

CUCKOO LONG CALLS

A knocking series or Rattle, either becoming a complex series or radically changing in tone quality.

All notes knocking; second series slower, made of couplets
Yellow-billed Cuckoo, p. 92

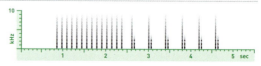

Knocking notes change gradually into clear notes
Black-billed Cuckoo, p. 93

Knocking notes change abruptly into nasal, quacking Barks
Mangrove Cuckoo, p. 94

DUCK LONG CALLS
A series of Quacks, notes usually loudest and longest near start.

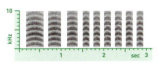

Usually long, loud, the Quacks slightly harsh
    Mallard, p. 48
    American Black Duck, p. 49
    Mottled Duck, p. 49
    Northern Shoveler, p. 51

Nasal, rather short
    Blue-winged Teal, p. 50

Nasal, usually 2–4 notes, not accelerating
    Northern Pintail, p. 52

Rather high, screechy, short
    Green-winged Teal, p. 53

Little-known, not often heard
    Gadwall, p. 46
    Cinnamon Teal, p. 51
    American Wigeon, p. 47
    Eurasian Wigeon, p. 47

A SCREECHY PHRASE
See 2- to 3-Syllabled Screechy Phrase, p. 554.

A SCREECHY CHORUS
A chorus of Screechy Phrases given by multiple birds at once.

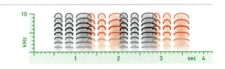

Plain Chachalaca Song, p. 66

RAIL LONG CALLS
A series of screechy Grunts, usually falling in pitch and often slightly accelerating. See also Grunt Series (p. 548).

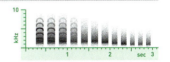

King Rail, p. 111
Clapper Rail, p. 110

Virginia Rail, p. 112

SMALL GULL AND TERN LONG CALLS
High, rather harsh grating notes in 1–3 consecutive series. Many patterns possible; some include short Keks or Kriks. All versions variable and plastic.

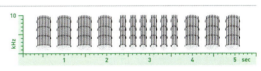

Most notes down- or under-slurred, about the same length
    Common Tern, p. 174

1–5 rapid Keks between grates
    Arctic Tern, p. 175
    Forster's Tern, p. 176

Roseate Tern, p. 177

Variable mix of long and short Grates
    Bonaparte's Gull, p. 158
    Black-headed Gull, p. 158
    Forster's Tern, p. 176

Longest notes upslurred
    Sabine's Gull, p. 157

Fairly low, first notes vaguely 2-syllabled
    American Coot, p. 117

Mix of grating notes and 2-syllabled Kideeks
    Least Tern, p. 178

A CROAKING PHRASE
A phrase of low, harsh croaking notes.

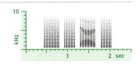

Crested Caracara, p. 245

GRATING, WAILING LONG CALLS
A series of Groans or Wails with burry or tremolo sections, often in plastic, unsynchronized chorus.

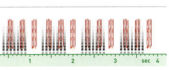

Red-necked Grebe Song, p. 81
Horned Grebe Song, p. 82
Red-throated Loon Song, p. 181

Sandhill Crane Long Call, p. 118
Whooping Crane Long Call, p. 119

Limpkin Wail Series, p. 120

Less grating, more whooping
    Tundra Swan Long Call, p. 43

A NASAL, HARSH, HARSH BURRY, OR KNOCKING COMPLEX SONG (INCLUDING "LONG CALLS"), CONTINUED

KUK-KUK-SCREECH AND SIMILAR PATTERNS
1 to many short Kuklike or Clucklike notes, then a longer, harsher, higher Screech or Shriek.

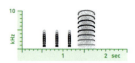

Cluck-cluck-screech
Red-billed Pigeon, p. 86

Kek-kek-shriek
Green Heron, p. 192

Grating Krik-krik-keer
Royal Tern, p. 172
Sandwich Tern, p. 173

Kuk-kuk-quack
Red-throated Loon, p. 181

(courting ducks may reproduce this pattern incidentally)

KEK-KEK-CHURR AND SIMILAR PATTERNS
1 to many short Keklike notes, then a longer, coarse harsh burry note, usually downslurred.

Noisy to screechy, unmusical
Northern Fulmar, p. 182
Helmeted Guineafowl, p. 65
Clapper Rail Kek-burr, p. 110
King Rail Kek-burr, p. 111
Wilson's Snipe Rattle, p. 142

First notes high, semimusical, more like Keeks
Virginia Rail Kee-kee-burr, p. 112
Black Rail Song, p. 109
Red-winged Blackbird Female Song, p. 468

In repeated grating series, often with other calls
Arctic Tern, p. 175
Forster's Tern, p. 176

TICK-TICK-SNARL
1 to many Ticklike notes, then a longer Snarl, given while swooping on potential predators.

Loud, often long
Common Tern, p. 174
Arctic Tern, p. 175

Rather soft
Tree Swallow, p. 304

A POLYPHONIC COMPLEX SONG

A POLYPHONIC PHRASE
A phrase of 4 or more different polyphonic notes.

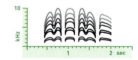

Blue-gray Gnatcatcher Short Song, p. 329

Eastern Bluebird Song, p. 341

A METALLIC SNARLING SONG
A short song ending in a loud harsh polyphonic Snarl or Groan.

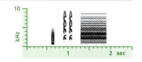

Yellow-headed Blackbird, p. 478

THRUSH SONG GROUP
Very intricate songs of burry polyphonic phrases, with a beautiful jingling metallic quality.

Spiraling upward (overslurred phrases in rising series)
Swainson's Thrush, p. 338

Spiraling downward (overslurred phrases in falling series)
Sprague's Pipit, p. 354
Veery, p. 335

Complex pattern, ending on a rising phrase
Bicknell's Thrush, p. 337

Complex pattern, usually ending on a falling phrase
Gray-cheeked Thrush, p. 336

Starts with monotone whistle
Hermit Thrush, p. 339

A low short series, a whistled phrase, then a high intricate polyphonic phrase
Wood Thrush, p. 340

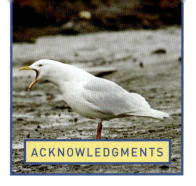

ACKNOWLEDGMENTS

This book was over a dozen years in the making, and countless people helped bring it to fruition. Unfortunately, I only have space to mention a few of them here.

I thank all those who contributed audio recordings. They are acknowledged on the website, www.petersonbirdsounds.com. Some, like Andrew Spencer, Tayler Brooks, Ian Cruickshank, and Bob McGuire, went out of their way to make recordings specifically for this project. Special thanks to Angelika Nelson of the Borror Laboratory of Bioacoustics, Carla Cicero of the Museum of Vertebrate Zoology at Berkeley, and Tom Webber of the Florida Museum of Natural History.

Many people at the Cornell Laboratory of Ornithology deserve thanks, including Jessie Barry, Tammy Bishop, Greg Budney, Greg Delisle, Jay McGowan, Matt Schloss, Mya Thompson, Brad Walker, and Michael Webster. Matt Medler arranged the digitization of old audio tapes, and helped me navigate the immense collection. Matt Young edited hundreds of sounds, hunted down hundreds more, and gave much expertise and encouragement. He is also responsible for the Red Crossbill maps.

I could not have completed the research for this project without the Xeno-Canto website, and many of the recordings come from Xeno-Canto contributors. I owe much to Willem-Pier Vellinga, Bob Planque, Jonathon Jongsma, and the rest of the Xeno-Canto team and recordist community.

Andrew Spencer contributed over 700 recordings, many hours of editing, and vast amounts of expertise and feedback from the very beginning of the project. Walter Szeliga volunteered his time to automate the spectrographic analysis of thousands of sounds. The contributions of Bill Evans, Andrew Farnsworth, and Michael O'Brien were essential to the treatment of nocturnal flight calls. For technical review, I am indebted to Andrew Spencer, Tayler Brooks, Paul Driver, Ted Floyd, Don Kroodsma, Arch McCallum, Michael O'Brien, and Chris Wood.

Others who helped with obtaining recordings or information include Patrik Aberg, Peter Adriaens, Chuck Aid, Nicholas Anich, Rick Baetsen, Mary Banker, Suzanne Beauchaine, Lance Benner, Magnus Bergsson, Kelly Bryan, Robin Carter, Kevin Colver, Ian Davies, Marco Dragonetti, Caroline Eastman, Bruce Falls, Matthias Feuersenger, Owen Fitzsimmons, Pat Gonzalez, Walter Graul, Larry Gregg, Manuel Grosselet, Wayne Hall, David Hof, Eric Hopps, Steve N. G.

Howell, Rich Hoyer, Paul Hurtado, Diana Iriarte, Stephanie Jones, Richard Kern, Niels Krabbe, Dan Lane, Albert Lastukhin, Ross Lein, Tony Leukering, Paul Marvin, Jarek Matusiak, Kevin McGowan, Martin Muller, Mike Nelson, Stein Nilsen, Stephen Nowicki, Ryan P. O'Donnell, Scott Olmstead, Brent Ortego, Leroy Overstreet, Ed Pandolfino, Chris Parrish, Iliana Pena, Susan Peters, Bill Pranty, Bruce Rideout, Andrew Rush, Michael Schroeder, Martin St. Michel, Ha-Cheol Sung, Patrick Turgeon, Rene Valdes, Benjamin Van Doren, Maarten van Kleinwee, Brad Walker, Paige Warren, Richard Webster, Russ Wigh, and Todd Wilson. My apologies and sincere thanks to all the others who helped out, but are not mentioned here.

My agent, Regina Ryan, was indefatigable and indispensable. My editor, Lisa White, was a pleasure to work with, as were many other people at Houghton Mifflin Harcourt, including Mary Dalton-Hoffman, Beth Burleigh Fuller, and Brian Moore. Loma Huh and Simone Payment copyedited and proofread, respectively, with sharp eyes, and Donna Riggs worked on the indexing. Eugenie S. Delaney's excellent work is evident in the layout and graphic design.

My family and my students workshopped parts of this book, from the initial proposal to the species account layouts. Thanks to all, especially my grandparents, my mother, my sister, my brother, and my sister-in-law. Last, a million thanks and a lifetime of love to Molly, for the many sacrifices she made for this project over the years I spent working on it.

I am very proud of this book and I could not have done it without you.

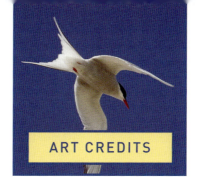

ART CREDITS

p. i: Andrew Spencer (Chestnut-collared Longspur)

pp. ii–iii: Jeffrey A. Gordon (Prairie Warbler)

p. vi: Andrew Spencer (Dunlin)

p. viii: Jeffrey A. Gordon (Red Knots and Ruddy Turnstones)

p. ix: Nathan Pieplow (Great-tailed Grackle)

p. x: Jeffrey A. Gordon (Barn Swallow)

p. 1: Andrew Spencer (Golden-winged Warbler)

pp. 34–35: Andrew Spencer (Red-necked Grebe)

p. 491: Andrew Spencer (Green-tailed Towhee)

pp. 492–493: Andrew Spencer (Philadelphia Vireo)

p. 494: Andrew Spencer (Lapland Longspur)

p. 495: Andrew Spencer (Stilt Sandpiper)

p. 579: Andrew Spencer (Glaucous Gull)

p. 581: Jeffrey A. Gordon (Artic Tern)

p. 583: Andrew Spencer(Sharp-tailed Grouse)

All paintings by Roger Tory Peterson, except:

Michael DiGiorgio: Egyptian Goose, p. 38, Domestic Chicken, p. 64, Indian Peafowl, p. 65, Helmeted Guineafowl, p. 65, Purple Swamphen, p. 114, Red-masked Parakeet, p. 247, White-eyed Parakeet, p. 247, Blue-crowned Parakeet, p. 247, Chestnut-fronted Macaw, p. 248, Orange-winged Parrot, p. 249, Red-vented Bulbul, p. 330.

Michael O'Brien: Cackling Goose, p. 40, Mitred Parakeet, p. 247, Northern Beardless-Tyrannulet, p. 251, Canyon Wren, p. 321, Cactus Wren, p. 327, Common Myna, p. 352, Hill Myna, p. 351, Northern Red Bishop, p. 358.

Drawings by Michael DiGiorgio: Northern Flicker, p. 240, Wood Stork, p. 183, Short-eared Owl p. 219, Ruffed Grouse, p. 72, Greater Prairie-Chicken, p. 76, Lesser Prairie-Chicken, p. 77, Gull Long Call, p. 155, Gull Mew Display, p. 155, Gull Head Toss, p. 155

All maps from the *Peterson Field Guide to Birds of Eastern and Central North America* by Roger Tory Peterson, created by Paul Lehman and Larry Rosche, except the following:

Nathan Pieplow: Egyptian Goose, Muscovy Duck, Montezuma Quail, Purple Swamphen, Willet (Western), Willet (Eastern), Antillean Nighthawk, Western Screech-Owl, Aplomado Falcon, Gray Vireo, Hutton's Vireo, Tamaulipas Crow, Marsh Wren (Western), Marsh Wren (Eastern), Red-whiskered Bulbul, Red-vented Bulbul, Common Myna, Hill Myna, Scaly-breasted Munia, Northern Red Bishop, Yellow-rumped Warbler (Myrtle), Yellow-rumped Warbler (Audubon's), Brewer's Sparrow, Black-throated Sparrow, Scott's Oriole.

Matt Young: Red Crossbill Type 1, Type 2, Type 3, Type 4, Type 10

Purchase Peterson Field Guide titles wherever books are sold.
For more information on Peterson Field Guides, visit **www.petersonfieldguides.com.**

PETERSON FIELD GUIDES®

Roger Tory Peterson's innovative format uses accurate, detailed drawings to pinpoint key field marks for quick recognition of species and easy comparison of confusing look-alikes.

BIRDS

Birds of North America

Birds of Eastern and Central North America

Western Birds

Eastern Birds

Feeder Birds of Eastern North America

Hawks of North America

Hummingbirds of North America

Warblers

Eastern Birds' Nests

PLANTS AND ECOLOGY

Eastern and Central Edible Wild Plants

Eastern and Central Medicinal Plants and Herbs

Western Medicinal Plants and Herbs

Eastern Forests

Eastern Trees

Western Trees

Eastern Trees and Shrubs

Ferns of Northeastern and Central North America

Mushrooms

North American Prairie

Venomous Animals and Poisonous Plants

Wildflowers of Northeastern and North-Central North America

MAMMALS

Animal Tracks

Mammals

Finding Mammals

INSECTS

Insects

Eastern Butterflies

Moths of Northeastern North America

REPTILES AND AMPHIBIANS

Eastern Reptiles and Amphibians

Western Reptiles and Amphibians

FISHES

Freshwater Fishes

SPACE

Stars and Planets

GEOLOGY

Rocks and Minerals

PETERSON FIRST GUIDES®

The first books the beginning naturalist needs, whether young or old. Simplified versions of the full-size guides, they make it easy to get started in the field, and feature the most commonly seen natural life.

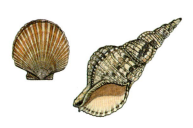

Astronomy

Birds

Butterflies and Moths

Caterpillars

Clouds and Weather

Fishes

Insects

Mammals

Reptiles and Amphibians

Rocks and Minerals

Seashores

Shells

Trees

Urban Wildlife

Wildflowers

PETERSON FIELD GUIDES FOR YOUNG NATURALISTS

This series is designed with young readers ages eight to twelve in mind, featuring the original artwork of the celebrated naturalist Roger Tory Peterson.

Backyard Birds

Birds of Prey

Songbirds

Butterflies

Caterpillars

PETERSON FIELD GUIDES® COLORING BOOKS®

Fun for kids ages eight to twelve, these color-your-own field guides include color stickers and are suitable for use with pencils or paint.

Birds

Butterflies

Dinosaurs

Reptiles and Amphibians

Wildflowers

Seashores

Shells

Mammals

PETERSON REFERENCE GUIDES®

Reference Guides provide in-depth information on groups of birds and topics beyond identification.

Behavior of North American Mammals

Birding by Impression

Molt in North American Birds

Owls of North America and the Caribbean

Seawatching: Eastern Waterbirds in Flight

PETERSON AUDIO GUIDES

Birding by Ear: Eastern/Central

Bird Songs: Eastern/Central

PETERSON FIELD GUIDE / *BIRD WATCHER'S DIGEST* BACKYARD BIRD GUIDES

Identifying and Feeding Birds

Hummingbirds and Butterflies

Bird Homes and Habitats

The Young Birder's Guide to Birds of North America

The New Birder's Guide to Birds of North America

DIGITAL

App available for Apple and Android.

Peterson Birds of North America

E-books

Birds of Arizona

Birds of California

Birds of Florida

Birds of Massachusetts

Birds of Minnesota

Birds of New Jersey

Birds of New York

Birds of Ohio

Birds of Pennsylvania

Birds of Texas

WHAT DID YOU HEAR?

A SINGLE NOTE
by itself, or repeated after a pause

A click, tap, or snap
(Bill snaps, wing claps, ticking calls, etc.)
p. 496

A Chip-like note
(Whits, Chips, Smacks, Tinks, Tsits, etc.)
pp. 496–499

A Peep- or Chirp-like note
(Pips, Peeps, Piks, Kips, Chirps, Cheeps, etc.)
pp. 499–501

A high Seet-like note
(Seets, Tseews, Seers, Dzits, etc.)
pp. 501–503

A whistle
(Medium- to high-pitched, musical)
pp. 504–508

A low-pitched note
(Whoops, Hoots, Coos, Growls, Groans, etc.)
pp. 508–509

A burry or buzzy note
(Musical to unmusical)
pp. 509–513

A short nasal note
(Squeaks, Keeks, Yelps, Yaps, Honks, etc.)
pp. 513–516

A longer nasal note
(Mews, Yanks, Wails, Squeals, Moans, etc.)
pp. 516–519

A short harsh note
(Clucks, Chits, Chuks, Chups, etc.)
pp. 519–520

A longer harsh note
(Barks, Grunts, Screeches, Hisses, etc.)
pp. 520–523

A harsh burry note
(Grates, Churrs, Quacks, Caws, Croaks, etc.)
pp. 523–527

A polyphonic note
(Metallic, dissonant, or whiny)
pp. 527–530

THE SAME NOTE REPEATED
without a significant pause

A ticking or tapping series
(Ticking trills, woodpecker drums, etc.)
pp. 531–533

A Chip series or unmusical trill
(Chipping series, smacking rattles, etc.)
pp. 533–535

A peeping series or rattle
(Peeping twitters, woodpecker rattles, etc.)
pp. 535–536

A very high series or trill
(Waxwing trills, Blackpoll Warbler song, etc.)
pp. 536–537

A musical series or trill
(Medium- to high-pitched)
pp. 537–539

A low-pitched series or trill
(Hooting, cooing, or growling series, etc.)
pp. 539–540

A series of burrs or buzzes
(Series of Dzik notes, Peents, etc.)
pp. 540–541

A series of short nasal notes
(Whinnies, Laughs, Gobbles, etc.)
pp. 542–545

A series of longer nasal notes
(Series of Screams, Squeals, Wails, Moans, etc.)
pp. 545–546

A series of short harsh notes
(Chatters, Chuckles, Bark Series,
Kek Series, etc.)
pp. 546–547

A series of longer harsh notes
(Series of Grunts, Screeches, etc.)
pp. 547–548

A harsh burry series
(Series of Quacks, Rasps, etc.)
pp. 548–549

A polyphonic series
(Series of Jits, metallic series, whiny Chatters, etc.)
p. 549